Plant Genetics and Evolution

Plant Genetics and Evolution

Edited by Ben Davies

SYRAWOOD
PUBLISHING HOUSE
New York

Published by Syrawood Publishing House,
750 Third Avenue, 9th Floor,
New York, NY 10017, USA
www.syrawoodpublishinghouse.com

Plant Genetics and Evolution
Edited by Ben Davies

International Standard Book Number: 978-1-68286-502-6 (Hardback)

Cataloging-in-Publication Data

Plant genetics and evolution / edited by Ben Davies.
p. cm.
Includes bibliographical references and index.
ISBN 978-1-68286-502-6
1. Plant genetics. 2. Plants--Evolution. 3. Evolution (Biology).
I. Davies, Ben.
QK981 .P53 2018
581.35--dc23

Printed in China.

TABLE OF CONTENTS

PREFACE

The world is advancing at a fast pace like never before. Therefore, the need is to keep up with the latest developments. This book was an idea that came to fruition when the specialists in the area realized the need to coordinate together and document essential themes in the subject. That's when I was requested to be the editor. Editing this book has been an honour as it brings together diverse authors researching on different streams of the field. The book collates essential materials contributed by veterans in the area which can be utilized by students and researchers alike.

Plant genetics is the study of the evolution and heredity of plants. Plant genetics follows two major procedures - gene gun method and the agrobacterium method. There is a great demand for genetically modified food. Crop yields have significantly altered since the introduction of genetically modified plants. Such plants remain fresh and pest free for long periods of time. This book on plant genetics and evolution discusses the vital concepts and methods related to this field. It will help new researchers by foregrounding their knowledge in this branch. A number of latest researches have been included herein to keep the readers up-to-date with the global concepts in this area of study.

Each chapter is a sole-standing publication that reflects each author´s interpretation. Thus, the book displays a multi-facetted picture of our current understanding of application, resources and aspects of the field. I would like to thank the contributors of this book and my family for their endless support.

Editor

Genome Analysis of *Staphylococcus aureus* ST291, a Double Locus Variant of ST398, Reveals a Distinct Genetic Lineage

Marc Stegger[1,2]*, Maliha Aziz[2], Tomasz Chroboczek[4], Lance B. Price[2,3], Troels Ronco[1], Kristoffer Kiil[1], Robert L. Skov[1], Frederic Laurent[4], Paal S. Andersen[1,2]

1 Microbiology and Infection Control, Statens Serum Institut, Copenhagen, Denmark, 2 Translational Genomics Research Institute, Pathogen Genomics Division, Flagstaff, Arizona, United States of America, 3 Department of Environmental and Occupational Health, George Washington University, Washington, D. C., United States of America, 4 National Reference Center for Staphylococci, Laboratory of Bacteriology, Hôpital de la Croix Rousse, Inserm U851, Lyon, France

Abstract

Staphylococcus aureus ST291 has been reported as a homologue recombinant double locus variant of the livestock associated *S. aureus* ST398. However, whole genome sequencing show that ST291 is a unique genetic lineage with highly variable content within its accessory genome compared to both human and livestock associated genome sequenced CC398s.

Editor: J. Ross Fitzgerald, University of Edinburgh, United Kingdom

Funding: The authors have no support or funding to report.

Competing Interests: The authors have declared that no competing interests exist.

* E-mail: mtg@ssi.dk

Introduction

Since its discovery, there has been a great interest in *Staphylococcus aureus* isolates related to clonal complex (CC) 398, both in terms of epidemiology and diversity of its genomic content [1–5]. To facilitate rapid detection of *S. aureus* CC398 isolates, we recently developed a PCR targeting the CC398-specific variant of *sau1-hsdS1*, a gene encoding the specificity subunit of the *S. aureus* restriction-modification system Sau1 [6]. *S. aureus* isolates of sequence type (ST) 291, have been assumed to be double locus variants (DLV) of CC398, potentially as a result of a homologue recombination event as indicated by multi locus sequence typing (MLST) [7]. ST291 has been reported to be sensitive to *Sma*I digest and thus different from the normal CC398 population by standard pulsed-field gel electrophoresis (PFGE) procedures, and also to exhibit major differences in *spa* repeats compared to other published CC398-associated *spa* types. It has also escaped detection by ourCC398 specific PCR [6]. In a recent study using whole genome sequence data to analyze the population structure of CC398, no ST291s were included to give insight into this specific subgroup of CC398 [5]. ST291 isolates have been scarcely described in the literature from areas associated with the presence of CC398, such as France, Italy, Switzerland, and the US, but recently more abundant in areas that have not previously been associated with CC398 such as England, India, Iran, Korea, Lebanon, Mali, and Tunisia [7–17], including reports of invasive, methicillin-susceptible as well as methicillin resistant, variants. To further understand and characterize how this reportedly DLV of ST398 differ from the general population structure of CC398 we performed whole genome sequencing (WGS) of two ST291 isolates.

Materials and Methods

MLST analysis

A population snapshot using eBURST at the MLST database (http://eburst.mlst.net/) identified STs either directly linked to CC398 or being DLV of sequence types linked to CC398 (data not shown). A maximum parsimony analysis using MEGA 5.05 [18] of the concatenated MLST alleles from the most closely related available *S. aureus* genome to ST398, MRSA252 (GenBank accession no. NC_002952), and the identified STs was used to evaluate the MLST allele based population structure of *S. aureus* CC398.

DNA sequencing and typing

Two invasive MSSA representatives of the ST291 genotype, HT20040853 and ST20090964, from France (*spa* type t3642) and Algeria (*spa* type t779), were sequenced by paired-end Illumina sequencing with 101-bp read lengths as previously described [5]. The two isolates were *in silico* MLST typed using assembled whole genome sequence data at http://cge.cbs.dtu.dk/services/MLST [19]. The Illumina whole genome sequence data for the two ST291 isolates are available at the Short Read Archive with accession ID SRA060904, with ~200 fold coverage. The accession numbers for the previously sequenced Illumina sequence data generated from 88 *S. aureus* CC398 isolates are available in the Sequence Read Archive under the following accession numbers: SRX129593 to SRX129632, SRX129682 to SRX129686, SRX129691, SRX129696, SRX129697, SRX129701, SRX129702, SRX129704 to SRX129707, SRX129714, SRX129718, SRX129758, SRX129763,

SRX129764, SRX129766, SRX129775, SRX129779, SRX129784, and SRX129816 to SRX129840.

De novo and reference assembly

Reference assembly against the MRSA livestock associated S0385 ST398 genome and the MSSA human associated ST398NM01/71193 ST398 genome (GenBank accession no. AM990992 and CP003045, respectively) was performed using CLCbio's Genomics Workbench 5.5.1 (CLCbio, Aarhus, Denmark). *De novo* assemblies of both the unmapped reads from the reference assemblies as well as on the entire data sets were also performed using the Genomics Workbench.

SNP calling

Single nucleotide polymorphisms (SNPs) was identified as described previously [5]. Briefly, in order to avoid false calls due to sequencing errors, SNP loci were excluded if they did not meet a minimum coverage of 10× and if the variant was present in less than 90% of the base calls for that position. SNP calls were combined for all of the sequenced genomes such that for the locus to be included in the final SNP matrix, it had to be present in all of the genomes. SNPs in the duplicated regions on the reference genome were discarded.

Genome analysis

Comparative genomics was performed using reference assemblies against the MRSA livestock associated S0385 ST398 genome and the MSSA human associated ST398NM01/71193 ST398 genome and gaps were identified in regions where there was no coverage (missing genes/genomic regions) and by subsequent *de novo* assembly of unmapped reads from the reference assemblies were performed using CLCbio's Genomics Workbench 5.5.1 (CLCbio) thereby identifying genes/genomic regions present in ST291 but absent in CC398. The distinct, non-ST398 mobile genetic elements (MGE) were annotated and compared using the annotation pipeline and sequence based comparison tool at RAST (http://rast.nmpdr.org) [20]. For visualization, the ST291 WGS data sets were aligned against the two *S. aureus* ST398 reference genomes (GenBank accession no. AM990992 and NC_017673), using BWA version 0.6.2-r126 and SAMtools version 0.1.18 [21,22]. BLAST atlases comparing the two reference sequences

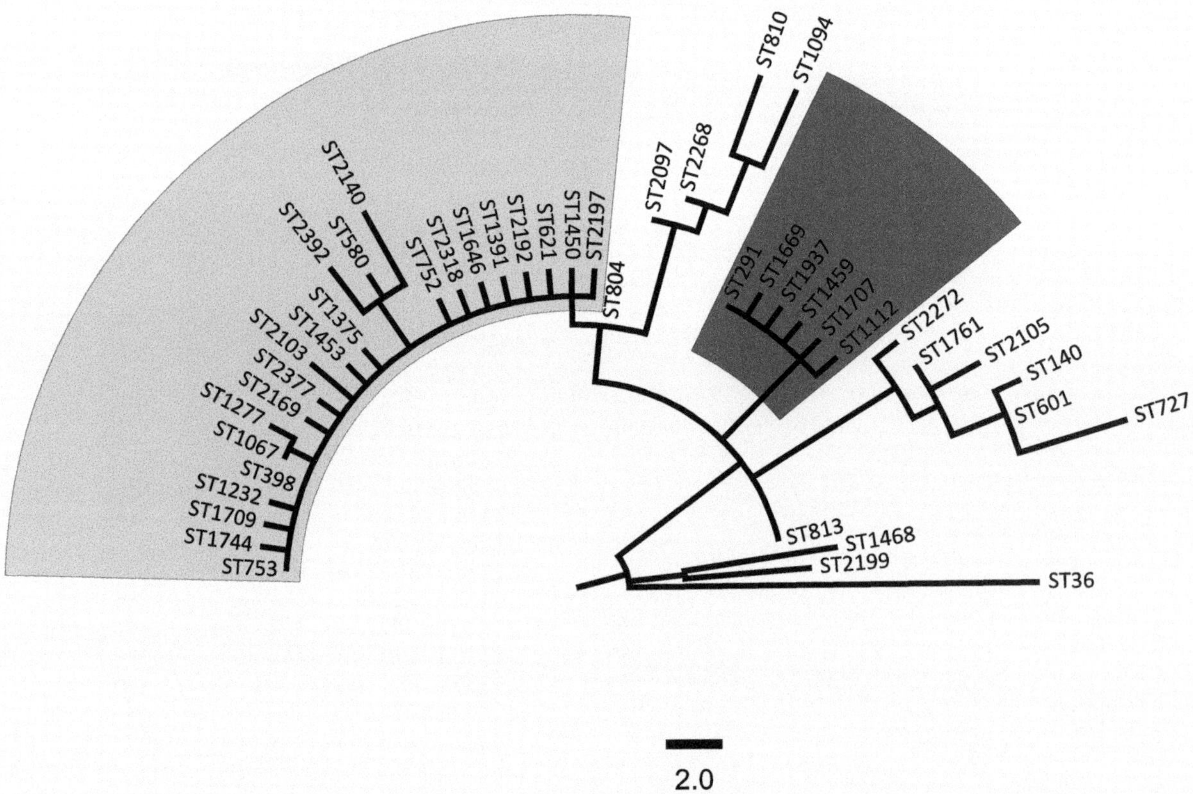

Figure 1. MLST based CC398 phylogeny. Maximum parsimony analysis of the concatenated allelic sequences from the 43 CC398 sequence types and rooted using MRSA252 (ST36). The cluster containing ST398 is highlighted in light grey and the ST291 cluster is dark grey. The scale indicates the number of SNP differences.

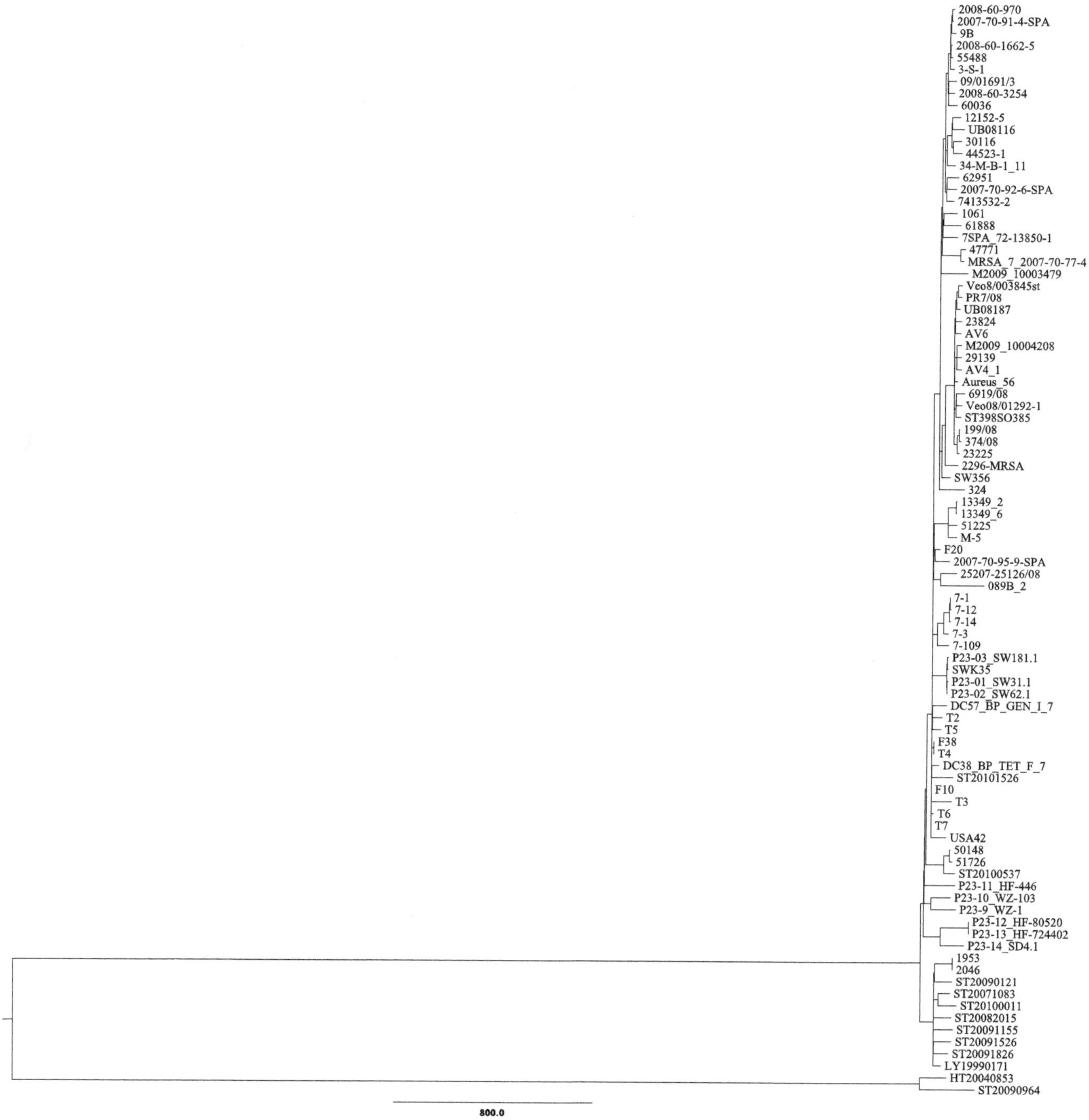

Figure 2. Whole genome sequence based analysis of divergence of ST291 and ST398. Maximum-parsimony tree of the two ST291 isolates (HT20040853 and ST20090964) and 89 CC398 isolates based on 12,735 SNPs total SNPs. The tree was rooted with MRSA252 (ST36), the closest non-CC398 available out-group. The scale indicates the number of SNP differences.

were generated using GeneWiz with an additional lane added to display the read depth of the aligned reads [23].

Results and Discussion

Phylogenetic analyses of the concatenated MLST alleles readily identified several sub-clusters within the *S. aureus* CC398 complex, with the presence of a distinct ST291 clade comprised of six different STs as seen in Figure 1. A total of 12,735 SNPs were identified among the entire 91 CC398 isolates, and with 11,744 identified SNPs after exclusion of the 123 kb horizontally acquired ST9 region. The maximum parsimony phylogenetic tree presented in Figure 2 based on the 11,744 SNPs revealed that ST291 formed a very distal cluster in relation to the previously published 89 CC398 genomes, with no evidence of any major horizontal gene transfers (HGTs) to account for the observed SNP diversity (see Figure 3). A total of 7,800 shared SNPs separated both ST291s from the S0385 CC398 reference genome, and only 329 SNPs differentiated the two ST291 isolates. In reference, the

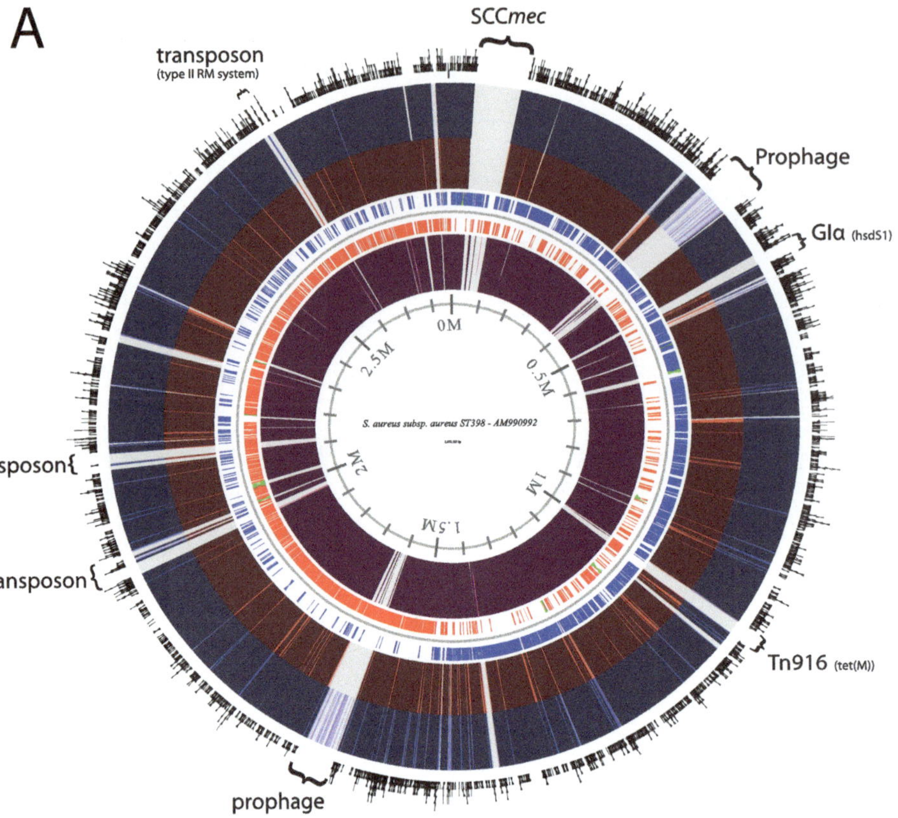
A
SCCmec
transposon
(type II RM system)
Prophage
Glα (hsdS1)
0M
0.5M
1M
1.5M
2M
2.5M
S. aureus subsp. aureus ST398 - AM990992
transposon
transposon
Tn916 (tet(M))
prophage

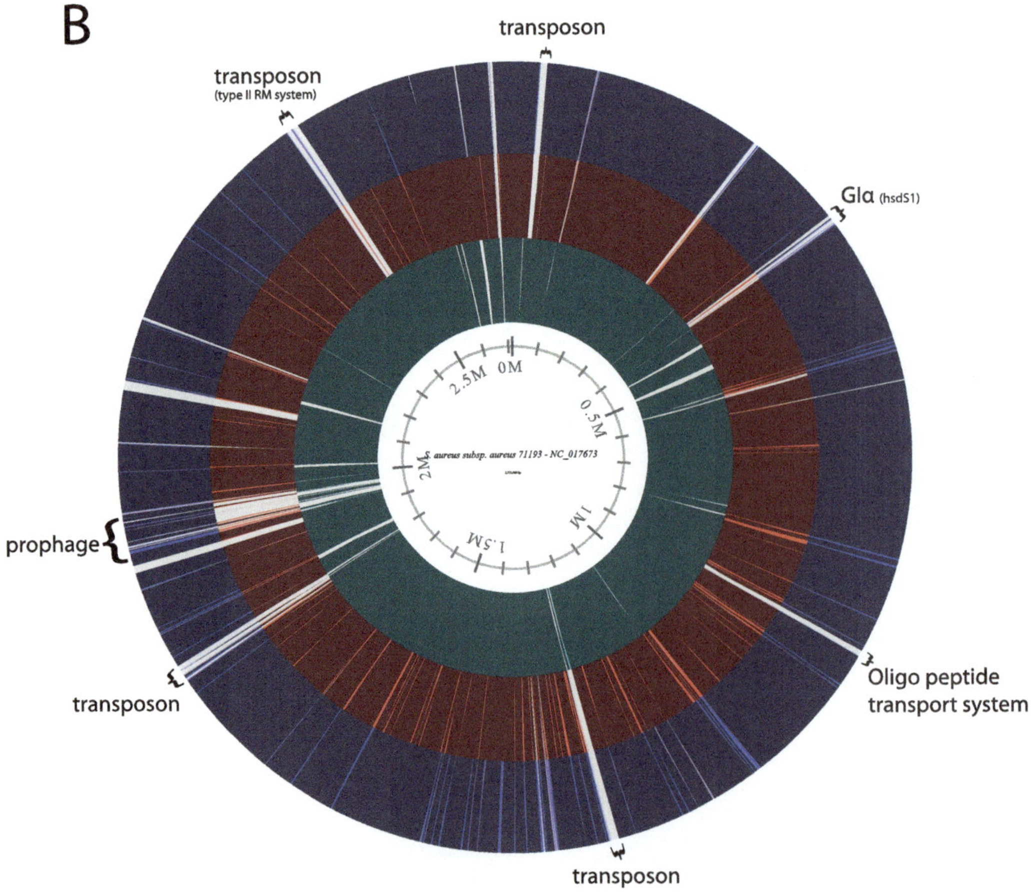
B
transposon
transposon
(type II RM system)
Glα (hsdS1)
0M
0.5M
1M
1.5M
2M
2.5M
S. aureus subsp. aureus 71193 - NC_017673
prophage
transposon
Oligo peptide
transport system
transposon

Figure 3. Diversity in *S. aureus* CC398 genome content. Comparison of the genome content of the livestock- and human-associated *S. aureus* CC398s with the two ST291s. Reference genomes are in A) S0385 the livestock associated CC398, and in B) ST398NM01/71193 the human associated CC398. Presented are analysis of ST398NM01/71193 (purple), S0385 (cyan), HT20040853 (blue) and ST20090964 (red), compared to the references. The outer circle in A represent the distribution of the 7,800 SNPs differentiating both ST291s from the SO385 reference genome.

phylogeny on the 89 previously published *S. aureus* CC398 genomes used to evaluate the overall population structure of *S. aureus* CC398 was based on a total of 4328 SNPs [5].

CC398 encodes a single *saul-hsdS1* gene located in the genomic island (GI) α, whereas most other *S. aureus* lineages encode two distinct variants of *saul-hsdS*; *saul-hsdS1* and *saul-hsdS2* located in GIα and GIβ, respectively [1,24]. ST291 also encode two specificity subunits, with the *saul-hsdS1* showing ~60% nucleotide similarity to the CC398 *saul-hsdS1*, and consequently the CC398 specific PCR do not identify these isolates as part of the CC398 cluster. The *sau1-hsdS1* has a very high sequence similarity (>99%) to a *sau1-HsdS* gene present in a non-GI associated restriction-modification system present in the genome sequence of a representative of the European community acquired ST80 lineage [25]. Additionally, ST291 isolates have been reported as being capable of *Sma*I restriction cleavage in pulsed-field gel electrophoresis analysis unlike CC398 [8].

Comparative genomics show that an extensive number of MGEs are absent from the two genome sequenced ST291 isolates compared to both the human and livestock associated ST398 (see Figure 3). These MGEs include both larger regions such as the SCC*mec*, multiple transposon and phage elements, but also "core genome" encoded adhesins, lipoproteins, various enzymes and a number of hypothetical proteins. A significant proportion of the genomic island α is also lacking from the ST291 genomes when compared to the two representative ST398s, including the gene encoding the von Willebrand factor-binding protein, in addition to harboring the divergent *saul-hsdS1* gene. A large proportion of the MGEs that seems to have been acquired at least in the representative livestock associated CC398 reference genome compared to reference human associated CC398 genome are also absent from our two ST291 genomes, see Figure 3A. We do however also see that multiple MGEs are unique in the human associated reference genome both compared to the livestock associated representative genome and to the ST291 genomes, see Figure 3B. *De novo* assembled contigs from unmapped ST291 reads from reference assemblies against the human associated ST398NM01/71193 ST398 genome sequence identified approximately 140 and 205 kb of novel content on 39 and 54 contigs >500 bp in ST20090964 and HT20040853 respectively, see Supplementary files S1 and S2. These *de novo* contigs encode a high number of phage related proteins, but also membrane transport systems, an ESAT-6 gene cluster, serine proteases, multiple components of restriction modification systems, and replication related proteins indicating at least one plasmid in each of the two isolates. The genomes of the two ST291 isolates contain the staphylococcal epidermal cell differentiation inhibitor (EDIN) exotoxin, which has been associated with promoting formation of infection foci in deep-seated tissues [26]; and both isolates encode SCIN, also suggesting them to belong to a human associated clonal lineage [5,27]. Additionally, the French ST291 isolate carries the *lukS/F-PV* operon encoding the Panton-Valentine Leukocidin.

ST291 is the fifth most abundant carrier clone in Mali [17], and searching the Ridom Spa Server (http://spa.ridom.de) for *spa* types related to t3642 (e.g. t937, t1149, t1614, t2313, t2993) show the widespread presence of both methicillin-susceptible and methicillin-resistant strains within the USA, Europe, and the Middle East. Identifying *spa* types related to t779 with its single repeat (r08), or linking them to both a group of STs or even specific clonal lineages may be associated with great error, but *spa* type t779 is reported via the Spa Server from throughout Central and Northern Europe.

Conclusions

Based on our analysis, ST291 isolates and likely isolates with the STs highlighted in dark grey in Figure 1, are not closely related to ST398 or to the livestock associated CC398 group in general. Concordantly, we suggest when reporting CC398s related to livestock, that isolates related to ST291 should not be grouped as part of this cluster. The reported ST291 furthermore shows a noticeably different geographic dispersion compared to ST398, indicating a distinct epidemiology possible linked to a community associated human lineage.

Acknowledgments

Tania Contente-Cuomo and Katerina Soldanova, Translational Genomics Research Institute, are thanked for excellent technical assistance.

Author Contributions

Conceived and designed the experiments: MS LP PSA. Performed the experiments: MS MA TR PA. Analyzed the data: MS PSA. Contributed reagents/materials/analysis tools: MA TC LP KK RLS FL PSA. Wrote the paper: MS PSA.

References

1. Schijffelen MJ, Boel CH, van Strijp JA, Fluit AC (2010). Whole genome analysis of a livestock-associated methicillin-resistant *Staphylococcus aureus* ST398 isolate from a case of human endocarditis. BMC Genomics 11:376.
2. Uhlemann AC, Porcella SF, Trivedi S, Sullivan SB, Hafer C, et al. (2012) Identification of a highly transmissible animal-independent *Staphylococcus aureus* ST398 clone with distinct genomic and cell adhesion properties. MBio 3:2.
3. McCarthy AJ, Witney AA, Gould KA, Moodley A, Guardabassi L, et al. (2011) The distribution of mobile genetic elements (MGEs) in MRSA CC398 is associated with both host and country. Genome Biol Evol 3:1164–74.
4. Wassenberg MW, Bootsma MC, Troelstra A, Kluytmans JA, Bonten MJ (2011) Transmissibility of livestock-associated methicillin-resistant *Staphylococcus aureus* (ST398) in Dutch hospitals. Clin Microbiol Infect 17:316–9.
5. Price LB, Stegger M, Hasman H, Aziz M, Larsen J, et al. (2012) *Staphylococcus aureus* CC398: host adaptation and emergence of methicillin resistance in livestock. MBio 3(1).
6. Stegger M, Lindsay JA, Moodley A, Skov R, Broens EM, et al. (2011) Rapid PCR detection of *Staphylococcus aureus* clonal complex 398 by targeting the restriction-modification system carrying *sau1-hsdS1*. J Clin Microbiol 49:732–4.

7. Mediavilla JR, Chen L, Uhlemann AC, Hanson BM, Rosenthal M, et al. (2012) Methicillin-susceptible *Staphylococcus aureus* ST398, New York and New Jersey, USA. Emerg Infect Dis 18:700–2.
8. Shambat S, Nadig S, Prabhakara S, Bes M, Etienne J, et al. (2012) Clonal complexes and virulence factors of *Staphylococcus aureus* from several cities in India. BMC Microbiol 12:64.
9. Cafiso V, Bertuccio T, Spina D, Campanile F, Bongiorno D, et al. (2010) Methicillin resistance and vancomycin heteroresistance in *Staphylococcus aureus* in cystic fibrosis patients. Eur J Clin Microbiol Infect Dis 29:1277–85.
10. Rasigade JP, Laurent F, Lina G, Meugnier H, Bes M, et al. (2010) Global distribution and evolution of Panton-Valentine leukocidin-positive methicillin-susceptible *Staphylococcus aureus*, 1981–2007. J Infect Dis 201:1589–97.
11. Sakwinska O, Kuhn G, Balmelli C, Francioli P, Giddey M, et al. (2009) Genetic diversity and ecological success of *Staphylococcus aureus* strains colonizing humans. Appl Environ Microbiol 75:175–83.
12. Green SM, Marsh P, Ahmad N, Jefferies JM, Clarke SC (2010) Characterization of community and hospital *Staphylococcus aureus* isolates in Southampton, UK. J Med Microbiol 59:1084–8.
13. Havaei SA, Vidovic S, Tahmineh N, Mohammad K, Mohsen K, et al. (2011) Epidemic methicillin-susceptible *Staphylococcus aureus* lineages are the main cause of infections at an Iranian university hospital. J Clin Microbiol 49:3990–3.
14. Moon SY, Lee HJ, Lee MS (2010) Molecular characteristics of methicillin-resistant *Staphylococcus aureus* blood isolates: clonal spread of staphylococcal cassette chromosome *mec* type IVA between the community and the hospital. Microb Drug Resist 16:217–22.
15. Tokajian S, Haddad D, Andraos R, Hashwa F, Araj G (2011) Toxins and Antibiotic Resistance in *Staphylococcus aureus* Isolated from a Major Hospital in Lebanon. ISRN Microbiology 2011: 812049.
16. Kechrid A, Pérez-Vázquez M, Smaoui H, Hariga D, Rodríguez-Baños M, et al (2011) Molecular analysis of community-acquired methicillin-susceptible and resistant *Staphylococcus aureus* isolates recovered from bacteraemic and osteomyelitis infections in children from Tunisia. Clin Microbiol Infect 17:1020–6.
17. Ruimy R, Maiga A, Armand-Lefevre L, Maiga I, Diallo A, et al (2008) The carriage population of *Staphylococcus aureus* from Mali is composed of a combination of pandemic clones and the divergent Panton-Valentine leukocidin-positive genotype ST152. J Bacteriol 190:3962–8.
18. Tamura K, Peterson D, Peterson N, Stecher G, Nei M, et al. (2011) MEGA5: molecular evolutionary genetics analysis using maximum likelihood, evolutionary distance, and maximum parsimony methods. Mol Biol Evol 28:2731–9.
19. Larsen MV, Cosentino S, Rasmussen S, Friis C, Hasman H, et al. (2012) Multilocus sequence typing of total-genome-sequenced bacteria. J Clin Microbiol 50:1355–61.
20. Aziz RK, Bartels D, Best AA, DeJongh M, Disz T, et al (2008) The RAST Server: rapid annotations using subsystems technology. BMC Genomics 9:75.
21. Li H, Durbin R (2009) Fast and accurate short read alignment with Burrows-Wheeler transform. Bioinformatics 25:1754–1760.
22. Li H, Handsaker B, Wysoker A, Fennell T, Ruan J, et al. (2009) The Sequence Alignment/Map format and SAMtools. Bioinformatics 25:2078–9
23. Hallin PF, Stærfeldt HH, Rotenberg E, Binnewies TT, Benham CJ, et al. (2009). GeneWiz browser: An Interactive Tool for Visualizing Sequenced Chromosomes. Stand Genomic Sci 25:204–15.
24. Waldron DE, Lindsay JA (2006) Sau1: a novel lineage-specific type I restriction-modification system that blocks horizontal gene transfer into *Staphylococcus aureus* and between *S. aureus* isolates of different lineages. J Bacteriol 188:5578–85.
25. Stegger M, Price LB, Larsen AR, Gillece JD, Waters AE, et al. (2012) Genome sequence of *Staphylococcus aureus* strain 11819-97, an ST80-IV European community-acquired methicillin-resistant isolate. (2012) J Bacteriol 194:1625–6.
26. Munro P, Benchetrit M, Nahori MA, Stefani C, Clément R, et al. (2010) The *Staphylococcus aureus* epidermal cell differentiation inhibitor toxin promotes formation of infection foci in a mouse model of bacteremia. Infect Immun 78:3404–11.
27. Lowder BV, Guinane CM, Ben Zakour NL, Weinert LA, Conway-Morris A, et al. (2009) Recent human-to-poultry host jump, adaptation, and pandemic spread of *Staphylococcus aureus*. Proc Natl Acad Sci U S A 106:19545–50.

2

Identification and Molecular Characterization of MYB Transcription Factor Superfamily in C_4 Model Plant Foxtail Millet (*Setaria italica* L.)

Mehanathan Muthamilarasan, Rohit Khandelwal, Chandra Bhan Yadav, Venkata Suresh Bonthala, Yusuf Khan, Manoj Prasad*

National Institute of Plant Genome Research, New Delhi, India

Abstract

MYB proteins represent one of the largest transcription factor families in plants, playing important roles in diverse developmental and stress-responsive processes. Considering its significance, several genome-wide analyses have been conducted in almost all land plants except foxtail millet. Foxtail millet (*Setaria italica* L.) is a model crop for investigating systems biology of millets and bioenergy grasses. Further, the crop is also known for its potential abiotic stress-tolerance. In this context, a comprehensive genome-wide survey was conducted and 209 MYB protein-encoding genes were identified in foxtail millet. All 209 *S. italica* MYB (SiMYB) genes were physically mapped onto nine chromosomes of foxtail millet. Gene duplication study showed that segmental- and tandem-duplication have occurred in genome resulting in expansion of this gene family. The protein domain investigation classified SiMYB proteins into three classes according to number of MYB repeats present. The phylogenetic analysis categorized SiMYBs into ten groups (I - X). *SiMYB*-based comparative mapping revealed a maximum orthology between foxtail millet and sorghum, followed by maize, rice and *Brachypodium*. Heat map analysis showed tissue-specific expression pattern of predominant SiMYB genes. Expression profiling of candidate MYB genes against abiotic stresses and hormone treatments using qRT-PCR revealed specific and/or overlapping expression patterns of *SiMYBs*. Taken together, the present study provides a foundation for evolutionary and functional characterization of MYB TFs in foxtail millet to dissect their functions in response to environmental stimuli.

Editor: Girdhar K. Pandey, University of Delhi South Campus, India

Funding: This work was financially supported by the core grant of National Institute of Plant Genome Research (NIPGR), New Delhi, India. The funder had no role in study design, data collection and analysis, decision to publish, or preparation of the manuscript.

* Email: manoj_prasad@nipgr.ac.in

Introduction

In plants, transcription factors (TFs) play crucial role in regulating the gene expression and thereby guiding various biological processes including growth, development, cell division and response to stresses. Deciphering the molecular and physiological roles of these TFs has become hottest topic of research since understanding the control of gene expression would enable researchers to develop elite varieties of crop plants with desired agronomic traits which will ensure global food security in the scenario of climate change. Hence research on TFs had gained momentum and numerous TFs regulating vital biological processes in plants have been identified. MYB (myeloblastosis) TFs are omnipresent in eukaryotic systems and in plants was first identified in *Zea mays* involving the regulation of anthocyanin biosynthesis [1].

MYB proteins possess a highly conserved MYB DNA-binding domain throughout the eukaryotes at N-terminus and 1–4 imperfect repeats (R0, R1, R2, and R3) [2]. Each of these repeats contains 50–53 amino acids and encodes three α-helices, with second and third helices forming a helix–turn–helix (HTH) structure. Upon binding to DNA, HTH intercalates in the major groove [2,3]. In addition, each MYB repeat comprises three regularly spaced tryptophan residues, which form a cluster in a hydrophobic core of the repeat and stabilize the structure of DNA-binding domain [4]. Contrarily, C-terminus is the activation domain and varies significantly between MYB proteins, which results in wide range of regulatory roles of MYB gene family [3,5,6].

There are numerous reports available demonstrating the role of MYB TFs in various plant processes such as secondary metabolism [7–10], hormone signal transduction [11,12], environmental stresses [13–15], cell structuring and organ development [16–19]. The emergence of next-generation sequencing technologies and high-throughput analysis platforms had expedited genome sequencing projects in crop plants [20]. The availability of genome sequence information facilitated the crop research community to perform genome-wide analysis of TFs which would be more informative in conducting further research on functional characterization of candidate TFs. Of the various TFs, MYBs are most extensively studied in almost all land plants whose genome sequence is available [21–25], though there was no such genome-wide study on MYB TFs conducted in foxtail millet (*Setaria italica* L.).

Foxtail millet is a member of Poaceae family and it is the second most-widely cultivated millets, next to pearl millet (FAOSTAT 2012; http://faostat.fao.org/). Being a C_4 panicoid crop, it is efficient in photosynthesis and better water use efficiency (WUE). Photosynthesis in C_4 crops is more efficient than C_3 plants even under high light intensity and high temperatures since phosphoenolpyruvate carboxylase (PEPC) helps in immediate uptake of carbon di oxide and delivering it to RuBisCO. The immediate quenching of carbon di oxide and delivery by PEPC does not require keeping stomata open for a long period and hence less water is lost by transpiration resulting in better WUE [26,27]. Moreover, foxtail millet has been considered as a model crop for studying systems biology of biofuel crops due to its close phylogenetic relationships with several biofuel crops such as switchgrass (*Panicum virgatum*), napier grass (*Pennisetum purpureum*) and pearl millet (*Pennisetum glaucum*) [26,27]. Hence considering its importance, BGI (Beijing Genomics Institute), China and the Joint Genome Institute (JGI) of Department of Energy, USA has sequenced the genome of two foxtail millet accessions [28,29]. This motivated the present study, to perform (i) genome-wide identification of MYB TFs encoded in foxtail millet genome, (ii) protein and gene structure analysis using proteomic tools as well as phylogenetic analysis, (iii) promoter and miRNA analysis, (iv) *in silico* expression profiling using RNA-Seq data, (v) ortholog identification in related grass genomes, (vi) evolutionary analysis of orthologs and paralogs, and (vii) expression analysis of candidate *MYB* genes using qRT-PCR.

Materials and Methods

Identification, chromosomal location and gene duplication analysis of MYB proteins

The Hidden Markov Model (HMM) profile of MYB DNA-binding domain (PF00249) downloaded from Pfam v27.0 (http://pfam.sanger.ac.uk/family/PF00249) [30] was queried against Phytozome v9.1 (www.phytozome.net) of *Setaria italica*. All hits with expected values less than 1.0 were retrieved and compared with MYB protein dataset available in Foxtail millet Transcription Factor Database (http://59.163.192.91/FmTFDb/) [31]. In addition, *S. italica* MYB (SiMYB) proteins were examined using CDD search (http://www.ncbi.nlm.nih.gov/Structure/cdd/wrpsb.cgi) for presence of one or more MYB DNA-binding domains. The chromosomal location and identification of alternate transcripts of SiMYBs were performed by executing BLASTP search against the foxtail millet protein sequences in Phytozome and redundant sequences with different identification numbers and same chromosome locus were discarded. Based on ascending order of chromosomal position from short arm telomere to long arm telomere, *SiMYBs* were annotated as *SiMYB001* to *SiMYB209* and physical map was constructed using MapChart v2.2 [32]. The map was manually inspected for identifying tandem duplications. Adjacent genes of same sub-family located within 10 predicted genes apart or within 30 kb of each-other are considered as tandem duplicated genes [33–36]. Segmental duplications were predicted using Multiple Collinearity Scan toolkit (MCScanX) [37] as described by Puranik et al. [35] and Mishra et al. [36].

Protein structure, gene organization and phylogenetic analysis

The presence of conserved domains in SiMYB proteins were analysed by employing hmmscan tool (http://hmmer.janelia.org/search/hmmscan). BioEdit tool (http://www.mbio.ncsu.edu/bioedit/bioedit.html) was used to perform multiple sequence alignment (MSA) of SiMYB amino acid sequences. The physicochemical properties of SiMYB proteins were then investigated using ExPASy ProtParam tool (http://web.expasy.org/protparam/). Gene Structure Display Server was used to identify the position of introns and exons in respective SiMYB genes (http://gsds.cbi.pku.edu.cn/) [38]. The SiMYB protein sequences were then imported into MEGA v6.06 [39] and MSA was performed using ClustalW with default parameters. The alignment file was used to construct neighbor-joining phylogenetic tree using a bootstrapping method with 1000 replicates.

Gene Ontology annotation, promoter analysis and prediction of miRNAs targeting SiMYBs

The amino acid sequences of SiMYB are loaded in Blast2GO suite [40] to perform BLASTP, against *Oryza sativa* protein sequences of NCBI, with default parameters. The hits were mapped to retrieve GO terms and annotation of GO terms associated with BLAST hits was executed. For promoter analysis, 2 kb upstream sequences of SiMYB genes were retrieved from Phytozome through in-house perl programming and *cis*-regulatory elements were identified using database of Plant *Cis*-acting Regulatory DNA Elements (PLACE) [41]. Moreover, *Setaria italica* miRNAs (Sit-miRs) targeting *SiMYB*s were predicted by aligning Sit-miRs [42,43] with *SiMYB* transcript sequences using psRNATarget [44].

In silico expression profiling and marker localization

Setaria italica Illumina RNA-HiSeq reads for 4 tissues namely spica, stem, leaf and root were retrieved from European Nucleotide Archive [SRX128226 (spica); SRX128225 (stem); SRX128224 (leaf); SRX128223 (root)] [45]. The RNA-seq data was then filtered by NGS toolkit (http://59.163.192.90:8080/ngsqctoolkit/) [46] to remove low quality reads and mapped onto the gene sequences of *S. italica* using Bowtie2 [47]. The number of reads mapped was normalized by RPKM (reads per kilobase per million) method. The heat map showing tissue specific expression was generated on RPKM values for each gene in all the tissue samples using TIGR MultiExperiment Viewer (MeV4) software [48,49]. The physical positions of DNA markers reported in foxtail millet [50] including simple sequence repeats (SSRs) [51,52], EST-derived SSRs (eSSRs) [53] and intron length polymorphic markers (ILPs) [54,55] were compared with the chromosomal location of SiMYB genes to identify the transcription factor gene-derived functional markers.

Identification of SiMYB orthologs in related grass genomes and evolutionary analysis of paralogs and orthologs

SiMYB protein sequences were BLASTP searched against the peptide sequences of sorghum, maize, rice and *Brachypodium* (http://www.gramene.org/; http://www.phytozome.net/) to predict the MYB orthologs in these grass species. A reciprocal BLAST (BLASTP) was also performed to ensure unique relationship between the orthologous genes and all hits with E-value≤1e-5 and at least 80% homology were considered significant. The orthologous relationships between foxtail millet and these grass species were then visualized using Circos v0.55 (http://circos.ca/). The synonymous (Ks) and non-synonymous (Ka) substitution rates were estimated for paralog and ortholog MYB gene pairs through CODEML program in PAML interface tool of PAL2NAL (http://www.bork.embl.de/pal2nal/) [56]. Time (million years

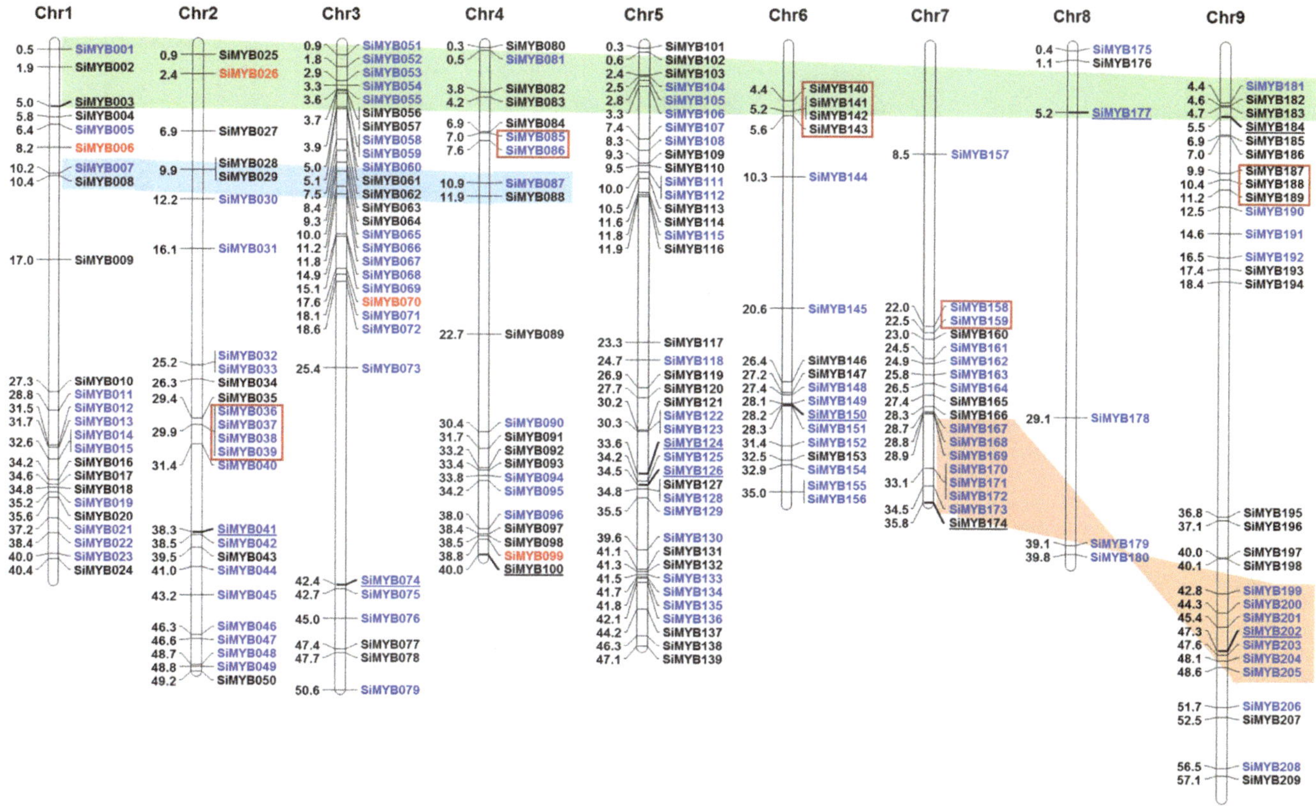

Figure 1. Physical map of 209 SiMYB genes. SiMYB genes were plotted onto nine chromosomes of foxtail millet. The numbers at left indicates the position of SiMYB genes, and the names of the respective genes are given in right. The 'MYB-Related' are given in black, 'MYB-R2R3' are in blue and 'MYB-R1R2R3' are in red. The SiMYB genes used for expression analysis through qRT-PCR are underlined. Segmentally duplicated gene-pairs are highlighted with different colours. Tandemly duplicated gene-pairs are enclosed within red boxes.

ago, Mya) of duplication and divergence of SiMYB gene pairs were estimated using a synonymous mutation rate of λ substitutions per synonymous site per year, as $T = Ks/2\lambda$, where, $\lambda = 6.5 \times 10^{-9}$ [35,57,58,59].

Plant materials and treatments

A salt and dehydration tolerant foxtail millet cultivar 'Prasad' [60,61] was chosen for analysing the expression pattern of candidate *SiMYB*s. The seedlings were grown in a plant growth chamber (PGC-6L; Percival Scientific Inc., USA) for 21 d after germination at 28±1°C day/23±1°C night/70±5% relative humidity with a photoperiod of 14 h and a photosynthetic photon flux density of 500 µmol m^{-2} s^{-1}. The plants were watered daily with one-third strength Hoagland's solution [62]. For stress and hormone treatments, 21-day-old seedlings were individually exposed to 250 mM NaCl (salinity), 20% PEG 6000 (dehydration), 100 µM abscisic acid (ABA), 100 µM methyl jasmonate (MJ) and 100 µM salicylic acid (SA) for 1 h (early) and 24 h (late) following previous reports [35,61,63]. After respective treatment, whole seedlings are immediately frozen in liquid nitrogen and stored at −80°C for RNA isolation. Untreated plants were maintained as control. The above experiments were repeated thrice to ensure precision and reproducibility.

RNA isolation and quantitative real-time PCR analysis

Total RNA was isolated by following the method described by Longeman et al. [64] and treated with RNase-free DNase I (50 U/µl; Fermentas, USA). Its quality and purity was ensured using NanoDrop Spectrophotometer (Thermo Fisher Scientific, USA) [OD_{260}: OD_{280} nm absorption ratio (1.8–2.0)] and integrity of RNA was checked by resolving on 1.2% agarose gel containing 18% formaldehyde. An amount of ~1 µg total RNA was reverse transcribed to first strand cDNA using random primers by Thermo Scientific Verso cDNA kit (Thermo Fisher Scientific, USA) following manufacturer's instructions. The qRT-PCR primers were designed using GenScript Real-time PCR Primer Design tool (https://www.genscript.com/ssl-bin/app/primer) (Table S1). qRT-PCR analysis was performed in three technical replicates for each biological duplicate by StepOne Real-Time PCR Systems (Applied Biosystems, USA). The PCR mixtures and reactions were used as described previously by Kumar et al. [65] Melting curve analysis (60 to 95°C after 40 cycles) and agarose gel electrophoresis were performed to check amplification specificity. A constitutive *Act2* gene-based primer was used as endogenous control [65]. The amount of transcripts accumulated for SiMYB genes normalized to the internal control *Act2* were analysed using $2^{-\Delta\Delta Ct}$ method cDNA synthesis [65]. The PCR efficiency was calculated as: Efficiency $= 10^{(-1/\text{slope})} - 1$ by the default software (Applied Biosystems).

Homology modeling and active site prediction

SiMYB protein sequences were searched against Protein Data Bank (PDB) (http://www.rcsb.org/pdb/) for predicting the best template with homologous amino acid sequence and known structure. The result was then imported in Phyre2 server (Protein Homology/AnalogY Recognition Engine; http://www.sbg.bio.ic.

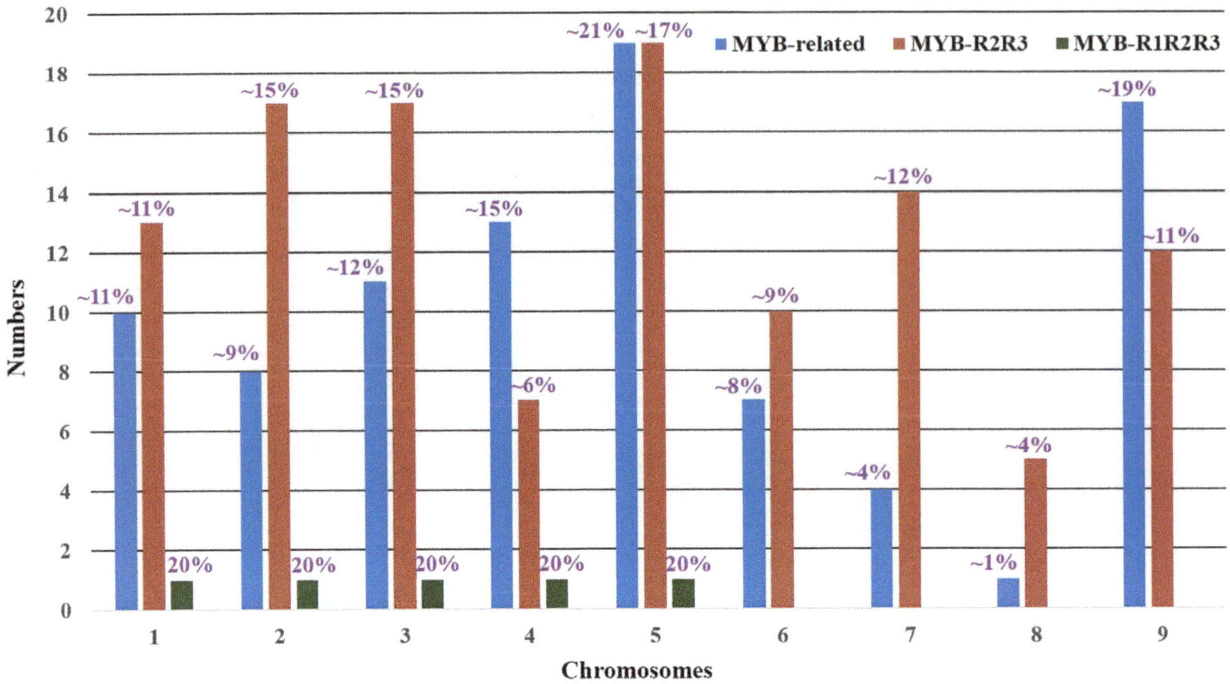

Figure 2. Chromosome-wise distribution of SiMYB genes. The distribution of three different SiMYB genes on individual chromosomes is shown.

ac.uk/phyre2) for predicting three-dimensional structures of proteins by homology modeling under 'normal' mode [66]. Active sites were precited using COACH server (http://zhanglab.ccmb.med.umich.edu/COACH/) and highlighted using UCSF Chimera 1.8.

Results and Discussion

Identification of SiMYB genes

The availability of foxtail millet whole genome sequence [28,29] facilitated genome wide identification of MYB transcription factor (TF) gene family. The HMM search showed the presence of 229 MYB genes in foxtail millet genome (Table S2). The result was then confirmed by comparing the data with Foxtail millet Transcription Factor Database (FmTFDb; http://59.163.192.91/FmTFDb/) [31]. Further, all 229 SiMYB genes were BLAST searched against foxtail millet genome in Phytozome to identify alternate transcripts and chromosomal location. The results showed that 209 *SiMYBs* were primary transcripts and 20 were alternate transcripts. SiMYB017 had a maximum of three alternate transcripts, whereas SiMYB080, SiMYB137, SiMYB143 and SiMYB185 have two splice variants. Thirteen *SiMYBs* were evidenced to have one splice variant each.

The 209 SiMYB genes reported in the present study is higher than the numbers reported in Arabidopsis (197) [67], grape (118) [68] and Populus (192) [69]. On contrary, this number is lesser than soybean (244) [23] and apple (229) [70] (Figure S1). Among sequenced grass genomes, foxtail millet was evidenced to encode for a highest number of MYB TFs. Its closest relative sorghum has 100 MYB genes, whereas maize, rice and *Brachypodium* has 132, 155 and 98 MYB genes, respectively (http://planttfdb.cbi.pku.edu.cn/; Figure S1). Of note, further experimentations are requisite to identify the pseudogenes among the 209 SiMYBs.

SiMYBs location and duplication in foxtail millet genome

All the 209 SiMYB genes were physically mapped onto nine chromosomes of foxtail millet (Figure 1). The map revealed an uneven distribution of SiMYB genes in foxtail millet genome (Table S2). Highest number of SiMYB genes were observed in chromosome 5 (39; ~19%) and lowest in chromosome 8 (6; ~3%), with a density of 0.8 and 0.1 MYB genes/Mb, respectively. The distribution pattern of *SiMYBs* on individual chromosomes also revealed certain physical regions with a relatively higher accumulation of gene clusters. For instance, *SiMYBs* located on chromosomes 1, 3, 4, 5, 6 and 9 appear to be clustered at both upper end and lower end of the arms. Moreover, examination of gene duplication showed that SiMYB genes underwent both tandem and segmental duplications (Figure 1). Twelve gene-pairs were evidenced to be segmentally duplicated, whereas 15 genes were tandemly duplicated.

Structural classification of SiMYB proteins

The CDD search demonstrated the presence of conserved motifs in SiMYB proteins (Figure S2). It also showed the presence of more than one MYB repeats in some of these proteins along with the presence of additional domains. The hmmscan analysis provided comprehensive information on functional domains present in SiMYB proteins (Table S3). Based on the presence of one, two and three MYB repeats, the SiMYBs are classified into 'MYB-related', 'MYB-R2R3' and 'MYB-R1R2R3' (Table S2). The multiple sequence alignment of other three types of SiMYBs confirmed the presence of MYB repeats in the protein sequences (Figure S3–S5). Of the three types, 'MYB-R2R3' genes were found to be in maximum (114; ~55%) followed by 'MYB-related' (90; ~43%). Only 5 'MYB-R1R2R3' were evidenced in foxtail millet distributed one MYB-R1R2R3 encoding gene each in chromosome 1 to 5 (Figs. 1, 2). The predominance of 'MYB-R2R3' type *SiMYBs* suggest that they could have been evolved

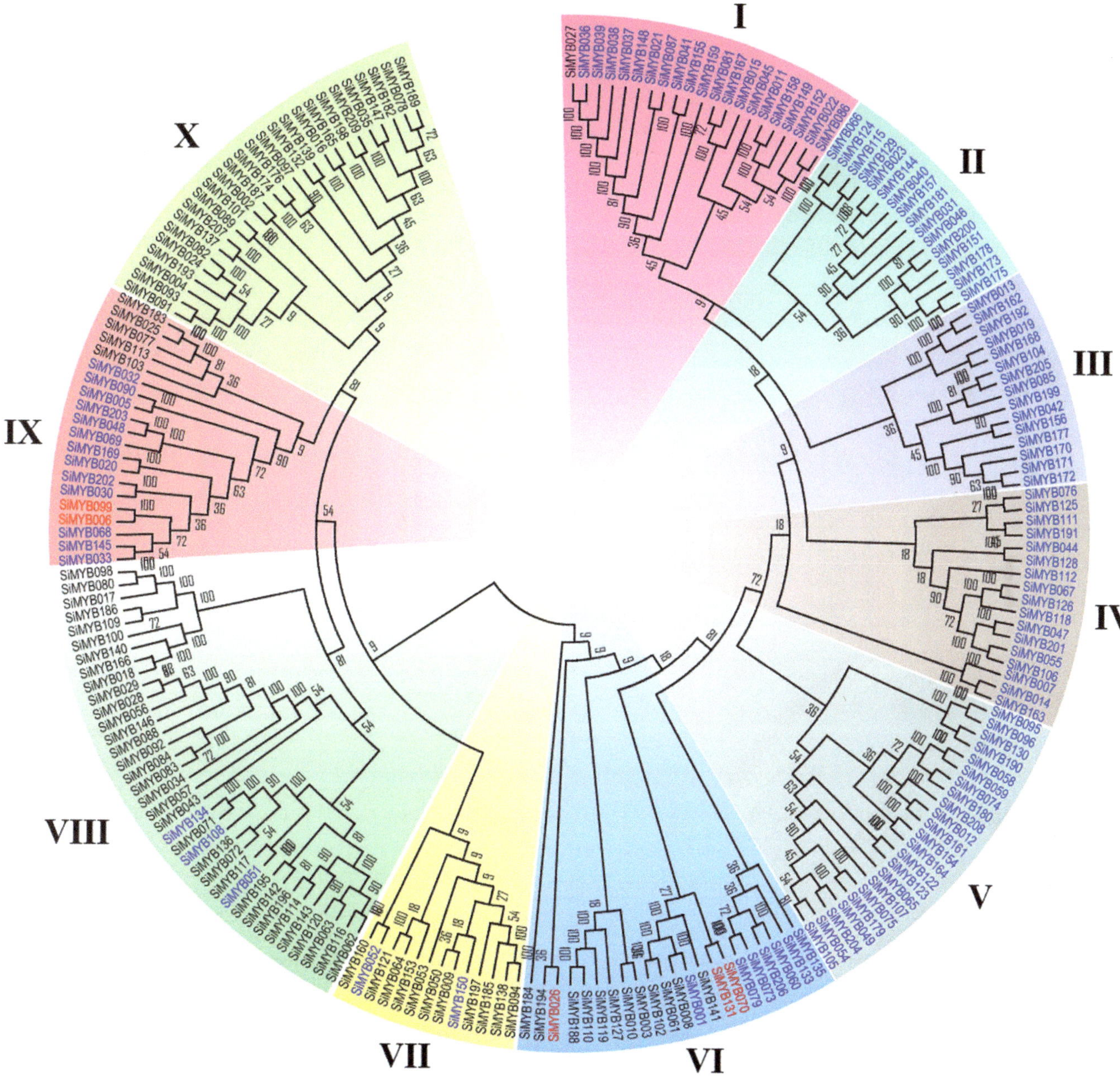

Figure 3. Phylogenetic relationships of foxtail millet MYB transcription factors. The unrooted phylogenetic tree was constructed by neighbor-joining method with 1000 bootstrap replicates. The bootstrap values are shown at the nodes. The tree was divided into ten phylogenetic cluster designated as I to X.

from MYB-R1R2R3 gene progenitor through loss of R1 repeat or from a MYB-R1 gene through duplication of R1 repeat [71,72]. Interestingly, no 'Atypical MYB' was identified in the foxtail millet, though one 'Atypical MYB' was reported in rice, two each in *Arabidopsis* and soybean [23,67]. Among 'MYB-R2R3' and 'MYB-related' genes, highest numbers were found to be encoded in chromosome 5 (19 each). The lowest number of these genes are in chromosome 8 (5; ~4% and 1; ~1%, respectively) (Figs. 1, 2).

Further, hmmscan analysis also revealed the presence of specific, functionally important domains present in C-terminal regions outside MYB domain which are involved in transcriptional activity and protein-protein interactions (Table S3). SiMYB003, SiMYB094, SiMYB160 and SiMYB185 possess SWIRM domain. It is a small alpha-helical domain of about 85 amino acid residues found in chromosomal proteins. It contains a helix-turn-helix motif and binds to DNA [73]. 'Response_reg' is 'response regulator receiver domain', which was first identified to receive signal from the sensor partner in bacterial two-component systems. SiMYB004, SiMYB024, SiMYB082, SiMYB091, SiMYB093, SiMYB137, SiMYB193 and SiMYB207 have this domain. It is usually found at N-terminal to a DNA binding effector domain and its role in plant systems remain elusive [74]. The domain 'P_C' in SiMYB205 denotes 'P protein C-terminus', which represents C-terminus of plant P proteins. The maize P gene is a transcriptional regulator of genes encoding enzymes for flavonoid biosynthesis in the pathway leading to production of a

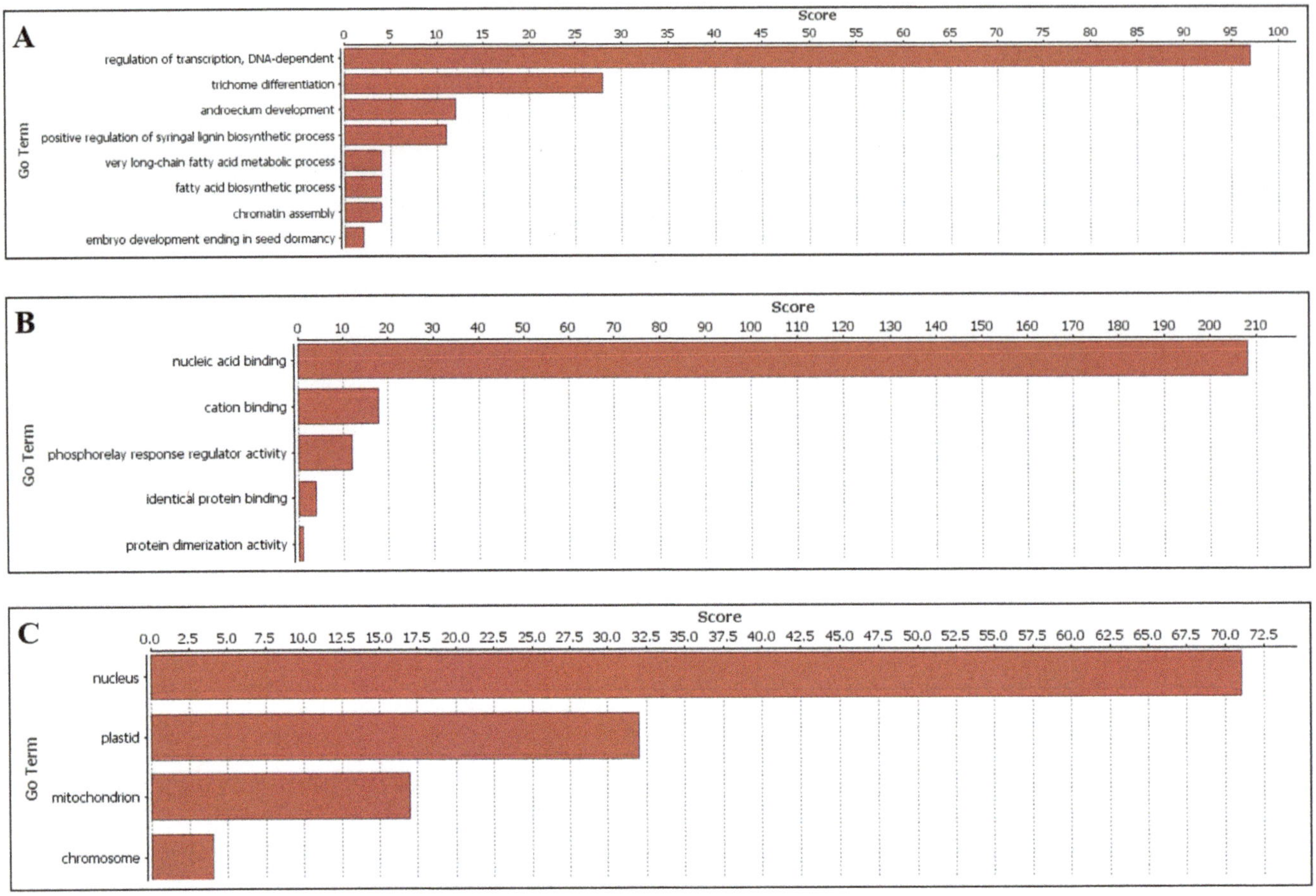

Figure 4. Gene Ontology distribution for SiMYB proteins. Blast2GO program defines the gene ontology under three categories. (**A**) biological processes, (**B**) molecular functions and (**C**) cellular component.

red phlobaphene pigment [75] and P proteins are homologous to DNA-binding domain of MYB-like transcription factors [76]. The 'ZZ' domain is ZZ type-Zinc finger whose binding to DNA is not yet identified [77]. This domain is present in SiMYB094, SiMYB184 and SiMYB185. It has 4–6 cysteine residues in its sequence, which are responsible for coordinating zinc ions and to reinforce the structure [78]. SiMYB153 was evidenced to have 'Bromodomain', which is reported to recognize acetylated lysine residues. This recognition is prerequisite for protein-histone association and chromatin remodelling [79]. The domain 'DUF3651' is present in SiMYB055 and SiMYB106, which denotes 'Domain of Unknown Function'. The 'Linker_histone' present in SiMYB110, SiMYB119, SiMYB127 and SiMYB188, which is an essential component of chromatin structure [80].

Except for the presence of conserved MYB domain, SiMYBs varied notably in size, sequence, intron-exon numbers and distribution and its physicochemical properties (Table S2). Gene structure analysis showed that 10 *SiMYB* were intron-less while *SiMYB026* and *SiMYB184* have maximum of 11 introns (Figure S6). The longest gene was *SiMYB094* with 8194 bases and the shortest was *SiMYB186* (348 bases). The longest protein sequence was SiMYB053 (1741 amino acids) and the least was SiMYB194 (98 amino acids) (Table S2). The enormous diversity with respect to MYB gene and proteins observed in foxtail millet was not reported in any other crop species [21–25], and this shows the presence of putative novel variants.

Phylogenetic classification of SiMYB proteins

The phylogenetic tree constructed using neighbour-joining (NJ) method for all 209 SiMYB proteins conformed to the phylogenetic classification enlisted in FmTFDb (Figure 3). The internal branches were detected to have high bootstrap values and this lead to the derivation of statistically reliable pairs of possible homologous proteins possessing similar functions from a common ancestor. Based on topology and clade support values, the phylogenetic tree was divided into ten clades (I to X) (Figure 3). Although SiMYB proteins tended to cluster together in the tree with respect to their type and they were not equally distributed in the clades which may be due to the occurrence of duplication and divergence of SiMYB genes (Figure 3).

Gene Ontology annotation

Gene Ontology (GO) annotation for all 209 SiMYB proteins was performed using Blast2GO. The SiMYB protein sequences were BLAST searched against *Oryza sativa* protein database and its biological processes, molecular functions and cellular components were identified (Figure 4). The results indicate that SiMYB proteins participate in diverse biological and molecular functions in the cell (Table S4). Noteworthy, predominant of SiMYBs were predicted to function in regulation of transcription in a DNA-dependent manner, followed by playing pivotal role in trichome differentiation and androecium development. SiMYBs are reported to least participate in breaking seed dormancy (Figure 4). The

Table 1. List of *Setaria italica* miRNAs targeting *SiMYB* transcripts.

miRNA ID	Target.	E-value	UPE	Target start	Target end	miRNA aligned fragment	Target aligned fragment	Inhibition
Sit-miR5568.58	SiMYB026	3.0	16.26	1923	1943	UUUCUAGGUUUAUAUCUUUUG	CACAAGAGAUAGAUCUAGAAA	Cleavage
Sit-miR2118d	SiMYB028	2.5	14.91	150	171	UUCCUGAUGCCUCUCAUUCCUA	UUGGAAGGAGAGGCAUCAGGUA	Cleavage
Sit-miR2118d	SiMYB029	2.5	13.89	152	171	UUCCUGAUGCCUCUCAUUCC	GGAAGGAGAGGCAUCAGGUA	Cleavage
Sit-miR159a	SiMYB130	2.5	16.88	960	979	UUUGGAUUGAAGGGAGCUCU	GGAGCUCCCUUCACUCCAAG	Cleavage
Sit-miR159a	SiMYB190	3.0	19.88	884	904	UUUGGAUUGAAGGGAGCUCUG	UGGAGCUCCCUUCAAACCAAU	Cleavage

UPE - unpaired energy.

biological function analysis also revealed that 68 (32.5%) SiMYB proteins function in response to stress (Table S4). The molecular role showed that most of SiMYBs have function in nucleic acid binding followed by cation binding. Cellular component data predicted that SiMYBs are nuclear localized. In addition, SiMYB proteins were also indicated to localize in plastids and mitochondria (Figure 4). In addition, Blast2GO was performed to correlate the domain composition of families/sub-families with functional classes, but no correlation was observed.

Cis-acting elements and *SiMYB* targeting miRNAs

The promoter analysis identified ~300 diverse *cis*-acting elements upstream of *SiMYB* genes (Table S5). Of these, ARR1AT, CAATBOX1, CACTFTPPCA1, DOFCOREZM, EBOXBNNAPA, GATABOX, GT1CONSENSUS, GTGANTG10, MYCCONSENSUSAT, POLLEN1LELAT52, WBOXNTERF3 and WRKY71OS were present in all 209 *SiMYB* genes. ARR1AT (NGATT) is a cytokinin response regulator binding motif [81] and CAATBOX1 (CAAT) are reported to regulate flowering [82]. CACTFTPPCA1 (CACGTG) is responsible for mesophyll-specific gene expression of C4 phosphoenolpyruvate carboxylase gene in C_4 plants [83] and DOFCOREZM (AAAG) is the binding cite of Dof transcription factors [84]. EBOXBNNAPA (CANNTG) is E-box sequence responsible for light responsiveness and is regulated by bHLH and MYB-transcription factor in directing tissue-specific expression [13]. GATABOX (GATA) is the binding site for transcription factors with a zinc finger motif [85], GT1CONSENSUS (GRWAAW) recognizes GT-1 proteins which have tri-helix DNA-binding domains [86] and GTGANTG10 (GTGA) is a pollen-specific *cis*-elements [87]. MYCCONSENSUSAT (CANNTG) is a MYC recognition site which regulates transcription of genes under cold conditions by a MYC-like bHLH transcriptional activator [88], POLLEN1LELAT52 (AGAAA) is a regulatory element responsible for pollen-specific activation of gene expression [89], WBOXNTERF3 (TGACY) is a W-box promoter motif functioning in response to wound signal [90] and WRKY71OS (TGAC) is a binding site of rice WRKY71, a transcriptional repressor of gibberellin signaling pathway [91]. In addition to these, some *cis*-elements were present unique to specific *SiMYBs* such as AAGACGTAGATACL12 in *SiMYB167*, ABADESI1 in *SiMYB025*, ABADESI2 in *SiMYB110*, ABRE3HVA22 in *SiMYB158* and ABREDISTBBNNAPA in *SiMYB122*. Further, *Setaria italica* miRNAs (Sit-miRs) targeting the *SiMYB* transcripts for post-transcriptional gene regulation were also identified. Three Sit-miRs namely Sit-miR5568.58, Sit-miR2118d and Sit-miR159a were evidenced to target *SiMYB026*, *SiMYB028*, *SiMYB029*, *SiMYB130* and *SiMYB190* (Table 1). This data would assist in deciphering post-transcriptional control of gene regulation during physiological and stress-induced cellular responses.

Heat map analysis and identification of transcription factor-derived markers

The publicly available Illumina RNA-Seq data for four tissues of foxtail millet root, leaf, stem and spica were used to analyse the expression pattern of SiMYB genes. The heat map showed differential expression of *SiMYBs* across the tissues (Figure 5). Predominant of *MYB* genes including *SiMYB003*, *SiMYB004*, *SiMYB005*, *SiMYB017*, *SiMYB025*, *SiMYB051*, *SiMYB064*, *SiMYB080*, *SiMYB089*, *SiMYB090*, *SiMYB0100* were predicted to be highly expressed in all the tissues. Interestingly, tissue-specific higher expression was also noted. *SiMYB007* and *SiMYB087* were highly expressed in leaf, while *SiMYB192* showed higher

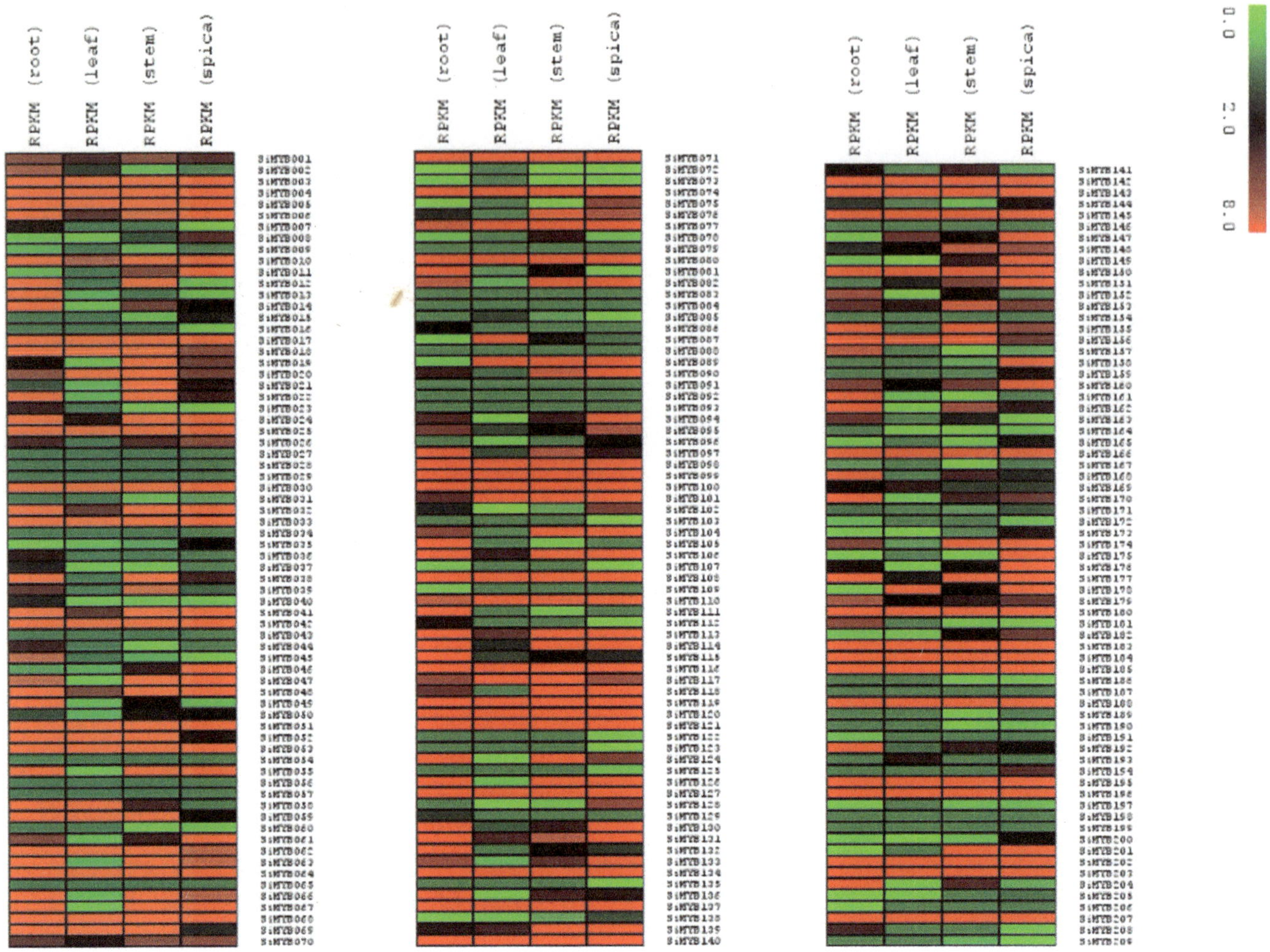

Figure 5. Heat map representation of SiMYB genes in four tissues. The Illumina RNA-seq data were re-analyzed and heat map was generated. Bar at the top represents log_2 transformed values, thereby values 0.0, 2.0 and 8.0 represent low, intermediate and high expression, respectively.

expression in root (Figure 5). This expression profiling would facilitate combinatorial usage of *SiMYBs* in transcriptional regulation of different tissues, whereas ubiquitously expressed *SiMYBs* might regulate the transcription of a broad set of genes. Further, this heat map data also enables selection and designing of studies to impart stress tolerance in both foxtail millet and related crop species.

In addition, molecular markers present in *SiMYBs* were also identified. Seventy three SiMYB genes (~35%) were evidenced to possess DNA markers (Table S6). Of the various types of markers reported in foxtail millet, presence of microsatellites (SSRs and eSSRs) in *SiMYB*s were found predominant (~90%) followed by ILP markers (~10%). Among the SSRs (63), tri-nucleotide repeats were maximum (41) followed by di-nucleotide repeats (20). Only two eSSRs namely SieSSR302 and SieSSR70 were found to be present in SiMYB030 and SiMYB094, respectively, and both the markers were of tri-nucleotide repeat type. SSRs SiGMS12780 and SiGMS8287 were a tetra- and hexa-nucleotide repeats, evidenced to be present in SiMYB179 and SiMYB125, respectively (Table S6). The functional relevance of these informative genic markers needs to be demonstrated by integrating agronomic trait association analysis with genetic mapping, differential expression profiling, protein modelling and haplotype gene evolution study in relation to *MYB* genes.

SiMYB orthologs in sorghum, maize, rice and *Brachypodium*

Of the 209 SiMYB genes, 72 (~34%) showed significant orthologous relationships with sorghum, 55 (~26%) with maize, 23 (~11%) with rice and 13 (~6%) with *Brachypodium* (Figure 6; Tables S7–S10). Foxtail millet chromosome 5 showed highest synteny with sorghum chromosome 3 (14; ~19%), and maize chromosomes 3 and 8 (11; 20%). Chromosome 3 of foxtail millet has highest orthologous pairs with rice chromosomes 4, 5 and 12 (~22%), and *Brachypodium* chromosomes 2, 4 and 5 (~38%). Except the presence of conserved nucleotide and amino acid sequences of MYB domain between foxtail millet-rice and foxtail millet-*Brachypodium*, the regions other than the MYB domain were highly diverse. The degree of decrease in orthologous relationships among foxtail millet-sorghum, -maize, -rice and *Brachypodium* is due to genetic relatedness of these genomes. Foxtail millet along with sorghum and maize belong to Panicoideae, whereas rice and *Brachypodium* belong to Ehrhartoideae and Pooideae, respectively. This pattern of decreasing

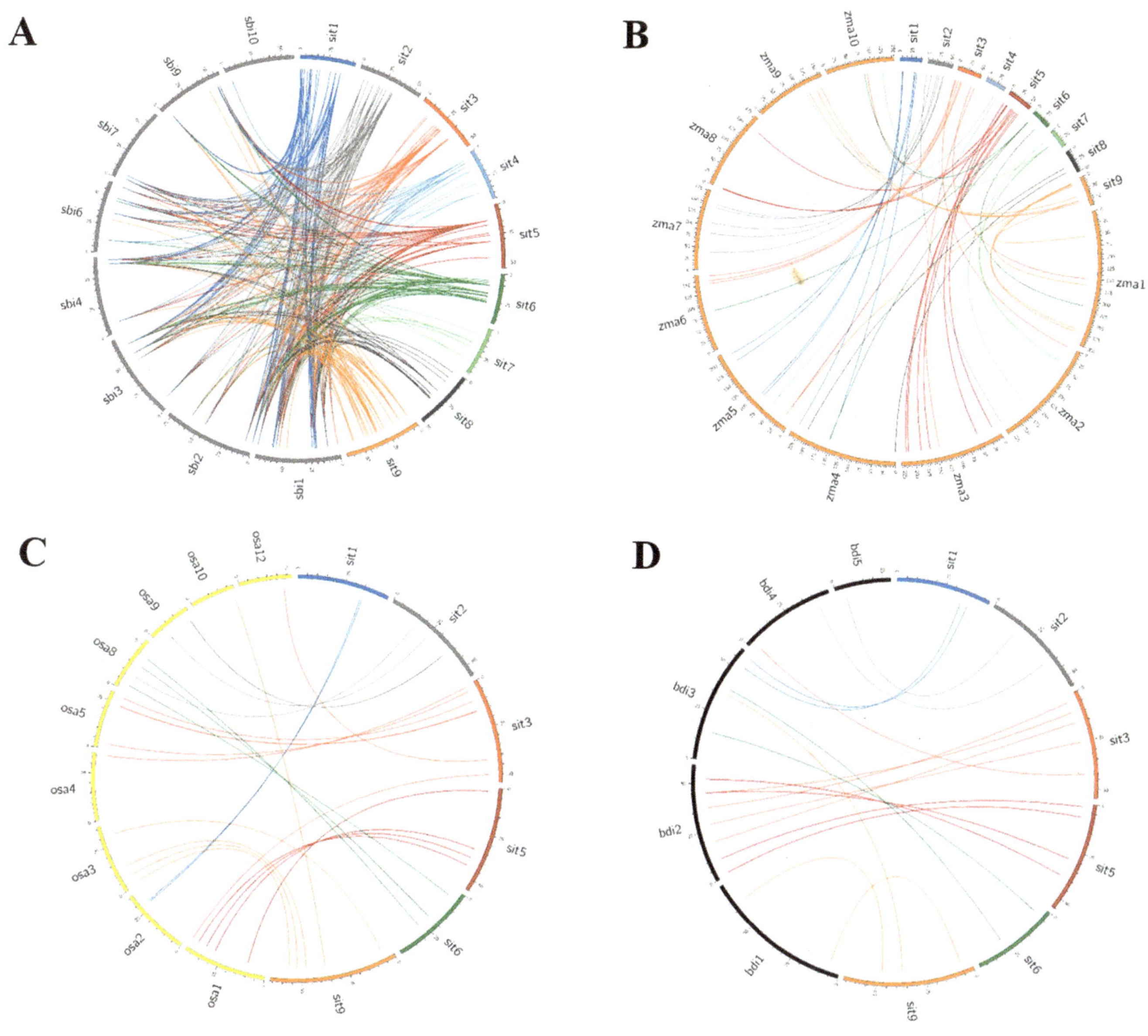

Figure 6. Comparative physical mapping revealed high degree of orthologous relationships of SiMYB genes located on nine chromosomes of foxtail millet. (**A**) *Sit* (foxtail millet) – *Sbi* (sorghum), (**B**) *Sit* – *Zma* (maize), (**C**) *Sit* – *Osa* (rice), and (**D**) *Sit* – *Bdi* (*Brachypodium*).

synteny was also observed in the comparative mapping of NAC TFs [35], WD40 genes [36], Dicer-like, Argonaute, RNA-dependent RNA polymerase genes [92], and C_2H_2-type zinc finger TFs [59]. This comparative map would enable the researchers to select candidate MYB genes from foxtail millet and utilize them in genetic enhancement of related grass family members.

Evolutionary aspects of duplication and divergence

The ratios of non-synonymous (Ka) versus synonymous (Ks) substitution rate (Ka/Ks) were estimated for duplicated and orthologous gene-pairs of *SiMYB*. The ratios of Ka/Ks for segmentally and tandemly duplicated gene-pairs ranged from 0.02 to 0.1 and 0.07 to 0.1, respectively (Tables S11–S12). The estimated Ka/Ks was <1, and this justified that the duplicated SiMYB genes were under strong purifying selection pressure. This is in agreement with previous genome-wide reports on NAC TFs by Puranik et al. [35], WD40 gene family by Mishra et al. [36], and C_2H_2-type zinc finger TFs by Muthamilarasan [59]. In addition, it was also estimated that the duplication event of these segmentally and tandemly duplicated genes occurred ~33 Mya (Figure 7). Although further experimentations are necessary to validate this, the present study suggests that these duplication events must be specific to *Setaria* since divergence of sorghum and maize had occurred ~13 Mya [28,29].

Among orthologous gene-pairs of *SiMYB*, average Ka/Ks value was maximum between foxtail millet and *Brachypodium* followed by foxtail millet and rice (Figure 7). The Ka/Ks was minimum for foxtail millet-sorghum gene-pairs. The relatively higher rate of synonymous substitution between *Brachypodium* and foxtail millet *MYB* genes suggested their earlier divergence around 42 Mya. The evolutionary analysis also predicts the divergence time of foxtail millet-rice as ~39 MYA, foxtail millet-maize as ~29 MYA and foxtail millet-sorghum as ~27 MYA (Figure 7). These Ka/Ks

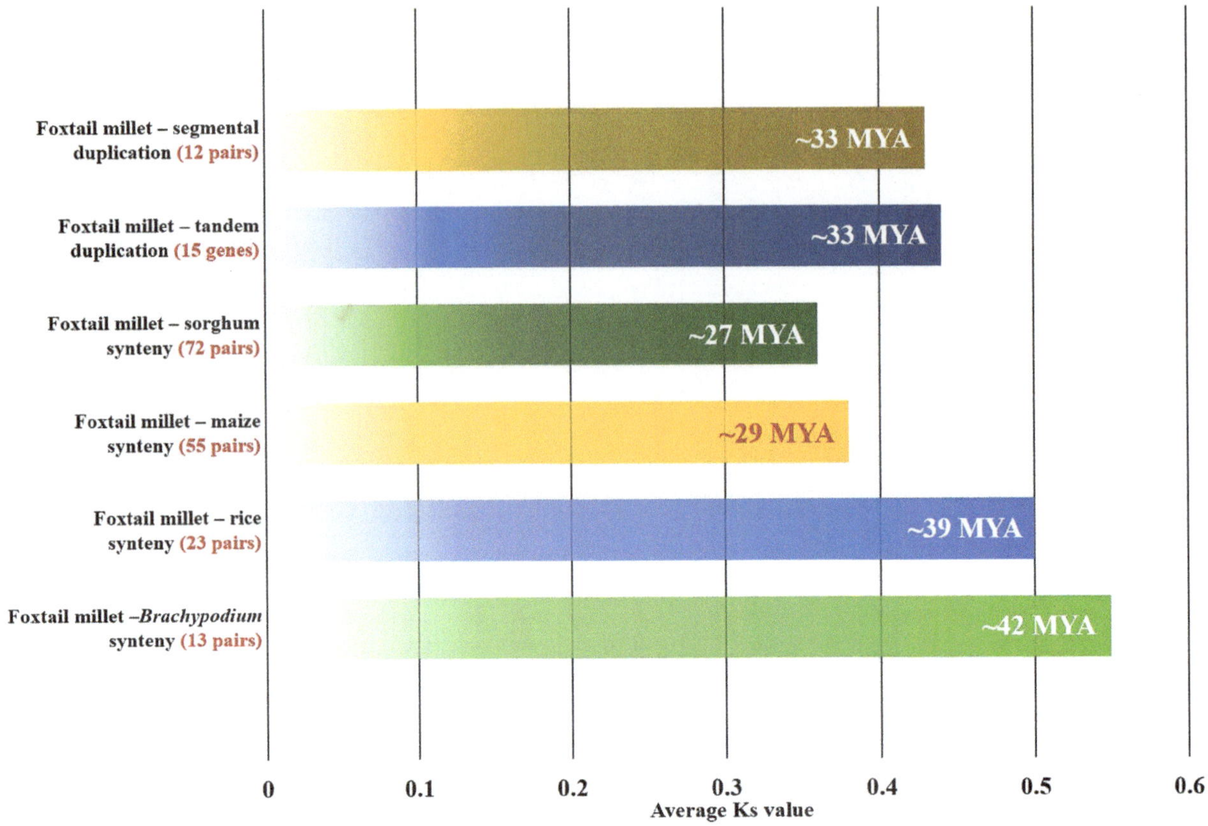

Figure 7. Time of duplication and divergence (MYA) based on synonymous substitution rate (Ks) estimated using duplicated SiMYB gene-pairs of foxtail millet and orthologous *MYB* gene pairs between foxtail millet- sorghum, -maize, -rice and *Brachypodium*.

dating data along with comparative map would assist in understanding the evolution of MYB genes in grasses genomes.

Expression pattern of *SiMYBs* under abiotic stress and hormone treatments

To study the expression patterns of identified *SiMYBs*, 11 candidate SiMYB genes were chosen representing all the nine chromosomes of foxtail millet and all the 10 clades of phylogenetic tree. The relative transcript abundance of these candidate *MYBs* at both early (1 h) and late (24 h) duration of abiotic stresses (salinity and dehydration), and hormone treatments (ABA, MJ, SA) were studied using qRT-PCR (Figure 8). During early phase of salinity stress *SiMYB003*, *SiMYB100*, *SiMYB124* and *SiMYB174* were highly up-regulated whereas *SiMYB041*, *SiMYB074*, *SiMYB126*, *SiMYB150* and *SiMYB202* were down-regulated. Though *SiMYB100* and *SiMYB174* showed significant up-regulation during early hours of salinity stress, a drastic down-regulation was observed at late phase. *SiMYB150* showed ~2 fold higher expression during late phase of salinity stress. In course of dehydration stress, up-regulation during early phase was evidenced in *SiMYB074*, *SiMYB100*, *SiMYB174* and *SiMYB202*. SiMYB124 showed higher expression at both early and late phases. A drastic up-regulation of SiMYB genes at 24 h after dehydrations stress was noticed in *SiMYB126, SiMYB110* and *SiMYB184*.

During hormone treatments, *SiMYB126*, *SiMYB150* and *SiMYB177* were down-regulated in all the treatments and time points, except a considerable higher expression of *SiMYB126* and *SiMYB177* at 24 h post ABA treatment. *SiMYB003* was down-regulated at both the phases of ABA treatment, while significant up-regulation at 24 h post ABA application was observed in *SiMYB184* and *SiMYB202*. The genes *SiMYB003*, *SiMYB100* and *SiMYB184* were up-regulated during early phase of MJ treatment, whereas up-regulation was observed at late phase in *SiMYB074* and *SiMYB202*. Up-regulation of *SiMYB100* and *SiMYB184* was evidenced in early phase of SA treatment. After 24 h, *SiMYB003* and *SiMYB202* showed higher expression.

Since foxtail millet is well known for its adaptability to abiotic stresses such as dehydration and salinity [60–62], deciphering the role of MYB TFs will provide novel insights into the stress tolerance mechanism of this model crop. This preliminary study on expression profiling of candidate SiMYB genes in response to different stress and hormone treatments would serve as a foundation for downstream characterization of these TFs.

Homology modelling of SiMYB proteins

Three dimensional protein models were constructed for all 209 SiMYBs and of note, the proteins were modelled at 90% confidence (Table S13). Further, its potential active sites were also identified (Figure S7). As mentioned earlier, the SiMYBs were classified into 'MYB-related', 'MYB-R2R3' and 'MYB-R1R2R3' based on the presence of one, two and three MYB repeats in the N-terminus. The homology modelling showed that each MYB repeat comprised of 52 amino acids and encoded three α-helices, with the second and third helices forming a helix-turn-helix (HTH) structure. Further, it included three regularly spaced tryptophan residues, forming a hydrophobic core in the three dimensional HTH structure (Figure S7). The third helix of each repeat formed

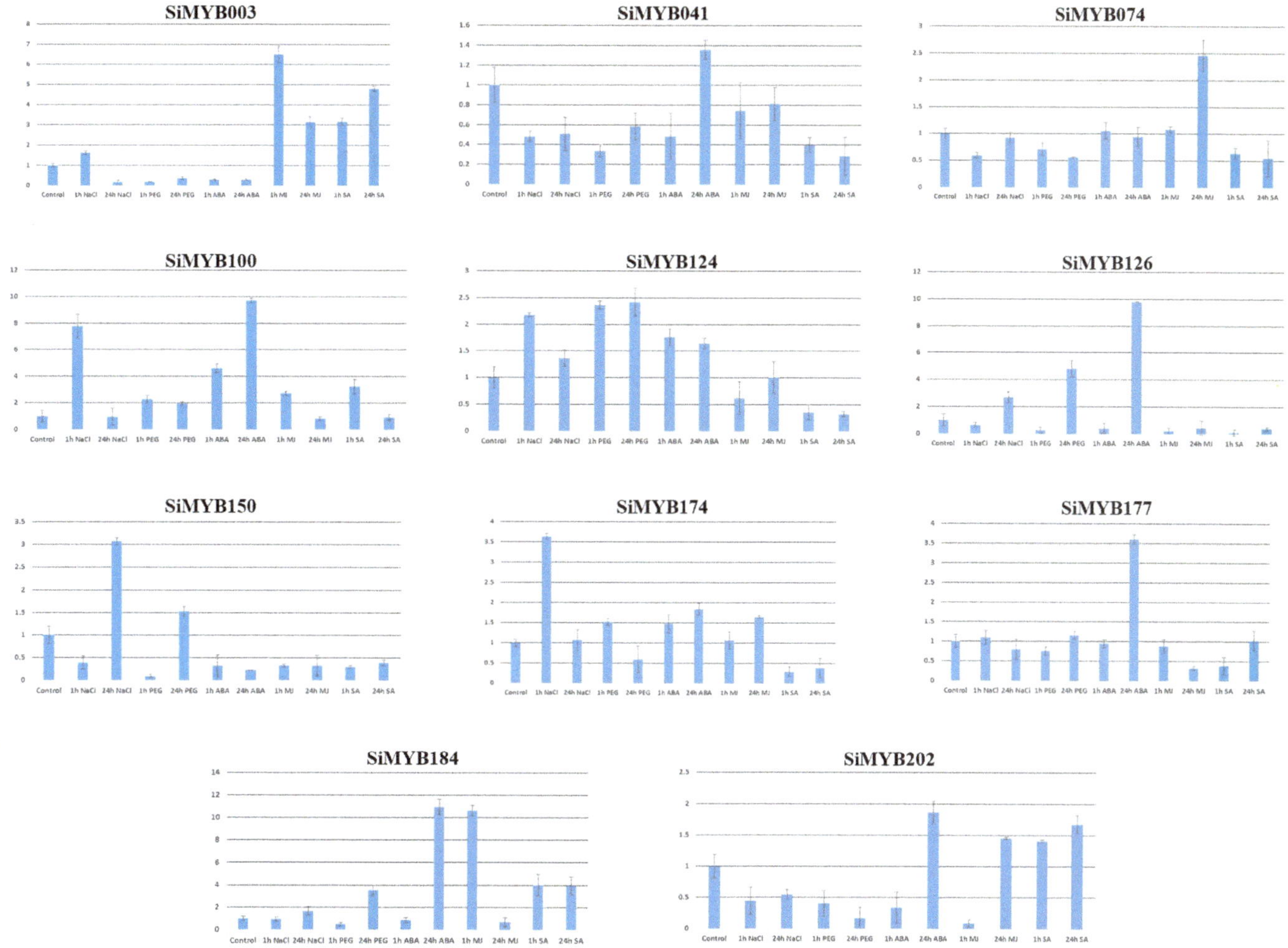

Figure 8. The relative expression ratio of 11 candidate SiMYB genes analysed using qRT-PCR under salinity stress, dehydration stress and hormone treatments for 1 h (early) and 24 h (late). The relative expression ratio of each gene was calculated relative to its expression in control sample (0 h). *Act2* was used as an internal control to normalize the data. Error bars representing standard deviation were calculated based on three technical replicates for each biological duplicate. *ABA - abscisic acid, MJ - methyl jasmonate, SA - salicylic acid.*

'recognition helix', which was reported to interact with DNA and intercalate in the major groove [6]. Contrarily, the C-terminus varied considerably between SiMYB proteins and this could be responsible for the wide range of regulatory roles of SiMYB gene family. This preliminary data would provide a base for further molecular structure analyses and interaction studies of SiMYB proteins.

In summary, the present study identified 209 MYB proteins in model crop foxtail millet and physically mapped them onto the genome. The protein structure and gene organization of individual SiMYBs were analysed and a phylogenetic tree was constructed. The domain analysis showed the presence of diverse protein domains in MYB proteins and the multiple sequence alignment revealed the conserved sequences in these domains. In addition, DNA markers present in *SiMYB* genes were also investigated. Candidate *SiMYB*s were chosen for expression profiling through qRT-PCR and the analysis showed the role of these genes in response to abiotic stresses and hormone treatments. The present study is the first report on genome-wide identification, characterization and expression profiling of MYB transcription factors in foxtail millet and promisingly, the data obtained from this study would contribute to a better understanding of the complexity of MYB TFs in plants and provide a useful reference for further functional analysis of MAPK genes in foxtail millet and related grasses.

Supporting Information

Figure S1 Phylogenetic relationships between all the sequenced plant species. The total number of MYB proteins found in each genome is indicated on the right. The data was retrieved from PlantTFDB (http://planttfdb.cbi.pku.edu.cn/). The data of foxtail millet (*Setaria italica*) excludes alternate transcripts.

Figure S2 Conserved domains in SiMYB proteins, identified using CDD database.

Figure S3 The multiple sequence alignment of 'MYB-related' proteins.

Figure S4 The multiple sequence alignment of 'MYB-R2R3' proteins.

Figure S5 The multiple sequence alignment of 'MYB-R1R2R3' proteins.

Figure S6 Gene structures of SiMYB proteins. Exons and introns are represented by green boxes and black lines, respectively.

Figure S7 Predicated three dimensional structures of all the 209 SiMYB proteins.

Table S1 Details of primers used for quantitative real-time PCR.

Table S2 Characteristic features of 209 MYB Transcription factor gene family members identified in *Setaria italica*.

Table S3 Summary of functional domains present in the 209 SiMYB proteins.

Table S4 Blast2GO annotation details of SiMYB protein sequences.

Table S5 Characteristics of the promoter region of SiMYB genes.

Table S6 Details of DNA markers present in SiMYB genes.

Table S7 The Ka/Ks ratio, estimated divergence time and percentage identity for orthologous SiMYB proteins between foxtail millet and sorghum.

Table S8 The Ka/Ks ratio, estimated divergence time and percentage identity for orthologous SiMYB proteins between foxtail millet and maize.

Table S9 The Ka/Ks ratio, estimated divergence time and percentage identity for orthologous SiMYB proteins between foxtail millet and rice.

Table S10 The Ka/Ks ratio, estimated divergence time and percentage identity for orthologous SiMYB proteins between foxtail millet and *Brachypodium*.

Table S11 The Ka/Ks ratios and estimated divergence time for segmentally duplicated SiMYB genes.

Table S12 The Ka/Ks ratios and estimated divergence time for tandemly duplicated SiMYB genes.

Table S13 List of templates used in homology modelling of 209 SiMYB proteins.

Acknowledgments

MM acknowledges the award of Junior Research Fellowship from University Grants Commission, New Delhi, India. Assistance from Ms Jananee Jaishankar and Mr Subodh Verma, NIPGR is greatly appreciated.

Author Contributions

Conceived and designed the experiments: MP. Performed the experiments: MM RK CBY VSB YK. Analyzed the data: MP MM. Wrote the paper: MM MP.

References

1. Paz-Ares J, Ghosal D, Wienand U, Peterson PA, Saedler H (1987) The regulatory c1 locus of *Zea mays* encodes a protein with homology to myb proto-oncogene products and with structural similarities to transcriptional activators. EMBO J 6: 3553–3558.
2. Lipsick JS (1996) One billion years of Myb. Oncogene 13: 223–235.
3. Stracke R, Werber M, Weisshaar B (2001) The R2R3-MYB gene family in *Arabidopsis thaliana*. Curr Opin Plant Bio l4: 447–456.
4. Ogata K, Morikawa S, Nakamura H, Hojo H, Yoshimura S, et al. (1995) Comparison of the free and DNA-complexed forms of the DNA-binding domain from c-Myb. Nat Struct Biol 2: 309–320.
5. Jin H, Martin C (1999) Multifunctionality and diversity within the plant MYB-gene family. Plant Mol Biol 41: 577–585.
6. Dubos C, Stracke R, Grotewold E, Weisshaar B, Martin C, et al. (2010) MYB transcription factors in Arabidopsis. Trends Plant Sci 15: 573–581.
7. Goicoechea M, Lacombe E, Legay S, Mihaljevic S, Rech P, et al. (2005) EgMYB2, a new transcriptional activator from Eucalyptus xylem, regulates secondary cell wall formation and lignin biosynthesis. Plant J 43: 553–567.
8. Mehrtens F, Kranz H, Bednarek P, Weisshaar B (2005) The Arabidopsis transcription factor MYB12 is a flavonol-specific regulator of phenylpropanoid biosynthesis. Plant Physiol 138: 1083–1096.
9. Bedon F, Grima-Pettenati J, Mackay J (2007) Conifer R2R3-MYB transcription factors: sequence analyses and gene expression in wood-forming tissues of white spruce (*Picea glauca*). BMC Plant Biol 7: 17.
10. Stracke R, Ishihara H, Huep G, Barsch A, Mehrtens F, et al. (2007) Differential regulation of closely related R2R3-MYB transcription factors controls flavonol accumulation in different parts of the *Arabidopsis thaliana* seedling. Plant J 50: 660–677.
11. Gocal GF, Poole AT, Gubler F, Watts RJ, Blundell C, et al. (1999) Long-day up-regulation of a GAMYB gene during *Lolium temulentum* inflorescence formation. Plant Physiol 119: 1271–1278.
12. Abe H, Urao T, Ito T, Seki M, Shinozaki K, et al. (2003) Arabidopsis AtMYC2 (bHLH) and AtMYB2 (MYB) function as transcriptional activators in abscisic acid signaling. Plant Cell 15: 63–78.
13. Hartmann U, Sagasser M, Mehrtens F, Stracke R, Weisshaar B (2005) Differential combinatorial interactions of cis-acting elements recognized by R2R3-MYB, BZIP, and BHLH factors control light-responsive and tissue-specific activation of phenylpropanoid biosynthesis genes. Plant Mol Biol 57: 155–171.
14. Jung C, Seo JS, Han SW, Koo YJ, Kim CH, et al. (2008) Overexpression of AtMYB44 enhances stomatal closure to confer abiotic stress tolerance in transgenic Arabidopsis. Plant Physiol 146: 623–635.
15. Adler G, Blumwald E, Bar-Zvi D (2010) The sugar beet gene encoding the sodium/proton exchanger 1 (BvNHX1) is regulated by a MYB transcription factor. Planta 232: 187–195.
16. Higginson T, Li SF, Parish RW (2003) AtMYB103 regulates tapetum and trichome development in *Arabidopsis thaliana*. Plant J 35: 177–192.
17. Perez-Rodriguez M, Jaffe FW, Butelli E, Glover BJ, Martin C (2005) Development of three different cell types is associated with the activity of a specific MYB transcription factor in the ventral petal of *Antirrhinum majus* flowers. Development 132: 359–370.
18. Baumann K, Perez-Rodriguez M, Bradley D, Venail J, Bailey P, et al. (2007) Control of cell and petal morphogenesis by R2R3 MYB transcription factors. Development 134: 1691–1701.
19. Pattanaik S, Kong Q, Zaitlin D, Werkman JR, Xie CH, et al. (2010) Isolation and functional characterization of a floral tissue-specific R2R3 MYB regulator from tobacco. Planta 231: 1061–1076.
20. Muthamilarasan M, Theriappan P, Prasad M (2013) Recent advances in crop genomics for ensuring food security. Curr Sci 105: 155–158.

21. Dias AP, Braun EL, McMullen MD, Grotewold E (2003) Recently duplicated maize R2R3 Myb genes provide evidence for distinct mechanisms of evolutionary divergence after duplication. Plant Physiol 131: 610–620.
22. Li Q, Zhang C, Li J, Wang L, Ren Z (2012) Genome-wide identification and characterization of R2R3MYB family in *Cucumis sativus*. PLoS ONE 7: e47576.
23. Du H, Yang SS, Feng BR, Liu L, Tang YX (2012) Genome-wide analysis of the MYB transcription factor superfamily in soybean. BMC Plant Biol 12: 106.
24. Du H, Feng BR, Yang SS, Huang YB, Tang YX (2012) The R2R3-MYB transcription factor gene family in maize. PLoS ONE 7: e37463.
25. Du H, Wang YB, Xie Y, Liang Z, Jiang SJ, et al. (2013) Genome-wide identification and evolutionary and expression analyses of MYB-related genes in land plants. DNA Res 20: 437–448.
26. Doust AN, Kellogg EA, Devos KM, Bennetzen JL (2009) Foxtail millet: A sequence-driven grass model system. Plant Physiol 149: 137–141.
27. Lata C, Gupta S, Prasad M (2013) Foxtail millet: a model crop for genetic and genomic studies in bioenergy grasses. Crit Rev Biotechnol 33: 328–343.
28. Zhang G, Liu X, Quan Z, Cheng S, Xu X, et al. (2012) Genome sequence of foxtail millet (*Setaria italica*) provides insights into grass evolution and biofuel potential. Nature Biotech 30: 549–554.
29. Bennetzen JL, Schmutz J, Wang H, Percifield R, Hawkins J, et al. (2012) Reference genome sequence of the model plant *Setaria*. Nature Biotechnol 30: 555–561.
30. Finn RD, Bateman A, Clements J, Coggill P, Eberhardt RY, et al. (2014) The Pfam protein families database. Nucleic Acids Res 42: D222–D230.
31. Bonthala VS, Muthamilarasan M, Roy R, Prasad M (2014) FmTFDb: a foxtail millet transcription factors database for expediting functional genomics in millets. Mol Biol Rep DOI: 10.1007/s11033-014-3574-y.
32. Voorrips RE (2002) MapChart: software for the graphical presentation of linkage maps and QTLs. J Hered 93: 77–78.
33. Du D, Zhang Q, Cheng T, Pan H, Yang W, et al. (2012) Genome-wide identification and analysis of late embryogenesis abundant (LEA) genes in *Prunus mume*. Mol Biol Rep 40: 1937–1946.
34. Shiu S-H, Bleecker AB (2003) Expansion of the Receptor-Like Kinase/Pelle Gene Family and Receptor-Like Proteins in *Arabidopsis*. Plant Physiol 132: 530–543.
35. Puranik S, Sahu PP, Mandal SN, B VS, Parida SK, et al. (2013) Comprehensive genome-wide survey, genomic constitution and expression profiling of the NAC transcription factor family in foxtail millet (*Setaria italica* L.). PLoS ONE 8: e64594.
36. Mishra AK, Muthamilarasan M, Khan Y, Parida SK, Prasad M (2014) Genome-wide investigation and expression analyses of WD40 protein family in the model plant foxtail millet (*Setaria italica* L.). PLoS ONE 9: e86852.
37. Wang Y, Tang H, DeBarry JD, Tan X, Li J, Wang X, et al. (2012) MCScanX: a toolkit for detection and evolutionary analysis of gene synteny and collinearity. Nucleic Acids Res 40: e49.
38. Guo AY, Zhu QH, Chen X, Luo JC (2007) GSDS: a gene structure display server. Yi Chuan 29: 1023–1026.
39. Tamura K, Stecher G, Peterson D, Filipski A, Kumar S (2013) MEGA6: Molecular Evolutionary Genetics Analysis version 6.0. Mol Biol Evol 30: 2725–2729.
40. Götz S, García-Gómez JM, Terol J, Williams TD, Nagaraj SH, et al. (2008) High-throughput functional annotation and data mining with the Blast2GO suite. Nucleic Acids Res 36: 3420–3435.
41. Higo K, Ugawa Y, Iwamoto M, Korenaga T (1999) Plant cis-acting regulatory DNA elements (PLACE) database: 1999. Nucleic Acids Res 27: 297–300.
42. Yi F, Xie S, Liu Y, Qi X, Yu J (2013) Genome-wide characterization of microRNA in foxtail millet (*Setaria italica*). BMC Plant Biol 13: 212.
43. Khan Y, Yadav A, Suresh BV, Muthamilarasan M, Yadav CB, et al. (2014) Comprehensive genome-wide identification and expression profiling of foxtail millet [*Setaria italica* (L.)] miRNAs in response to abiotic stress and development of miRNA database. Plant Cell Tiss Organ Cult 118: 279–292.
44. Dai X, Zhao PX (2013) psRNATarget: a plant small RNA target analysis server. Nucleic Acids Res 39: W155–W159.
45. Cochrane G, Alako B, Amid C, Bower L, Cerdeño-Tárraga A, et al. (2013) Facing growth in the European Nucleotide Archive. Nucleic Acids Res 41: D30–D35.
46. Patel RK, Jain M (2012) NGS QC Toolkit: A toolkit for quality control of next generation sequencing data. PLoS ONE 7: e30619.
47. Langmead B, Salzberg SL (2012) Fast gapped-read alignment with Bowtie 2. Nat Methods 9: 357–359.
48. Saeed AI, Bhagabati NK, Braisted JC, Liang W, Sharov V, et al. (2006) TM4 microarray software suite. Methods Enzymol 411: 134–193.
49. Saeed AI, Sharov V, White J, Li J, Liang W, et al. (2003) TM4: a free, open-source system for microarray data management and analysis. Biotechniques 34: 374–378.
50. Suresh BV, Muthamilarasan M, Misra G, Prasad M (2013) FmMDb: a versatile database of foxtail millet markers for millets and bioenergy grasses research. PLoS ONE 8: e71418.
51. Gupta S, Kumari K, Sahu PP, Vidapu S, Prasad M (2013) Sequence-based novel genomic microsatellite markers for robust genotyping purposes in foxtail millet [*Setaria italica* (L.) P. Beauv]. Plant Cell Rep 31: 323–337.
52. Pandey G, Misra G, Kumari K, Gupta S, Parida SK, et al. (2013) Genome-wide development and use of microsatellite markers for large-scale genotyping applications in foxtail millet [*Setaria italica* (L.)]. DNA Res 20: 197–207.
53. Kumari K, Muthamilarasan M, Misra G, Gupta S, Subramanian A, et al. (2013) Development of eSSR-markers in *Setaria italica* and their applicability in studying genetic diversity, cross-transferability and comparative mapping in millet and non-millet species. PLoS ONE 8: e67742.
54. Gupta S, Kumari K, Das J, Lata C, Puranik S, et al. (2011) Development and utilization of novel intron length polymorphic markers in foxtail millet (*Setaria italica* (L.) P. Beauv.) Genome 54: 586–602.
55. Muthamilarasan M, Venkata Suresh B, Pandey G, Kumari K, Parida SK, et al. (2014b) Development of 5123 intron-length polymorphic markers for large-scale genotyping applications in foxtail millet. DNA Res 21: 41–52.
56. Suyama M, Torrents D, Bork P (2006) PAL2NAL: robust conversion of protein sequence alignments into the corresponding codon alignments. Nucleic Acids Res 34: W609–W612.
57. Lynch M, Conery JS (2000) The evolutionary fate and consequences of duplicate genes. Science 290: 1151–1155.
58. Yang Z, Gu S, Wang X, Li W, Tang Z, et al. (2008) Molecular evolution of the cpp-like gene family in plants: insights from comparative genomics of *Arabidopsis* and rice. J Mol Evol 67: 266–277.
59. Muthamilarasan M, Bonthala VS, Mishra AK, Khandelwal R, Khan Y, et al. (2014) C_2H_2 type of zinc finger transcription factors in foxtail millet define response to abiotic stresses. Funct Integr Genomics 14: 531–543.
60. Lata C, Sahu PP, Prasad M (2010) Comparative transcriptome analysis of differentially expressed genes in foxtail millet (*Setaria italica* L.) during dehydration stress. Biochem Biophys Res Commun 393: 720–727.
61. Puranik S, Jha S, Srivastava PS, Sreenivasulu N, Prasad M (2011) Comparative transcriptome analysis of contrasting foxtail millet cultivars in response to short-term salinity stress. J Plant Physiol 168: 280–287.
62. Lata C, Bhutty S, Bahadur RP, Majee M, Prasad M (2011) Association of an SNP in a novel DREB2-like gene SiDREB2 with stress tolerance in foxtail millet (*Setaria italica* (L.)). J Exp Bot 62: 3387–3401.
63. Le DT, Nishiyama R, Watanabe Y, Mochida K, Yamaguchi-Shinozaki K, et al. (2011) Genome-wide survey and expression analysis of the plant-specific NAC transcription factor family in soybean during development and dehydration stress. DNA Res 18: 263–276.
64. Longeman J, Schell J, Willmitzer L (1987) Improved method for the isolation of RNA from plant tissues. Anal Biochem 163: 16–20.
65. Kumar K, Muthamilarasan M, Prasad M (2013) Reference genes for quantitative Real-time PCR analysis in the model plant foxtail millet (*Setaria italica* L.) subjected to abiotic stress conditions. Plant Cell Tiss Organ Cult 115: 13–22.
66. Kelley LA, Sternberg MJE (2009) Protein structure prediction on the Web: a case study using the Phyre server. Nature Protocols 4: 363–371.
67. Katiyar A, Smita S, Lenka SK, Rajwanshi R, Chinnusamy V, et al. (2012) Genome-wide classification and expression analysis of MYB transcription factor families in rice and Arabidopsis. BMC Genomics 13: 544.
68. Matus JT, Aquea F, Arce-Johnson P (2008) Analysis of the grape MYB R2R3 subfamily reveals expanded wine quality-related clades and conserved gene structure organization across Vitis and Arabidopsis genomes. BMC Plant Biol 8: 83.
69. Wilkins O, Nahal H, Foong J, Provart NJ, Campbell MM (2009) Expansion and diversification of the Populus R2R3–MYB family of transcription factors. Plant Physiol 149: 981–993.
70. Cao ZH, Zhang SZ, Wang RK, Zhang RF, Hao YJ (2013) Genome wide analysis of the apple MYB transcription factor family allows the identification of MdoMYB121 gene confering abiotic stress tolerance in plants. PLoS ONE 8: e69955.
71. Jiang C, Gu J, Chopra S, Gu X, Peterson T (2004) Ordered origin of the typical two– and three-repeat Myb genes. Gene 326: 13–22.
72. Rosinski JA, Atchley WR (1998) Molecular evolution of the Myb family of transcription factors: evidence for polyphyletic origin. J Mol Evol 46: 74–83.
73. Da G, Lenkart J, Zhao K, Shiekhattar R, Cairns BR, et al. (2006) Structure and function of the SWIRM domain, a conserved protein module found in chromatin regulatory complexes. Proc Natl Acad Sci U S A 103: 2057–2062.
74. Pao GM, Saier MH (1995) Response regulators of bacterial signal transduction systems: selective domain shuffling during evolution. J Mol Evol 40: 136–154.
75. Chopra S, Athma P, Peterson T (1996) Alleles of the maize P gene with distinct tissue specificities encode Myb-homologous proteins with C-terminal replacements. Plant Cell 8: 1149–1158.
76. Grotewold E, Drummond BJ, Bowen B, Peterson T (1994) The myb-homologous P gene controls phlobaphene pigmentation in maize floral organs by directly activating a flavonoid biosynthetic gene subset. Cell 76: 543–553.
77. Ponting CP, Blake DJ, Davies KE, Kendrick-Jones J, Winder SJ (1996) ZZ and TAZ: new putative zinc fingers in dystrophin and other proteins. Trends Biochem Sci 21: 11–13.
78. Roberts RG (2001) Dystrophins and dystrobrevins. Genome Biol 2: REVIEWS3006.
79. Dhalluin C, Carlson JE, Zeng L, He C, Aggarwal AK, et al. (1999) Structure and ligand of a histone acetyltransferase bromodomain. Nature 399: 491–496.
80. Ramakrishnan V, Finch JT, Graziano V, Lee PL, Sweet RM (1993) Crystal structure of globular domain of histone H5 and its implications for nucleosome binding. Nature 362: 219–223.

81. Sakai H, Aoyama T, Oka A (2000) Arabidopsis ARR1 and ARR2 response regulators operate as transcriptional activators. Plant J 24: 703–711.
82. Wenkel S, Turck F, Singer K, Gissot L, Le Gourrierec J, et al. (2006) CONSTANS and the CCAAT box binding complex share a functionally important domain and interact to regulate flowering of Arabidopsis. Plant Cell 18: 2971–2984.
83. Gowik U, Burscheidt J, Akyildiz M, Schlue U, Koczor M, et al. (2004) cis-Regulatory elements for mesophyll-specific gene expression in the C4 plant *Flaveria trinervia*, the promoter of the C4 phosphoenolpyruvate carboxylase gene. Plant Cell 16: 1077–1090.
84. Yanagisawa S, Schmidt RJ (1999) Diversity and similarity among recognition sequences of Dof transcription factors. Plant J 17: 209–214.
85. Reyes JC, Muro-Pastor MI, Florencio FJ (2004) The GATA family of transcription factors in Arabidopsis and rice. Plant Physiol 134: 1718–1732.
86. Villain P, Mache R, Zhou DX (1996) The mechanism of GT element-mediated cell type-specific transcriptional control. J Biol Chem 271: 32593–32598.
87. Hobo T, Suwabe K, Aya K, Suzuki G, Yano K, et al. (2008) Various spatiotemporal expression profiles of anther-expressed genes in rice. Plant Cell Physiol 49: 1417–1428.
88. Chinnusamy V, Ohta M, Kanrar S, Lee BH, Hong X, et al. (2003) ICE1: a regulator of cold-induced transcriptome and freezing tolerance in Arabidopsis. Genes Dev 17: 1043–1054.
89. Filichkin SA, Leonard JM, Monteros A, Liu PP, Nonogaki H (2004) A novel endo-beta-mannanase gene in tomato LeMAN5 is associated with anther and pollen development. Plant Physiol 134: 1080–1087.
90. Nishiuchi T, Shinshi H, Suzuki K (2004) Rapid and transient activation of transcription of the ERF3 gene by wounding in tobacco leaves: possible involvement of NtWRKYs and autorepression. J Biol Chem 279: 55355–55361.
91. Zhang ZL, Xie Z, Zou X, Casaretto J, Ho TH, et al. (2004) A rice WRKY gene encodes a transcriptional repressor of the gibberellin signaling pathway in aleurone cells. Plant Physiol 134: 1500–1513.
92. Yadav CB, Muthamilarasan M, Pandey G, Prasad M (2014) Identification, characterization and expression profiling of Dicer-like, Argonaute and RNA-dependent RNA polymerase gene families in foxtail millet. Plant Mol Biol Rep DOI: 10.1007/s11105-014-0736-y.

3

Accelerating the Switchgrass (*Panicum virgatum* L.) Breeding Cycle Using Genomic Selection Approaches

Alexander E. Lipka[1¤a*], Fei Lu[1], Jerome H. Cherney[2], Edward S. Buckler[1,3,4], Michael D. Casler[5,6], Denise E. Costich[1,3¤b]

1 Institute for Genomic Diversity, Cornell University, Ithaca, New York, United States of America, **2** Department of Crop and Soil Sciences, Cornell University, Ithaca, New York, United States of America, **3** Agricultural Research Service, United States Department of Agriculture, Ithaca, New York, United States of America, **4** Department of Plant Breeding and Genetics, Cornell University, Ithaca, New York, United States of America, **5** Agricultural Research Service, United States Department of Agriculture, Madison, Wisconsin, United States of America, **6** Department of Agronomy, University of Wisconsin–Madison, Madison, Wisconsin, United States of America

Abstract

Switchgrass (*Panicum virgatum* L.) is a perennial grass undergoing development as a biofuel feedstock. One of the most important factors hindering breeding efforts in this species is the need for accurate measurement of biomass yield on a per-hectare basis. Genomic selection on simple-to-measure traits that approximate biomass yield has the potential to significantly speed up the breeding cycle. Recent advances in switchgrass genomic and phenotypic resources are now making it possible to evaluate the potential of genomic selection of such traits. We leveraged these resources to study the ability of three widely-used genomic selection models to predict phenotypic values of morphological and biomass quality traits in an association panel consisting of predominantly northern adapted upland germplasm. High prediction accuracies were obtained for most of the traits, with standability having the highest ten-fold cross validation prediction accuracy (0.52). Moreover, the morphological traits generally had higher prediction accuracies than the biomass quality traits. Nevertheless, our results suggest that the quality of current genomic and phenotypic resources available for switchgrass is sufficiently high for genomic selection to significantly impact breeding efforts for biomass yield.

Editor: David D. Fang, USDA-ARS-SRRC, United States of America

Funding: This work was supported by Department of Energy-United States Department of Agriculture Plant Feedstock Genomics for Bioenergy Program Project Number DE-A102-07ER64454 (http://genomicscience.energy.gov/research/DOEUSDA/) and United States Department of Agriculture—Agricultural Research Services (http://www.ars.usda.gov/main/main.htm). The funders had no role in study design, data collection and analysis, decision to publish, or preparation of the manuscript.

Competing Interests: The authors have declared that no competing interests exist.

* Email: alipka@illinois.edu

¤a Current address: Department of Crop Sciences, University of Illinois, Urbana, Illinois, United States of America
¤b Current address: International Maize and Wheat Improvement Center (CIMMYT), Texcoco, Mexico

Introduction

Switchgrass (*Panicum virgatum* L.) is undergoing development as a biofuel feedstock due to its high biomass yield, broad adaptation, perennial growth habit, and long-standing presence in the seed industry [1]. Once inhabiting prairie and savanna ecosystems from Canada to Mexico and east of the Rocky Mountains, native switchgrass is now confined to thousands of prairie and savanna remnants that range in size from a few plants to a few hundred hectares [2]. Driven largely by photoperiod and temperature, latitude is the principal source of adaptive phenotypic variability across a broad landscape [3,4].

Switchgrass contains three principal taxa: a tetraploid ($2n = 4x = 28$) lowland ecotype, a tetraploid upland ecotype, and an octoploid ($2n = 8x = 56$) upland ecotype [5]. Upland ecotypes originated from upland prairie and savanna habitats that were frequently exposed to drought, especially toward the western portion of the range [6]. Lowland ecotypes originated in low-lying riverine or lacustrine habitats that were exposed to seasonal wet periods [6]. Upland ecotypes tend to be more northern adapted, while lowland ecotypes tend to be more southern adapted, with a transition zone where both can be found, sometimes within a single prairie or savanna remnant [5]. Upland and lowland ecotypes are highly cross-fertile and significant gene flow has occurred between the ecotypes during glacial maxima of the past million years [7]. Ploidy is the secondary taxonomic division within the species, primarily within the upland ecotype; lowland plants at the octoploid level are rare [8]. Gene flow has occurred between tetraploid and octoploid levels, largely by 2n gametes (4x to 8x) or haploidy (8x to 4x), but at relatively low frequencies due to the role of ploidy as a hybridization barrier [9].

Since the establishment of switchgrass as the herbaceous model species for cellulosic biofuel feedstock development in 1992 [1], a total of 12 breeding programs have been developed in North America [5]. Due to phenotypic differences among the three principal taxa and to the magnitude of adaptive phenotypic variation for flowering time and temperature (cold and heat) tolerance, there is very little overlap or duplication among these breeding programs. Collectively, their target population of environments covers eastern North America, but their individual target regions are realistically broken down into a minimum of eight regional gene pools or cultivar deployment zones [5]. Because adaptive phenotypic variation is a strong driver of both adaptation and production traits, genotype × environment interactions are a dominant force and individual cultivars are rarely adapted to more than three hardiness zones, as defined by [10].

Due to the length of the breeding cycle and the need for frequent (perhaps constant) phenotypic assessment of adaptive traits, few cultivars have been developed with documented improvement in biomass production traits. 'Liberty' is the most notable example, demonstrating both an increase in biomass yield and broader adaptation into USDA hardiness zone 3 [11]. Recent advances in the development of genomic tools for measuring and quantifying DNA marker diversity and sophisticated statistical tools to associate marker variation with phenotypic variation have the potential to revolutionize switchgrass breeding methodology [12]. Switchgrass breeding is complicated by the perennial nature of the species and the need for accurate measurement of biomass yield on a per-hectare basis, the single trait that is most limiting for sustainable and economically viable biomass production [13]. Simple-to-measure surrogate traits are needed to speed up the breeding cycle. Genomic selection [14,15] offers such an opportunity by developing predictive equations that allow breeders to measure DNA markers on seedlings and to predict which seedlings will have the highest biomass yield potential as adult plants [12].

The potential of genomic selection for improving the effectiveness of breeding programs has been successfully demonstrated in livestock [16–18], annual crops [19–23], and forest trees [24–26]. In these species, genomic selection has been shown to increase selection accuracy, reduce evaluation cost per genotype, and reduce breeding cycle time compared to phenotypic selection. More specifically, a recent evaluation of genomic selection methods concluded that genomic selection for perennial biofuel crops, such as switchgrass, is most advantageous when biomass yield on a per-hectare basis is difficult or expensive to measure, when it is difficult or impossible to apply meaningful selection pressure on plants within families, and when cycle times are >5 years, which is typically the case with switchgrass [12].

The purpose of this study was to explore the potential for genomic selection to increase the breeding cycle in switchgrass, particularly for seven morphological traits and 13 biomass quality traits. For most of these traits, reasonably high prediction accuracies were obtained. Our analysis was conducted within an association panel of 515 genotypes defined as a random sample of switchgrass from the northern USA gene pools. The population was evaluated using a set of 16,669 single nucleotide polymorphisms (SNPs) obtained using genotyping by sequencing (GBS) techniques [27,28] that were subsequently mapped to the recently available *Panicum virgatum* genome sequence v1.1 reference genome [29].

Materials and Methods

Germplasm

We analyzed the switchgrass association panel described in [27]. Briefly, this panel included 66 diverse switchgrass populations derived from predominantly northern adapted upland germplasm. Both tetraploid and octoploid germplasm were included. This panel was grown from seed planted at the greenhouse in the USDA-ARS Dairy Forage Research Center in Madison, Wisconsin in 2007. Ten clones or genotypes from each population were vegetatively propagated, then planted in Ithaca, New York in 2008 in a randomized complete block design with two replicates. Subsequently, a total of 540 plants from the Ithaca location were used for genotypic and phenotypic evaluation.

Morphological traits

The association panel was evaluated for seven morphological traits in 482 of the plants grown in Ithaca, NY during the 2009, 2010, and 2011 field seasons. These traits included anthesis date, heading date, standability, leaf length, leaf width, plant height, and total plant height. Descriptions of how each of these traits was measured are presented in Table 1, and the tools used to obtain the measurements are described at http://www.maizegenetics.net/phenotyping-tools [30]. Prior to subsequent analysis, the heading and anthesis dates were converted to growing degree days (GDD) as follows:

1) The first day in which GDD was recorded occurs the day after the first five consecutive days where the average temperature is >32° F.
2) After this day, GDD for a single day is calculated as:

$$[(Adj.Min + Adj.Max)/2] - 32$$

where *Adj. Min* is the maximum of the minimum daily temperature and 32°F, and *Adj.Max* is the minimum of the of the maximum daily temperature and 86°F. Intuitively, *Adj. Min* and *Adj.Max* limits the recorded minimum and maximum daily temperatures to 32°F and 86°F, respectively.

3) For each day after the first day in which GDD is recorded, the cumulative GDD is also recorded. The cumulative GDD is used to record heading date and anthesis date.

Biomass Quality traits

Near-infrared reflectance spectroscopy (NIRS, described in [31]) was used to estimate 42 biomass quality traits for a total 515 genotypes grown during two field seasons at the Ithaca, NY location. Samples were ground in Ithaca, NY, shipped to Madison, WI, and scanned on an NIRS unit at the U.S. Dairy Forage Research Center, as described in [31]. A total of 42 biomass quality traits were predicted using equations developed by [31], but only 13 of those traits were analyzed in this study due to their direct relevance and practical value in a breeding program focused on improving conversion efficiency, and to minimize redundancy from collinear traits. Specifically, these traits include acid detergent lignin, minerals (total ash), carbon, high heating value, cell wall concentration, ethanol/g dry forage, etherified ferulates, in vitro dry matter digestibility, pentose sugars release/g dry forage, total soluble carbohydrates, starch, sucrose, and total sugar. No sample had an H-statistic >3.0, indicating that none of the samples could be classified as outliers.

Description of SNPs

The Universal Network-Enabled Analysis Kit (UNEAK) discovery pipeline [27] was used to generate 29,221 SNPs with a minimum call rate of 0.5 and minimum minor allele frequency (MAF) of 0.05 among the 540 plants grown at the Ithaca location. These SNPs were then aligned to the *Panicum virgatum* genome sequence v1.1 [29]. The resulting 16,669 uniquely aligned SNPs were used for subsequent analysis.

Table 1. Phenotyping protocol for seven morphology traits measured in three summer environments, in Ithaca, NY across three years.

Trait Name (units)	Trait Description	Measured in Following Years
Anthesis Date	50% of panicles have 50% open florets	2009–2011
Heading Date	at least 50% of stems are 50% emerged (panicle branches still upright, just starting to spread)	2009–2011
Standability	0 = prostrate	2010–2011
	10 = upright	
Leaf length (mm)	Leaf below flag; base to tip	2009–2011
Leaf width (mm)	Leaf below flag; widest part	2009–2011
Plant Height (cm)	Base of longest flowering stem to the node at the base of the panicle	2009–2011
Total Plant Height (cm)	Base of the longest flowering stem to the tip of the panicle	2009–2011

-

Phenotypic evaluation

A subset of the 540 plants that yielded sufficient biomass for at least one field season was evaluated for morphological and quality traits. Specifically, 482 plants were evaluated for seven morphological traits and 515 plants were evaluated for 13 quality traits. All 20 traits were examined for outliers using Studentized deleted residuals [32] from a mixed linear model including year, field, block, and population as random effects in SAS version 9.3 [33]. For each trait, best linear unbiased predictors (BLUPs) were obtained for each line across years and replicates, using a mixed linear model fitted in ASReml version 3.0 [34]. Details of the model fitting procedure have been described in [35]. The relationship between each of these BLUPs was then evaluated using the Pearson correlation coefficient (r). Variance component estimates from the model used to obtain BLUPs were also used to estimate repeatability on a clone mean basis ($\hat{h}^2_l$) [36,37]. These repeatability estimates are upper bounds of the heritabilities for each trait. The delta method was used to approximate the standard error of the repeatability estimates [36]. Finally, the Box-Cox procedure [38] was implemented to find the optimal transformation of the BLUPs, as described in [39].

Genomic Selection

Prior to evaluating the genomic selection models, missing allelic values among the 16,669 SNPs anchored to the *Panicum virgatum* genome sequence v1.1 reference genome were imputed using fastPhase version 1.4.0 [40]. The allele frequencies of these SNPs were calculated among the 482 plants evaluated for the morphology traits and again among the 515 plants evaluated for the quality traits. Within each of these two subsets, SNPs with MAF <0.05 were removed. Consequently, 11,857 SNPs were used in the genomic selection models for the morphology traits, and 12,180 SNPs were used in the models for the quality traits.

To assess the capability of our imputed markers to predict morphological and quality trait values, three genomic selection approaches were tested, namely ridge regression-best linear unbiased prediction (RR-BLUP) [14], least absolute shrinkage and selection operator (LASSO) [41], and elastic net [42]. Although these three approaches have been shown to produce similar results in practice (e.g., [21]), the performance of each approach could depend on the genetic architecture of the evaluated traits. Specifically, RR-BLUP should theoretically outperform LASSO for complex traits, while LASSO should be superior for simpler traits. The elastic net, whose penalty is a weighted average of the penalties of RR-BLUP and LASSO, is considered to be a compromise between the two approaches. In this study, the mixing parameter for the elastic net was set to $\alpha = 0.5$, meaning that the RR-BLUP and LASSO penalties were given equal weights. The RR-BLUP approach was conducted using the rrBLUP package [43] in the R programming language

Table 2. Means and ranges for best linear unbiased predictors (BLUPs) of seven morphological traits evaluated on a switchgrass association panel, and estimated repeatability on a clone-mean basis in three summer environments, in Ithaca, NY across three years.

Trait	No. Lines	BLUP Mean	BLUP SD[b]	BLUP Range	Repeatability	Repeatability SE[c]
Anthesis Date (GDD[a])	481	3840.53	450.21	2630.25–5272.48	0.93	0.01
Heading Date (GDD)	482	2870.47	343.81	2111.75–4547.04	0.91	0.01
Standability (0–10 scale)	481	5.36	1.60	1.47–8.59	0.88	0.01
Leaf Length (mm)	482	528.88	73.30	294.62–708.48	0.85	0.02
Leaf Width (mm)	482	13.32	1.91	6.56–20.75	0.82	0.02
Plant Height (cm)	482	88.78	16.45	44.75–146.16	0.75	0.03
Total Plant Height (cm)	482	162.22	20.24	105.43–222.81	0.85	0.02

[a]GDD, Growing degree dates
[b]SD, Standard deviation
[c]SE, Standard error

Table 3. Means and ranges for best linear unbiased predictors (BLUPs) of 13 quality traits evaluated on a switchgrass association panel, and estimated repeatability on a clone-mean basis in two summer environments, in Ithaca, NY, across two years.

Trait (µg/g)	No. Lines	BLUP Mean	BLUP SD[a]	BLUP Range	Repeatability	Repeatability SE[b]
Acid detergent lignin	514	75.62	5.45	61.47–90.02	0.81	0.02
Minerals (total ash)	514	69.20	4.60	54.47–83.62	0.67	0.03
Carbon	514	443.86	2.19	438.78–452.85	0.67	0.03
High Heating Value	514	4182.59	17.6	4136.14–4237.54	0.76	0.02
Cell wall concentration	514	673.11	47.31	564.12–832.92	0.87	0.01
Ethanol/g dry forage	514	82.73	7.46	60.83–106.78	0.78	0.02
Etherified ferulates	514	0.88	0.10	0.64–1.28	0.83	0.48
In vitro dry matter digestibility	514	410.54	35.16	311.86–494	0.82	0.01
Pentose sugars release/g dry forage	515	191.29	8.30	167.13–218.05	0.77	0.02
Total soluble carbohydrates	514	51.27	8.58	29.22–74.16	0.71	0.03
Starch	514	6.35	2.72	0.67–17.27	0.59	0.04
Sucrose	514	28.29	5.71	13.71–45.13	0.72	0.02
Total sugar	514	625.44	19.01	572.91–691.56	0.79	0.02

[a]SD, Standard deviation
[b]SE, Standard error

[44], while LASSO and elastic net were conducted using the glmnet R package [45].

For any genomic selection model, it is important to ensure that SNP effects arising from overall differences in population structure are factored out [17]. Given the genetic differences attributable to the observed ecotypes and ploidies in our association panel, it is crucial to account for such SNP effects prior to conducting our genomic selection study. Based on the results presented in [27], we hypothesized that the first two principal components (PCs) of a principal component analysis (PCA) of the 16,669 SNPs imputed with fastPhase would sufficiently account for these genetic differences. Accordingly, we fitted a model to each trait where the trait was the response variable and the first two PCs from the PCA of these SNPs were the explanatory variables. The residuals from each model were used for genomic selection.

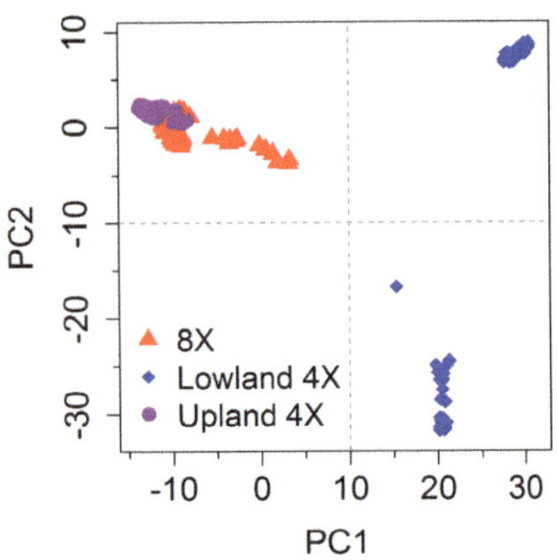

Figure 1. First two principal components of 16,669 single nucleotide polymorphisms separate plants by ploidy and ecotype. The first and second principal components (x- and y-axis, respectively) from a principal component analysis of 540 switchgrass closes separate octoploid (8X) and upland tetraploid (4X) accessions from lowland 4X accessions. The lowland accessions are also separated into two distinct clusters.

The performance of each model was assessed through ten-fold cross validation, as described in [46]. Briefly, the association panel was partitioned into ten equally-sized subgroups. Nine of the ten subgroups (i.e., the training set) were used to fit each prediction model while the remaining subgroup (the prediction set) was used to assess the correlation between the observed and predicted trait values. This process was repeated ten times, with each subgroup being the prediction set exactly once. For each trait, prediction accuracies were calculated by dividing the average Pearson's correlation coefficient across the ten folds by the square root of the repeatability [25]. To prevent inflated prediction accuracies arising from clones nested within populations, the data were partitioned for ten-fold cross validation so that none of the populations were in both the training and prediction sets. All phenotypic and genotypic data used to conduct this analysis are included in File S1.

Results

Extensive Phenotypic Variability among Clones

Substantial variation was observed for each of the seven morphological traits, with differences between minimum and maximum values of each trait ranging from 2-fold for anthesis date to 5.84-fold for standability (Table 2). In general, the majority of the morphological traits were highly correlated, with the strongest Pearson's correlation being between heading date and anthesis date ($r = 0.92$; Table S1). High correlations between leaf width, plant height, and total plant height were also observed (Pearson correlations ranging from $r = 0.53$ to $r = 0.88$). The average repeatability among the seven morphological traits was 0.86, with a range from 0.75 for plant height to 0.93 for anthesis date (Table 2). These high repeatabilities suggest that the majority of the phenotypic variation is attributable to genetic effects, and that genomic selection could be a useful breeding approach for morphological traits in switchgrass.

In comparison to the morphological traits, a greater range of fold differences between the minimum and maximum values of each trait was observed for the quality traits (Table 3). Although many of the correlations between the quality traits were generally

Table 4. Prediction accuracies of seven morphological traits in a switchgrass association panel.

Trait	Mean prediction accuracy	Prediction accuracy: RR-BLUP[a]	Prediction accuracy: Elastic net	Prediction accuracy: LASSO[b]
Anthesis Date	0.44	0.55 (0.21)	0.38 (0.23)	0.38 (0.23)
Heading Date	0.36	0.39 (0.14)	0.34 (0.20)	0.34 (0.19)
Standability	0.52	0.51 (0.27)	0.53 (0.19)	0.52 (0.19)
Leaf Length	0.40	0.55 (0.21)	0.34 (0.30)	0.32 (0.29)
Leaf Width	0.19	0.32 (0.29)	0.13 (0.24)	0.13 (0.24)
Plant Height	0.25	0.34 (0.18)	0.21 (0.26)	0.20 (0.26)
Total Plant Height	0.15	0.30 (0.28)	0.09 (0.19)	0.06 (0.19)

Standard errors of prediction accuracies are provided in parentheses.
Mean prediction accuracies were obtained by averaging results across ridge regression best linear unbiased prediction (RR-BLUP), least absolute shrinkage and selection operator (LASSO), and elastic net analysis.
[a]RR-BLUP, Ridge regression-best linear unbiased prediction
[b]LASSO, Least absolute shrinkage and selection operator

lower than those between the morphology traits, some individual quality traits were strongly correlated. For example, a Pearson correlation coefficient of $r = 0.95$ was observed between sucrose and total soluble carbohydrates (Table S2). Although lower than observed among the morphology traits, the estimated repeatabilities of the quality traits were sufficiently high enough to merit investigation into the utility of genomic selection.

First Two Principal Components of SNPs Sufficiently Account for Ploidy and Ecotype Differences

The first two PCs of the imputed GBS markers subdivided the plants used in this study into three genetically distinct subgroups (Figure 1). Specifically, the octoploid and upland tetraploid plants were clustered into one group, while the lowland tetraploid plants were subdivided into two distinct clusters. Collectively, these results suggest that the first two PCs of the SNPs capture a substantial amount of the major genetic differences between the ploidies and ecotypes of the plants included in our association panel. Moreover, these results justify our use of the first two PCs to factor out the SNP effects arising from overall population structure differences prior to conducting our genomic selection study.

Genomic Selection has Considerable Potential in Switchgrass

As expected, the observed prediction accuracies were similar across the three GS approaches (Tables 4–5). The predictive ability of the morphological traits were generally higher than those of the quality traits, with the highest prediction accuracies

Table 5. Prediction accuracies of 13 quality traits in a switchgrass association panel.

Trait	Mean prediction accuracy	Prediction accuracy: RR-BLUP[a]	Prediction accuracy: Elastic net	Prediction accuracy: LASSO[b]
Acid detergent lignin	0.34	0.41 (0.25)	0.31 (0.21)	0.30 (0.21)
Minerals (total ash)	−0.08	−0.09 (0.18)	−0.06 (0.13)	−0.10 (0.15)
Carbon	0.12	0.21 (0.25)	0.09 (0.27)	0.07 (0.27)
High Heating Value	0.22	0.26 (0.14)	0.21 (0.16)	0.20 (0.17)
Cell wall concentration	0.23	0.30 (0.23)	0.21 (0.19)	0.19 (0.18)
Ethanol/g dry forage	0.43	0.46 (0.20)	0.42 (0.20)	0.41 (0.21)
Etherified ferulates	0.22	0.27 (0.23)	0.20 (0.16)	0.19 (0.15)
In vitro dry matter digestibility	0.35	0.43 (0.27)	0.32 (0.25)	0.30 (0.25)
Pentose sugars release/g dry forage	0.06	0.15 (0.20)	0.03 (0.26)	0.01 (0.25)
Total soluable carbohydrates	0.30	0.39 (0.21)	0.26 (0.23)	0.25 (0.23)
Starch	0.08	0.19 (0.26)	0.03 (0.16)	0.03 (0.15)
Sucrose	0.32	0.44 (0.20)	0.26 (0.24)	0.25 (0.24)
Total sugar	0.04	0.16 (0.17)	0.00 (0.14)	−0.03 (0.17)

Standard errors of prediction accuracies are provided in parentheses.
Mean prediction accuracies were obtained by averaging results across ridge regression best linear unbiased prediction (RR-BLUP), least absolute shrinkage and selection operator (LASSO), and elastic net analysis.
[a]RR-BLUP, Ridge regression-best linear unbiased prediction
[b]LASSO, Least absolute shrinkage and selection operator

(averaged across all three tested GS models) obtained for standability (0.52), anthesis date (0.44), ethanol/g dry forage (0.43), leaf length (0.40), and heading date (0.36). We also obtained relatively strong positive Spearman's rank correlation coefficients between repeatabilities and unstandardized prediction accuracies for both the morphology ($r_{SP} = 0.61$) and the quality ($r_{SP} = 0.44$) traits. Consistent with the findings of previous studies (e.g., [25]), this result suggests that all three GS approaches successfully use the larger genetic contribution to phenotypic variability of the more heritable traits to obtain higher prediction accuracies.

Discussion

We evaluated the ability of three popular genomic selection approaches to predict the phenotypic values of seven morphological traits and 13 quality traits in a switchgrass association panel. Such a study is important because the successful application of genomic selection to switchgrass could significantly reduce the breeding cycle of this important biofuel feedstock. In general, our prediction accuracies are comparable to those reported in previous studies (e.g., [17] and [23]) that identified quantifiable advantages of genomic selection compared to traditional breeding programs. For perennial grasses such as switchgrass, one important quantity to consider is the expected genetic gain per unit of time. Because genomic selection does not require on-site phenotyping to identify accessions with superior trait values, multiple cycles of breeding could be completed with a genomic selection breeding program during the same amount of time required to achieve one cycle of breeding using traditional breeding programs [17,23]. For instance, it is demonstrated in [23] that it is possible for genomic selection breeding programs in maize and winter wheat to respectively achieve three cycles and two cycles of breeding during the same amount of time to complete one cycle of marker-assisted selection breeding. Moreover, the same study concluded that the expected genetic gain per year from a genomic selection breeding program will exceed that of a marker-assisted selection breeding program for traits with prediction accuracies as low as 0.20 in maize and 0.30 in winter wheat. Because many of our tested traits had prediction accuracies that exceed these thresholds, we believe that it is possible for similar advantages in expected genetic gain per unit of time to be achieved in switchgrass genomic selection breeding programs.

To our knowledge, the genetic architectures of the traits we evaluated are unknown in switchgrass. In particular, little is known about the number of genes underlying each trait. Therefore, we used three genomic selection models that have been hypothesized to perform differently under various genetic architectures. In general, we obtained similar prediction accuracies for all three genomic selection models. This result is especially apparent if we consider the standard errors of the prediction accuracies. Suppose we use the prediction accuracies and their standard errors from each genomic selection model (presented in Tables 4 and 5) to construct 95% confidence intervals. For each trait, the confidence intervals from the three genomic selection approaches overlap. This suggests that there are no discernible differences in prediction accuracies among the three genomic selection models. Indeed, this finding has been reported in other studies (e.g. [46]) and is theoretically justified in [47]. Nevertheless, we recommend repeating our study because we anticipate that the sampling, genotyping, and phenotypic resources available to the switchgrass community will continue to expand and improve, and it is imperative to confirm that these conclusions still hold given the new information we expect to obtain from these resources.

We observed a wide range of prediction accuracies across the traits. We suspect that this result was obtained because our markers provided incomplete coverage of the switchgrass genome, and it is likely that they tagged only a subset of the loci underlying the genetic sources of variation for each trait. It is therefore plausible that traits with higher prediction accuracies have causal loci that were in higher linkage disequilibrium with our markers compared to traits with lower prediction accuracies. Nonetheless, the prediction accuracies for many of the studied traits were suitably high enough to justify further investigation into the application of genomic selection to switchgrass breeding programs. Indeed one major factor contributing to our observed prediction accuracies was the availability of the *Panicum virgatum* genome sequence v1.1 reference genome. Because of this reference genome, we were able to use genotypic information from neighboring markers to impute missing genotypic data, and ultimately obtain substantial increases in the predictive abilities of our genomic selection models. Thus, we strongly recommend that switchgrass genomic selection breeding programs only use markers that are anchored to a reference genome. This will enable accurate imputation of missing data, and should ultimately result in genomic selection models with higher predictive abilities.

In general, lower prediction accuracies were obtained for the biomass quality traits relative to the morphology traits. We suspect that this result could have arisen from two different sources. In contrast to the morphological traits, the process of obtaining the quality traits was a lengthy procedure that was conducted in the laboratory. As such, it is possible that a greater amount of variability was introduced into the quality traits, which ultimately resulted in lower prediction accuracies. Factors such as spatial variability in the field, diurnal variation in biomass quality traits manifested by variation in sampling time, variation in grinding time and blade sharpness, and moisture content of the samples may all introduce variability to the measurement of biomass quality traits.

Our study suggests that the implementation of genomic selection approaches to switchgrass breeding programs will be highly beneficial. We believe that such an approach will revolutionize switchgrass breeding programs just as it has in at least four dairy cattle breeding programs [17]. Indeed, the large body of theoretical and empirical studies conducted in plant and animal species [15,22,25,46,48,49] suggests that genomic selection is a cost-effective approach that will substantially speed up breeding cycles, and we expect that these advantages will significantly benefit the development of switchgrass as a biofuel feedstock. As high as the prediction accuracies were in our study, we expect them to increase as more attention is focused on the characterization and exploitation of switchgrass phenotypic and genotypic resources. Specifically, we believe that increased prediction accuracies will arise from improvements to the switchgrass reference genome, improvements in phenotyping techniques, and the development of markers with higher levels of genomic coverage and density.

Supporting Information

Table S1 Correlation matrix for untransformed BLUPs of the seven morphological traits. Pearson correlation coefficients are presented in the upper triangle, and the P-values for the significance of associations are in the lower triangle.

Table S2 Correlation matrix for untransformed BLUPs of the 13 quality traits. Pearson correlation coefficients are presented in the upper triangle, and the P-values for the significance of associations are in the lower triangle.

File S1 Data files used to conduct analysis. All files used to conduct the genomic selection analysis are included in this file.

Acknowledgments

We would like to thank the propagators on our team, Ken Paddock (Cornell) and Nick Baker (USDA-ARS, Madison, Wisconsin), as well as the legions of undergraduate students who assisted in growing and phenotyping switchgrass, and preparing samples for both phenotypic and genotypic analyses. Nick Lepak (USDA-ARS, Ithaca, New York) provided valuable support at the Ithaca field site and contributed to the GDD calculations and selection of the appropriate formula. We also thank the Cornell University Genomics Core Facility and the staff at the Institute for Genomic Diversity (Cornell) for their support: they made the development of our GBS markers possible. AEL acknowledges the USDA-ARS for providing a Postdoctoral Fellowship to pursue this research. Prepublication data for the *Panicum virgatum* genome sequence v1.1 reference genome were provided by the Department of Energy Joint Genome Institute.

Author Contributions

Conceived and designed the experiments: ESB DEC JHC MDC. Performed the experiments: FL DEC JHC MDC. Analyzed the data: AEL FL. Contributed reagents/materials/analysis tools: FL DEC MDC JHC. Wrote the paper: AEL MDC DEC.

References

1. Sanderson MA, Adler PR, Boateng AA, Casler MD, Sarath G (2006) Switchgrass as a biofuels feedstock in the USA. Can J Plant Sci 86: 1315–1325.
2. Stubbendieck JL, Hatch SL, Butterfield CH (1992) North American range plants: U of Nebraska Press.
3. Casler M, Vogel K, Taliaferro C, Ehlke N, Berdahl J, et al. (2007) Latitudinal and longitudinal adaptation of switchgrass populations. Crop Sci 47: 2249–2260.
4. Casler M, Vogel K, Taliaferro C, Wynia R (2004) Latitudinal adaptation of switchgrass populations. Crop Sci 44: 293–303.
5. Casler MD (2012) Switchgrass breeding, genetics, and genomics. Switchgrass: Springer. pp. 29–53.
6. Porter CL Jr (1966) An analysis of variation between upland and lowland switchgrass, Panicum virgatum L., in central Oklahoma. Ecology: 980–992.
7. Zhang Y, Zalapa J, Jakubowski AR, Price DL, Acharya A, et al. (2011) Natural hybrids and gene flow between upland and lowland switchgrass. Crop Sci 51: 2626–2641.
8. Zhang Y, Zalapa JE, Jakubowski AR, Price DL, Acharya A, et al. (2011) Post-glacial evolution of Panicum virgatum: centers of diversity and gene pools revealed by SSR markers and cpDNA sequences. Genetica 139: 933–948.
9. Martinez-Reyna J, Vogel K (2002) Incompatibility systems in switchgrass. Crop Sci 42: 1800–1805.
10. Cathey HM (1990) USDA plant hardiness zone map.
11. Vogel KP, Mitchell R, Casler M, Sarath G (2014) Registration of 'Liberty' switchgrass. J Plant Regist.
12. Simeão Resende RM, Casler MD, Vilela de Resende MD (2014) Genomic selection in forage breeding: accuracy and methods. Crop Sci 54: 143–156.
13. Perrin R, Vogel K, Schmer M, Mitchell R (2008) Farm-scale production cost of switchgrass for biomass. Bioenerg Res 1: 91–97.
14. Meuwissen T, Hayes B, Goddard M (2001) Prediction of total genetic value using genome-wide dense marker maps. Genetics 157: 1819–1829.
15. de los Campos G, Hickey JM, Pong-Wong R, Daetwyler HD, Calus MP (2013) Whole-genome regression and prediction methods applied to plant and animal breeding. Genetics 193: 327–345.
16. Schaeffer L (2006) Strategy for applying genome-wide selection in dairy cattle. J Anim Breed Genet 123: 218–223.
17. Hayes B, Bowman P, Chamberlain A, Goddard M (2009) Invited review: Genomic selection in dairy cattle: Progress and challenges. J Dairy Sci 92: 433–443.
18. VanRaden P, Van Tassell C, Wiggans G, Sonstegard T, Schnabel R, et al. (2009) Invited Review: Reliability of genomic predictions for North American Holstein bulls. J Dairy Sci 92: 16–24.
19. Asoro FG, Newell MA, Beavis WD, Scott MP, Jannink JL (2011) Accuracy and training population design for genomic selection on quantitative traits in elite North American oats. Plant Gen 4: 132–144.
20. Zhao Y, Gowda M, Liu W, Würschum T, Maurer HP, et al. (2012) Accuracy of genomic selection in European maize elite breeding populations. Theor Appl Genet 124: 769–776.
21. Riedelsheimer C, Czedik-Eysenberg A, Grieder C, Lisec J, Technow F, et al. (2012) Genomic and metabolic prediction of complex heterotic traits in hybrid maize. Nat Genet 44: 217–220.
22. Heffner EL, Sorrells ME, Jannink JL (2009) Genomic selection for crop improvement. Crop Sci 49: 1–12.
23. Heffner EL, Lorenz AJ, Jannink JL, Sorrells ME (2010) Plant breeding with genomic selection: gain per unit time and cost. Crop Sci 50: 1681–1690.
24. Resende MD, Resende MF, Sansaloni CP, Petroli CD, Missiaggia AA, et al. (2012) Genomic selection for growth and wood quality in Eucalyptus: capturing the missing heritability and accelerating breeding for complex traits in forest trees. New Phytol 194: 116–128.
25. Resende MF, Muñoz P, Resende MD, Garrick DJ, Fernando RL, et al. (2012) Accuracy of genomic selection methods in a standard data set of loblolly pine (Pinus taeda L.). Genetics 190: 1503–1510.
26. Resende M, Munoz P, Acosta J, Peter G, Davis J, et al. (2012) Accelerating the domestication of trees using genomic selection: accuracy of prediction models across ages and environments. New Phytologist 193: 617–624.
27. Lu F, Lipka AE, Glaubitz J, Elshire R, Cherney JH, et al. (2013) Switchgrass genomic diversity, ploidy, and evolution: novel insights from a network-based SNP discovery protocol. PLoS Genet 9: e1003215.
28. Elshire RJ, Glaubitz JC, Sun Q, Poland JA, Kawamoto K, et al. (2011) A robust, simple genotyping-by-sequencing (GBS) approach for high diversity species. PLoS One 6: e19379.
29. Joint Genome Institute (2014) Phytozome 10. Panicum virgatum v1.1 DOE-JGI. Available: http://phytozome.jgi.doe.gov.
30. Buckler Lab for Maize Genetics and Diversity (nd) Buckler lab phenotyping tools. Available: http://www.maizegenetics.net/phenotyping-tools.
31. Vogel KP, Dien BS, Jung HG, Casler MD, Masterson SD, et al. (2011) Quantifying actual and theoretical ethanol yields for switchgrass strains using NIRS analyses. Bioenerg Res 4: 96–110.
32. Kutner MH, Nachtsheim CJ, Neter J, Li W (2004) Applied Linear Statistical Models. Boston, MA: McGraw-Hill.
33. SAS Institute (2012) The SAS system for Windows. Release 9.3. SAS Institute. Institute SAS: Sas Inst.
34. Gilmour AR, Gogel B, Cullis B, Thompson R, Butler D, et al. (2009) ASReml user guide release 3.0. VSN International Ltd, Hemel Hempstead, UK.
35. Chandler K, Lipka AE, Owens BF, Li H, Buckler ES, et al. (2013) Genetic Analysis of Visually Scored Orange Kernel Color in Maize. Crop Sci 53: 189–200.
36. Holland JB, Nyquist WE, Cervantes-Martínez CT (2003) Estimating and interpreting heritability for plant breeding: An update. Plant Breed Rev 22: 9–112.
37. Hung HY, Browne C, Guill K, Coles N, Eller M, et al. (2012) The relationship between parental genetic or phenotypic divergence and progeny variation in the maize nested association mapping population. Heredity (Edinb) 108: 490–499.
38. Box GEP, Cox DR (1964) An Analysis of Transformations. J Roy Stat Soc B Met 26: 211–252.
39. Lipka AE, Gore MA, Magallanes-Lundback M, Mesberg A, Lin H, et al. (2013) Genome-wide association study and pathway-level analysis of tocochromanol levels in maize grain. G3: Genes, Genomes, Genet 3: 1287–1299.
40. Scheet P, Stephens M (2006) A fast and flexible statistical model for large-scale population genotype data: applications to inferring missing genotypes and haplotypic phase. Am J Hum Genet 78: 629–644.
41. Tibshirani R (1996) Regression shrinkage and selection via the lasso. J Roy Stat Soc B Met: 267–288.
42. Zou H, Hastie T (2005) Regularization and variable selection via the elastic net. J Roy Stat Soc B Met 67: 301–320.
43. Endelman JB (2011) Ridge regression and other kernels for genomic selection with R package rrBLUP. Plant Gen 4: 250–255.
44. Team RC (2012) R: A language and environment for statistical computing.
45. Friedman J, Hastie T, Tibshirani R (2010) Regularization paths for generalized linear models via coordinate descent. J Stat Softw 33: 1.
46. Heslot N, Yang HP, Sorrells ME, Jannink JL (2012) Genomic selection in plant breeding: a comparison of models. Crop Sci 52: 146–160.
47. Gianola D (2013) Priors in whole-genome regression: the Bayesian alphabet returns. Genetics 194: 573–596.
48. Jannink J, Lorenz AJ, Iwata H (2010) Genomic selection in plant breeding: from theory to practice. Brief Funct Genomics 9.2 166–177.
49. Wolc A, Stricker C, Arango J, Settar P, Fulton JE, et al. (2011) Breeding value prediction for production traits in layer chickens using pedigree or genomic relationships in a reduced animal model. Genet Sel Evol 43.

4

Complete Chloroplast Genome Sequence of Omani Lime (*Citrus aurantiifolia*) and Comparative Analysis within the Rosids

Huei-Jiun Su[1], Saskia A. Hogenhout[2], Abdullah M. Al-Sadi[3], Chih-Horng Kuo[4,5,6*]

1 Institute of Ecology and Evolutionary Biology, National Taiwan University, Taipei, Taiwan, **2** Department of Cell and Developmental Biology, John Innes Centre, Norwich, United Kingdom, **3** Department of Crop Sciences, Sultan Qaboos University, Al Khoud, Oman, **4** Institute of Plant and Microbial Biology, Academia Sinica, Taipei, Taiwan, **5** Molecular and Biological Agricultural Sciences Program, Taiwan International Graduate Program, National Chung Hsing University and Academia Sinica, Taipei, Taiwan, **6** Biotechnology Center, National Chung Hsing University, Taichung, Taiwan

Abstract

The genus *Citrus* contains many economically important fruits that are grown worldwide for their high nutritional and medicinal value. Due to frequent hybridizations among species and cultivars, the exact number of natural species and the taxonomic relationships within this genus are unclear. To compare the differences between the *Citrus* chloroplast genomes and to develop useful genetic markers, we used a reference-assisted approach to assemble the complete chloroplast genome of Omani lime (*C. aurantiifolia*). The complete *C. aurantiifolia* chloroplast genome is 159,893 bp in length; the organization and gene content are similar to most of the rosids lineages characterized to date. Through comparison with the sweet orange (*C. sinensis*) chloroplast genome, we identified three intergenic regions and 94 simple sequence repeats (SSRs) that are potentially informative markers with resolution for interspecific relationships. These markers can be utilized to better understand the origin of cultivated *Citrus*. A comparison among 72 species belonging to 10 families of representative rosids lineages also provides new insights into their chloroplast genome evolution.

Editor: Tongming Yin, Nanjing Forestry University, China

Funding: Funding for this work was provided by research grants from Biotechnology and Biological Sciences Research Council (BB/J004553/1) and the Gatsby Charitable Foundation to SAH, Sultan Qaboos University (SR/AGR/CROP/13/01) to AMA, and the Institute of Plant and Microbial Biology at Academia Sinica to CHK. The funders had no role in study design, data collection and analysis, decision to publish, or preparation of the manuscript.

* Email: chk@gate.sinica.edu.tw

Introduction

Citrus is in the family of Rutaceae, which is one of the largest families in order Sapindales. Flowers and leaves of *Citrus* are usually strong scented, the extracts from which contain many useful flavonoids and other compounds that are effective insecticides, fungicides and medicinal agents [1–3]. *Citrus* is of great economic importance and contains many fruit crops such as oranges, grapefruit, lemons, limes, and tangerines. However, due to a long cultivation history, wide dispersion, somatic bud mutation, and sexual compatibility among *Citrus* species and related genera, the taxonomy of *Citrus* remains controversial [4,5] and the origination of many *Citrus* species and hybrids is still unresolved [6,7].

The chloroplast (cp) genome sequence contains useful information in plant systematics because of its maternal inheritance in most angiosperms [8,9] and its highly conserved structures for developing promising genetic markers. The only complete cp genome available in *Citrus* is sweet orange (*Citrus sinensis*) [10], which has provided valuable information to the position of Sapindales in rosids. Although a genome sequencing project is in progress for *C. clementine*, its complete chloroplast genome sequence is not available yet. To identify the cp genome regions that are polymorphic and may be used as molecular markers for resolving the evolutionary relationships among *Citrus* species, a second cp genome within the genus is necessary for comparative analysis. For this purpose, the major aim of this study is to determine the complete cp genome sequence of *C. aurantiifolia*.

C. aurantiifolia, which is commonly known as Key lime, Mexican lime, Omani lime, Indian lime, or acid lime, is native to Southeast Asia and widely cultivated in tropics and subtropics. Oman is known to be a transit country for lime, from which lime spread to Africa and the New World [11]. In Oman, Omani lime is considered the fourth most important fruit crop in terms of cultivated area and production. The products of Omani lime can be used for beverage, food additives and cosmetic industries [12]. Omani lime is sensitive to several biotic agents, the most serious of which is '*Candidatus* Phytoplasma aurantifolia', the cause of witches' broom disease of lime (WBDL). Recent studies on WBDL focused on effect of genetic diversity of Omani limes on the disease [13], transcriptome and proteomic analysis of lime response to infection by phytoplasma [14–16] and effect of phytoplasma on seed germination, growth and metabolite content in lime [17,18].

Here, we present the complete chloroplast genome sequence of Omani lime (*C. aurantiifolia*). To identify loci of potential utility for the molecular identification and phylogenetic analyses of *Citrus* cultivars and species, we compared the intergenic regions and SSRs in the cp genomes of *C. aurantiifolia* and *C. sinensis*. Furthermore, we performed phylogenetic analyses to infer the history of gene losses in the cp genome evolution among representative rosids lineages.

Materials and Methods

Sample Preparation and Sequencing

The Omani lime leaves were collected from a 5-year-old lime tree at a private farm located in the Omani territory of Madha (GPS coordinates: 25.276318, 56.318909). This farm is owned by one of the co-authors of this work, Dr. Abdullah M. Al-Sadi, whom should be contacted for future permissions. This study does not involve endangered or protected species and does not require specific permission from regulatory authority concerned with protection of wildlife. The sample was stored in a cool box and transported to the Plant Pathology Research Laboratory at Sultan Qaboos University (Al Khoud, Oman) for DNA extraction

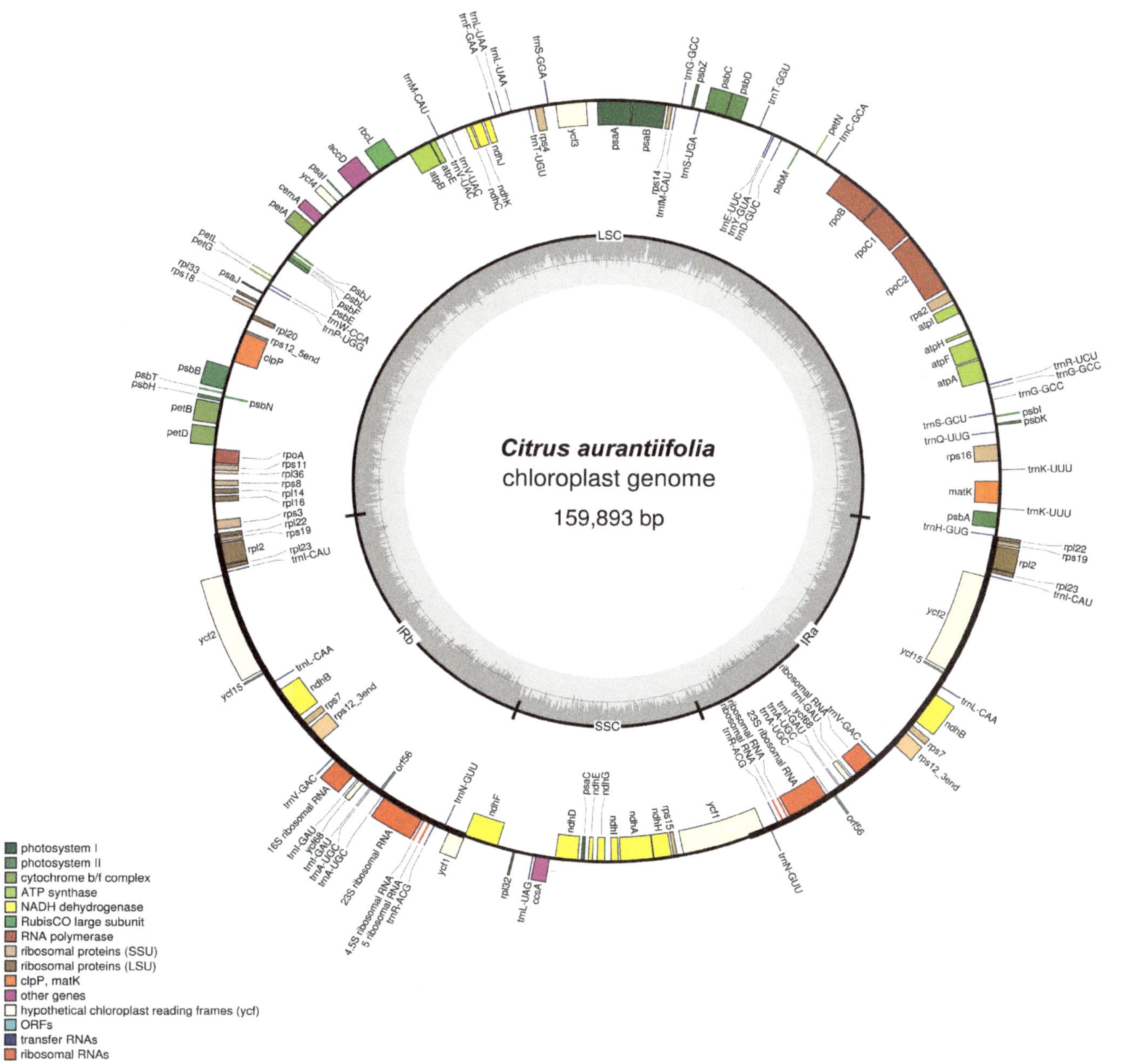

Figure 1. Chloroplast genome map of *Citrus aurantiifolia*. Gene drawn inside the circle are transcribed clockwise, whereas those outside are counterclockwise. The within-genome GC content variation is indicated in the middle circles.

following a protocol of Maixner et al. [19]. The leaves were washed with clear water before the isolation procedure. 1 g of leaves were used and crushed in 3 ml CTAB extraction buffer (2% CTAB, 1.4 M NaCl, 500 mM EDTA pH8, 1 M Tris-HCl pH8 and 0.2% beta-mercaptol). 1.5 ml of the leave extract was transferred to a 2 ml tube and incubated in a water-bath at 65°C for 15 min. The tube was turned up and down twice during incubation, centrifuged at 960 g for 5 min, and the supernatant was subsequently transferred to a clean eppendorf tube. An equal volume of chloroform-isoamyl alcohol mix (24:1) was added and the tube was centrifuged at 21000 g for 20 min. The supernatant was transferred to a new tube and then 0.6 volume of isopropanol was added to the supernatant and incubated at −20°C for 30 min. The DNA pellet was collected by centrifugation at 21000 g for 20 min and then washed with 1 ml of 70% ethanol. The final DNA was resuspended in 100 µl TE (Tris 10 mM, EDTA 1 mM pH8) and was stored at −80°C until used.

The library construction and sequencing were done at the Genome Analysis Centre (Norwich, UK). The Illumina TruSeq DNA Sample Preparation v2 Kit was used to prepare an indexed library. The DNA sample was sheared to a fragment size of 500–600 bp using a sonicator, followed by end-repair and the addition of a single A base for binding of the indexed adapter. The appropriate sized library (500 bp) was selected by gel electrophoresis, followed by PCR enrichment. The 251 bp paired-end sequencing run was performed on an Illumina MiSeq instrument using the SBS chemistry and Illumina software MCS v2.3.0.3 and RTA v1.18.42. The raw reads were deposited at the NCBI Sequence Read Archive under the accession number SRR1611615.

Genome Assembly and Analyses

The procedures for genome assembly and annotation were based on our previous studies of cp genomes [20,21]. In addition to the standard *de novo* assembly approach by using Velvet v1.2.10 [22] with the k-mer size set to 243, a reference-based approach for assembly as described below was used in parallel. All of the raw reads were initially mapped onto the published cp genome of *C. sinensis* [10] using BWA v0.6.2 [23]. The sequence variations were identified with SAMtools v0.1.19 [24] and visually inspected using IGV v2.3.25 [25]. The variants were corrected with the raw reads and the regions without sufficient coverage were converted into gaps. This corrected sequence was then used as the new draft reference for the next iteration of verification. Gaps were filled using the reads overhang at margins and the process was repeated until the reference was fully supported by all mapped raw reads. The final assembly, which was supported by our *de novo* and reference-based approaches, resulted in an average of 1,441-fold coverage of paired-end reads with a mapping quality of 60 and the region with the lowest coverage is 506-fold.

The preliminary annotations of the *C. aurantiifolia* cp genome were performed online using the automatic annotator DOGMA [26] and verified using BLASTN [27,28] searches (e-value cutoff = 1e-10) against other land plant cp genomes. Each annotated gene was manually compared with *C. sinensis* cp genome for start and stop codons or intron junctions to ensure accurate annotation. The codon usage was analyzed by using the seqinr R-cran package [29]. A circular map of genome was produced using OGDRAW [30].

To identify the differences between *C. aurantiifolia* and *C. sinensis,* the two sequences were aligned using Mauve v2.3.1 [31] and the result was analyzed using custom Perl scripts. Intergenic gene regions were parsed out from the two *Citrus* cp genomes and aligned using MUSCLE v3.8.31 [32] with the default settings. The pairwise distances were calculated using the DNADIST program in the PHYLIP package v3.695 [33].

The positions and types of simple sequence repeats (SSRs) in the two *Citrus* cp genomes were detected using MISA (http://pgrc.ipk-gatersleben.de/misa/). The minimum number of repeats were set to 10, 5, 4, 3, 3, and 3 for mono-, di-, tri-, tetra-, penta-, and hexanucleotides, respectively. For long repeats, the program REPuter [34] was used to identify the number and location of direct and inverted (i.e., palindromic) repeats. A minimum repeat

Table 1. Summary of the *Citrus* chloroplast genome characteristics.

Attribute	*C. aurantiifolia* (KJ865401)	*C. sinensis* (NC_008334)
Size (bp)	159,893	160,129
overall GC content (%)	38.4	38.5
LSC size in bp (% total)	87,148 (54.5%)	87,744 (54.8%)
SSC size in bp (% total)	18,763 (11.7%)	18,393 (11.5%)
IR size in bp (% total)[a]	26,991 (16.9%)	26,996 (16.9%)
Protein-coding regions size in bp (% total)	81,468 (51.0%)	79,773 (49.8%)
rRNA and tRNA size in bp (% total)	11,850 (7.5%)	11,850 (7.4%)
Introns size in bp (% total)	17,129 (10.7%)	18,252 (11.4%)
Intergenic spacer size in bp (% total)	49,446 (30.9%)	50,254 (31.4%)
Number of different genes	115	113[b]
Number of different protein-coding genes	81	79[b]
Number of different rRNA genes	4	4
Number of different tRNA genes	30	30
Number of different genes duplicated by IR	22	20
Number of different genes with introns	17	17

[a]Each cp genome contains two copies of inverted repeats (IRs).
[b]According to the original annotation, not including *orf56*.

Table 2. Differences between the *C. aurantiifolia* and *C. sinensis* cp genomes.

Indel				
		Length (bp)	Count	
		1	43	
		2–10	20	
		11–100	18	
		101–1,000	3	
	Sum	1,780	116	Percentage[a]: 1.11%
Substitution				
		Type	Count	
		A <-> T	34	
		C <-> G	15	
		A <-> C	81	
		T <-> C	64	
		A <-> G	51	
		T <-> G	85	
	Sum		330	Percentage[a]: 0.21%
10 most divergent intergenic regions				
		Region	Length[b] (bp)	Pairwise distance
		rps3 - rpl22 (LSC)	234	0.027
		ndhE - ndhG (SSC)	276	0.018
		psaC - ndhE (SSC)	231	0.017
		psbH - petB (LSC)	118	0.017
		trnY-GUA-trnE-UCC (LSC)	59	0.017
		trnH-GUG - psbA (LSC)	449	0.016
		rpl32 - trnL-UAG (SSC)	1,141	0.015
		psbT-psbN (LSC)	66	0.015
		trnG-GCC-trnR-UCU (LSC)	204	0.015
		trnD-GUC-trnY-GUA (LSC)	469	0.013

[a]Relative to the length of *C. aurantiifolia*.
[b]Length in C. aurantiifolia.

size of 30 bp and sequence identity greater than 90% setting were used according to the study of *C. sinensis* cp genome [10]. The redundant or overlapping repeats were identified and filtered manually.

Phylogenetic Inference

Phylogenetic analysis of the representative rosids lineages with complete cp genomes available was performed using PhyML v20120412 [35] with the GTR+I+G model. A total of 72 rosids species were chosen as the ingroups and *Vitis venifera* was included as the outgroup, the accession numbers were provided in Table S1. The protein-coding and rRNA genes were parsed from the selected cp genomes and clustered into ortholog groups using OrthoMCL [36]. The presence/absence of orthologous genes in each genome was examined and further verified using TBLASTN [27,28] searches (e-value cutoff= 1e-10). The nucleotide sequences of the conserved genes were aligned individually by using MUSCLE with the default settings. The concatenated alignment was used to infer a maximum likelihood phylogeny as described above. The bootstrap supports were estimated from 1,000 resampled alignments generated by the SEQBOOT program in the PHYLIP package.

Investigations of *orf56* and *ycf68*

To investigate the presence/absence of *orf56* and *ycf68* in the selected cp genomes, the gene sequences from *C. aurantiifolia* was used as the queries to perform BLASTN [27,28] searches (e-value cutoff= 1e-10). The significant hits were examined to investigate the presence of intact open reading frames (ORFs). Phylogenetic analysis of the cp *orf56* genes and the homologous mitochondrial sequences was performed as described above. The final alignment contains 190 aligned nucleotide sites and a total of 70 sequences, including two sequences of *Amborella* as the outgroup.

Results and Discussion

General Features of the Omani Lime Chloroplast Genome

The complete cp genome of *C. aurantiifolia* (Christm.) Swingle (GenBank accession number KJ865401.1) is 159,893 bp in length, including a large single copy (LSC) region of 87,148 bp, a small single copy (SSC) region of 18,763 bp, and a pair of inverted repeats (IRa and IRb) of 26,991 bp each (Figure 1 and Table 1). A total of 137 different genes, including 93 protein-coding genes, 30 tRNA genes, and four rRNA genes, were annotated (Table S2). Among these, 12 protein-coding genes and 7 tRNA genes are duplicated in the IR regions. Most of the protein-coding genes are

Table 3. List of simple sequence repeats.

Repeat unit	Length (bp)	Number of SSRs	Start position[a]
A	10	6	4512; 47812; 53871; **72614**; 121748; **159288**
	11	6	**6866**; 10130; 69481; 71892; 117725; **134802**
	12	9	**8332**; 31399; 47307; 63928; **111804**; **113977 (*ycf1*)**; 118367; 140302 (*ycf68*); 144255
	13	2	10107; 84557
	14	1	**385**
	15	1	32360
	16	2	69965; 118302
	17	3	7620; 39139; 74176
	19	1	**12023**
	22	1	70289
T	10	10	**2424 (*matK*)**; **19786**; **26964 (*rpoB*)**; **37622**; 46938; **63632**; **87731**; 117742; 117871; 118851
	11	11	9401; 10416; **17001**; 30912; 46021; 63530; 112216; 117988; 118224; **121703**; **131189** (*ycf1*)
	12	6	**14722**; 29024; 102773; 106715 (*ycf68*); **133040 (*ycf1*)**; **135213**
	13	2	**73946**; 80423
	14	2	1776; 85274
	15	2	54209; 57817
	17	1	45965
	18	3	52748; 68339; 81409
	20	1	49202
	23	2	23694; **33282**
C	10	2	**28769**; **104247**
G	10	1	**142772**
AT	10	4	**20631 (*rpoC2*)**; **33636**; 11817; **121517 (*ndhD*)**
AAG	12	1	**97331**
AAT	12	2	**38604**; **122629**
ATA	12	1	70220
ATT	12	3	**10283**; **53810**; **54088**
	18	1	**1760**;
CTT	12	2	**37353 (*psbC*)**; **149686**
TAA	12	2	**30250**; 61945
TAT	12	1	**83297**
TTC	12	1	**73084**
TAAA	12	2	4866; **45088**
AAAT	12	3	**30423**; **32502**; **71394**
ATAC	12	1	**51167**
ATTT	12	1	49193
	20	1	**117168**
TTAA	12	2	39175; 39188
TTAG	12	1	**61483**
TTTC	12	1	**14352**
TCTT	12	1	**46961**
AATAA	20	1	144226
TTTTA	20	1	102781
TTCAAA	18	1	63817

[a]The SSR-containing coding regions are indicated in parentheses. SSRs that are identical in the *C. sinensis* chloroplast genome are highlighted in bold; SSRs that are conserved but with different lengths are highlighted by underline.

Table 4. List of long repeat sequences.

Repeat size	Type[a]	Start position of 1st repeat	Start position the repeat found in other region	Location[b]	Region
30	D	1759	1762	IGS (*psbA-trnK-UUU*)	LSC
30	P	1771	12015	IGS (*psbA-trnK-UUU, atpA-atpF*)	LSC
30	P	8231	37726, 47606	**IGS (*trnS-GCU, trnS-UGA, trnS-GGA*),**	LSC
30	D	23226	85067	**intron (*rpoC1*), IGS (*rpl16-rps3*)**	LSC
30	D	23686	52733	intron (*rpoC1*), IGS (*ndhK-ndhC*)	LSC
30	P	23687	70291	intron (*rpoC1*), IGS (*trnP-UGG-psaJ*)	LSC
30	D	23692	33280	intron (*rpoC1*), IGS (*trnE-UUC-trnT-GGU*)	LSC
30	D	49192	117171	IGS (*psbA-trnK-UUU*), IGS (*atpA-atpF*)	LSC, IR
30	D, P	49197	102764, 144233	IGS (*trnT-UGU-trnL-UAA*), IGS (*rps12-trnV-GAC*)	LSC, IR
30	D, P	51215	102768, 144229	IGS (*trnF-GAA-ndhJ*), IGS (*rps12-trnV-GAC*)	LSC, IR
30	P	71344	71344	**IGS (*rpl33-rps18*)**	LSC
30	D, P	102768	102773, 144224	IGS (*rps12-trnV-GAC*)	IR
30	D	144225	144230	IGS (*trnV-GAC-rps12*)	IR
31	P	4492	117868	IGS (*trnK-UUU-rps16*), IGS (*rpl32-trnL-UAG*)	LSC, IR
31	P	10106	49188	IGS (*trnG-GCC-trnR-UCU, trnT-UGU-trnL-UAA*)	LSC
31	P	29811	29811	**IGS (*petN-psbM*)**	LSC
31	P	33281	70282	**IGS (*trnE-UUC-trnT-GGU, trnP-UGG-psaJ*)**	LSC
31	P	119977	119977	intron (*ccsA*)	IR
32	D	7615	74171	IGS (*psbK-psbI*), intron (*clpP*)	LSC
32	P	39166	39166	IGS (*trnG-GCC-trnfM-CAU*)	LSC
34	P	38774	38782	**IGS (*psbZ-trnG-GCC*)**	LSC
34	P	49186	70288	rps4, IGS (*trnP-UGG-psaJ*)	LSC
34	D, P	111432	111464, 135529, 135561	**IGS (*rrn4.5-rrn5*)**	IR
35	P	10097	49193	IGS *trnG-GCC-trnR-UCU, trnT-UGU-trnL-UAA*)	LSC
36	P	27648	27648	IGS (*rpoB-trnC-GCA*)	LSC
40	P	77776	77776	IGS (*psbT-psbN*)	LSC
41	D	41294	43518	***psaB, psaA***	LSC
41	D, P	102353	124945, 144633	**IGS (*rps12-trnV-GAC*), intron (*ndhA*)**	IR
48	P	30626	30626	**IGS (*petN-psbM*)**	LSC
50	D	39020	39044	**IGS (*trnG-GCC-trnfM-CAU*)**	LSC
51	P	9984	9984	IGS (*trnG-GCC-trnR-UCU*)	LSC
53	P	8869	31095	**IGS** (*trnS-GCU-trnG-GCC*, ***psbM-trnD-GUC***)	LSC
54	P	441	441	IGS (*trnH-GUG-psbA*)	LSC

[a]D: direct repeat; P: palindrome inverted repeat.
[b]IGS: intergenic spacer region. Sequences conserved in the *C. sinensis* chloroplast genome are highlighted in bold.

composed of a single exon, while 14 contain one intron and three contain two introns. The gene *rps12* was predicted to undergo trans-splicing, with the 5′ exon located in the LSC region and the other two exons located in the IR regions.

The protein-coding regions contain a total of 27,159 codons (Table S3). Isoleucine and cysteine are the most and least frequent amino acids and have 2,892 (10.7%) and 359 (1.2%) codons, respectively. The codon usage is biased towards a high ratio of A/T at the third position, which is also observed in many land plant cp genomes [37].

Sequence Comparisons with Sweet Orange

The general characteristics of the two *Citrus* cp genomes are summarized in Table 1, overall the compositions are quite similar. The GC content of these *Citrus* cp genomes is approximately 38.5%, which is slightly higher than the average of the 72 representative rosids lineages (36.7%). In these two *Citrus* cp genomes, the genic regions, introns, and intergenic regions account for ca. 58%, 11%, and 31%, respectively (Table 1).

The pairwise sequence alignment between the two *Citrus* cp genomes revealed approximately 1.3% sequence divergence (Table 2), including 1,780 indels (1.11%) and 330 substitutions (0.21%). The LSC region contains more sequence polymorphisms than expected by its size, including 1,360 (76.4%) indels and 235 (71.2%) substitutions. In contrast, the two IR regions account for ca. 34% of the cp genome yet contain only 16 (0.9%) indels and 12 (3.6%) substitutions. The size differences in the LSC and SSC regions between these two cp genomes are mostly explained by one large indel in each region. The LSC sizes differ by 596 bp and a 523-bp indel was found in the spacer between *rps16* and *trnQ-*

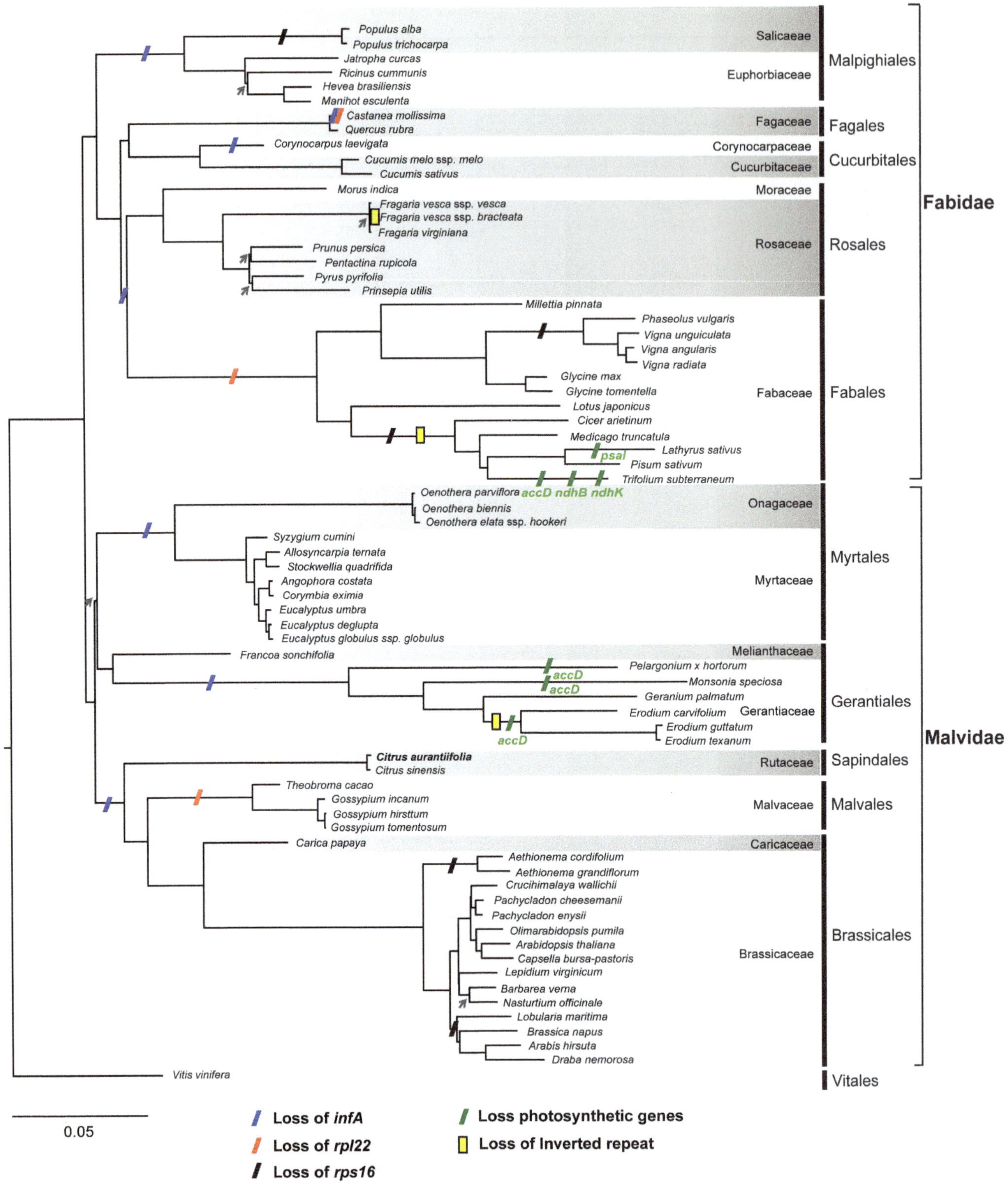

Figure 2. Maximum likelihood phylogeny of the representative rosids lineages. The common grape vine (*Vitis vinifera*) is included as the outgroup to root the tree. The concatenated alignment includes 62 conserved chloroplast genome genes and 54,689 aligned nucleotide sites. Nodes received <70% bootstrap support are indicated by gray arrows. The putative events of gene losses are inferred based on the most parsimonious scenario.

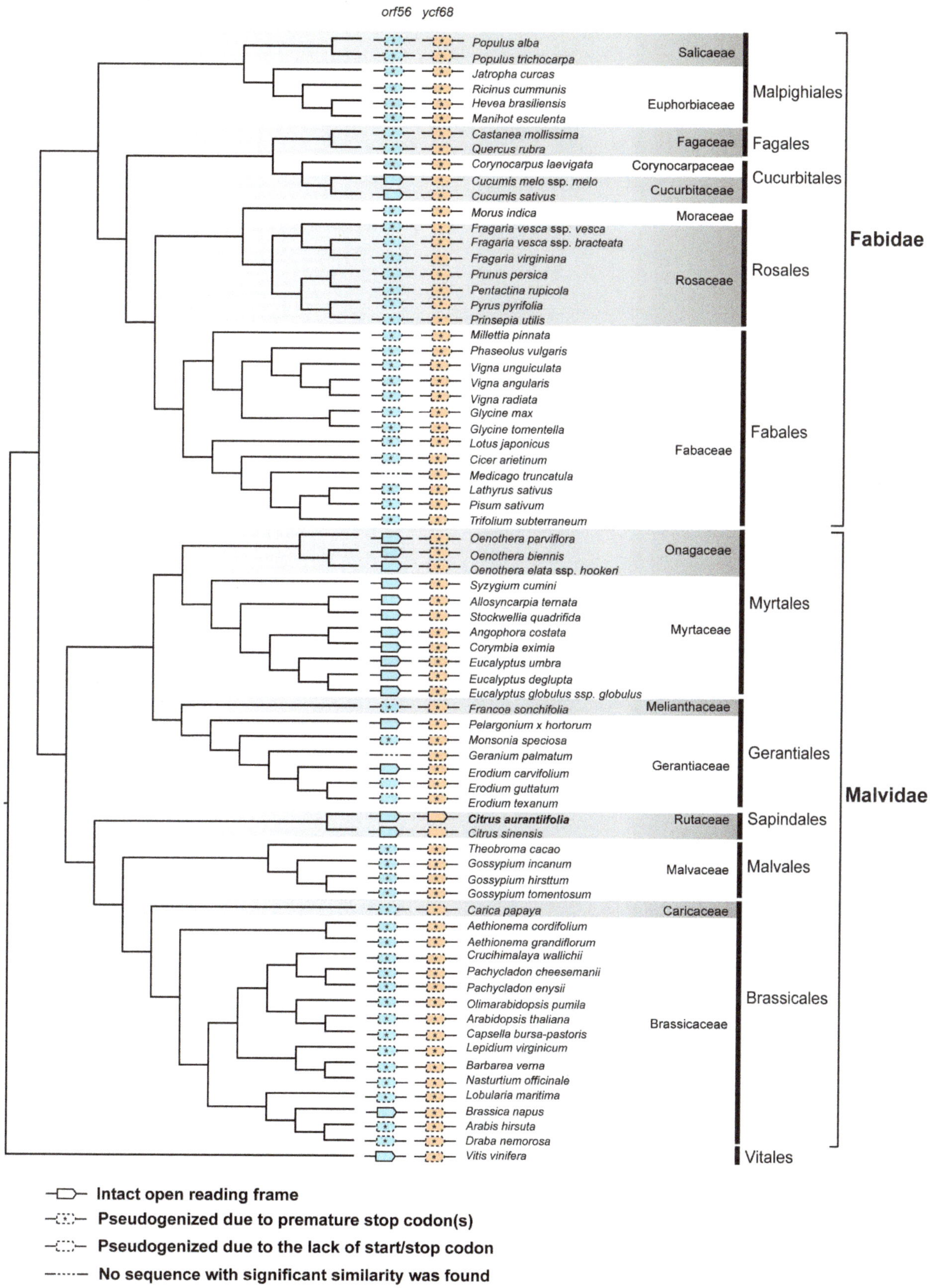

Figure 3. The phylogenetic distribution patterns of *orf56* and *ycf68*.

UUG. The SSC sizes differ by 370 bp and a 354-bp indel was found in the spacer between *rpl32* and *trnL-UAA*.

To identify the intergenic regions that may be useful for phylogenic analysis or molecular identification, we searched for the spacers that are >400 bp in length and exhibit above-average sequence divergence between the two *Citrus* species (i.e., >1.3%). A total of three regions satisfied these criteria, including the spacer between *trnH-GUG* and *psbA* (449 bp, 1.6% divergence), the spacer between *rpl32* and *trnL-UAG* (1141 bp, 1.5% divergence), and the spacer between *trnD-GUC* and *trnY-GUA* (469 bp, 1.3% divergence).

The junctions between the IR, LSC, and SSC regions in *C. aurantiifolia* are similar to that of *C. sinensis* except for the LSC-IRb boundary. A total of 23 indels and five substitutions were found at this region, resulting in one copy of *rpl22* spanning across the LSC-IRb junction in *C. aurantiifolia*. Comparing the IR junctions of *Citrus* with *Theobroma* and *Gossypium* in Malvaceae [38], it was found that the IRs in *Citrus* have expanded to include *rps19* and 252 nt of *rpl22*, whereas in Malvaceae, *rps19* is located in LSC and *rpl22* was missing [38–40].

Analyses of Repetitive Sequences

A total of 109 SSR loci were found in the cp genome of *C. aurantiifoliaa*, accounting for 1,352 bp of the total sequence (ca. 0.8%). Among these, 94 were also found in *C. sinensis* and 42 exhibit length polymorphism (Table 3). Most SSRs are located in intergenic regions, but some were found in coding genes such as *matK* and *ycf1*. Concerning the controversial status of *Citrus* taxonomy, the SSRs identified in this study may provide new perspective to refine the phylogeny and elucidate the origin of the cultivars. Furthermore, these SSRs may be used as molecular markers for population studies.

In addition, 62 large repeats that are longer than 30 bp were found in the *C. aurantiifolia* cp genome (Table 4). Most of these repeats are located in intergenic spacers, except for three that are located in the coding regions of *rps4*, *psaA* and *psaB*. Twelve of these long repeats were also found in *C. sinensis,* indicating that these repeats might be widespread in the genus.

Gene Content Analyses within the Rosids

A maximum likelihood phylogenetic analysis of 72 representative rosids lineages was conducted based on a concatenated alignment of four rRNA and 58 protein-coding genes with 54,689 sites (Figure 2). *Citrus* represents Sapindales and is sister to the clade containing Malvales and Brassicales. These relationships are congruent with the previous reports [10,41–43]. Based on this phylogeny and the gene content, we inferred the gene loss events during the cp genome evolution in rosids.

The translation initiation factor gene *infA* in cp has been lost independently at least 24 times in angiosperms and evidence provided from some cases suggested functional replacement by a nucleus copy [44]. Although the majority of *infA* in our selected cp genomes were found to be pseudogenized or completely lost, an intact *infA* was found in *Quercus*, *Francoa*, and two *Cuscumis* species.

The *rpl22* were found to be lost in Fabaceae [45] and *Castanea* of Fagaceae [46] following independent transfers to nucleus. Furthermore, another putative loss of *rpl22* was detected in *Passiflora* [46]. The *rpl22* in Malvaceae, including *Theobroma* and three *Gossypium* species, were found to be pseudogenized in our analysis. In *Citrus*, the ORF of *rpl22* was shortened to 252–264 nt compared to the typical length of 399–489 nt in other rosids [10,46]. However, compared with the pseudogenized *rpl22* found in Malvalvace, the *rpl22* homologs in *Citrus* still show high sequence conservation. Additionally, the *rpl22* transcripts can be identified in the EST database for various *Citrus* species (data not shown). Taking account into the above consideration, we did not annotate *rpl22* as a pseudogene in *Citrus*.

The parallel losses of *rps16* were found in several rosids lineages (Figure 2), including one time in Salicaceae, two times in Fabaceae and another two times in Brassicaceae. The loss of *rps16* in *Medicago* and *Populus* was found to be substituted by a nuclear-encoded copy that transferred from the mitochondrion (mt) [47]. Because the nuclear-encoded RPS16 was found to target both mt and cp in *Arabidopsis*, *Lycopersicon*, and *Oryza* [47], it is possible that the cp genome-encoded *rps16* would not be maintained by selection and will eventually become lost in these lineages.

There are only a few gene loss events of photosynthetic genes found in rosids. In addition to the loss of *psaI* in *Lathyrus sativus* [48], the *accD* seems to be lost independently in *Trifolium subterraneum* and several Gerantiaceae species except for *Geranium palmatum*. In *Trifolium,* a nuclear-encoded *accD* copy has been reported [48], which presented another example of horizontal gene transfer from cp to nucleus. Successful gene transfers from cp to the nucleus in angiosperms are rare and have been only documented for four genes in rosids. Other than the three genes described above (i.e., *infA*, *rpl22*, and *accD*), the *rpl32* in *Populus* (Salicaceae) is the fourth example [49–51].

The IR has been reported to be independently lost at least five times among seed plants, two of which are within rosids [51]. In addition to the inverted repeat lacking clade (IRLC) of papilionoid Fabaceae [52] and *Erodium* of Gerantiaceae [53,54], the IR was found to be lost in two lineages of *Fragaria* (Rosaceae), which are *F. vesca* ssp. *bracteatea* and *F. mandschurica* (accession: NC_018767, not shown in Figure 2). Based on the *Fragaria* phylogeny shown in a previous study [55], it seems that IR loss was not a single event in *Fragaria*.

Molecular Evolution of *orf56* and *ycf68* within the Rosids

In the comparison of gene content between the two *Citrus* cp genomes, *C. aurantiifolia* was found to contain two additional protein-coding genes. The first gene, *orf56*, is located in the *trnA-UGC* intron that contains one sequence homologous to previously recognized mitochondrial *ACRS* (ACR-toxin sensitivity gene) in *Citrus* [56]. In addition to the 171-bp identical sequences between cp *orf56* and the ORF sequences of *ACRS* in mt, the full length of 355-bp region of *ACRS* that conferred sensitivity to ACR-toxin in *E. coil* are also identical. Furthermore, the whole *trnA-UGC* among two *Citrus* cp regions and *C. jambhiri* mitochondrial *ACRS* shared more than 96% identity (Figure S1), which highlight the conservation of this region between cp and mt.

The gene *orf56* has also been included in the annotation of complete cp genomes of *Calycanthus* [57] and *Pelargonium* [58]. Our BLAST search against the rosids genome database revealed that in addition to *Citrus* and *Pelargonium*, all of the species examined in Cucurbitaceae and Myrtales also contain an intact *orf56* (Figure 3). Moreover, an intact *ACRS* ORF is also present in the mt genomes of *Liriodendron* [59] and *Silene* [60] and the ORF sequences between cp and mt are identical. Goremykin et al. [57] suggested that the *ACRS* gene was relative recently transferred from cp to mt. Based on the phylogeny containing the cp *orf56* and the mt *ACRS* (Figure S2), it appears that *orf56* has been independently transferred from cp to mt in different lineages.

The second gene, *ycf68*, is located in the *trnI-GAU* intron. A nearly identical sequence was found in *C. sinensis* but an additional T insertion near the C-terminus abolished the stop codon at the corresponding position. The intact *ycf68* can be detected in several monocots and Nymphaeaceae [61,62]. However, in the majority of other rosids (Figure 3) and the rest of the eudicots [61], the *ycf68* homologs all contain premature stop codons. Although Raubeson et al. [61] argued that *ycf68* is not a protein-coding gene based on the lack of intron-folding pattern, the high levels of sequence conservation among the ORFs of identified homologs suggest that the true identity and functionality of this putative gene remains to be further investigated.

Conclusions

We reported the complete cp genome sequence of *Citrus aurantiifolia* (Rutaceae) in this study. The genome organization and gene content is typical of most angiosperms and highly similar to that of *C. sinensis* (i.e., 98.7% identical at the nucleotide level). The only difference in the gene content between the two *Citrus* cp genomes is the *C. aurantiifolia*-specific presence of a protein-coding gene (*ycf68*) in the *trnI-GAU* intron. Notably, three long intergenic spacers with high sequence divergence and 94 shared SSR regions were identified in the *C. aurantiifolia-C. sinensis* comparison. These regions may provide phylogenetic utility at low taxonomic levels and could be applied to the molecular identification of *Citrus* cultivars. Finally, our comparative analysis of gene content among 72 representative rosids lineages highlighted multiple events of gene losses within this group.

Acknowledgments

We thank Sam T. Mugford and Allyson M. MacLean for help with purification and quality controls of DNA samples.

Author Contributions

Conceived and designed the experiments: HJS CHK. Performed the experiments: SAH AMA. Analyzed the data: HJS CHK. Contributed reagents/materials/analysis tools: HJS AMA CHK. Contributed to the writing of the manuscript: HJS SAH AMA CHK.

References

1. Mabberley DJ (2004) *Citrus* (Rutaceae): a review of recent advances in etymology, systematics and medical applications. Blumea 49: 481–198.
2. Tripoli E, Guardia ML, Giammanco S, Majo DD, Giammanco M (2007) *Citrus* flavonoids: molecular structure, biological activity and nutritional properties: a review. Food Chem 104: 466–479.
3. Ezeabara CA, Okeke CU, Aziagba BO, Ilodibia CV, Emeka AN (2014) Determination of saponin content of various parts of six *Citrus* species. Int Res J Pure Appl Chem 4: 137–143.
4. Nicolosi E, Deng ZN, Gentile A, La Malfa S, Continella G, et al. (2000) *Citrus* phylogeny and genetic origin of important species as investigated by molecular markers. Theor Appl Genet 100: 1155–66.
5. Hynniewta M, Malik SK, Rao SR (2014) Genetic diversity and phylogenetic analysis of Citrus (L) from north-east India as revealed by meiosis, and molecular analysis of internal transcribed spacer region of rDNA. Meta Gene 2: 237–251.
6. Liu Y, Heying E, Tanumihardjo SA (2012) History, global distribution, and nutritional importance of citrus fruits. Compr Rev Food Sci Food Saf 11: 530–545.
7. Penjor T, Yamamoto M, Uehara M, Ide M, Matsumoto N, et al. (2013) Phylogenetic relationships of *Citrus* and its relatives based on *matK* gene sequences. PLoS ONE 8: e62574.
8. Corriveau JL, Coleman AW (1988) Rapid screening method to detect potential biparental inheritance of plastid DNA and results for over 200 angiosperm species. Am J Bot 75: 1443–1458.
9. Zhang Q, Liu Y, Sodmergen (2003) Examination of the cytoplasmic DNA in male reproductive cells to determine the potential for cytoplasmic inheritance in 295 angiosperm species. Plant Cell Physiol 44: 941–951.
10. Bausher MG, Singh ND, Lee SB, Jansen RK, Daniell H (2006) The complete chloroplast genome sequence of *Citrus sinensis* (L.) Osbeck var 'Ridge Pineapple': organization and phylogenetic relationships to other angiosperms. BMC Plant Biol 6: 21.
11. Davies FS, Albrigo LG (1994). Citrus. CABI International, Wiltshire, UK. 1–2.
12. Vand SH, Abdullah TL (2012) Identification and introduction of Thornless Lime (*Citrus aurantifolia*) in Hormozgan, Iran. Indian J Sci Technol 5: 3670–3673.
13. Al-Sadi AM, Al-Moqbali HS, Al-Yahyai RA, Al-Said FA (2012) AFLP data suggest a potential role for the low genetic diversity of acid lime (*Citrus aurantifolia* Swingle) in Oman in the outbreak of witches' broom disease of lime. Euphytica 188: 285–297.
14. Taheri F, Nematzadeh G, Zamharir MG, Nekouei MK, Naghavi M, et al. (2011) Proteomic analysis of the Mexican lime tree response to "*Candidatus* Phytoplasma aurantifolia" infection. Mol Biosyst 7: 3028–3035.
15. Zamharir MG, Mardi M, Alavi SM, Hasanzadeh N, Nekouei MK, et al. (2011) Identification of genes differentially expressed during interaction of Mexican lime tree infected with "*Candidatus* Phytoplasma aurantifolia". BMC Microbiol 11: 1.
16. Monavarfeshani A, Mirzaei M, Sarhadi E, Amirkhani A, Khayam Nekouei M, et al. (2013) Shotgun proteomic analysis of the Mexican lime tree infected with "*Candidatus* Phytoplasma aurantifolia." J Proteome Res 12: 785–795.
17. Faghihi MM, Bagheri AN, Bahrami HR, Hasanzadeh H, Rezazadeh R, et al. (2011) Witches'-broom disease of lime affects seed germination and seedling growth but is not seed transmissible. Plant Disease 95: 419–422.
18. Zafari S, Niknam V, Musetti R, Noorbakhsh SN (2012) Effect of phytoplasma infection on metabolite content and antioxidant enzyme activity in lime (*Citrus aurantifolia*). Acta Physiol Plant 34: 561–568.
19. Maixner M, Ahrens U, Seemüller E (1995) Detection of the German grapevine yellows (Vergilbungskrankheit) MLO in grapevine, alternative hosts and a vector by a specific PCR procedure. Eur J Plant Pathol 101: 241–250.
20. Ku C, Hu J-M, Kuo C-H (2013) Complete plastid genome sequence of the basal asterid *Ardisia polysticta* Miq. and comparative analyses of asterid plastid genomes. PLoS ONE 8: e62548.
21. Ku C, Chung W-C, Chen L-L, Kuo C-H (2013) The complete plastid genome sequence of Madagascar periwinkle *Catharanthus roseus* (L.) G. Don: plastid genome evolution, molecular marker identification, and phylogenetic implications in asterids. PLoS ONE 8: e68518.
22. Zerbino DR, Birney E (2008) Velvet: algorithms for *de novo* short read assembly using de Bruijn graphs. Genome Res 18: 821–829.
23. Li H, Durbin R (2009) Fast and accurate short read alignment with Burrows-Wheeler transform. Bioinformatics 25: 1754–1760.
24. Li H, Handsaker B, Wysoker A, Fennell T, Ruan J, et al. (2009) The Sequence Alignment/Map format and SAMtools. Bioinformatics 25: 2078–2079.
25. Robinson JT, Thorvaldsdottir H, Winckler W, Guttman M, Lander ES, et al. (2011) Integrative genomics viewer. Nat Biotechnol 29: 24–26.
26. Wyman SK, Jansen RK, Boore JL (2004) Automatic annotation of organellar genomes with DOGMA. Bioinformatics 20: 3252–3255.
27. Altschul SF, Gish W, Miller W, Myers EW, Lipman DJ (1990) Basic local alignment search tool. J Mol Biol 215: 403–410.
28. Camacho C, Coulouris G, Avagyan V, Ma N, Papadopoulos J, et al. (2009) BLAST+: architecture and applications. BMC Bioinformatics 10: 421.
29. Charif D, Lobry JR (2007) SeqinR 1.0–2: a contributed package to the R project for statistical computing devoted to biological sequences retrieval and analysis. In: Bastolla DU, Porto PDM, Roman DHE, Vendruscolo DM, editors.

Structural Approaches to Sequence Evolution. Biological and Medical Physics, Biomedical Engineering. Springer Berlin Heidelberg. 207–232.
30. Lohse M, Drechsel O, Bock R (2007) OrganellarGenomeDRAW (OGDRAW): a tool for the easy generation of high-quality custom graphical maps of plastid and mitochondrial genomes. Curr Genet 52: 267–274.
31. Darling ACE, Mau B, Blattner FR, Perna NT (2004) Mauve: multiple alignment of conserved genomic sequence with rearrangements. Genome Res 14: 1394–1403.
32. Edgar RC (2004) MUSCLE: multiple sequence alignment with high accuracy and high throughput. Nucl Acids Res 32: 1792–1797.
33. Felsenstein J (1989) PHYLIP - Phylogeny Inference Package (version 3.2). Cladistics 5: 164–166.
34. Kurtz S, Schleiermacher C (1999) REPuter: fast computation of maximal repeats in complete genomes. Bioinformatics 15: 426–427.
35. Guindon S, Gascuel O (2003) A simple, fast, and accurate algorithm to estimate large phylogenies by maximum likelihood. Syst Biol 52: 696–704.
36. Li L, Stoeckert CJ, Roos DS (2003) OrthoMCL: Identification of ortholog groups for eukaryotic genomes. Genome Res 13: 2178–2189.
37. Clegg MT, Gaut BS, Learn GH, Morton BR (1994) Rates and patterns of chloroplast DNA evolution. Proc Natl Acad Sci U S A 91: 6795–6801.
38. Kane N, Sveinsson S, Dempewolf H, Yang JY, Zhang D, et al. (2012) Ultra-barcoding in cacao (*Theobroma* spp.; Malvaceae) using whole chloroplast genomes and nuclear ribosomal DNA. Am J Bot 99: 320–329.
39. Lee SB, Kaittanis C, Jansen RK, Hostetler JB, Tallon LJ, et al. (2006) The complete chloroplast genome sequence of *Gossypium hirsutum*: organization and phylogenetic relationships to other angiosperms. BMC Genomics 7: 61.
40. Xu Q, Xiong G, Li P, He F, Huang Y, et al. (2012) Analysis of complete nucleotide sequences of 12 *Gossypium* chloroplast genomes: origin and evolution of allotetraploids. PLoS ONE 7: e37128.
41. Bremer B, Bremer K, Chase MW, Fay MF, Reveal JL, et al. (2009) An update of the Angiosperm Phylogeny Group classification for the orders and families of flowering plants: APG III. Bot J Linn Soc 161: 105–121.
42. Worberg A, Alford MH, Quandt D, Borsch T (2009) Huerteales sister to Brassicales plus Malvales, and newly circumscribed to include *Dipentodon*, *Gerrardina*, *Huertea*, *Perrottetia*, and *Tapiscia*. Taxon 58: 468–478.
43. Ruhfel BR, Gitzendanner MA, Soltis PS, Soltis DE, Burleigh JG (2014) From algae to angiosperms-inferring the phylogeny of green plants (*Viridiplantae*) from 360 plastid genomes. BMC Evol Biol 14: 23.
44. Millen RS, Olmstead RG, Adams KL, Palmer JD, Lao NT, et al. (2001) Many parallel losses of *infA* from chloroplast DNA during angiosperm evolution with multiple independent transfers to the nucleus. Plant Cell 13: 645–658.
45. Gantt JS, Baldauf SL, Calie PJ, Weeden NF, Palmer JD (1991) Transfer of *rpl22* to the nucleus greatly preceded its loss from the chloroplast and involved the gain of an intron. EMBO J 10: 3073–3078.
46. Jansen RK, Saski C, Lee S-B, Hansen AK, Daniell H (2011) Complete plastid genome sequences of three rosids (*Castanea*, *Prunus*, *Theobroma*): evidence for at least two independent transfers of *rpl22* to the nucleus. Mol Biol Evol 28: 835–847.
47. Ueda M, Nishikawa T, Fujimoto M, Takanashi H, Arimura S, et al. (2008) Substitution of the gene for chloroplast RPS16 was assisted by generation of a dual targeting signal. Mol Biol Evol 25: 1566–1575.
48. Magee AM, Aspinall S, Rice DW, Cusack BP, Sémon M, et al. (2010) Localized hypermutation and associated gene losses in legume chloroplast genomes. Genome Res 20: 1700–1710.
49. Cusack BP, Wolfe KH (2007) When gene marriages don't work: divorce by subfunctionalization. Trends Genet 23: 270–272.
50. Ueda M, Fujimoto M, Arimura S, Murata J, Tsutsumi N, et al. (2007) Loss of the *rpl32* gene from the chloroplast genome and subsequent acquisition of a preexisting transit peptide within the nuclear gene in *Populus*. Gene 402: 51–56.
51. Jansen RK, Ruhlman TA (2012) Plastid genomes of seed plants. In: Bock R, Knoop V, editors. Genomics of chloroplasts and mitochondria. Advances in photosynthesis and respiration. Springer Netherlands. 103–126.
52. Wojciechowski MF, Lavin M, Sanderson MJ (2004) A phylogeny of legumes (Leguminosae) based on analysis of the plastid *matK* gene resolves many well-supported subclades within the family. Am J Bot 91: 1846–1862.
53. Blazier JC, Guisinger MM, Jansen RK (2011) Recent loss of plastid-encoded *ndh* genes within *Erodium* (Geraniaceae). Plant Mol Biol 76: 263–272.
54. Guisinger MM, Kuehl JV, Boore JL, Jansen RK (2011) Extreme reconfiguration of plastid genomes in the angiosperm family Geraniaceae: rearrangements, repeats, and codon usage. Mol Biol Evol 28: 583–600.
55. Njuguna W, Liston A, Cronn R, Ashman TL, Bassil N (2013) Insights into phylogeny, sex function and age of *Fragaria* based on whole chloroplast genome sequencing. Mol Phylogenet Evol 66: 17–29.
56. Ohtani K, Yamamoto H, Akimitsu K (2002) Sensitivity to *Alternaria alternata* toxin in citrus because of altered mitochondrial RNA processing. Proc Natl Acad Sci U S A 99: 2439–2444.
57. Goremykin V, Hirsch-Ernst KI, Wölfl S, Hellwig FH (2003) The chloroplast genome of the "basal" angiosperm *Calycanthus fertilis* – structural and phylogenetic analyses. Plant Syst Evol 242: 119–135.
58. Chumley TW, Palmer JD, Mower JP, Fourcade HM, Calie PJ, et al. (2006) The complete chloroplast genome sequence of *Pelargonium* × *hortorum*: organization and evolution of the largest and most highly rearranged chloroplast genome of land plants. Mol Biol Evol 23: 2175–2190.
59. Richardson AO, Rice DW, Young GJ, Alverson AJ, Palmer JD (2013) The "fossilized" mitochondrial genome of *Liriodendron tulipifera*: ancestral gene content and order, ancestral editing sites, and extraordinarily low mutation rate. BMC Biol 11: 29.
60. Sloan DB, Müller K, McCauley DE, Taylor DR, Šorchová H (2012) Intraspecific variation in mitochondrial genome sequence, structure, and gene content in *Silene vulgaris*, an angiosperm with pervasive cytoplasmic male sterility. New Phytol 196: 1228–1239.
61. Raubeson LA, Peery R, Chumley TW, Dziubek C, Fourcade HM, et al. (2007) Comparative chloroplast genomics: analyses including new sequences from the angiosperms *Nuphar advena* and *Ranunculus macranthus*. BMC Genomics 8: 174.
62. Ahmed I, Biggs PJ, Matthews PJ, Collins LJ, Hendy MD, et al. (2012) Mutational dynamics of aroid chloroplast genomes. Genome Biol Evol 4: 1316–1323.

5

Direct Chloroplast Sequencing: Comparison of Sequencing Platforms and Analysis Tools for Whole Chloroplast Barcoding

Marta Brozynska, Agnelo Furtado, Robert James Henry*

Queensland Alliance for Agriculture and Food Innovation, The University of Queensland, Brisbane, Queensland, Australia

Abstract

Direct sequencing of total plant DNA using next generation sequencing technologies generates a whole chloroplast genome sequence that has the potential to provide a barcode for use in plant and food identification. Advances in DNA sequencing platforms may make this an attractive approach for routine plant identification. The HiSeq (Illumina) and Ion Torrent (Life Technology) sequencing platforms were used to sequence total DNA from rice to identify polymorphisms in the whole chloroplast genome sequence of a wild rice plant relative to cultivated rice (cv. Nipponbare). Consensus chloroplast sequences were produced by mapping sequence reads to the reference rice chloroplast genome or by *de novo* assembly and mapping of the resulting contigs to the reference sequence. A total of 122 polymorphisms (SNPs and indels) between the wild and cultivated rice chloroplasts were predicted by these different sequencing and analysis methods. Of these, a total of 102 polymorphisms including 90 SNPs were predicted by both platforms. Indels were more variable with different sequencing methods, with almost all discrepancies found in homopolymers. The Ion Torrent platform gave no apparent false SNP but was less reliable for indels. The methods should be suitable for routine barcoding using appropriate combinations of sequencing platform and data analysis.

Editor: Hikmet Budak, Sabanci University, Turkey

Funding: The research was funded by Rural Industries Research and Development Corporation (Australia). The funders played no role in study design, data collection, decision to publish or preparation of the manuscript.

Competing Interests: This research was supported by a grant from Rural Industries Research and Development Corporation, Australian Government.

* Email: robert.henry@uq.edu.au

Introduction

A universal plant barcoding method based upon sequencing a specific gene or combination of specific genes has proven elusive. Analysis of different plant groups has required the use of different genetic loci. The emergence of next generation sequencing (NGS) technology allowing whole chloroplast genomic DNA sequencing has provided the opportunity to use the whole chloroplast genome sequence as a barcode [1]. The chloroplast has unique features found in all green plants, conserved sufficiently to be readily aligned for comparison of different samples and large enough to contain variation, which allow species or sub species distinction across the seed plants [2]. Analysis of complex mixtures of plants has applications in many ecological studies [3].

A generic method for rapid and cost effective DNA-based identification of plants at this level (species or sub-species) will find wide application in industry and research [4,5]. Industrial applications will include the identification of plant components in whole and processed foods and the management of food processing to ensure food safety and authenticity of labelling [6]. This will complement the use of next generation sequencing to screen for the presence of pathogenic microbes [7]. Protection of intellectual property rights associated with plant varieties will also be simplified by a standard approach to plant identification. These tools will also support protection of biosecurity and management of wild plant populations of rare or threatened plants.

Early analysis relied on either specific amplification of the chloroplast genome or separation of chloroplasts from nuclear and mitochondrial DNA before chloroplast DNA isolation and sequencing the amplified or cloned product derived from PCR amplification [8]. However, the universal amplification of chloroplast sequences from all species has proven difficult and chloroplast purification is laborious and not perfect [9,10]. Recently a simple approach involving NGS of a total genomic DNA preparation was proposed [11]. This method relies on successful extraction of chloroplast genome sequence reads from total genomic DNA reads and their assembly to form a consensus sequence for the sample that can be used as a barcode. Reference-guided mapping of sequence reads and *de novo* assembly have been used to generate these whole chloroplast barcodes [12].

DNA sequencing platforms differ in their throughput and accuracy [13]. Accuracy of a DNA sequence assembled using NGS reads is dependent on the read length, sequencing depth, sequence coverage or width and evenness of coverage but also on the accuracy of the sequencing platform and the assembly and analysis pipeline [14]. We compared two different sequencing technologies, the Illumina platform that uses sequencing-by-syntheses (SBS) chemistry and the Ion Torrent that is based on semiconductor sequencing technology. Sequencing on both platforms commences with library preparation, which involves fragmentation of total genomic DNA, purifying to uniform and desired fragment size and ligation to sequencing adapters specific to the platform [15,16]. In Illumina systems the fragments are subsequently denatured and fixed on the surface of a proprietary glass flowcell, followed by solid-phase amplification (bridge amplification). As a result clusters are created which contain clonally amplified DNA templates. Next the clusters are sequenced in parallel using four fluorescently labelled nucleotides with reversible dye terminators. After every sequencing cycle and base incorporation to the template the polymerization terminates, a CCD (charge-coupled device) captures the fluorescent signal and identifies the base. After the base call the dye is cleaved and the sequencing process continues. This technology was first implemented in an instrument called Genome Analyzer (GA) with subsequent releases of improved series up to GAIIx. The next major system launched by Illumina was the HiSeq2000 with improved output and read length. Data from both GAIIx and HiSeq was used in the current study.

In Ion Torrent systems the prepared libraries are immobilized to beads and amplified using emulsion PCR which takes place within microdroplets of aqueous solution and oil [17]. During the sequencing individual bases are incorporated by the action of a DNA polymerase. As a result of this reaction a proton is released and the resulting change in pH is measured. The reaction occurs in a proprietary chip, which acts as a pH meter. Unlike Illumina, this system does not use fluorescent dyes and light detection, shortening the sequencing time. The Ion Torrent instrument used in the study was a Personal Genome Machine (PGM).

We examined the reliability of these two different sequencing platforms coupled with different data analysis approaches when applied to obtaining the chloroplast consensus sequences of a wild rice genotype and a cultivated rice genotype. Although the Ion Torrent platform is widely used for sequencing and limited published data is available on its application for transcript [18] and amplicon [19] sequencing in plants, it has not previously been applied to plant genomic DNA analysis. The most common applications of the Ion Torrent system have been to small genomes (bacterial samples) and targeted sequencing of amplicons or transcripts, whereas the Illumina has been widely used for applications including whole human genome re-sequencing, *de novo* plant and animal genome sequencing, exome, transcriptome (RNA-Seq) and metagenomics investigations. The chloroplast genome represents a specific sequencing challenge with the presence of an inverted repeat. This study compares total DNA sequencing with the Illumina and Ion Torrent platforms and options for analysis of the sequences generated to produce a whole chloroplast genome sequence as a barcode for plant identification. Whole genome NGS was carried out on two rice genotypes: *Oryza sativa* spp. *japonica* var. Nipponbare referred to here as the reference-rice-genotype (R-rice-genotype) as its chloroplast sequence (reference-sequence) is publicly available and an Australian wild rice genotype sample (W-rice-genotype).

Materials and Methods

Plant materials

Seeds of rice (*Oryza sativa* cv. Nipponbare), referred to as reference-rice-genotype (R-rice-genotype), were germinated and entire seedlings were used for DNA extraction. Leaf tissue from a wild rice plant, referred to as wild-rice-genotype (W-rice-genotype) collected from a field containing a population of perennial wild rice (*Oryza rufipogon*-like) at Abattoir Creek in North Queensland, was used for DNA extraction [20].

DNA extraction

DNA from leaf tissue of the W-rice-genotype was extracted using modification of the CTAB method [21]. DNA from seedlings of the R-rice-genotype was extracted as described for the W-rice-genotype but with further purification as described by [11].

Sequencing

The Illumina and the Ion Torrent sequencing platforms were used for the shotgun sequencing of total genomic DNA of both rice samples. The wild rice sample was sequenced on the Ion Torrent (200 bp reads) and the Illumina HiSeq platform (100 bp paired end reads). The cultivated rice genotype was subjected to sequencing using the Ion Torrent (200 bp reads) platform while Illumina reads (36 bp paired-end), generated on the GAII, were sourced from available archived data [11]. Sequencing on the Illumina HiSeq 2000 (Illumina, San Diego, CA, USA), to generate 100 bp paired end reads with an average library insert size of 500–600 bp, was outsourced to the Australian Genome Research Facility (AGRF, Melbourne, Australia). For sequencing using the Ion Torrent platform, genomic DNA was sheared using the Covaris S220 instrument (www.covarisinc.com) and used for preparing sequencing libraries according to the standard Ion Torrent PGM protocol. The resulting individual DNA libraries were quality checked and quantified on the Agilent 2100 Bioanalyzer using the High Sensitivity DNA kit (Agilent). Following template amplification and enrichment on the Ion OneTouch2 (Ion OneTouch 200 Template Kit v2; #4478316) and OneTouch2 ES instruments, each sample was loaded onto one PGM #318 chip and sequenced using Ion PGM Sequencing 200 Kit v.2 (#4482006) according to manufacturer's protocol.

Data analysis – consensus

Raw reads from both sequencing platforms, PGM Ion Torrent and Illumina, were imported to CLC Genomics Workbench 6.0 (CLC-GW) (www.clcbio.com) and read statistics assessed using sequencing data quality control, followed by read trimming for quality, length and presence of ambiguous bases. Ion Torrent reads were trimmed with the quality score limit set to 0.05 (which corresponded to PHRED quality value >15) and a minimum read length of 30 bp. Reads from both Illumina platforms, GAII and HiSeq, were trimmed with a quality score limit of 0.01 (which corresponded to PHRED >22) and the same minimum read length of 30 bp. Ion Torrent data was trimmed at a slightly lower quality value due to the lower average quality compared to the Illumina reads (Table S1; avg. PHRED for Ion Torrent reads: 25 and 26, and avg. PHRED for Illumina: 29 and 32). The Illumina HiSeq platform generates sequencing data fifty to sixty times more in order of magnitude as compared to the Ion Torrent data. Thus, in order to compare Ion Torrent and Illumina HiSeq platforms at similar coverage, a subset of HiSeq Illumina reads for the wild rice sample and matching the Ion Torrent read numbers for that same sample, was randomly extracted from the whole reads set. Ion Torrent raw reads were also alternatively trimmed using the Torrent Suite Software version 3.6 using default analysis settings with the following modifications: the quality-based trimming of the sequencing reads was adjusted to predicted-PHRED-score 17 over the sliding window of 20 bp (default values 15, and 30 respectively) and the minimal length filter for reported reads was set to 20 (default 8).

Consensus sequence generation using the mapping tool was carried out as follows. The CLC-GW was used to map trimmed reads to the chloroplast genome sequence of *Oryza sativa* spp. *japonica* var. Nipponbare (GenBank accession – GU592207) used as the reference-sequence. The mapping process involved the following parameters: mismatch cost 2, insertion cost 3, deletion cost 3, length fraction 0.9, similarity fraction 0.9, and global alignment setting. Non-specific matches were also mapped randomly to the reference genome. Consensus sequences were extracted from all mapping runs giving five sequences in total; two consensus sequences for *O. sativa* Nipponbare each derived from the Illumina GAII reads and the Ion Torrent reads, and three consensus sequences for the wild rice sample each derived from the Illumina HiSeq sequence reads, Illumina HiSeq sequence subset-reads and the Ion Torrent reads. Conflicts between reads were resolved by voting for the majority of the reads at the given position. Consensus sequence were aligned to the reference sequence and analysed for SNPs and indels.

Consensus sequence generation using the assembly tool was carried out as follows using both the CLC-GW and the Torrent Suite Software. The CLC trimmed reads from both platforms were assembled in CLC-GW using the *de novo* assembly tool with the following parameters: automatic word and bubble sizes with 200 bp for minimum contig length. Moreover the reads were mapped back to contigs (mismatch cost 2, insertion cost 3, deletion cost 3, length fraction 0.9, and similarity fraction 0.9) and contigs were updated based on the results. The trimmed Ion Torrent reads were also assembled in the Torrent Suite Software using the Assembler plugin v.3.4.2 utilizing the MIRA assembly algorithm (http://sourceforge.net/projects/mira-assembler/). The assembly was performed in the reference-assisted mode using the reference-sequence. All contigs generated were aligned to the reference-sequence using NUCmer alignment generator from MUMmer package [22]. Owing to the chloroplast genome structure (two long inverted repeats) the option to use all matches regardless of the uniqueness was enabled in the script. All other parameters were left as default. Based on the MUMer output the longest contigs with the best mapping results (query coverage, precent identity and alignment length) were chosen as scaffolds and manually joined together by overlapping the remaining short contigs. In cases when there was not enough contigs to reconstruct the chloroplast genome sequence, *de novo* assembly was repeated with different parameter for word size in the CLC-GW and subjected again to the alignment tool. Consensus sequences from contigs mapping and joining were also screened for SNPs and indels based on the reference-sequence.

Data analysis – variant detection

The consensus sequences generated by both mapping and *de novo* assembly of the reads were imported into Geneious 6.1.6 (www.geneious.com) and each of them was separately mapped to reference-sequence. To generate the alignments the Geneious Map to Reference tool was used with the default parameters. Subsequently the alignments were screened to find variations/SNPs using the tool available in Geneious package.

Validation of the most probable variant

The wild rice chloroplast consensus sequences derived from reads mapping from Illumina and Ion Torrent platforms were aligned and scanned for SNPs and indels (as described above). If a variant was found in one consensus but was not present in the other it was reported as a discrepancy and closely studied. To confirm which of the disagreeing variants was more likely to be true the following procedure was adopted: (1) visual investigation of mapping results in CLC-GW and (2) aligning the conflicting regions from mapping derived consensus with contigs sequences originated from *de novo* assemblies.

Data deposition

The chloroplast genome for Australian wild rice genotype sample (W-rice-genotype) has been deposited in GenBank with the sequence accession number KF428978.

Results

Mapping and *de novo* assembly of sequence reads

The CLC Genomics Workbench was used to trim sequence reads from both the Illumina and Ion Torrent platforms and then processed, using the mapping and *de novo* assembly mode, to generate a mapping-consensus or *de novo*-consensus chloroplast sequences, respectively. The Torrent Suite Software, a dedicated analysis tool for Ion Torrent reads, was used to trim sequence reads from the Ion Torrent platform and then processed using the reference-guided *de novo* assembly mode (the only available mode) to generate a reference-guided *de novo*-consensus chloroplast sequence. The published chloroplast sequence of the R-rice-genotype (*Oryza sativa* spp. *japonica* var. Nipponbare) was used as a reference chloroplast genome sequence (reference-sequence) for mapping of reads or contigs. Sequencing data of the two rice genotypes, generated by both sequencing platforms, was of good quality with average PHRED scores ranging above 25 (Table S1).

In case of the R-rice-genotype, CLC trimmed-reads from both sequencing platforms mapped to the entire reference-sequence indicating complete reference-sequence coverage. However, the mapping-consensus chloroplast sequences varied in length based on sequencing platform reads. Similar results were obtained for the W-rice-genotype (Table S2).

Using the *de novo* mode in the CLC-GW and reads from both sequencing platforms, a number of contigs were generated with a subset of these mapping to the reference-sequence (Table S3). Similar results were obtained with the Ion Torrent reads when using the reference-assisted *de novo* mode on the Torrent Suite Software (Table S3). For both rice genotypes, contigs, which mapped to the reference-sequence, were checked if contiguous/overlapping to generate a *de novo*-consensus sequence. Using the *de novo* mode in the CLC-GW, respective *de novo*-consensus chloroplast sequences were generated for the reference-rice-genotype and the wild-rice-genotype but only from their respective Ion Torrent reads and not their Illumina reads. Using the reference-assisted *de novo* mode, reference-guided *de novo*-consensus chloroplast sequences were generated for both genotypes.

Thus, all contigs generated from Ion Torrent reads using both assembly algorithms (in the CLC-GW and Torrent Suite Software) could be joined together based on their position and overlapping information creating full length consensus sequences for the chloroplast genome (Table S3). Consensus sequences could not be created from contigs generated in the CLC-GW from Illumina reads for either of the rice genotypes, possibly because of the limitations of read length (Table S3). Although these contigs were well mapped to the reference, there were gaps between some of them that prevented their joining.

Analysis of reference-rice-genotype

The mapping-consensus chloroplast genome sequence (134,551 bp) generated by mapping the Illumina reads from [11] to the reference-sequence in the CLC-GW was, as expected, identical to the reference-sequence (Table 1). The mapping-

Table 1. Comparison of chloroplast consensus sequences of the cultivated reference rice genotype (*Oryza sativa* Nipponbare).

Sequencing platform	Source	Sequence length (bp)	Variants	Deletions	Insertions	Mismatches
Used available reference	GenBank (GU592207)	134,551	-	-	-	-
Illumina [11]	mapping- consensus	134,551	0	0	0	0
Ion Torrent	mapping- consensus	134,520	31	30	1	0
	de novo- consensus (CLC-GW	134,503	54	48	6	0
	reference-assisted *de novo*- consensus (Torrent Suite)	134,581	50	9	41	0

consensus sequence generated by mapping of the Ion Torrent reads to the reference-sequence in the CLC-GW was shorter than the reference-sequence by 31 nucleotides with no base mismatches, but diverged at 31 positions comprising of 30 deletions and 1 insertion and where all but one of the polymorphisms were found in homopolymer regions of length from 2 to 9 bases. The *de novo*-consensus sequence built by mapping and joining the contigs from *de novo* assembly of Ion Torrent reads (performed in CLC-GW) was shorter than the reference-sequence by 48 bp with no base mismatches, but diverged at 54 positions comprising of 48 deletions and 6 insertion and where all but two of the polymorphisms (indels) were found in homopolymer regions of length from 2 to 17 identical bases.

The reference-guided *de novo*-consensus chloroplast sequence obtained from contigs generated in Torrent Suite Software using Ion Torrent reads was longer than the reference-sequence by 30 bases with no base mismatches, but diverged at 50 positions comprising of 9 deletions and 41 insertions and where all polymorphisms (indels) were found in homopolymer regions of length from 2 to 12 identical bases.

The variants in the mapping-derived consensus sequences generated by the CLC-GW and using the Illumina or Ion Torrent reads were predominantly deletions, which is in contrast to the reference-guided *de novo*-consensus chloroplast sequence generated by the Torrent Suite Software using Ion Torrent reads, where the variants were predominantly insertions. Overall, 135 variants were detected using these different approaches. Among them only 8 were common for all of the methods (the same position, length and type). The rest of the variants were either present in two of the approaches or were unique to only one of them.

The analysis of regions in the chloroplast genome where the indels were found in the Ion Torrent data (both mapping and assembly consensus) revealed biases in terms of polymorphism nature, nucleotide type and location. Data from all approaches were merged and duplicated regions were removed. The number of indels that involved C/G bases was slightly higher than A/T bases (53 and 44, respectively). It is worth noting that the GC content of the chloroplast genome is 39%. Also, more deletions than insertions were observed (except for the consensus from the Torrent Suite Software assembly) and the majority were of G/C nucleotides in short homopolymeric regions (2–5 bp long with the vast majority of these variants occurring in 3 bp long homopolymer) (Figure 1). On the other hand, insertions were found mostly in longer homopolymers (5–17 bp long) and involved almost only A/T nucleotides.

Analysis of wild-rice-genotype

The reference-sequence was used to generate consensus chloroplast sequences using mapping or *de novo* approaches. For variant analysis, consensus sequences were compared to the reference-sequence unless indicated otherwise. The mapping-consensus sequence generated by mapping the Illumina reads of the wild rice plant using the CLC-GW was shorter than the reference-sequence by 20 bases (Table 2). In comparison, there were 128 variants found between the two sequences, which included 18 deletions, 13 insertions and 97 mismatches. Among mismatches 92 were single-nucleotide variants (SNPs) (48 transitions and 44 transversions) and 5 were multi-nucleotide variants (MNVs). The mapping-consensus obtained by mapping a subset of Illumina reads differed from the previous sequence by one deletion of 2 bases (Table 2), although, this variation was not found in the reference-sequence.

The consensus sequence generated by mapping Ion Torrent reads of the wild-rice-genotype using the CLC-GW was shorter than the reference-sequence by 26 bases and shorter by 6 bases than the consensus sequence from Illumina sequencing. The two sequences differed by 139 variants, which included 30 deletions, 14 insertions and 95 mismatches. Among the mismatches 90 were SNPs (47 transitions and 43 transversions) and 5 were MNVs.

As was observed for the R-rice-genotype, the de novo-consensus sequence differed from the reference-sequence at more positions than the mapping-consensus sequence (Table 1 and Table 2). In addition, a higher number of deletions were observed in consensus sequences obtained from the CLC-GW analysis as compared to the Torrent Suite Software analysis where more insertions were observed (Table 1 and Table 2).

Comparison of sequencing platforms

The mapping-consensus chloroplast sequences of the wild-rice-genotype generated in the CLC-GW using the Illumina and Ion Torrent reads were used to compare sequence platforms. When comparing variations found in the wild-rice-genotype by both sequencing technologies, 20 discrepancies (variant in one consensus not present in the other) were observed (Table 3). However, in the Ion Torrent consensus sequence there were fifteen variants and in the Illumina consensus there were only three. Moreover, two variants found at the same position differed in nucleotide composition in both consensus sequences.

Read mapping files derived from both mapping-consensus sequences were visually inspected to check for reads mapped to all 20 positions to determine sequence read error or for mapping error. The region in the Ion Torrent mapping-consensus sequence with the substitution of TT → AA (Table 3, variant #1) had several mapped reads with many mismatches. Reads from this region when extracted and blasted against the nucleotide sequence collection (nr/nt) at NCBI website (http://blast.ncbi.nlm.nih.gov) gave hits to the chloroplast but also to the nuclear DNA of *Oryza* species. In addition, this variant was not observed in the corresponding region of the contig sequences obtained from *de novo* assembly (from both analysis software), indicating that the

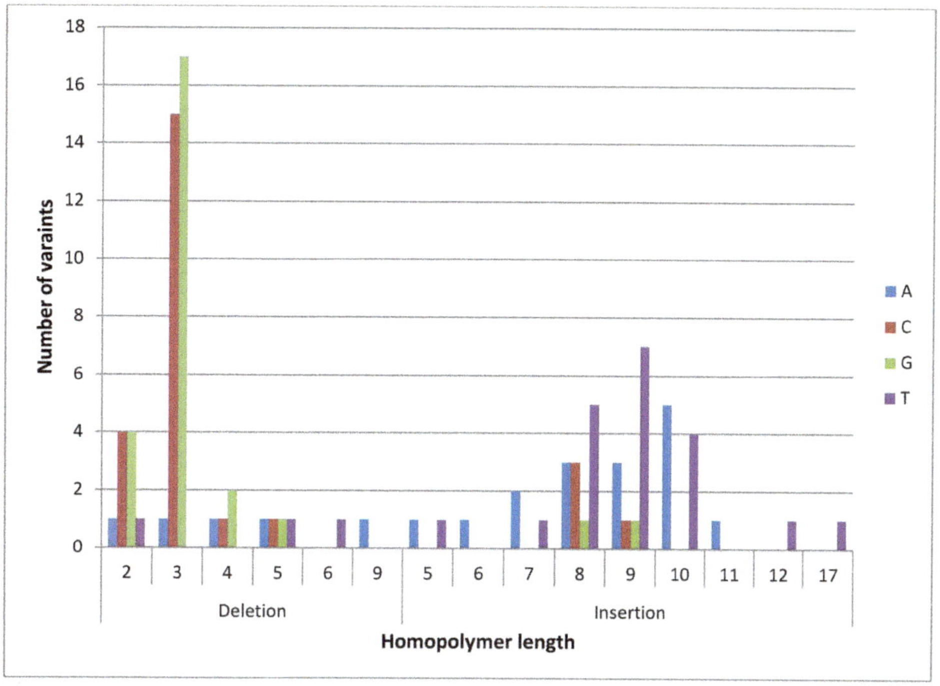

Figure 1. Variants in indels in cultivated (cv. Nipponbare) rice chloroplast consensus. Sequences generated by mapping and assembly of Ion Torrent reads to the available chloroplast sequence in GenBank for this genotype. The number of variants is shown with respect to its type (deletion or insertion) and position (the length of homopolymer region where the variants were found).

variant was not due to sequencing error but due to mapping artefact. Similarly, mapping error and not sequence read error was the reason for variants detected at several positions in one or the other mapping-consensus sequences (Table 3: for # 2 see Figure 2, for #3 and #4 see Figure 3), as these variants were not observed in the corresponding region of the contig sequences obtained from *de novo* assembly (from both analysis software). One of the variants (#7), an insertion of an A in the Ion Torrent consensus and AA in the Illumina consensus, was found in a long homopolymer stretch of 10 A's. The location of the variation suggested that in both cases it could be an error. Variations at this position in contigs sequences varied from 2 to 3 A's insertions. This polymorphism was not called in the consensus from the subset of Illumina data. Comparison of other chloroplast genomes, known to have been sequenced on the Illumina platform (GAII) [11], showed some to have this insertion, namely Australian *Oryza rufipogon* (GenBank accession – JN005833), Asian *Oryza rufipogon* (GenBank accession – JN005832) and *Oryza meridionalis* (GenBank accession – JN005831). Interestingly, the insertion was not present in the chloroplast genome sequence of *Oryza sativa* spp. *japonica* var. Nipponbare (reference-sequence), *Oryza sativa* ssp. *indica* isolate 93-11 (GenBank accession – AY522329) and *Oryza nivara* (GenBank accession – AP006728) which were sequenced by Sanger technology. Thus for this variant we cannot conclude with certainty if this discrepancy is due to mapping artefact or due to read error. However, all other errors found in Ion Torrent sequence (Table 3, #8 to #20) were not due mapping errors but read errors from either deletions or insertions in homopolymer regions.

Based on the analysis and all the findings using various approaches outlined above, an amended wild-rice-genotype chloroplast consensus sequence was created incorporating the most probable variants from Table 3. To assess the correctness of this amended chloroplast sequence paired-end reads from Illumina

Table 2. Comparison of chloroplast consensus sequences of the wild rice (*Oryza rufipogon*-like).

Sequencing platform	Source	Sequence length (bp)	Variants	Deletions	Insertions	Mismatches
Used available reference	GenBank (GU592207)	134,551	–	–	–	–
Illumina	mapping- consensus	134,531	128	18	13	97
	mapping-consensus (subset of reads)	134,529	129	19	13	97
Ion Torrent	mapping consensus	134,525	139	30	14	95
	de novo- consensus (CLC-GW)	134,521	155	43	17	95
	reference-assisted *de novo*- consensus (Torrent Suite)	134,554	147	23	28	96

Table 3. Inconsistent variations found in wild rice chloroplast mapping-consensus sequences and their validation.

Variations		Reference-sequence		Mapping-consensus sequence		The most probable variant
No	**Type**	**Position**	**Allele**	**HiSeq**	**Ion Torrent**	
#1	MNV	57,036	TT	TT	AA	TT
#2	ins	65,465^65,466	–	–	TCCTATTTAATA	TTCCTATTTAATA
#3	MNV	66,897	CGAT	TAGA	CGAT	TAGAAATAAAAAATTCTAA
#4	SNP	66,902	C	A	C	TAGAAATAAAAAATTCTAA
#5	SNP	17,366	T	A	T	T
#6	SNP/del	17,368	C	A	–	–
#7	ins	3,545^3,546	–	AA	A	–
#8	del	21,808	C	C	–	C
#9	del	57,027	T	T	–	T
#10	del	81,342	G	G	–	G
#11	del	91,427	C	C	–	C
#12	del	91,589	C	C	–	C
#13	del	97,135	G	G	–	G
#14	del	111,639	G	G	–	G
#15	del	116,139	C	C	–	C
#16	del	118,025	C	C	–	C
#17	del	119,245	G	G	–	G
#18	del	122,914	C	C	–	C
#19	del	123,568	G	G	–	G
#20	del	133,816	C	C	–	C

Variations derived by Illumina and Ion Torrent sequencing.
SNP – single-nucleotide variant, MNV – multi-nucleotide variant, ins – insertion, del – deletion.

sequencing platform were re-mapped to this new wild rice chloroplast genome. Illumina reads as against Ion Torrent reads of the W-rice-genotype were selected for remapping as even short Illumina reads (36 bases) of the R-rice-genotype provided a consensus matching the reference-sequence (Table 1) [11], as against an inaccurate consensus when Ion Torrent reads were used (Table 1) due to indel-associated errors. The new mapping results for the wild rice show only one discrepancy described earlier (Table 3, #7) in a long homopolymer region. All other sites were identical giving an additional assurance of the correctness of the final consensus.

Discussion

Various NGS technologies are now available for the rapid sequencing of whole genomes [17,23,24]. The choice of selecting one or more NGS technologies depends on the yield of data required, read length required, cost per data point and accuracy of the sequence data [25]. Systematic as against random errors can compromise the use of sequence data [26,27]. In this study we analysed NGS reads of two rice genotypes obtained from the Ion Torrent platform and compared it for accuracy to those obtained from the Illumina platform. A comparison of the three sequencing systems is presented in Table 4.

The Ion Personal Genome Machine (PGM) is characterised by long read length, very short run times and inexpensive consumables [28]. The drawbacks include low output, high cost per Gb (gigabase) of sequence [25] and the longer time required for the library preparation relative to that required for the sequencing run. Moreover, the PGM produces biased sequences from high homopolymer regions, which result in indel errors [27]. However, this is a developing platform with high potential for improvement in accuracy, read length and cost per read and per Gb. As the machine is the chip, it can be easily upgraded by the release of new chips. The PGM was the first next-generation sequencer, which price that dropped below $100 K. At the time of the study the PGM cost was around $80 K, which included the instrument, temple preparation system, enrichment system and server. The main advantages of HiSeq2000 system are the high throughput, large output data volumes per run and low cost of reagents per Gb [29]. The shortcomings of the platform embrace short read length and the need for advanced computational resources to process and store the enormous data volumes created from each run. Furthermore, the run time is considerably longer than that for the PGM, and the Illumina has a much higher initial capital cost. The specification of the GAIIx system was very similar to the HiSeq2000 with smaller output per run, higher cost per Gb and significantly lower instrument cost. The sequencing accuracy on both Illumina platforms is high with the most commonly encountered errors being substitutions. A wide range of Illumina instruments are now available with differing data volumes, running times and costs.

We selected the cultivated rice genotype (cv. Nipponbare) as one of the samples as its chloroplast sequence, derived by Sanger sequencing and considered as accurate, was available and could be used as reference sequence to check for errors in consensus sequences derived from NGS sequencing on both platforms. The aim was to generate both mapping and *de novo*-consensus sequences to determine read accuracy of both platforms.

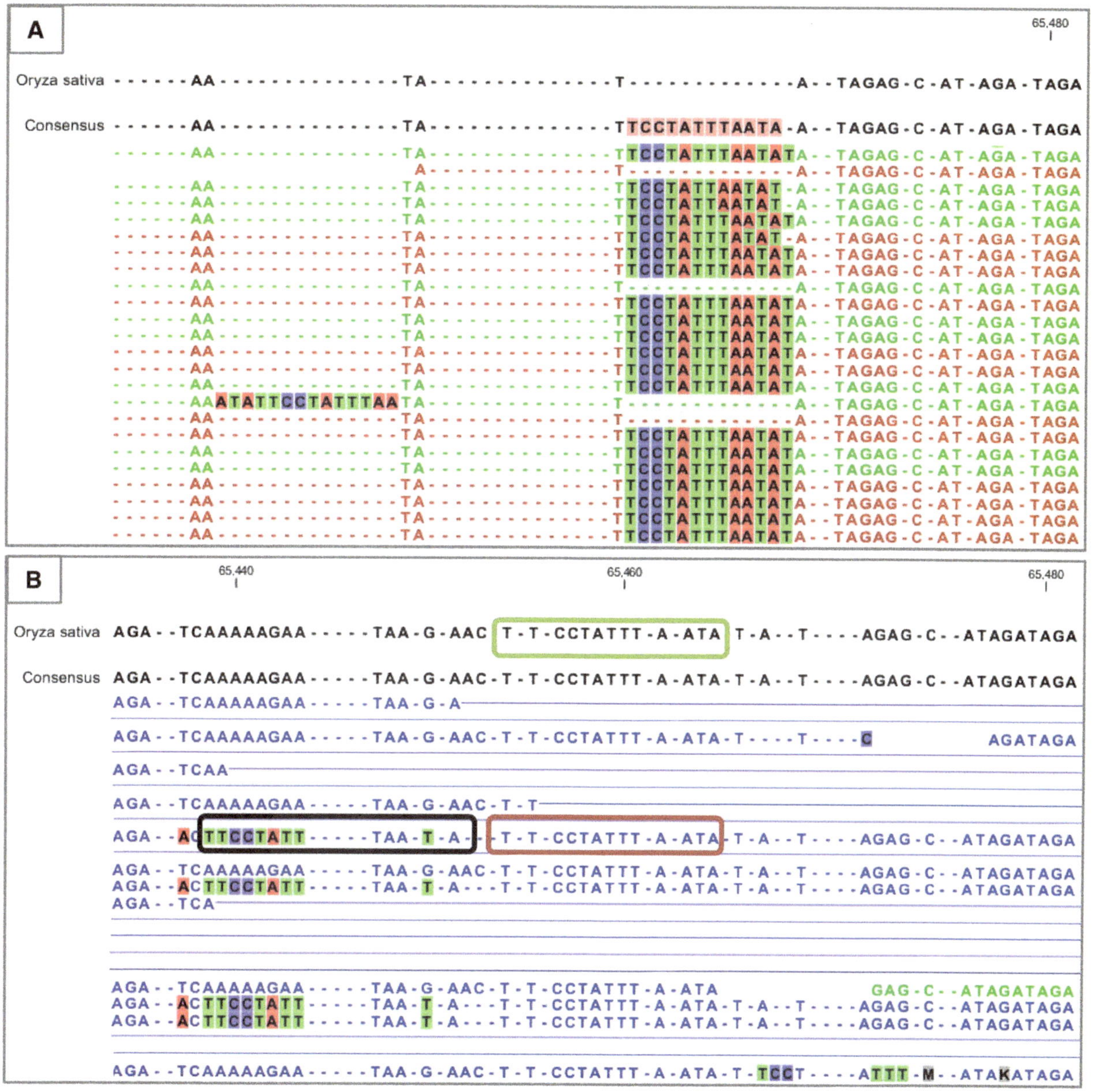

Figure 2. Snapshot of mapping results of wild rice Ion Torrent (A) and Illumina (B) reads. Reads were mapped to the chloroplast reference of *Oryza sativa* cv. Nipponbare. In the mapping of Ion Torrent reads there was a long insertion (TCCTATTTAATA) reported in the consensus sequence of wild rice chloroplast ((A), marked with orange background colour). This insertion was missed in the mapping of Illumina reads, although it was present in the reads ((B), example of the read sequence marked in black rectangle). The nucleotides in the insertion were duplicated in wild rice (sequence marked in red rectangle), and not in the reference genome where only one copy of these nucleotides was present (marked in green rectangle). The duplicated region was a probable cause of the misalignment of reads. Oryza sativa – fragment of chloroplast sequence of *Oryza sativa* spp. *japonica* var. Nipponbare; Consensus – consensus sequence of wild rice chloroplast sequence derived by mapping reads from Illumina (A) and Ion Torrent (B) platforms to the reference. Nucleotides with background colours represent the mismatches between reads and the reference sequence; paired end reads are shown in blue; single reads are shown in green and red (in forward and reverse orientation, respectively).

We were unable to obtain *de novo*-consensus sequences for both genotypes with Illumina reads when processed by CLC-GW as the contigs generated were non-contiguous when mapped to the reference-sequence. In the case of reference-genotype, read coverage depth and read length contributed to non-contiguous contigs as the read depth in regions corresponding to gaps was between 7 and 30, when examined for base calling and read mapping depth to the mapping-consensus, which is well below the average mapping read coverage depth of 127.7 (Table S2). In the case of wild-rice-genotype with longer read length (100 bases), read length but not read coverage depth may have contributed to non-contiguous contigs as the read depth in regions corresponding to the gaps was between 7,000 to 11,000, when examined for base calling and read mapping depth to the in the mapping-consensus, which is well above the average mapping read coverage depth of 8,143.2 (Table S2). Interestingly, *de novo*-consensus, although not accurate (Table 1) was generated for the reference-genotype using CLC-GW and the longer (200 bases) Ion Torrent reads. We can

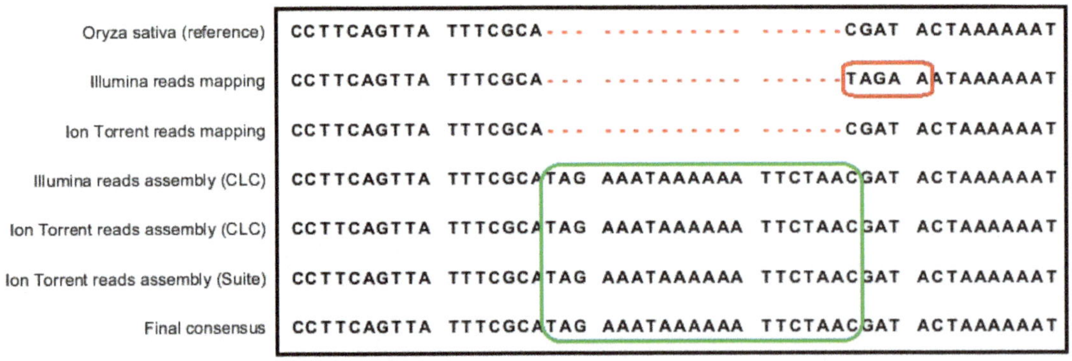

Figure 3. Alignment of regions #3 and #4 from Table 3 showing discrepancies in consensus sequences. The fragment circled in red shows false called SNPs (#3 and #4, Table 3, Illumina consensus); these SNPs were incorrect because of the long insertion present in wild rice sequence but not in the reference. The fragments circled in green illustrate this long insertion found in wild rice chloroplast genome by means of reads assembly from both platforms and both assembly tools. Final sequence was created based on this information. Oryza sativa (reference) – region 66860.66940 from chloroplast sequence of *Oryza sativa* spp. *japonica* var. Nipponbare; Illumina reads mapping and Ion Torrent reads mapping – regions from consensus sequence generated by mapping wild rice Illumina and Ion Torrent reads, respectively, to the reference sequence; Illumina reads assembly and Ion Torrent reads assembly – regions from contigs generated by assembly of reads from Illumina and Ion Torrent platforms, respectively; CLC – assembly performed in CLC Genomic Workbench; Suite – assembly performed in Torrent Suite Software; Final consensus – final wild rice chloroplast genome sequence (GenBank accession – KF428978).

thus conclude that read lengths of more than 100 bases would be required to successfully obtain a *de novo*-consensus of the chloroplast genome.

In case of the reference-genotype, mapping-consensus sequences were generated using reads from both platforms, with no mismatches. However, mapping-consensus sequences showed several indels due to inaccuracies in the Ion Torrent reads mainly in homopolymer regions. In case of the short Illumina reads (36 bases) of the R-rice-genotype, neither the read length (36 bases) nor the sequence yield (minimum reference-sequence read mapping coverage of 7) was a limitation in generating an accurate mapping-consensus when processed by the CLC-GW, indicating the accuracy of the Illumina sequencing platform. In case of the longer Ion Torrent reads (200 bases), when processed by the CLC-GW and the Torrent Suite Software, consensus-sequences with no mismatches were generated but with several variants all of which were indels and predominantly in homopolymer regions (Table 1). These results indicate the actual presence of these anomalous indels in the Ion Torrent reads which contributed to the inaccuracy in the consensus sequence generated and to the unreliability of any indels calls made when using these reads. It is known that the Ion Torrent sequencing platform generates anomalous indels [27]. Interestingly, the preference of the CLC-GW and the Torrent Suite analysis tool in filtering out predominantly either insertions or deletion, respectively, as observed in both rice genotypes (Tables 1 and 2), indicates the

Table 4. The comparison between the tree sequencing systems utilised in the study.

	Illumina GAIIx	Illumina HiSeq2000	Ion Torrent PGM –318 chip
Sequencing method	Synthesis (light detection)	Synthesis (light detection)	Synthesis (proton detection)
Amplification method	Bridge PCR	Bridge PCR	Emulsion PCR
Read length[a]	Up to 2×150 bp	1×50 bp, 2×50 bp, 2×100 bp	~200 bp, ~400 bp
Paired reads	Yes	yes	Yes
Insert size	Up to 700 bp	Up to 700 bp	Up to 250 bp
Output data/run	30 Gb	600 Gb	Up to 2 Gb
Time/run[b]	10–14 days	8–11 days	4–7 hours
Cost/Gb[c]	$148	$41	$1000
Instrument cost	$256 K	$654 K	$80 K
Accuracy	>99.9%	>99.9%	99%
Error rate[d]	≥0.1%	≥0.1%	~1%
Primary errors	Substitutions	Substitutions	Indels
DNA requirements	0.05–1 ug	0.05–1 ug	0.1–1 ug

Gb – gigabase, bp – base pair, K – thousand, uq - microgram.
[a] Annotation '2 x' refers to paired end reads and '1 x' to single reads.
[b] Run time from minimum to maximum read lengths.
[c] Includes one sample and one sequencing kit per run.
[d] percentage of errors per base in single read.

possibility of these analysis tools to be amended to deal with anomalous indels in the Ion Torrent reads. We also observed a bias for indels in homopolymer regions, where deletions occurred in shorter regions while insertions occurred in longer regions. These findings are inconsistent with a recent study on Ion Torrent sequencing bias where it was reported that an increased deletion rate was positively correlated with increased lengths of homopolymer regions and that insertions were independent of homopolymer length [27,30]. This is probably because of the different genomes used or that the chloroplast genome is AT-rich and the longest C/G stretches found are 9 bp long (2 regions), 8 bp (4 regions) and 7 bp (9 regions).

The systematic pattern of bias, specific types of nucleotide variants in long versus short repeats, in the single base indels in homopolymers (Figure 1) is likely to be a feature of chloroplasts with their specific patterns of homopolymer repeat frequency. However, systematic errors reported in the Ion Torrent data in earlier studies [27] were not found in this analysis of the chloroplast possibly because we used a more recent version of the technology. We can conclude that both sequencing platforms produce high quality data with relatively low rates of discrepancy in the calling of polymorphisms, especially SNPs, but the Ion Torrent data would be less reliable for indel calls.

We compared the two sequencing platforms in generating an accurate consensus chloroplast sequence by comparing the mapping-consensus sequences generated for both genotypes. For the R-rice genotype, no mismatches were observed indicating the two platforms to be comparable for SNP calls as has been reported for microbial genomes [25]. When comparing the mapping-consensus sequences of the W-rice-genotype we identified twenty discrepancies. It must be noted that some of the discrepancies were caused by misalignment of reads and as has been previously recognized [31] there is an important issue with mismatches in close proximity of indel events. Some read mapping tools can have problems with these regions resulting in false SNPs calling. However, for both the rice genotypes we observed erroneous indel calls with Ion Torrent reads mainly in homopolymer regions.

The importance of obtaining a *de novo* sequence was clear from our study as discrepancies observed in the mapping sequences of the W-rice-genotype were curated using the *de novo* contigs. The *de novo* assembly of chloroplast genomes using the tools applied in this study was easier with the longer Ion Torrent data than with the Illumina data. Although increasing number of reference genome sequences are being generated to support chloroplast genome analysis, the ability to generate whole chloroplast genome sequences *de novo* will find wide application. The first report of total plant DNA analysis for chloroplast sequencing [11] used short Illumina reads (36 bp) and relied on a reference genome for successful assembly. More recent studies based upon longer reads [12,23] have reported more success with *de novo* assembly.

Conclusions

This study demonstrates that *de novo* assembly of an accurate whole chloroplast genome sequence will be possible for routine plant barcoding. Analysis of plants based upon appropriate and careful analysis of shotgun sequencing of total DNA promises to provide a barcode that will have wide application. Having a well curated and reliable consensus chloroplast sequence for a plant sample provides greater certainty of obtaining reliable results in many critical studies such as plant identification, purity assessment, phylogenetic analysis and heteroplasmy analysis. Continuing improvements in sequencing platforms and analysis tools will make this method more reliable and cost effective for a wide range of research and industrial applications. The significant advances in these competing sequencing platforms that have been foreshadowed by the manufactures promise dramatic reductions in cost in the near future. This would make barcoding by sequencing the whole chloroplast in this way the preferred option for plant identification in many research and industrial applications. A comparison of the platforms at the same level of coverage is useful in providing a guide to likely comparative performance as the number of reads on these platforms increases in the future. Costs will change for each platform as read length and read volume are increased.

However cost is not the only advantage of this approach. The use of total DNA from the sample without amplification simplifies the analysis but also provides greater opportunity to measure in an unbiased way the contribution of different genotypes in complex mixtures. The risks of preferential amplification or enrichment of chloroplasts from a specific genotype that complicated earlier analysis will be avoided.

Acknowledgments

We thank the local support team from Life Technologies for technical assistance with this project.

Author Contributions

Conceived and designed the experiments: MB AF RH. Performed the experiments: MB. Analyzed the data: MB AF. Contributed to the writing of the manuscript: MB AF RH.

References

1. Straub SCK, Parks M, Weitemier K, Fishbein M, Cronn RC, et al. (2012) Navigating the tip of the genomic iceberg: Next-generation sequencing for plant systematics. Am J Bot 99: 349–364.
2. Parks M, Cronn R, Liston A (2009) Increasing phylogenetic resolution at low taxonomic levels using massively parallel sequencing of chloroplast genomes. BMC Biol 7: 84–84.
3. Taberlet P, Coissac E, Pompanon F, Brochmann C, Willerslev E (2012) Towards next-generation biodiversity assessment using DNA metabarcoding. Mol Ecol 21: 2045–2050.
4. Henry RJ (2001) Plant genotyping: the DNA fingerprinting of plants. New York: CABI Pub.
5. Henry RJ (2013) Molecular Markers in Plants. Hoboken, NJ: John Wiley & Sons.

6. Galimberti A, De Mattia F, Losa A, Bruni I, Federici S, et al. (2013) DNA barcoding as a new tool for food traceability. Food Res Int 50: 55–63.
7. Diaz-Sanchez S, Hanning I, Pendleton S, D'Souza D (2013) Next-generation sequencing: the future of molecular genetics in poultry production and food safety. Poult Sci 92: 562–572.
8. Jansen RK, Raubeson LA, Boore JL, dePamphilis CW, Chumley TW, et al. (2005) Methods for Obtaining and Analyzing Whole Chloroplast Genome Sequences. In: Elizabeth AZ, Eric HR, editors. Methods in Enzymology: Academic Press. pp. 348–384.
9. Lang BF, Burger G (2007) Purification of mitochondrial and plastid DNA. Nat Protoc 2: 652–660.
10. Cronn R, Liston A, Parks M, Gernandt DS, Shen R, et al. (2008) Multiplex sequencing of plant chloroplast genomes using Solexa sequencing-by-synthesis technology. Nucleic Acids Res 36: e122.
11. Nock CJ, Waters DLE, Edwards MA, Bowen SG, Rice N, et al. (2011) Chloroplast genome sequences from total DNA for plant identification. Plant Biotechnol J 9: 328–333.
12. McPherson H, Rossetto M, van der Merwe M, Delaney SK, Edwards MA, et al. (2013) Capturing chloroplast variation for molecular ecology studies: a simple next generation sequencing approach applied to a rainforest tree. BMC Ecol 13: 8–18.
13. Loman NJ, Misra RV, Dallman TJ, Constantinidou C, Gharbia SE, et al. (2012) Performance comparison of benchtop high-throughput sequencing platforms. Nat Biotechnol 30: 434–434.
14. Sims D, Sudbery I, Ilott NE, Heger A, Ponting CP (2014) Sequencing depth and coverage: key considerations in genomic analyses. Nat Rev Genet 15: 121–132.
15. Illumina Inc website. Available: http://res.illumina.com/documents/products/techspotlights/techspotlight_sequencing.pdf. Accessed 2014 September 8.
16. Illumina Inc website. Available: http://res.illumina.com/documents/products/illumina_sequencing_introduction.pdf. Accessed 2014 September 8.
17. Rothberg JM, Edwards M, Hoon J, Simons JF, Marran D, et al. (2011) An integrated semiconductor device enabling non-optical genome sequencing. Nature 475: 348–352.
18. Lee KH, Park J, Williams DS, Xiong Y, Hwang I, et al. (2013) Defective chloroplast development inhibits maintenance of normal levels of abscisic acid in a mutant of the Arabidopsis RH3 DEAD-box protein during early post-germination growth. Plant J 73: 720–732.
19. Hartwig B, James GV, Konrad K, Schneeberger K, Turck F (2012) Fast isogenic mapping-by-sequencing of ethyl methanesulfonate-induced mutant bulks. Plant Physiol 160: 591–600.
20. Sotowa M, Ootsuka K, Kobayashi Y, Hao Y, Tanaka K, et al. (2013) Molecular relationships between Australian annual wild rice, Oryza meridionalis, and two related perennial forms. Rice 6: 26.
21. Furtado A (2014) DNA extraction from vegetative tissue for next-generation sequencing. In: Henry RJ, Furtado A, editors. Cereal Genomics. Methods in Molecular Biology. New York: Springer Science+Business Media. pp. 1.
22. Kurtz S, Phillippy A, Delcher AL, Smoot M, Shumway M, et al. (2004) Versatile and open software for comparing large genomes. Genome Biol 5: 12–R12.
23. Ferrarini M, Cestaro A, Sargent DJ, Moretto M, Ward JA, et al. (2013) An evaluation of the PacBio RS platform for sequencing and de novo assembly of a chloroplast genome. BMC Genomics 14: 670–670.
24. Shendure J, Ji H (2008) Next-generation DNA sequencing. Nat Biotechnol 26: 1135–1145.
25. Quail MA, Smith M, Coupland P, Otto TD, Harris SR, et al. (2012) A tale of three next generation sequencing platforms: comparison of Ion Torrent, Pacific Biosciences and Illumina MiSeq sequencers. BMC Genomics 13: 341–341.
26. Meacham F, Boffelli D, Dhahbi J, Martin DIK, Singer M, et al. (2011) Identification and correction of systematic error in high-throughput sequence data. BMC Bioinformatics 12: 451–451.
27. Bragg LM, Stone G, Butler MK, Hugenholtz P, Tyson GW (2013) Shining a light on dark sequencing: characterising errors in Ion Torrent PGM data. Plos Comput Biol 9: e1003031.
28. Liu L, Li Y, Li S, Hu N, He Y, et al. (2012) Comparison of next-generation sequencing systems. J Biomed Biotechnol 2012: 251364.
29. Glenn TC (2011) Field guide to next-generation DNA sequencers. Molecular ecology resources 11: 759–769.
30. Ross MG, Russ C, Costello M, Hollinger A, Lennon NJ, et al. (2013) Characterizing and measuring bias in sequence data. Genome Biol 14: R51.
31. Albers CA, Lunter G, MacArthur DG, McVean G, Ouwehand WH, et al. (2011) Dindel: accurate indel calls from short-read data. Genome Res 21: 961–973.

6

Whole Genome Association Mapping of Plant Height in Winter Wheat (*Triticum aestivum* L.)

Christine D. Zanke[1], Jie Ling[1], Jörg Plieske[2], Sonja Kollers[3], Erhard Ebmeyer[3], Viktor Korzun[3], Odile Argillier[4], Gunther Stiewe[5], Maike Hinze[5], Kerstin Neumann[1], Martin W. Ganal[2], Marion S. Röder[1]*

1 Leibniz Institute of Plant Genetics and Crop Plant Research (IPK), Gatersleben, Germany, **2** TraitGenetics GmbH, Gatersleben, Germany, **3** KWS LOCHOW GMBH, Bergen, Germany, **4** Syngenta France S.A.S., Orgerus, France, **5** Syngenta Seeds GmbH, Bad Salzuflen, Germany

Abstract

The genetic architecture of plant height was investigated in a set of 358 recent European winter wheat varieties plus 14 spring wheat varieties based on field data in eight environments. Genotyping of diagnostic markers revealed the *Rht-D1b* mutant allele in 58% of the investigated varieties, while the *Rht-B1b* mutant was only present in 7% of the varieties. *Rht-D1* was significantly associated with plant height by using a mixed linear model and employing a kinship matrix to correct for population stratification. Further genotyping data included 732 microsatellite markers, resulting in 770 loci, of which 635 markers were placed on the ITMI map plus a set of 7769 mapped SNP markers genotyped with the 90 k iSELECT chip. When Bonferroni correction was applied, a total of 153 significant marker-trait associations (MTAs) were observed for plant height and the SSR markers ($-\log_{10}$ (P-value) ≥4.82) and 280 ($-\log_{10}$ (P-value) ≥5.89) for the SNPs. Linear regression between the most effective markers and the BLUEs for plant height indicated additive effects for the MTAs of different chromosomal regions. Analysis of syntenic regions in the rice genome revealed closely linked rice genes related to gibberellin acid (GA) metabolism and perception, i.e. GA20 and GA2 oxidases orthologous to wheat chromosomes 1A, 2A, 3A, 3B, 5B, 5D and 7B, *ent*-kaurenoic acid oxidase orthologous to wheat chromosome 7A, *ent*-kaurene synthase on wheat chromosome 2B, as well as GA-receptors like DELLA genes orthologous to wheat chromosomes 4B, 4D and 7A and genes of the GID family orthologous to chromosomes 2B and 5B. The data indicated that besides the widely used GA-insensitive dwarfing genes *Rht-B1* and *Rht-D1* there is a wide spectrum of loci available that could be used for modulating plant height in variety development.

Editor: Pilar Hernandez, Institute for Sustainable Agriculture (IAS-CSIC), Spain

Funding: This research was funded by the Plant Biotechnology program of the German Federal Ministry of Education and Research (BMBF) within the frame of the projects GABI-Wheat and VALID (project numbers 0315067 and 0315947). The funders had no role in study design, data collection and analysis, decision to publish, or preparation of the manuscript.

Competing Interests: SK,VK and EE are employed by the company KWS-LOCHOW GMBH, OA is employed at Syngenta France SAS, GS and MH are employed by Syngenta Seeds GmbH and MG, JP are employed by the company TraitGenetics GmbH. The companies have commercial interest in the results for application in variety development and for the provision of genotyping services. The authors declare that the research was conducted in the absence of any commercial or financial relationships that could be construed as a potential conflict of interest.

* Email: roder@ipk-gatersleben.de

Introduction

Plant height (PH) is an intrinsic component of plant architecture with effects on lodging stability, harvest index and yield. The introduction of dwarfing genes was a major factor during the 'Green Revolution' in the 1970 s, which led to a dramatic increase of production in the staple crops rice, wheat and maize at reduced costs [1,2].

In wheat, the major dwarfing genes are *Rht-B1* and *Rht-D1* (old nomenclature *Rht1* and *Rht2*), which reduce plant height by reducing the response to gibberellin (gibberellin insensitive dwarfing genes) with pleiotropic effects on grain number and yield. The molecular identification of the *Rht-B1* and *Rht-D1* genes in wheat [3] confirmed them as orthologs to the gibberellin response regulator *GAI* in Arabidopsis [4] and the dwarfing genes *dwarf-8* (*d8*) and *dwarf-9* (*d9*) in maize [5]. Mutations in the DELLA-domain of these proteins often cause a dominant gibberellin-insensitive phenotype by increasing the stability of this negative regulator of gibberellin signal transduction [6]. Other orthologous DELLA genes are the *Slender* rice 1 [7] and the barley Slender1 [8,9] mutants. The gibberellin-insensitive dwarf mutant *Dwarf 1* in rice encodes the α-subunit of GTP-binding protein which may be associated with gibberellin signal transduction [10]. The *GID*1 family in rice encodes a soluble receptor for gibberellin [11], while the *gid2* mutant is caused by the loss of function of a putative F-box protein [12].

Besides the gibberellin insensitive dwarfing genes also gibberellin sensitive dwarfing genes exist. A prominent example is the semi-dwarf (sd-1) "green revolution" rice, which contains a defective gibberellin 20-oxidase gene [13,14]. Mutations in *GA20ox1* also caused semi-dwarf phenotypes in Arabidopsis [15] and an orthologous gene is likely to cause the *sdw1/denso* mutant in barley [16]. Like the gibberellin 20-oxidase mutants, the maize

Dwarf3 gene, encoding a protein of the cytochrome P450 superfamily, is involved in gibberellin biosynthesis [17].

A different mechanism for height reduction was described in the maize *brachytic2* (*br2*) mutants, where the loss of a P-glycoprotein modulated polar auxin transport in the stalk [18]. An orthologous mutant *dwarf3* was discovered in sorghum and the ABCB1 P-glycoprotein is a likely candidate for the *d2* dwarfing gene in pearl millet [19].

In wheat, besides the dwarfing genes with known function, e.g. *Rht-B1* and *Rht-D1* [20], a number of additional dwarfing genes were mapped in bi-parental populations [21–23]. In the current study, we were interested to gain a comprehensive picture about the loci involved in the regulation of plant height in European winter wheat, by applying a whole genome association mapping approach. Association mapping can be carried out with candidate genes [24,25] or with anonymous markers covering the genome [26,27]. In fact, the associations of the maize *Dwarf8* polymorphisms with variation in flowering time and plant height were regarded as first association mapping approach reported for crop plants [28], though the validity of this assumption recently has been questioned [29]. In comparison to bi-parental mapping populations, association panels have the advantage that the genetic diversity of larger germplasm panels can be monitored and a refined resolution can be obtained by monitoring the "historical" recombinations, which occurred during evolution or breeding of a line [30]. Association mapping and genomic selection approaches for plant height were conducted in barley [31], sorghum [32], maize [33] and wheat [34].

The goals of the current study were (i) to establish marker-trait associations (MTAs) for plant height in a panel of European winter wheat varieties, (ii) to compare the MTAs obtained with SSR markers to those obtained with SNP markers, and (iii) to connect the obtained MTAs to known genes of the gibberellic acid (GA) metabolism by exploiting the synteny to the annotated genome sequence of rice [35]. Our wheat association panel previously was analyzed for resistance to fungal pathogens [36–38] and heading date [39].

Materials and Methods

Plant material and phenotyping

No specific permits were required for the described field trials. The field trials were performed by companies. The field trials did not involve endangered or protected species. The plant material, consisting of 358 European winter wheat varieties plus 14 spring wheat varieties as a outgroup, is described in more detail in [36]. Field trials were conducted in the season 2008/2009 in Andelu/France (09.AND), Seligenstadt/Germany (09.SEL) and Wohlde/Germany (09.WOH) and in the season 2009/2010 in Andelu/France (10.AND), Janville/France (10.JAN), Saultain/France (10.SAU), Seligenstadt/Germany (10.SEL) and Wohlde/Germany (10.WOH) by applying an alpha design with two replications per site. Plot sizes were 5 to 6.75 m^2 and climatic factors are described in Table S1. Both winter and spring varieties were sown in autumn and common agricultural practices were applied, which included the applications of growth regulators (Cycocel CL at a rate of 2.5L/ha 09.AND. 10.AND, 10.JAN and 10.SAU; CCC720 at a rate of 1.5L/ha at 09.WOH and 10.WOH; 1.0 L/ha CCC720 and 0.3 L/ha Moddus at 09.SEL and 1.3 L/ha CCC720 at 10.SEL). The active substance for CCC720 is Chlormequatchlorid (720 g/L), for Moddus it is Trinexapacethyl (250 g/L) and for Cycocel CL it is Chlormequat (460 g/L) and Imazaquine (10 g/L). Cycocel is based on the action of chlormequat chloride which is an inhibitor of *ent*-kaurene synthase involved in the early steps of GA-biosynthesis [40]. As untreated control, all varieties were grown in a nursery in rows with 6 plants in Gatersleben 2012 (12.GAT). Average plant height of each plot was measured before harvest with exclusion of awns.

Molecular data analysis and genetic mapping

For marker-trait analysis a set of 732 microsatellite markers, resulting in 770 different loci spred across all chromosomes of wheat was used. Of these 770 loci, 635 loci were mapped and 135 loci were unmapped. Since the microsatellites are multi-allelic, they amounted to 3176 alleles. More details about this data set and the description of linkage disequilibrium (LD) and population structure are provided in [36]. For SNP-analysis, all 372 varieties were genotyped on a novel 90 k Infinium chip (90k iSELECT) [41,42]. This resulted in a total of 21742 scorable and polymorphic markers on our association panel by considering all polymorphic markers with a minor allele frequency (MAF) >0.03. Of these markers, only the 7769 mapped markers were included in the association analysis, while the unmapped markers were not used for association analysis. The SSR-markers were mapped on the ITMI-population (International Triticeae Initiative) based on recombinant inbred lines between the parents W7984 (synthetic wheat) × Opata M85 [43,44], while the SNP markers were mapped on 138 lines of a newly created doubled-haploid population of the same parents [45,46]. Map construction was performed using the software package Joinmap 4.1. Both maps have different recombination values, and currently only few common markers are available, which makes comparisons difficult. For display a reduced version of the SNP-map was used containing most relevant markers with MTAs for PH. As candidate genes the photoperiodism gene *Ppd-D1* [47] and the dwarfing genes *Rht-B1* and *Rht-D1* [48] were genotyped on all varieties.

Statistical analysis and association mapping

Each year-location combination was considered as an environment in our study. For each environment, adjusted entry mean of the genotypes were estimated using GenStat 13th edition and following model:

$$y = \mu + replication + genotype + block + e$$

with replication and genotype as fixed factors and block as random effect nested within replication; μ represents an overall mean and e is a residual term; y represents the single plot values within each environment.

Adjusted entry means were used to calculate Best Linear Unbiased Estimators (BLUEs) across all eight environments using the software package GenStat 14th edition [VSN International, Hemel Hempstead, Hertfordshire, UK] as described in [36]:

$$y = \mu + genotype + environment + e$$

with genotype and environment as fixed effects; μ represents an overall mean and e is a residual term.

For calculating genotype-phenotype associations a minor allele frequency (MAF) threshold of 3% (equalling 11 varieties) was used for all markers. A mixed model for association mapping was applied based on the BLUEs across environments using the software package GenStat 14th edition as described in [36] by applying a kinship matrix as relationship model:

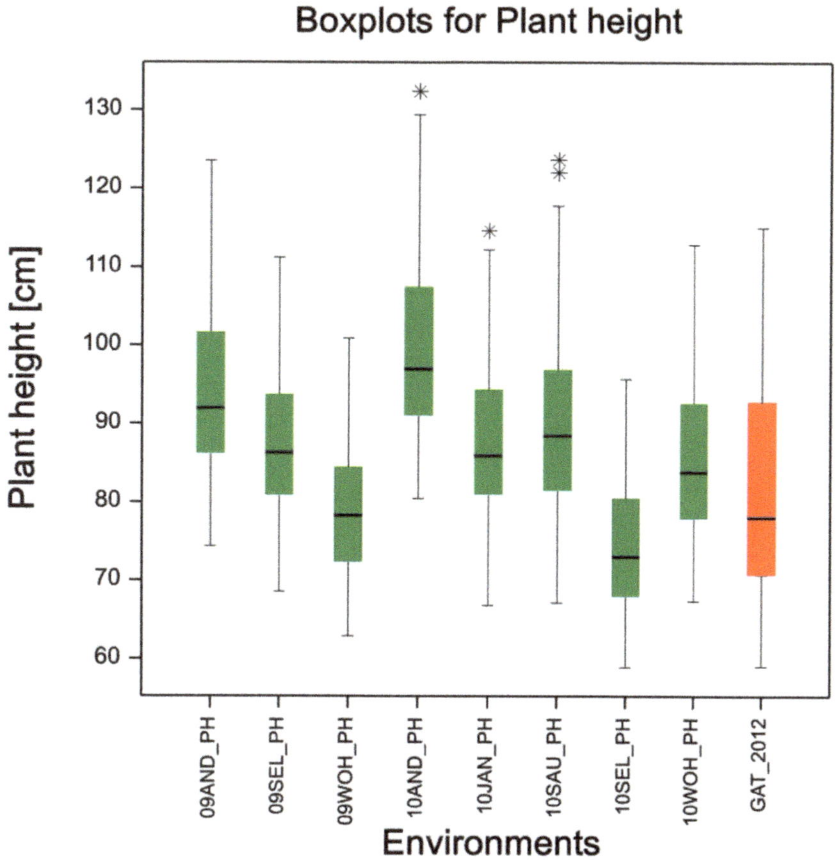

	09AND_PH	09SEL_PH	09WOH_PH	10AND_PH	10JAN_PH	10SAU_PH	10SEL_PH	10WOH_PH	GAT_2012
Min.	74.4	68.6	62.9	80.4	66.8	67.1	75.8	67.3	59.0
Max.	123.5	111.2	100.9	132.4	114.6	123.6	74.0	112.8	115.0
Mean	94.2	87.5	78.7	99.7	88.0	89.6	74.5	85.5	81.5

Figure 1. Boxplots for plant height. Boxplots for eight field environments are depicted in green, the untreated control in a nursery is shown in red. Asterisks mark outlier varieties.

$$y=\mu+marker+genotype+e$$

with $genotype \sim N(0,2K\sigma^2_{genotype}), error \sim N(0,\sigma^2_e)$.

Marker refers to a fixed effect of every marker, μ represents an overall mean, e is a residual term, and K denotes the kinship matrix among all genotypes. The Loiselle kinship matrix was calculated for 155 SSR markers, equally distributed on the genome, by using the software package SPAGeDi [49]. This kinship matrix was applied to correct for false positives when calculating MTAs with SSR as well as with SNP markers as described by [50]. The threshold of Bonferroni correction for multiple testing was calculated by dividing $P<0.01$ with the number of SSR or SNP markers used for the analysis. Marker effects (r^2) were estimated using the software package TASSEL 3.0. The additive effects were calculated by using GenStat 14th edition, where for the bi-allelic SNP-markers the most frequent alleles was set to zero and the difference of the phenotypic effect towards the less frequent allele was calculated. Therefore negative additive effects mean a decrease in plant height and positive additive effects mean an increase in plant height.

Spearman rank order correlations and ANOVA using the adjusted means of the eight environments were calculated with the software package SigmaPlot 11.0. The broad sense heritability was calculated from the variance components according to the formula:

$$H^2=Var(genotype)/(Var(genotype)+Var(error)/number\_of\_locations)$$

with variance components calculated with the software package SPSS v. 19. This software was also used to conduct a trait Posthoc test according to Tukey B.

Results

Description of phenotypic data

The phenotypic data were based on eight field environments with various distributions of the trait plant height (Figure 1). The

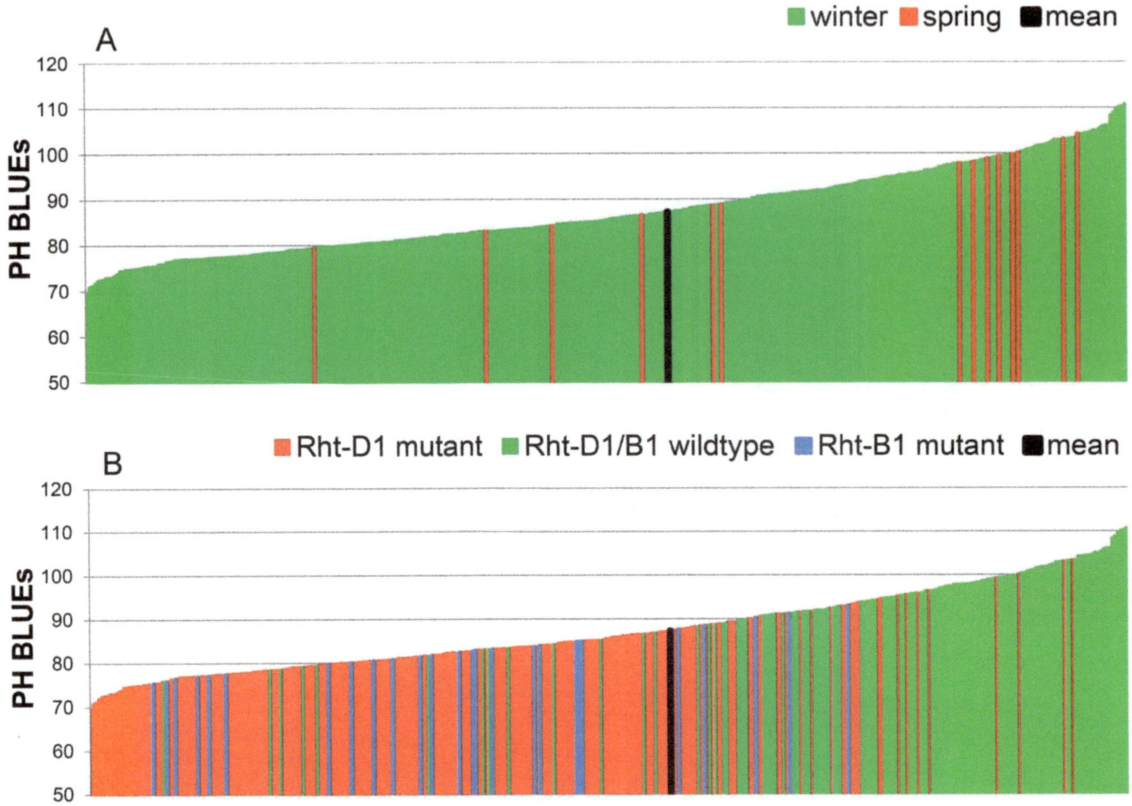

Figure 2. Phenotypic distribution of plant height BLUEs in 372 wheat varieties. The BLUEs for plant height were calculated across eight environments. Plant height BLUEs were arranged according to growth habit (A) or the distribution of the mutant alleles of the dwarfing genes *Rht-B1* and *Rht-D1* (B).

BLUEs for PH across the eight environments ranged from 69.5 cm to 110.7 cm with a mean of 87.2 (Table S2; Figure 2). The untreated control 12.GAT had very similar values ranging from 59.0 cm to 115.0 cm (Figure 1). The broad sense heritability among the eight field environments was 0.89. An ANOVA test revealed a significant effect of environment and a Tukey B test discriminated seven phenotypic classes (Table S3). The Spearman rank order correlations among the eight field environments ranged from 0.862 to 0.943 indicating a good reproducibility of the ranking of the varieties in the different environments (Table S4). The correlation between the BLUEs of the eight field environments and the untreated control 12.GAT was 0.882. The analysis of variance was significant for genotype as well as environment (Table S5).

Analysis of marker trait associations

For analysis of MTAs, the eight field environments and the BLUEs across these eight environments were taken into account. By considering all MTAs found for all eight environments with $-\log_{10}$ (P-value) $\geq$4.0, in total 1873 MTAs were found for the SSRs and 6245 for the SNPs, of those 239 MTAs were detected for calculating with BLUEs across all environments in the SSRs, and 788 MTAs detected in the SNPs (Table 1; Tables S6, S7). After applying Bonferroni-correction, $-\log_{10}$ (P-value) $\geq$4.82 was significant for the SSRs and $-\log_{10}$ (P-value) $\geq$5.89 was significant for SNPs. Then the numbers shifted from 1104 significant MTAs for SSRs associated with BLUEs to 153 MTAs, and from 1904 significant MTAs for the SNPs associated with BLUEs to 280 MTAs (Figure 3). However, many markers detected MTAs in several environments and also especially the significant SNPs co-segregated or were closely linked in many cases, so that the number of real loci was significantly lower (Figure S1). By considering significant MTAs for BLUEs only and by combining marker loci with distance in the genetic map of $\leq$6 cM, the number of combined marker loci was estimated to be 109 for the SSRs and 87 for the SNPs (Tables S6B, S7B, S8).

Analysis of candidate genes

The *Rht-D1b* mutant allele on chromosome 4D was detected in 216 varieties, while only 26 *Rht-B1b* mutant alleles on chromosome 4B were present (Table S2). No double-dwarfs were found. In the spectrum of varieties the *Rht-D1b* and *Rht-B1b* mutant alleles were mainly represented in the first half of the distribution containing the shorter varieties (Figure 2). The MTA with *Rht-D1* candidate gene was highly significant in all environments (Table S9). No significant MTAs with $-\log_{10}$ (P-value) $\geq$3.0 were detected for *Rht-B1*, which may be due to the low representation of the mutant allele in the germplasm.

The varieties were also genotyped for the *Ppd-D1* gene on chromosome 2DS regulating the photoperiodic sensitivity of the lines [46]. The mutant allele *Ppd-D1a* rendering a photoperiod insensitive phenotype was detected in 53 varieties, which mostly originated from South France. The *Ppd-D1* candidate gene was significantly associated with plant height in all environments. A possible reason is that *Ppd-D1* is closely linked to the dwarfing gene *Rht8* [51]. Allelic variants of microsatellite marker GWM261 were reported as diagnostic for *Rht8* [22,52]. The 192-bp allele of GWM261, which is considered to be diagnostic for reduced height

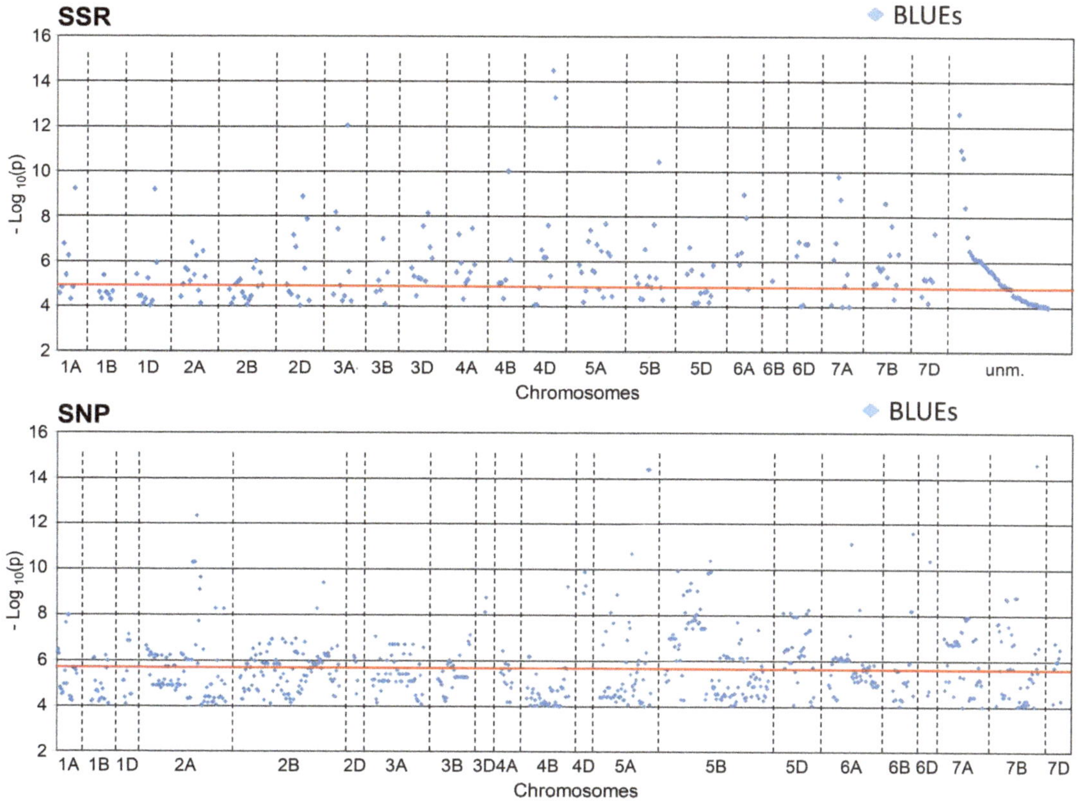

Figure 3. Manhattan plots of SSR or SNP marker alleles associated with plant height. The plot presents significant MTAs at a threshold of $-\log_{10}$ (P-value) ≥4.0 for BLUEs sorted according to their chromosomal location. The red line indicates the threshold for Bonferroni correction, $-\log_{10}$ (P-value) ≥4.82 (SSR) and ≥5.89 (SNP).

at the *Rht8* locus, was present in 35 varieties (Table S2). Moderately significant MTAs for PH were found with $-\log_{10}$ (P-value) ≥3.0 in four environments and with $-\log_{10}$ (P-value) ≥ 4.0 in two environments (Table S9) for this marker. However, no linkage disequilibrium based on r^2 was detected between the *Ppd-D1* candidate gene and marker GWM261, indicating that the significant MTA detected for *Ppd-D1* is independent of the effect based on *Rht8* (Figure S2). The Spearman rank order correlation between BLUEs for PH and BLUEs for heading date in the same environments was moderately significant with $r = 0.158$ ($P = 0.00231$).

Additive effects of SSR and SNP markers

The varieties carried between 19 and 170 PH reducing SSR alleles and between 4 to 101 PH promoting alleles (Figure S3). The Spearman Rank order correlation between the number of PH

Table 1. Number of MTAs per environment for the SSR markers and the SNPs of the 90K iSelect chip.

	SSR		90K iSelect	
Environments	**$-\log_{10}$ (P-value) ≥4.0**	**$-\log_{10}$ (P-value) ≥4.82**	**$-\log_{10}$ (P-value) ≥4.0**	**$-\log_{10}$ (P-value) ≥5.89**
Andelu (2009)	209	127	660	176
Seligenstadt (2009)	225	132	761	229
Wohlde (2009)	220	118	686	195
Andelu (2010)	221	130	705	273
Janvielle (2010)	169	87	566	134
Saultain (2010)	163	105	620	154
Seligenstadt (2010)	208	120	707	229
Wohlde (2010)	219	132	752	234
BLUEs	239	153	788	280
Sum	**1873**	**1104**	**6245**	**1904**

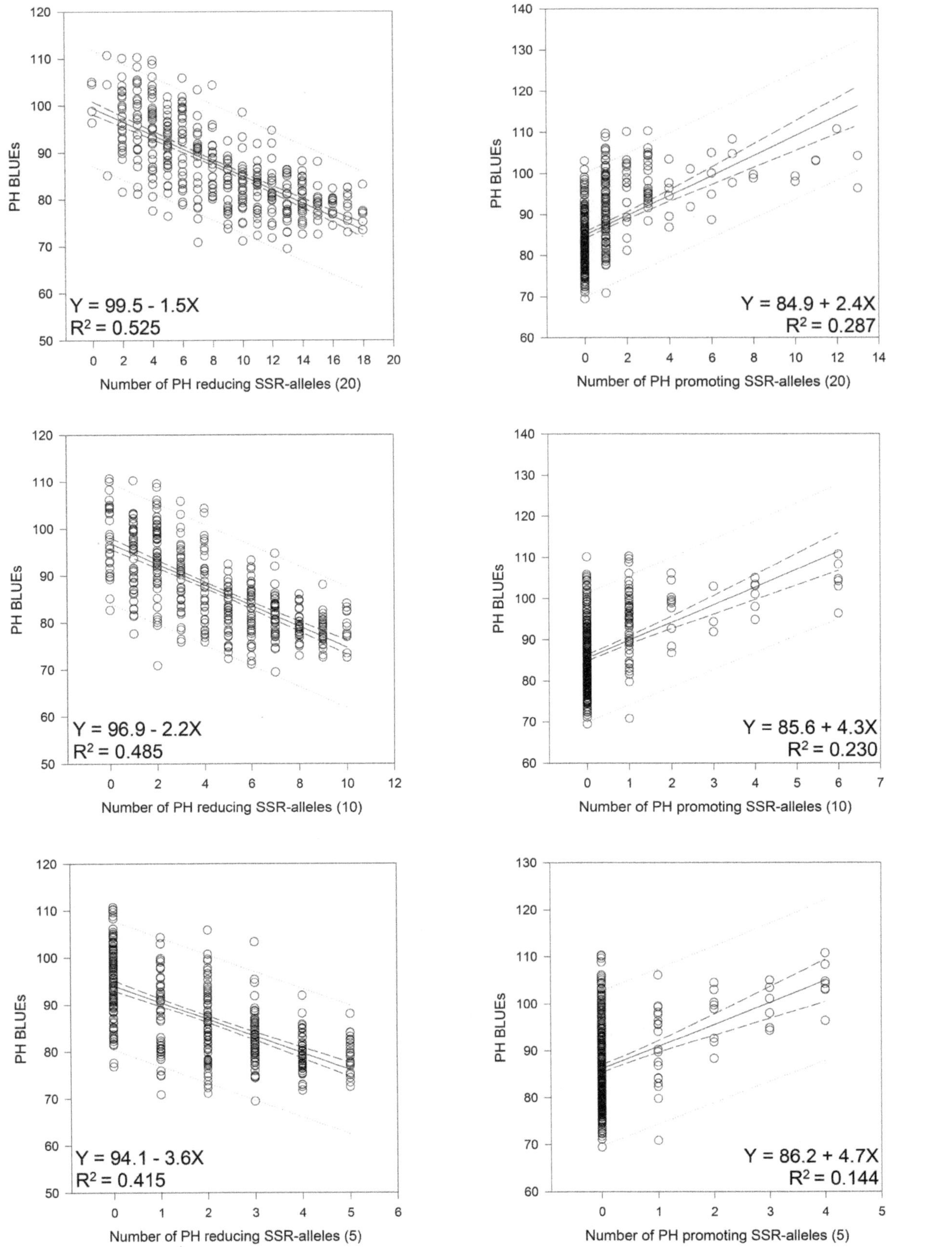

Figure 4. Linear regressions of the most PH reducing ("best") and the most PH promoting ("worst") SSR-alleles with PH-BLUEs. Linear regression resulted in a relationship between PH-BLUEs and the 20, 10 or 5 "best" or "worst" SSR-alleles in 372 varieties.

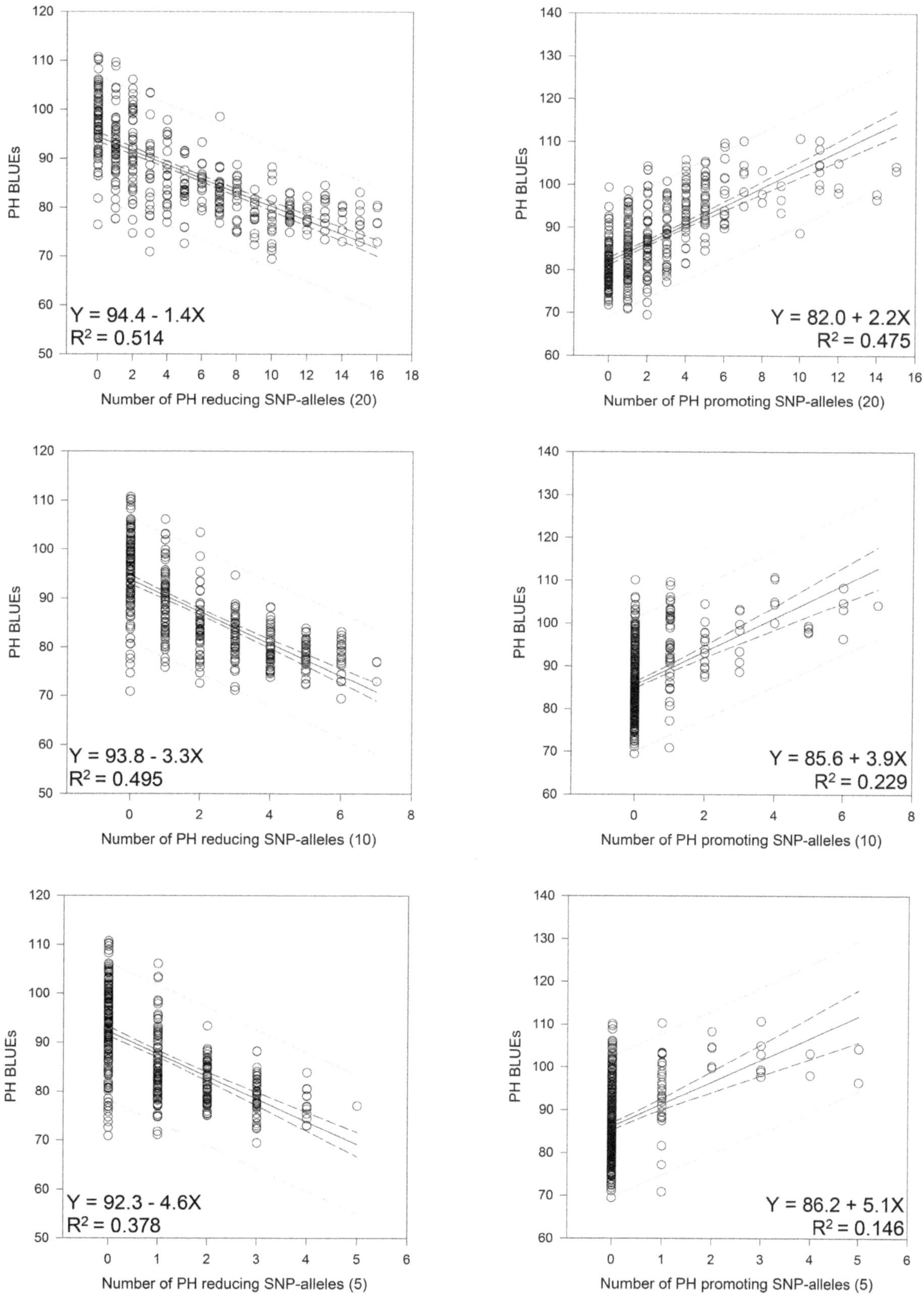

Figure 5. Linear regressions of the most PH reducing ("best") and the most PH promoting ("worst") SNP-alleles with PH-BLUEs. Linear regression resulted in a relationship between PH-BLUEs and the 20, 10 or 5 "best" or "worst" SNP-alleles from different loci in 372 varieties.

Table 2. List of the most PH reducing ("best") and most PH enhancing ("worst") SSR-alleles.

Marker alleles	Chromosome (linked gene)	Position (cM)	alleles belong to the					
			20 best	10 best	5 best	20 worst	10 worst	5 worst
GWM0357_128	1A	92.3	x	x				
GDM0126_196	1B	124.9				x		
GWM4780b_140	1D	89.1	x					
GWM4795_195	1D	130.8	x					
GWM1053_134	2A	0.2				x		
WMC0522_204	2A	88.7				x		
WMC0154_179*	2B	0.4	x					
GWM1419_197	2D	118.4				x		
GWM0539_151	2D	136.0				x	x	
GWM1204_279	2D	140.4	x					
GWM0349_232	2D	192.8				x	x	x
GWM1235_122	2D	215.0	x	x				
GWM1507_135*	3A	61.0	x	x	x			
CFA2193_239	3A	155.5	x	x	x			
GWM0938_144	3B	176.5				x	x	x
GWM1564_235	3B	238.5				x		
GWM0383a_183	3D	131.1				x	x	x
WMC0262_188	4A (*KAO-B1*)	165.1	x					
BARC0163_161	4B	81.2				x	x	x
GWM0819_183*	4D (*Rht-D1*)	34.8	x					
GWM4346_219*	4D (*Rht-D1*)	41.5	x	x	x			
GWM4001a_211	4D	100.6	x	x	x			
BARC0100b_162*	5A	88.1				x		
CFA2163a_187	5A (*Rht9*)	163.3				x	x	
WMC0415b_176*	5B	97.6				x	x	x
GWM1475a_122*	5B	108.2	x	x				
BARC0107_187*	6A	93.5	x					
GWM1241_126*	6D	138.5				x	x	
GWM1207_206	7A	230.3				x		
GWM0400_137	7B	3.4				x		
BARC0267a_145	7B	36.0	x	x				
GWM1650a_191	7D	29.1				x	x	
GWM0885_172	7D	52.1				x		
BARC0149_148	unm.	13.1	x					

Table 2. Cont.

Marker alleles	Chromosome (linked gene)	Position (cM)	alleles belong to the					
			20 best	10 best	5 best	20 worst	10 worst	5 worst
BARC1144_268	unm.	31.0				x	x	
CFD0045_177	unm.	36.0	x	x				
WMC0032_null	unm.	50.0	x					
WMC0054_162	unm.	51.4	x					
WMC0410_115	unm.	74.1	x	x	x			
WMJ3103b_164	unm.	113.1				x		

*Coincides with meta-QTL according to Griffiths et al. (2012) [54].

reducing alleles per variety and the PH BLUEs was −0.750 (P = 0.0000002), and for the number of PH promoting alleles it was 0.808 (P = 0.0000002).

We sorted the significant SSR alleles with $-\log_{10}$ (P-value) ≥4.0 according to their additive effects and tested the 20, 10 and 5 most significant alleles for their effects on the PH BLUEs (Table 2). Linear regression showed a dependence of the PH BLUEs from the number of PH reducing alleles with Y = 99.5–1.5X with R^2 = 0.525; for the PH promoting alleles the regression formula was Y = 84.9+2.4X with R^2 = 0.287 for the 20 most significant SSR-markers based on the additive effects (Figure 4). Even with only 5 alleles, an R^2 = 0.415 was found for the PH reducing alleles. Similar observations were found for the SNP-markers, where the 20 "best" and "worst" markers were selected which represented different combined loci, i.e. cluster of markers in LD were represented by only one marker (Table 3, Figure 5).

Analysis of syntenic regions in the rice genome

We developed a strategy to search for potential candidate genes for plant height regulation in the complete set of 21742 scorable wheat SNPs based on synteny to the annotated rice genome. For this purpose, a keyword search was conducted with keywords 'gibberellin', 'auxin', 'cytochrome P450', 'kaurene', 'kaurenoic' and 'DELLA' in the MSU rice database (http://rice.plantbiology.msu.edu/). The key words were based on gene ontologies of known dwarfing genes from the literature. We then used the BlastX-annotations in rice of the wheat SNP markers, as provided by [42], to relate the rice loci detected in the keyword search to the complete set of wheat SNPs. A total of 694 hits for all keywords were found in the MSU rice database (Table 4). Of those 190 hits were present, scorable and polymorphic in the data of the 90 k iSELECT chip and 44 hits were significant for PH-BLUEs with $-\log_{10}$ (P-value) ≥3.0 (Table 4, Table S10). The significant markers included 20 hits for 'cytochrome P450', 14 hits for 'auxin' and 10 hits for 'gibberellin'. Most significant with $-\log_{10}$ (P-value) ≥11.0 was wheat marker Kukri_c82296_367 on chromosome 5B with synteny to rice locus LOC_Os09g37500, annotated as OsSAUR55, an auxin responsive SAUR gene family. The SAUR (= small auxin-up RNAs) gene family in rice represents early auxin-responsive genes with largely unknown function [53]. Additionally, nine wheat markers with homology to auxin responsive factors on chromosomes 2B, 3A, 3B, 5A, 6A, 7A and 7D were significant. The significant hits for 'gibberellin' included a gibberellin response modulator protein on chromosome 5B and a chitin-inducible gibberellin-responsive protein on chromosome 3A (Table S10).

Out of 84 hits for the keyword 'gibberellin' in the rice database only 20 hits were present and polymorphic in our 90 k iSELECT chip. Therefore, we wanted to test if any rice-genes from the keyword search 'gibberellin' are linked to highly significant genomic regions in the wheat-genome. For this purpose, the syntenic rice loci were extracted from [42] for the significant wheat SNP-markers of Table S7B and this list was compared to the rice loci of the keyword search 'gibberellin'. The results of coinciding genomic regions are depicted in Table 5 and Table S11. The list of genes included several GA20 and GA2 oxidases orthologous to wheat chromosomes 1A, 2A, 3A, 3B, 5B, 5D and 7B, and other genes involved in GA synthesis, like ent-kaurenoic acid oxidase orthologous to wheat chromosome 7A and *ent*-kaurene synthase orthologous to wheat chromosome 2B. But there are also potential GA-receptors, like DELLA-genes orthologous to wheat chromosomes 4B, 4D and 7A and genes of the GID family orthologous to wheat chromosomes 2B and 5B (Table 5, Figure S1).

Table 3. List of the most PH reducing ("best") and most PH enhancing ("worst") SNP-alleles.

SNP-Marker	Chromosome (linked orthologous locus in rice)	Position (cM)	alleles belong to the					
			20 best	10 best	5 best	20 worst	10 worst	5 worst
BS00086680_51	1A	62.6	x	x	x			
Kukri_c54467_100	1A	65.9	x					
tplb0025b13_2687	1D	3.4				x	x	x
wsnp_Ex_c1358_2602235	1D	12.1				x	x	
wsnp_CAP8_c2677_1394934	2A	66.6				x		
BS00022896_51	2A	69.8	x					
RAC875_c38018_278	2A	69.8	x	x				
wsnp_CAP11_c1711_934478	2A (*OsGA 20 oxidase 2*)	95.8	x					
RAC875_c4609_1756	2A	105.3	x					
BS00023075_51	2A	117.2	x					
Ra_c23048_474	2B	66.0	x					
BobWhite_rep_c64068_241	2B	131.7				x		
Tdurum_contig91865_242	3A	38.3	x	x	x			
Kukri_c20889_526	3A (*OsGA2ox3**)	61.4				x	x	
BobWhite_c19725_1329	3B	57.4				x	x	
RAC875_c403_2247	3B (*OsGA2ox2**)	66.8				x	x	x
Jagger_c3839_60	3D	147.1	x					
IAAV7221	4B	2.2	x	x	x			
RAC875_c1357_860	4B	66.4				x		
RFL_Contig5341_816	4B	79.1				x	x	x
wsnp_BE444644A_Ta_2_1	5A	11.0	x					
wsnp_BF293620A_Ta_2_3	5A	57.7	x	x				
wsnp_Ex_c23795_33033959	5A	101.2				x		
RAC875_c25756_279	5B (*Os GA receptor GID1L2*)	50.5	x					
IAAV7207	5B	57.4				x	x	x
Excalibur_c5329_1335	5B	104.1				x		
wsnp_Ex_c65985_64188864	5D	27.9				x		
wsnp_Ex_c1278_2449191	5D	81.7	x	x				
BS00089597_51	5D (*OsGA20ox1**)	166.2	x	x				
BS00036878_51	6A	57.0				x		
BS00084314_51	6B	42.8	x	x	x			
RAC875_rep_c71463_98	6B	95.7	x	x				
Kukri_c73802_205	6D	2.3				x	x	
Excalibur_c15844_1470	6D	19.3				x		

Table 3. Cont.

SNP-Marker	Chromosome (linked orthologous locus in rice)	Position (cM)	alleles belong to the					
			20 best	10 best	5 best	20 worst	10 worst	5 worst
Ex_c101666_634	7B	42.9				x	x	
wsnp_Ex_c24376_33618864	7B	50.6				x		
BS00024215_51	7B	54.5	x					
Tdurum_contig42584_1190	7B	140.4	x	x	x			
Excalibur_c22419_460	7D	56.8				x	x	x
Ku_c884_791	7D	83.9				x		

*Nomenclature according to Sakamoto et al. 2004 [58].

Discussion

The pattern of inheritance for plant height was characterized by many closely linked markers with high LD, especially for the SNPs. When these closely linked marker loci were combined, we estimated 109 loci for the SSRs and 87 for the SNPs. The higher numbers for SSRs may be caused by a better coverage of the D-genome for this marker type. Eventually, further loci could be combined. A meta-QTL analysis of four doubled-haploid populations in European elite winter wheat germplasm estimated sixteen meta-QTL loci [54]. Several of these meta-QTL coincided with our detected MTAs (Table 6). In the population Avalon × Cadenza, a QTL distal to marker GWM261 on chromosome 2DS was described [54], which may coincide with the MTAs detected by markers GWM4815 and GWM1418 in our study. This QTL could be due to the effect of various *Rht8* alleles [22,52] or be a pleiotropic effect of *Ppd-D1* with a reduced time to flowering resulting in reduced plant height [55]. *Rht-D1* was included in a block of highly significant markers in LD including BARC98, WMC473b, GWM819, GDM129, GWM4346 for the SSRs and RAC875_rep_c105718_304 for the SNPs (Figure 6) and coincided with the respective meta-QTL on chromosome 4D [54]. Several of the markers coinciding with meta-QTL regions were found in our list of the 20 most PH reducing or PH promoting alleles based on the additive effects. These included besides the markers linked to *Rht-D1*, GWM4346 and GWM819, WMC154 on chromosome 2B, GWM1507 on 3A, BARC100 on 5A, WMC415b and GWM1475a on 5B, BARC107 on 6A and GWM1241 on chromosome 6D (Table 2).

Of the known mapped *Rht*-genes in wheat, besides *Rht-D1* and *Rht8*, possibly an allelic effect of *Rht4* was discovered in our study. *Rht4* was mapped on chromosome arm 2BL with marker WMC317 explaining most of the phenotypic variance in the cross Vigour18 × Burt *ert* (*Rht4*) [23]. WMC317 was not significant in our association study, however significant MTAs were found for the flanking markers GWM1399, GWM619 and GWM3038 (Figure S1). The *Rht5*-gene was mapped on chromosome 3BS in the Vigour18 × Marfed M (*Rht5*) population and linked to marker BARC102. BARC102 is located in a distance of 1.2 cM (http://wheat.pw.usda.gov/GG2/index.shtml) from marker WMC808, which was significant in our study. Therefore the MTAs discovered by WMC808 and GWM4145 are possibly caused by an allele of *Rht5*. Gene *Rht9* was mapped to chromosome arm 5AL in the Chuan Mai 18 (*Rht8*) × Mara (*Rht8Rht9*) population and linked to marker BARC151 [23]. While BARC151 was not significant in our association panel, the neighboring markers GWM1342 and CFA2163 discovered highly significant MTAs (Figure S1), which may be an allelic effect of *Rht9*. The *Rht13* gene was reported to be linked to marker GWM577 on the distal long arm of chromosome 7B [23]. GWM577 was not significant in our association panel, but marker BARC182 in a distance of 12 cM was highly significant, an effect which may be caused by an allele of *Rht13*.

For a number of gibberellin metabolic pathway genes the chromosomal location is known in wheat [56]. Three *ent*-kaurenic acid oxidase (KAO) genes were mapped to the distal ends of chromosome arms 7AS, 4AL and 7DS corresponding to the 7BS/4AL translocation region [57]. The significant markers GWM681 and GWM735 correspond to the location of *KAO-A1* south of marker GWM471 on chromosome arm 7AS, while the significant markers WMC262 and GWM1251 correspond to *KAO-B1* on chromosome 4AL north of marker GWM160. WMC262 is listed among the 20 most effective SSR-markers for plant height reduction based on the additive effects (Table 2). The significant

Table 4. Keyword search in rice for PH candidate genes.

	Keywords in rice				
	gibberellin	auxin	cytochrome P450	kaurene, kaurenoic	DELLA
Number of hits in MSU rice database	84	177	421	9	3
Number of syntenic wheat SNPs present and polymorphic in the 90 k iSELECT chip	20	61	103	6	0
Number of wheat SNPs significant for PH-BLUEs with LOD >3.0	10	14	20	0	0

MTAs at both sites, on chromosomes 7AS and 4AL, may be caused by the respective *KAO*-genes, since a *KAO*-gene has been described as *Dwarf3* gene in maize [17]. On chromosome 7AS a large significant cluster of SNP markers was detected which had synteny to the rice gene *OsKAO* [58] on rice chromosome 6 (Table 5, Figure S1), the syntenic locus to *KAO-A1* in wheat.

Three GA 20-oxidase genes, *TaGA20ox1*, were located on the distal long arms of chromosomes 5B and 5D [59,56] and chromosome arm 4AL in the region of the known 5AL/4AL translocation [60]. These loci may correspond to the significant MTAs detected in our study on chromosome arm 5DL by markers BARC177, WMC765, BARC144, GWM1454, GWM272, on chromosome arm 5BL by marker WMC 783 and on chromosome 4A by marker BARC 343. This assumption is supported by the fact that a cluster of SNP-markers on chromosome arm 5DL (wsnp_RFL_Contig2402_1924907, Excalibur_c24145_1643, wsnp_Ex_c11055_179228283, wsnp_Ex_c24145_33394561, wsnp_Ex_c24145_33394644) and an SNP marker on chromosome 5BL (Kukri_c1214_2565) spanned a region on rice chromosome 3, where *OsGA20ox1*, the syntenic locus to *TaGA20ox1*, is located (Table 5, Figure S1, Table S11).

Unfortunately, the 21742 scorable SNPs in our study did only include 20 hits for the keyword 'gibberellin' based on synteny to the rice genome (Table 4) and no GA20-oxidases were present. Therefore, we could only rely on indirect evidence based on mapping in similar genomic regions to assess the impact of gibberellin pathway genes (Table 5). Syntenic relationships to rice genomic regions containing GA oxidases were found for chromosomes 1A, 3A, 3B and 5B (Table 5). The regions on chromosomes 3A and 3B corresponded to *OsGA2ox3* and *OsGA20ox2* [58], both on rice chromosome 1. For both chromosomal regions large clusters of SNP markers were significant (Figure S1) and representative markers (Kukri_c20889_526 on chromosome 3A and RAC875_c403_2247 on chromosome 3B) were in the list of the makers with highest additive effects (Table 3). The semidwarfing gene *sdw1/denso* on chromosome 3HL in barley was described as most likely orthologous gene of the *sd1* gene in rice, which carries the mutation in the gene *OsGA20ox2* [16,61] encoding an oxidase enzyme involved in the biosynthesis of gibberellin [13,14]. Therefore, the highly significant SNP cluster on chromosome 3BL corresponding to *OsGA20ox2* in our study may be the syntenic locus to gene *sdw1/denso* in barley. Since co-location of loci is not a direct proof for the involvement of the mentioned genes in regulating PH, further sequence analysis and SNP-discovery in the wheat genes themselves is necessary to estimate their contribution to the expression of PH.

Syntenic relationships were also discovered for rice genomic regions carrying genes for GA perception, like DELLA proteins on wheat chromosomes 4B, 4D and 7A and GID receptors on wheat chromosomes 2B and 5B (Table 5). The cluster of SNP-markers

Table 5. List of orthologous loci in rice with genes of the GA metabolism or related to dwarfing phenotypes.

Wheat SNPs, chromosomal position	Orthologous locus in rice	Gene annotation in rice
1A; 69.8 cM	LOC_Os05g34854	OsGA2ox4*; GA 20 oxidase 2
2A; 95.8–98.1 cM	LOC_Os04g55070	GA 20 oxidase 2
2B; 64.3 cM	LOC_Os07g44850	GA receptor GID1L2
2B; 75.7–85.1 cM	LOC_Os04g52230	Ent kaurene synthase
3A; 59–61.4 cM	LOC_Os01g55240	OsGA2ox3*; GA 2-beta dioxygenase
3B; 59.7–66.8 cM	LOC_Os01g66100	OsGA20ox2*; GA 20 oxidase 2
4B; 50.8 cM	LOC_Os03g49990	DELLA, SLR1
4D; 47 cM	LOC_Os03g49990	DELLA, SLR1
5B; 45.8–50.5 cM	LOC_Os09g28230	GA receptor GID1L2
5B; 95.4–96.2 cM	LOC_Os03g42130	GA 20 oxidase 2
5B; 165 cM	LOC_Os03g63970	OsGA20ox1*; GA 20 oxidase 1
5D; 163.9–167.8 cM	LOC_Os03g63970	OsGA20ox1*; GA 20 oxidase 1
7A; 1.6–5.7 cM	LOC_Os06g02019	OsKAO*; ent-kaurenoic acid oxidase (cytochrom P450)
7A; 80.3 cM	LOC_Os06g02019	DELLA, RGL3
7B; 68.6 cM	LOC_Os08g44590	GA 20 oxidase 2

*Nomenclature according to Sakamoto et al. 2004 [58]

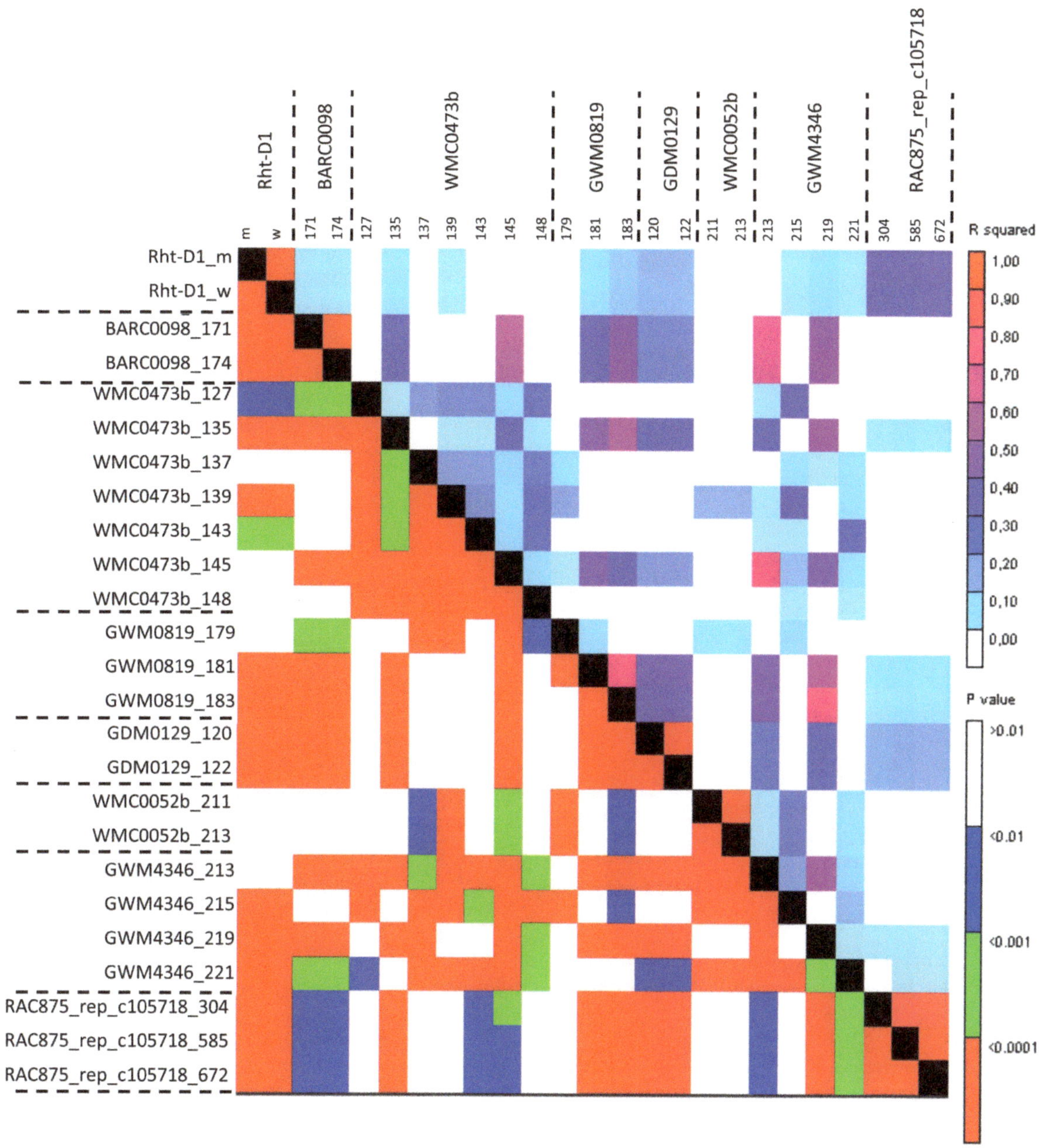

Figure 6. Linkage disequilibrium in the region of *Rht-D1*. The LD around *Rht-D1* on chromosome 4D extended to linked SSR and SNP markers with significant MTAs for plant height. In the upper triangle r^2 and in the lower triangle the P-values are depicted.

on wheat chromosome 4D with synteny to *OsDELLA* on rice chromosome 3 corresponds to the *Rht-D1* gene, since the respective SNP-Markers (RAC875_rep_c105718_304, RAC875_rep_c105718_672, RAC875_rep_c105718_585) are in LD with *Rht-D1* (Figure 6).

The significant candidate genes of the keyword search (Table 4) included 20 hits for 'cytochrome P450'. Cytochrom P450 is part of *ent*-kaurenoic acid oidase, an enzyme of the GA-metabolism, but it also is present in enzymes of other functions including the involvement in brassinosteroid biosynthesis [62]. Highly significant hits were found for auxin-related genes as well. These included markers with homology to auxin responsive factors and the SAUR gene family, the function of which is largely unknown in both cases [53,63].

Our genotyping results indicated, that the *Rht-D1b* mutant allele on chromosome 4D was present in 58% of the varieties analyzed, while the *Rht-B1b* mutant allele on chromosome 4B was found in only 7% of our varieties. The detection method employed [48] did not distinguish from further alleles, such as *Rht-B1b* from *Rht-B1d*, and *Rht-D1b* from *Rht-D1c* and *Rht-D1d* [64,65]. The high percentage of *Rht-D1* mutant allele in our germplasm set contradicted the results of [66], where 12% *Rht-B1b* and 16% *Rht-D1b* mutant alleles were reported in 103 French varieties, while in the total set of 368 worldwide accessions for both loci, *Rht-B1* and *Rht-D1*, a frequency of 12% mutant alleles was found. In contrast to our study, where only registered varieties were investigated, the study of [66] included landraces. Another study reported frequencies of 40% for *Rht-B1b* and 22% for *Rht-D1b* in

Table 6. Coincidences of significant MTAs in our study with QTL or meta-QTL regions described by Griffiths et al. 2012 [54].

Significant markers in our study	Chromosome	Coincidences with QTL regions described in [54]
GWM164	1A	WMC336 - PSP3027
GWM337	1D	GWM337 - WMC36
GWM359	2A	wPt6207 - WMC827
WMC154	2B	Stm17tcac - WMC154
GMW4815, GWM1418	2D	CFD36 - GWM261
GWM1507, GWM4018	3A	BARC45 - WMC269
GWM4660, GWM251, BARC163	4B	GWM149 - GWM6
BARC98, WMC473b, GWM819, GDM129, GWM4346	4D	WMC285 - *Rht-D1*
BARC4, BARC100	5A	wPt3794 - GWM617
GWM1475, WMC415b, BARC109a, GWM1165, GWM1108, GWM335	5B	GWM213 - GWM408
BARC107, GWM4608, WMC672	6A	BARC23 - GWM570
GWM1241	6D	CFD42 - GWM325
GWM1397, GWM3062, GWM4335	7D	wPt6263 - GWM44

a 172 genotypes originating from 20 countries [67]. The data indicated, that *Rht-D1b* may be prevalent in the Middle European wheat varieties, while *Rht-B1b* seems to occur mainly in the world wide germplasm including landraces.

The *Rht1* mutant phenotypes were found to increase harvest index and grain yield [68] and were therefore introduced in many commercial wheat varieties around the world. But also some negative effects, like decreased seedling vigour and shorter coleoptiles were reported for these GA-insensitive dwarfing genes [69,70]. Beneficial effects for the *Rht1* wild type alleles were reported in drought environments [71]. Therefore, the use of alternative height reducing genes was suggested [70,72]. Our data indicate a wealth of alternative loci that could be employed for modulating plant height reduction and which are already present in the germplasm. With only five SNP- or SSR-markers obvious additive effects for PH BLUEs were observed (Figures 4 and 5). Similar results were reported for pyramiding of elite alleles for plant height in two winter wheat populations [73].

Supporting Information

Figure S1 Chromosomal location of marker-trait associations for plant height. The linkage map for the SNP-markers of the 90 k iSELECT chip based on the ITMI-DH population is depicted on the left side, while the linkage map for the SSR-markers based on the ITMI-RIL population is shown on the right side for each linkage group. Each MTA for a single environment or the BLUEs are depicted by an icon as explained on the first page of the file. The locations of known *Rht* genes and known mapping positions for gibberellin metabolism genes for wheat are indicated in the SSR maps. Orthologous sites to rice (*Oryza sativa* = *Os*) loci with genes related to GA metabolism or perception are indicated in the SNP maps.

Figure S2 No Linkage disequilibrium was detected between SSR-marker GWM0261 and the candidate gene *Ppd-D1* on chromosome 2D.

Figure S3 Frequency of (A) PH reducing and (B) PH promoting SSR-alleles per variety.

Table S1 Climatic factors at trial sites.

Table S2 List of varieties, genotyping data of candidate genes and phenotypic data.

Table S3 Estimation of variance components and broad sense heritability and estimation of differences between groups (= environments) using ANOVA and a Tukey B test.

Table S4 Spearman rank order correlation of PH scores in 372 varieties among eight environments and the BLUEs. GAT_2012 represents an untreated control environment in a plant nursery.

Table S5 Analysis of variance (ANOVA) of plant height in 372 varieties in eight environments.

Table S6 List of significant ($-\log_{10}$(P) >4.0) marker trait associations of the SSR markers. Sheet 6A: List of significant ($-\log_{10}$(P) >4.0) marker trait associations of the SSR markers. Columns F to U describe the additive effect and the R^2 for each environment. Colum V indicates the R2 of the BLUEs values. Positive additive effects promote plant height and negative additive effects reduce plant height. **Sheet 6B: List of significant ($-\log_{10}$(P) >4.0) marker trait associations of the SSR markers for BLUEs only.** Markers with distances <6.0 cM were combined by colour coding. Columns F to U describe the additive effect and the R^2 for each environment. Colum V indicates the R^2 of the BLUEs values. Positive additive effects promote plant height and negative additive effects reduce plant height.

Table S7 List of significant ($-\log_{10}$(P) >4.0) marker trait associations of the SNP markers. Sheet 7A: List of significant ($-\log_{10}$(P) >4.0) marker trait associations of the SNP markers. Columns G to V describe the additive effect and the R^2 for each environment. Colum W indicates the R^2 of the BLUEs values. Positive additive effects promote plant height and negative additive effects reduce plant height. **Sheet 7B: List of significant (–$\log_{10}$(P) >4.0) marker trait associations of the SNP markers for the BLUEs only.** Markers with distances <6.0 cM were combined by colour coding. Columns G to V describe the additive effect and the R^2 for each environment. Colum W indicates the R^2 of the BLUEs values. Positive additive effects promote plant height and negative additive effects reduce plant height.

Table S8 Number of MTAs per chromosomes for the SSR and SNP markers.

Table S9 List of significant (–$\log_{10}$(P) >2.0) marker trait associations of the candidate genes. Columns F to U describe the additive effect and the R^2 for each environment. Colum V indicates the R^2 of the BLUEs values. Positive additive effects promote plant height and negative additive effects reduce plant height.

Table S10 List of significant (–$\log_{10}$(P) >3.0) marker trait associations of the candidates from the keyword search. Columns Y to AF describe the additive effect for each environment. Column AG indicates the additive effect of the BLUEs values. Positive additive effects promote plant height and negative effects reduce plant height.

Table S11 List of wheat SNP with GA related orthologous loci in rice.

Author Contributions

Conceived and designed the experiments: MWG EE VK OA GS MH MSR. Performed the experiments: EE OA VK JP SK. Analyzed the data: CDZ JL KN. Contributed to the writing of the manuscript: MSR CDZ.

References

1. Khush GS (2001) Green revolution: the way forward. Nat Rev Genetics 2: 815–822.
2. Hedden P (2003) The genes of the Green Revolution. Trends Genet 19: 5–9.
3. Peng J, Richards DE, Hartley NM, Murphy GP, Devos KM, et al. (1999) 'Green revolution' genes encode mutant gibberellin response modulators. Nature 400: 256–261.
4. Peng J, Carol P, Richards DE, King KE, Cowling RJ, et al. (1997) The Arabidopsis GAI gene defines a signaling pathway that negatively regulates gibberellin responses. Genes Dev 11: 3194–3205.
5. Winkler RG, Freeling M (1994) Physiological genetics of the dominant gibberellin-nonresponsive maize dwarfs, *Dwarf8* and *Dwarf9*. Planta 193: 341–348.
6. Lawit SJ, Wych HM, Xu D, Kundu S, Tomes DT (2010) Maize DELLA proteins dwarf plant8 and dwarf plant9 as modulators of plant development. Plant Cell Physiol 51: 1854–1868.
7. Ikeda A, Ueguchi-Tanaka M, Sonoda Y, Kitano H, Koshioka M, et al. (2001) *Slender* rice, a constitutive gibberellin response mutant, is caused by a null mutation of the SLR1 gene, an ortholog of the height-regulating gene *GAI/RGA/RHT/D8*. Plant Cell 13: 999–1010.
8. Chandler PM, Marion-Poll A, Ellis M, Gubler F (2002) Mutants at the *Slender*1 locus of barley cv. Himalaya. Molecular and Physiological characterization. Plant Phys 129: 181–190.
9. Chandler PM, Harding CA (2013) 'Overgrowth' mutants in barley and wheat: new alleles and phenotypes of the 'Green revolution' DELLA gene. Journal Experimental Botany 64: 1603–1613.
10. Ashikari M, Wu J, Yano M, Sasaki T, Yoshimura A (1999) Rice gibberellin insensitive dwarf mutant gene Dwarf 1 encodes the α-subunit of GTP-binding protein. Proc Natl Acad Sci USA 96: 10284–10289.
11. Ueguchi-Tanaka M, Ashikari M, Nakajima M, Itoh H, Katoh E, et al. (2005) *GIBBERELLIN INSENSITIVE DWARF1* encodes a soluble receptor for gibberellin. Nature 437: 693–698.
12. Sasaki A, Itoh H, Gomi K, Ueguchi-Tanaka M, Ishiyama K, et al. (2003) Accumulation of phosphorylated repressor for gibberellin signaling in an F-box mutant. Science 299: 1896–1898.
13. Spielmeyer W, Ellis MH, Chandler PM (2002) Semidwarf (*sd-1*), "green revolution" rice, contains a defective gibberellin 20-oxidase gene. Proc Natl Acad Sci USA 99: 9043–9048.
14. Sasaki A, Ashikari M, Ueguchi-Tanaka M, Itoh H, Nishimura A, et al. (2002) A mutant gibberellin-synthesis gene in rice. Nature 416: 701–702.
15. Barboza L, Effgen S, Alonso-Blanco C, Kooke R, Keurentjes JJB, et al. (2013) *Arabidopsis* semidwarfs evolved from independent mutations in *GA20ox1*, ortholog to green revolution dwarf alleles in rice and barley. Proc Natl Acad Sci USA 110: 15818–15823.
16. Jia Q, Zhang J, Westcott S, Zhang X-Q, Bellgard M, et al. (2009) GA-20 oxidase as a candidate for the semidwarf gene *sdw1/denso* in barley. Funct Integr Genomics 9: 255–262.
17. Winkler RG, Helentjaris T (1995) The maize Dwarf3 gene encodes a cytochrome P450-mediated early step in gibberellin biosynthesis. Plant Cell 7: 1307–1317.
18. Multani DS, Briggs SP, Chamberlin MA, Blakeslee JJ, Murphy AS, et al. (2003) Loss of an MDR transporter in compact stalks of maize *br2* and sorghum *dw3* mutants. Science 302: 81–84.
19. Parvathaneni RK, Jakkula V, Padi FK, Faure S, Nagarajappa N, et al. (2013) Fine-mapping and identification of a candidate gene underlying the *d2* dwarfing phenotype in pearl millet, *Cenchrus americanus* (l.) Morrone. G3-Genes/Genomes/Genetics 3: 563–572.
20. Börner A, Röder M, Korzun V (1997) Comparative molecular mapping of GA insensitive *Rht* loci on chromosomes 4B and 4D of wheat (*Triticum aestivum* L.). Theor Appl Genet 95: 1133–1137.
21. Korzun V, Röder M, Worland AJ, Börner A (1997) Intrachromosomal mapping of the dwarfing (*Rht12*) and vernalisation response (*Vrn1*) genes in wheat by using RFLP and microsatellite markers. Plant Breed 116: 227–232.
22. Korzun V, Röder MS, Ganal MW, Worland AJ, Law CN (1998) Genetic analysis of the dwarfing gene (*Rht8*) in wheat. Part I. Molecular mapping of the *Rht8* on the short arm of chromosome 2D of bread wheat (*Triticum aestivum* L.). Theor Appl Genet 96: 1104–1109.
23. Ellis MH, Rebetzke GJ, Azanza F, Richards RA, Spielmeyer (2005) Molecular mapping of gibberellin-responsive dwarfing genes in bread wheat. Theor Appl Genet 111: 423–430.
24. Harjes CE, Rocheford TR, Bai L, Brutnell TP, Kandianis CB, et al. (2008) Natural genetic variation in *Lycopene epsilon cyclase* tapped for maize biofortification. Science 319: 330–333.
25. Matthies IE, Weise S, Röder MS (2009) Association of haplotype diversity in the α-amylase gene *amy*1 with malting quality parameters in barley. Mol Breed 23: 139–152.
26. Waugh R, Jannink JL, Muehlbauer GL, Ramsay L (2009) The emergence of whole genome association scans in barley. Curr Opin Plant Biol 12: 1–5.
27. Rafalski JA (2010) Association genetics in crop improvement. Curr Opin Plant Biol 13: 1–7.
28. Thornsberry JM, Goodman MM, Doebley J, Kresovich S, Nielsen D, et al. (2001) *Dwarf*8 polymorphisms associate with variation in flowering time. Nat Genet 28: 286–289.
29. Larsson SJ, Lipka AE, Buckler ES (2013) Lessons from Dwarf8 on strengths and weaknesses of structured association mapping. PLOS Genetics 9: e1003246.
30. Hamblin MT, Buckler ES, Jannink J-L (2011) Population genetics of genomics-based crop improvement methods. Trends in Genet 27: 98–106.
31. Pasam RK, Sharma R, Malosetti M, van Eeuwijk FA, Haseneyer G, et al. (2012) Genome-wide association studies for agronomical traits in a world wide spring barley collection. BMC Plant Biol 12: 16.
32. Brown PJ, Rooney WL, Franks C, Kresovich S (2008) Efficient mapping of plant height quantitative trait loci in a Sorghum association population with introgressed dwarfing genes. Genetics 180: 629–637.
33. Weng J, Xie C, Hao Z, Wang J, Liu C (2011) Genome-wide association study identifies candidate genes that affect plant height in Chinese elite maize (*Zea mays* L.) inbred lines. PLoS ONE 6: e29229.
34. Zhao Y, Mette MF, Gowda M, Longin CFH, Reif JC (2014) Bridging the gap between marker-assisted and genomic selection of heading time and plant height in hybrid wheat. Heredity 112: 638–645.
35. International Rice Genome Project (2005) The map-based sequence of the rice genome. Nature 436: 793–800.
36. Kollers S, Rodemann B, Ling J, Korzun V, Ebmeyer E, et al. (2013) Whole genome association mapping of *Fusarium* head blight resistance in European winter wheat (*Triticum aestivum* L.). Plos One 8: e57500.

37. Kollers S, Rodemann B, Ling J, Korzun V, Ebmeyer E, et al. (2013) Genetic architecture of resistance to *Septoria tritici* blotch (*Mycosphaerella graminicola*) in European winter wheat. Mol Breed 32: 411–423.
38. Kollers S, Rodemann B, Ling J, Korzun V, Ebmeyer E, et al. (2014) Genome wide association mapping of tan spot resistance (*Pyrenophora tritici-repentis*) in European winter wheat. Mol Breed 34: 363–371.
39. Zanke C, Ling J, Plieske J, Kollers S, Ebmeyer E, et al. (2014) Genetic architecture of main effect QTL for heading date in European winter wheat. Frontiers in Plant Science 5: 217.
40. Rademacher W (2000) Growth Retardants: Effects on gibberellin biosynthesis and other metabolic pathways. Ann Rev Plant Phys Plant Mol Biol 51: 501–531.
41. Cavanagh CR, Chao S, Wang S, Huang BE, Stephen S, et al. (2013) Genome-wide comparative diversity uncovers multiple targets of selection for improvement in hexaploid wheat landraces and cultivars. Proc Natl Acad Sci USA 110: 8057–8062.
42. Wang S, Wong D, Forrest K, Allen A, Chao S, et al. (2014) Characterization of polyploid wheat genomic diversity using a high-density 90,000 SNP array. Plant Biotechnology Journal 12: 787–796.
43. Röder MS, Korzun V, Wendehake K, Plaschke J, Tixier M-H, et al. (1998) A microsatellite map of wheat. Genetics 149: 2007–2023.
44. Ganal MW, Röder MS (2007) "Microsatellite and SNP markers in wheat breeding", in Genomics Assisted Crop Improvement: Vol. 2: Genomics Applications in Crops, ed RK Varshney and R Tuberosa, pp. 1–24.
45. Sorrells ME, Gustafson JP, Somers D, Chao S, Benscher D, et al. (2011) Reconstruction of the synthetic W7984 × Opata M85 wheat reference population. Genome 54: 875–882.
46. Poland JA, Brown PJ, Sorrells ME, Jannink J-L (2012) Development of high-density genetic maps for barley and wheat using a novel two-enzyme genotyping-by-sequencing approach (2012) PLoS One 7: 1–8.
47. Beales J, Turner A, Griffiths S, Snape J, Laurie DA (2007) A pseudo-response regulator is misexpressed in the photoperiod insensitive *Ppd-D1a* mutant of wheat (*Triticum aestivum* L.). Theor Appl Genet 115: 721–733.
48. Ellis MH, Spielmeyer W, Gale KR, Rebetzke GJ, Richards RA (2002) "Perfect" markers for the *Rht-B1b* and *Rht-D1b* dwarfing genes in wheat. Theor Appl Genet 105: 1038–1042.
49. Hardy OJ, Vekemans X (2002) SPAGEDi: a versatile computer program to analyse spatial genetic structure at the individual or population levels. Mol Ecol Notes 2: 618–620.
50. Matthies IE, van Hintum T, Weise S, Röder MS (2012) Population structure revealed by different marker types (SSR or DArT) has an impact on the results of genome-wide association mapping in European barley cultivars. Mol Breed 30: 951–966.
51. Pestsova E, Röder MS (2002) Microsatellite analysis of wheat chromosome 2D allows the reconstruction of chromosomal inheritance in pedigrees of breeding programmes. Theor Appl Genet 106: 84–91.
52. Worland AJ, Korzun V, Röder MS, Ganal MW, Law CN (1998) Genetic analysis of the dwarfing gene *Rht8* in wheat. Part II. The distribution and adaptive significance of allelic variants at the *Rht8* locus of wheat as revealed by microsatellite screening. Theor Appl Genet 96: 1110–1120.
53. Jain M, Tyagi AK, Khurana JP (2006) Genome-wide analysis, evolutionary expansion, and expression of early auxin-responsive *SAUR* gene family in rice (*Oryza sativa*). Genomics 88: 360–371.
54. Griffiths S, Simmonds J, Leverington M, Wang Y, Fish L, et al. (2012) Meta-QTL analysis of the genetic control of crop height in elite European winter wheat germplasm. Mol Breed 29: 159–171.
55. Börner A, Worland AJ, Plaschke J, Schumann E, Law CN (1993) Pleiotropic effects of genes for reduced height (Rht) and day-length insensitivity (Ppd) on yield and its components for wheat grown in middle Europe. Plant Breed 111: 204–216.
56. Spielmeyer W, Ellis M, Robertson M, Ali S, Lenton JR, et al. (2004) Isolation of gibberellin metabolic pathway genes from barley and comparative mapping in barley, wheat and rice. Theor Appl Genet 109: 847–855.
57. Khlestkina EK, Kumar U, Röder MS (2010) Ent-kaurenoic acid oxidase genes in wheat. Mol Breed 25: 251–258.
58. Sakamoto T, Miura K, Itoh H, Tatsumi T, Ueguchi-Tanaka M, et al. (2004) An overview of gibberellin metabolism enzyme genes and their related mutants in rice. Plant Phys 134: 1642–1653.
59. Appleford NEJ, Evans DJ, Lenton JR, Gaskin P, Croker SJ, et al. (2006) Function and transcript analysis of gibberellin-biosynthetic enzymes in wheat. Planta 223: 568–582.
60. Linkiewicz AM, Qi LL, Gill BS, Ratnasiri A, Echalier B, et al. (2004) A 2500-locus bin map of wheat homoeologous group 5 provides insights on gene distribution and colinearity with rice. Genetics 168: 665–676.
61. Jia Q, Zhang X-Q, Westcott S, Broughton S, Cakir M, et al. (2011) Expression level of a gibberellin 20-oxidase gene is associated with multiple agronomic and quality traits in barley. Theor Appl Genet 122: 1451–1460.
62. Chapple C (1998) Molecular-genetic analysis of plant cytochrome P450-dependent monooxygenases. Ann Rev Plant Physiol Plant Mol Biol 49: 311–343.
63. Okushima Y, Overvoorde PJ, Arima K, Alonso JM, Chan A, et al. (2005) Functional genomic analysis of the *AUXIN RESPONSE FACTOR* gene family members in *Arabidopsis thaliana*: Unique and overlapping functions of *ARF7* and *ARF19*. Plant Cell 17: 444–463.
64. Pestsova EG, Korzun V, Börner A (2008) Validation and utilization of Rht dwarfing gene specific markers. Cereal Res Commun 36: 235–246.
65. Wilhelm EP, Mackay IJ, Saville RJ, Korolev AV, Balfourier F, et al. (2013) Haplotype dictionary for the *Rht-1* loci in wheat. Theor Appl Genet 126: 1733–1747.
66. Wilhelm EP, Boulton MI, Al-Kaff N, Balfourier F, Bordes J, Greenland AJ, Powell W, Mackay IJ (2013) *Rht-1* and *Ppd-D1* associations with height, GA sensitivity, and days to heading in a worldwide bread wheat collection. Theor Appl Genet 126: 2233–2243.
67. Tosovic-Maric B, Kobiliskij B, Obreht D, Vapa L (2008) Evaluation of wheat *Rht* genes using molecular markers. Genetika 40: 31–38.
68. Flintham JE, Börner A, Worland AJ, Gale MD (1997) Optimizing wheat grain yield: effects of Rht (gibberellin-insensitive) dwarfing genes. Journal of Agricultural Science 128: 11–15.
69. Rebetzke GJ, Appels R, Morrison AD, Richards RA, McDonald G, et al. (2001) Quantitative trait loci on chromosome 4B for coleoptile length and early vigour in wheat (*Triticum aestivum* L.). Australian Journal of Agricultural Research 52: 1221–1234.
70. Ellis MH, Rebetzke GJ, Chandler P, Bonnett D, Spielmeyer W, et al. (2004) The effect of different height reducing genes on the early growth of wheat. Functional Plant Biology 31: 583–589.
71. Zhang J, Dell B, Biddulph B, Drake-Brockman F, Walker E, et al. (2013) Wild-type alleles of *Rht-B1* and *Rht-D1* as independent determinants of thousand-grain weight and kernel number per spike in wheat. Mol Breed 32: 771–783.
72. Rebetzke GJ, Richards RA (2000) Gibberellic acid-sensitive dwarfing genes reduce plant height to increase kernel number and grain yield of wheat. Australian Journal of Agricultural Research 51: 235–246.
73. Zhang B, Shi W, Li W, Chang X, Jing R (2013) Efficacy of pyramiding elite alleles for dynamic development of plant height in common wheat. Mol Breed 32: 327–338.

7

Development and Validation of a 20K Single Nucleotide Polymorphism (SNP) Whole Genome Genotyping Array for Apple (*Malus* × *domestica* Borkh)

Luca Bianco[1], Alessandro Cestaro[1], Daniel James Sargent[1], Elisa Banchi[1], Sophia Derdak[2], Mario Di Guardo[1,3], Silvio Salvi[4], Johannes Jansen[5], Roberto Viola[1], Ivo Gut[2], Francois Laurens[6], David Chagné[7], Riccardo Velasco[1], Eric van de Weg[3], Michela Troggio[1]*

1 Research and Innovation Centre, Fondazione Edmund Mach, San Michele all'Adige, Trento, Italy, **2** CNAG – Centro Nacional de Análisis Genómico, Parc Científic de Barcelona, Barcelona, Spain, **3** Wageningen UR Plant Breeding, Wageningen University and Research Centre, Wageningen, The Netherlands, **4** DipSA - University of Bologna, Bologna, Italy, **5** Biometris, Wageningen University and Research Centre, Wageningen, The Netherlands, **6** INRA, UMR1345 Institut de Recherche en Horticulture and Semences, Beaucouzé, France, **7** Plant & Food Research, Palmerston North Research Centre, Palmerston North, New Zealand

Abstract

High-density SNP arrays for genome-wide assessment of allelic variation have made high resolution genetic characterization of crop germplasm feasible. A medium density array for apple, the IRSC 8 K SNP array, has been successfully developed and used for screens of bi-parental populations. However, the number of robust and well-distributed markers contained on this array was not sufficient to perform genome-wide association analyses in wider germplasm sets, or Pedigree-Based Analysis at high precision, because of rapid decay of linkage disequilibrium. We describe the development of an Illumina Infinium array targeting 20 K SNPs. The SNPs were predicted from re-sequencing data derived from the genomes of 13 *Malus* × *domestica* apple cultivars and one accession belonging to a crab apple species (*M. micromalus*). A pipeline for SNP selection was devised that avoided the pitfalls associated with the inclusion of paralogous sequence variants, supported the construction of robust multi-allelic SNP haploblocks and selected up to 11 entries within narrow genomic regions of ±5 kb, termed focal points (FPs). Broad genome coverage was attained by placing FPs at 1 cM intervals on a consensus genetic map, complementing them with FPs to enrich the ends of each of the chromosomes, and by bridging physical intervals greater than 400 Kbps. The selection also included ~3.7 K validated SNPs from the IRSC 8 K array. The array has already been used in other studies where ~15.8 K SNP markers were mapped with an average of ~6.8 K SNPs per full-sib family. The newly developed array with its high density of polymorphic validated SNPs is expected to be of great utility for Pedigree-Based Analysis and Genomic Selection. It will also be a valuable tool to help dissect the genetic mechanisms controlling important fruit quality traits, and to aid the identification of marker-trait associations suitable for the application of Marker Assisted Selection in apple breeding programs.

Editor: Pankaj Jaiswal, Oregon State University, United States of America

Funding: This work has been funded under the EU seventh Framework Programme by the FruitBreedomics project N°. 265582: Integrated Approach for increasing breeding efficiency in fruit tree crops. The views expressed in this work are the sole responsibility of the authors and do not necessary reflect the views of the European Commission. The funders had no role in study design, data collection and analysis, decision to publish, or preparation of the manuscript.

Competing Interests: The authors have declared that no competing interests exist.

* Email: michela.troggio@fmach.it

Introduction

Cultivated apple (*Malus* × *domestica*) is the most economically important deciduous fruit tree crop worldwide [1]. Breeding of novel cultivars with superior fruit quality characteristics is a slow and costly process, because of the extended juvenility period of the species. Typically, the selection process takes on average more than 20 years from seed to introduction, even when rootstocks conferring precocious flowering are exploited. Breeding for disease resistance takes even longer, requiring introgression of novel resistance genes from wild germplasm, followed by successive generations of back-crossing to restore fruit quality to a commercially acceptable level. The use of molecular markers could accelerate and enhance the breeding process, particularly for traits that are difficult to select for phenotypically, such as pyramided disease resistances, or for traits that are expressed only in mature trees, such as fruit characteristics.

Recent advances in genomics technologies have enabled the sequencing of the 'Golden Delicious' genome [2] and the subsequent development of a whole genome genotyping (WGG) (micro-) array for the species [3]. The International RosBREED Single nucleotide polymorphism (SNP) Consortium (IRSC) WGG array contains 7,867 *Malus* SNP markers as well as 921 SNPs derived from *Pyrus* [4]. The IRSC array greatly facilitated the development of high density linkage maps for segregating apple progenies [5–7]. It was used in a Genome-Wide Association Study (GWAS) of the genetic control of several significant fruit traits [8], for the implementation of Genomic Selection (GS) [9] and for resolving pedigrees [10]. However, these 8,788 potential genetic markers are not sufficient to perform GWAS in wider germplasm sets, or to perform Pedigree-Based Analysis (PBA) [11] with high levels of precision. This is due to the rapid linkage disequilibrium decay in apple [12], the limited proportion of robust, easy to score makers included on the IRSC SNP array [6,10], and their uneven distribution across the genome. A higher density array, with robust genome-wide markers is therefore required to perform such studies successfully.

In this investigation, we describe the development of such a high density WGG array for apple, using a focal point approach and stringent selection criteria built from experience of the analysis of the apple IRSC array [3,5,6]. A haplotype-targeting strategy similar to that adopted to design the IRSC array was implemented for the design of the array to combine information from individual SNPs into haploblocks and provide fully informative multi-allelic markers. We also summarize available metadata on the application of the array to a separate genetic mapping study.

Materials and Methods

SNP Discovery Panel and Re-sequencing

To enable the identification of SNPs, a discovery panel comprising the following 13 apple cultivars, including some of the core European apple breeding founder varieties [13–15], was re-sequenced using short-read sequencing technology: 'Braeburn', 'Cox's Orange Pippin', 'Common Antonovka', 'Delicious', 'Dr. Oldenburg', F2-26829-2-2, 'Fuji', 'Jonathan', 'Lady Williams', 'McIntosh', 'Macoun', 'Priscilla-NL' and 'Worcester Pearmain-USA'. Additionally, a scab-resistant accession of *M. micromalus* and two *M.* × *domestica* double haploid (DH) accessions, X9273 and X9748, which were derived from 'Golden Delicious', were included [16]. Leaf material was procured from various institutions (Table 1). For 'Priscilla-NL' and 'Worcester Pearmain-USA', the country of origin of the leaf material is included, to distinguish them from other genotypes with the same cultivar name [17]. The DHs were included to help identify pseudo-SNPs created from paralogous sequences of the apple genome that are erroneously assembled into a single locus, or that are located at different segments/chromosomes but that are targeted by the same Illumina probes. Since the DH lines are homozygous across their entire genomes, any heterozygous calls in these genotypes were thus considered evidence of paralogous sequences. DNA was extracted from freeze-dried, newly emerged leaf material using a phenol-chloroform isoamyl alcohol extraction method [10] and quantified with a Qubit fluorometer (Life Technologies). Sequencing libraries were constructed according to the TruSeq DNA sample preparation protocol (Illumina) with minor modifications, in particular employing double size selection steps. Two micrograms of genomic DNA were fragmented with a Covaris E210 and size selected to 300–600 bps. The resulting fragments were end-repaired, adenylated and ligated to Illumina paired-end adaptors. The size of the library was confirmed on the BioAnalyzer 2100 (Agilent) and the library was sequenced on an Illumina HiSeq 2000 platform with paired end runs of 2×101+7 bps. Base calling and quality control were performed using the Illumina RTA sequence analysis pipeline.

Read Alignment, Variant Detection and Quality Filtering

A schematic representation of all the steps in the pipeline for array development is given in Figure 1. As the first step, reads were sequentially aligned to the primary assembly and three alternative assemblies of the 'Golden Delicious' genome v2.0 (http://www.rosaceae.org/species/malus/malus_x_domestica/genome_v2.0) allowing up to seven mismatches in a four-step exhaustive alignment procedure using GEM [18] and BFAST [19]. Version 2.0 of the apple genome was created by removing 34,882 problematic contigs from the previous version [2]. The primary assembly, representing about 80% of the assembled and anchored genome, and the three alternative assemblies were produced following the NCBI AGP v2.0 specifications (http://www.ncbi.nlm.nih.gov/projects/genome/assembly/agp/AGP_Specification_v2.0.shtml).

Read pairs that always mapped to a single genomic location were kept in the alignment file for that reference at each step, while unaligned read pairs were submitted to the next mapping round. Pairs that did not align uniquely to any of the references were discarded (Figure 1). As a result, four .bam files were generated for each sample from the four alternative assemblies. Variant calling was performed for each of these .bam files separately, applying samtools and bcftools (version 0.1.18) [20] using default parameters. Positions with significant strand bias (p-value<0.001), significant tail distance bias (p-value<0.05) or within regions of low mapping confidence [21] were excluded. The variant-calling pipeline was set to produce genotype calls for each variant for all the re-sequenced cultivars. Calls with a support value of at least 10 reads were retained in the .vcf file, while the others were considered unreliable and therefore set to empty. To avoid the inclusion of potential paralogy-related SNPs, variants with a heterozygous genotype in the DHs were filtered out. Finally, known ambiguous bases in the primary reference were annotated.

A quality filter was then applied to remove potentially unreliable variants. Custom scripts written in Python (www.python.org) were developed to remove variants with low phred-scaled quality scores (i.e. below 20); a high combined read depth (i.e. higher than 2,000); and a low single-cultivar read depth (i.e. lower than 10) in more than 50% of the cultivars. A minimum phred-scaled quality score of 20 was chosen to ensure that only SNPs with a probability of less than 1% for the alternative allele being called wrongly were selected, and the maximum read depth value was used to ensure the removal of SNPs derived from paralogous regions rather than true heterozygous regions as done previously in the development of the IRSC apple array [3] and the 9 K peach array [22]. Additionally, a cut-off of at least ten reads per single cultivar in more than 50% of the cultivars was used to distinguish real variants from potential sequencing errors. The 50% cut-off was chosen since some cultivars, such as 'Common Antonovka' and F2-26829-2-2, and the accession of *M. micromalus*, were derived from different genetic backgrounds compared to the other accessions, and absence of sequence coverage in those regions could have been due to genomic variation. Finally, all SNPs with an allele frequency (AF) = 1 were discarded since this equated to all the re-sequenced varieties carrying an allele that was different from the 'Golden Delicious' reference genome. Such SNPs were discarded as they were likely to represent potential false SNPs resulting from sequencing errors in the 'Golden Delicious' reference sequence, or rare alleles derived from 'Golden Delicious'. For the same reason, rare alleles derived from other cultivars were discarded during the SNP selection phase described below.

Illumina Specific Filtering and SNP Selection

Quality filtered variants from the pipeline described above were then processed to meet Illumina Infinium II array design requirements (http://res.illumina.com/documents/products/technotes/technote_iselect_design.pdf). This third step removed

Table 1. Origin of samples and re-sequencing statistics for the 14 genotypes of the discovery panel and the two double haploids.

Sample name	Source of leaf material*	Tot read pairs	Mean percent unique reads	Mean coverage
'Dr Oldenburg'	JKI	134,605,846	55.98	43.13
'Fuji'	JKI	116,541,162	56.82	37.95
'Lady Williams'	JKI	115,667,659	56.17	38.12
F2-26829-2-2	UNIBO	110,254,636	55.14	35.52
'Macoun'	JKI	95,279,675	55.35	30.77
'Cox's Orange Pippin'	WUR	98,094,673	56.53	32.61
'Worchester Pearmain-USA' PI 206035	PGRU-Geneva	104,600,260	54.67	34.30
'Jonathan'	WUR	137,126,235	56.00	45.45
Malus micromalus	WUR	117,046,241	50.11	34.65
'McIntosh'	INRA	135,946,435	51.10	44.33
'Delicious'	INRA	139,815,864	57.49	45.88
'Braeburn'	KUL	122,735,693	50.48	39.37
Common Antonovka	VNIIISPK	133,423,835	50.73	43.32
'Priscilla-NL'	WUR	127,017,433	55.02	41.17
X9748	INRA	197,011,579	59.17	66.84
X9273	INRA	207,626,734	58.81	69.70

The 16 genomes were sequenced on 8 Illumina HiSeq 2000 lanes. Double haploids (DHs, acc. X9748 and X9273) feature almost a double coverage compared with the other genomes; the total number of read pairs ranges from 95 Million to 207 Million. The percentage of uniquely aligning reads is evenly spread over each genome and ranges from 50 to 60% of the total. The mean coverage similarly ranges from 30 to 45X, while it is higher for the DHs.
*(JKI) Julius Kühn-Institut; (UNIBO) Università di Bologna; (WUR) Wageningen University and Research centre; (PGRU-Geneva) Plant Genetic Resources Unit; (INRA) Institut National de la Recherche Agronomique; (KUL) Katholieke Universiteit Leuven; (VNIIISPK) The All Russian Research Institute of Horticultural Breeding.

variants that were either indels, A/T or C/G transversions, or tri-allelic SNPs. Additionally, variants located at sites containing additional high quality SNPs in either the up/down-stream 50 bp; or that contained both up-stream and down-stream sequence that appeared multiple times across the genome in the directly flanking 24 bp were removed [6]. Approximately one million SNPs obtained in this way were subsequently submitted to the Illumina Assay Design Tool (http://support.illumina.com/array/array_software/assay_design_tool.ilmn) for a preliminary estimation of the conversion rate (the Illumina SNP_Score). Following Illumina guidelines, markers with a SNP_Score below the threshold of 0.7 were discarded, while others were kept as high quality SNPs for the final selection of the 20,000 targets to be included on the 20 K Infinium array. In addition to these newly identified and validated SNPs from the discovery panel re-sequencing, 3,670 validated SNPs from the previously developed apple IRSC SNP WGG array were added to this selection. The SNPs selected from the IRSC array were chosen as they were robustly positioned in different genetic linkage maps, including those reported in [3] and [5].

The focal point strategy adopted for the design of this array followed that of the IRSC array design [3], with some modifications (Figure 2). The size of focal point intervals was reduced from 50 K bp, to 5 K bp up- or down-stream of the focal point itself to provide greater robustness in haplotype building (fewer recombination events within a focal point), and the number of SNPs within each FP was increased where feasible. Focal points evenly spread across the genome were first identified on the basis of the genetic positions available from previous mapping studies [3,5], which resulted in a total of 718 FPs. A further 1,184 FPs were then added to cover physical intervals of an average of 400 Kbps to reach an average genetic distance of one FP per centiMorgan (cM) (using an expected Kbp/cM ratio of 440 Kbps based on a previous estimate by [2]). Finally, an additional 120 FPs were selected to enrich the ends of each of the 17 *Malus* chromosomes, making the number of FPs selected 2,022. Through a SNP short-listing process, up to 11 markers for each FP were selected. Entries featuring the same genotyping profile across the 14 re-sequenced accessions within a given FP were removed to avoid having too many SNPs at a single point potentially derived from the same haplotype. Additionally, SNPs displaying a percentage of heterozygous genotypes in the re-sequencing panel greater than 65% were also excluded as they were most likely to be the result of paralogy instead of true heterozygosity. In cases where 11 or fewer SNPs remained within a given FP, these were all selected for inclusion on the array. Where more than 11 SNPs remained, a step-wise inclusion procedure was followed. Firstly, up to five SNPs were selected that were polymorphic in 35–60% of the panel members, which generally corresponds to 5–8 members. Their heterozygous profiles had to be complementary as far as possible, and panel members had to be more or less equally represented, thus balancing the number of heterozygous SNPs across all (diploid) panel members. Secondly, this first set of highly polymorphic SNPs was complemented with SNPs that were polymorphic in 10–35% of the panel members, usually corresponding to 2–4 members, following the same principles on complementarity and representation. Thirdly, the remaining positions were filled up with SNPs that were polymorphic in just one panel member. This meant that complementarity within any given FP was accounted for as far as possible. This selection

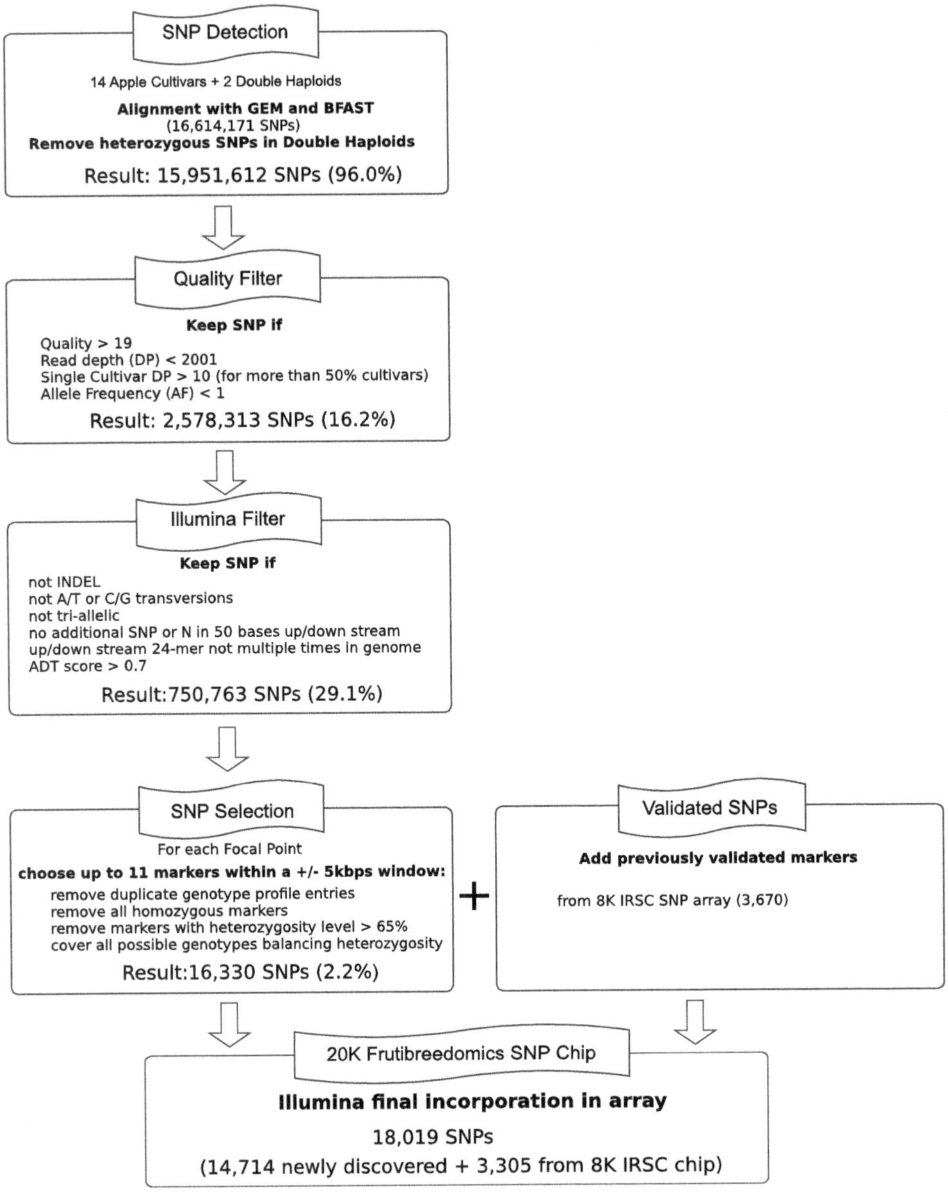

Figure 1. The SNP selection workflow. The SNP selection process was a cascade involving four steps: Detection, Quality Filtering, Illumina specific Filtering and SNP Selection, with the number of SNPs reduced after each step. The specific filtering criteria have been reported for each filtering stage, as well as the number of resulting SNPs, with the corresponding percentage survival, relative to the total number of SNP markers after the previous step. From the 20,000 selected, Illumina successfully incorporated a total of 18,019 SNP probes into the new array, 14,714 of which were newly discovered and 3,305 previously validated.

process was automated through a customized script that is available upon request from the corresponding author. Finally, the single member heterozygotes were manually scrutinized across all FPs, to ensure that FPs in the same region did not have a single heterozygous SNP for the same panel member. Also, *M. micromalus* specific SNPs were maintained at reduced frequency because of the currently limited use of this species in breeding (Figure 2).

SNP Validation

The discovery panel was genotyped with the 20 K SNP array following the standard Illumina protocol detailed in the IRSC apple array and its application papers [3,5]. In the framework of an ongoing PBA-based QTL mapping study of the FruitBreedomics project, over 1,600 seedlings from 21 full-sib families and their direct parents and additional progenitors were also genotyped and used for the construction of genetic maps. Although this work is not part of the current paper, we will present some of its outcomes, as they relate to the performance of the array. The full-sib families on which these data are based are listed in Table S1.

Genotyping data from the discovery panel and the full-sib families were analyzed using the GenomeStudio software (Illumina) with a GenCall threshold of 0.15 and a SNP filtering pipeline [23]. This employs the multipoint maximum likelihood mapping algorithm approach for cross pollinators in JoinMap 4.1 ([24,25] E. Van de Weg, unpublished data).

Genotype calls from the 14 accessions of the discovery panel were compared with those obtained through high-coverage

SNP #	LG	Position (bps)	HET	Calls of diploid discovery panel members														Di-haploids	
				Braeburn'	Common Antonovka'	Cox's Orange Pippin'	Delicious'	Dr. Oldenburg'	F2-26829-2-2	Fuji'	Jonathan'	Lady Williams'	Macoun'	*M.Micromalus*	McIntosh'	Priscilla'	Worcester Pearmain-USA'	X9273	X9748
1	1	633,002	7	0/1	1/1	0/0	0/1	0/0	0/0	0/1	0/0	0/1	1/1	0/1	0/1	0/0	0/1	0/0	0/0
2	1	***629,347***	6	0/1	1/1	0/0	0/1	0/0	0/0	0/1	0/0	0/1	1/1	0/0	0/1	0/0	0/1	0/0	0/0
3	1	**633,660**	6	1/2.	1/1	2/2	1/1	1/2.	2/2	1/2.	2/2	1/2.	1/1	1/1	1/2.	2/2	1/2.	1/1	1/1
4	1	632,163	3	1/1	1/1	1/1	1/1	1/1	1/2.	1/2.	2/2	1/2.	1/1		1/1	2/2	1/1	1/1	1/1
5	1	632,821	1	0/0	0/0	0/0	0/0	0/0	0/0	0/0	0/0	0/0	0/0	0/1	0/0	0/0	0/0	0/0	0/0
6	1	629,904	6	0/1	1/1	0/0	0/1	0/0	0/0	0/1	0/0	0/1	1/1	1/1	0/1	0/0	0/1	0/0	0/0
		Representation	*All*	*4*			*3*	*1*	*1*	*5*		*5*		*2*	*4*		*4*		
		heterozygotes	*Selected*	*3*			*2*	*1*	*1*	*4*		*4*		*2*	*3*		*3*		
7	9	39,002,373	7	1/2.	1/1	1/1	1/2.	1/1	1/2.	1/1	1/1	1/2.	1/2.	1/1	1/2.	1/1	1/2.	2/2	1/1
8	9	39,000,404	5	0/0	1/1		0/1	1/1	0/0		0/1	0/1	0/0	0/0	0/1	1/1	0/1	0/0	0/0
9	9	39,002,203	5	0/0	0/0	0/0	0/1	0/1	0/1	0/1	0/1	0/0	0/0	0/0	0/0	0/0	0/0	0/0	0/0
10	9	39,002,529	5	1/1	2/2	1/1	1/1	1/2.	1/1	1/2.	2/2	1/2.	1/2.	1/1	1/1	1/2.	1/1	1/1	2/2
11	9	39,004,315	5	0/1	0/1	1/1	0/0	0/0	0/1	0/0	1/1	0/0	0/1	0/0	0/0	0/0	0/1	0/0	0/0
12	9	38,999,475	4	0/1	0/0	0/1	0/1	0/1	0/0		0/0	0/0	0/0		0/0	0/0	0/0	0/0	
13	9	39,002,125	4	0/0	0/0	0/1	0/0	0/0	0/0	0/0	0/1	0/0	0/0	0/0	0/1	0/1		0/0	0/0
14	9	39,002,242	4	0/0	0/1	0/0	0/0	0/1	0/0	0/0	0/0	0/1	0/0	0/0	0/0	0/1	0/0	0/0	0/0
15	9	**39,004,972**	4	0/1	0/1	0/1	0/0	0/0	0/0	0/0	1/1	0/0	0/1	0/0	0/0	0/0	0/0	0/0	0/0
16	9	38,999,410	3	0/0	0/0	0/0	0/0	0/0	0/0	0/0	0/0	0/0	0/0		0/1	0/1	0/1	0/0	
17	9	39,001,465	2	1/1	1/1	1/1	1/1	1/1	1/1	1/2.	1/2.	1/1	1/1	2/2	1/1	1/1	1/1	1/1	2/2
18	9	38,999,308	8	0/1	0/1	1/1	0/1	0/1	0/0	1/1	1/1	0/0	0/1		0/1	0/1	0/1	0/0	
19	9	***38,998,668***	6	0/1	1/1	1/1	0/1	1/1	0/0		1/1	0/1	0/1		0/1	1/1	0/1	0/0	
20	9	38,998,717	6	0/1	0/1	1/1	0/1	1/1	0/0	1/1	0/0	0/1	0/0		0/1	1/1	0/1	0/0	
21	9	39,002,445	6	1/2.	1/1	1/1	1/2.	1/1	1/2.	1/1	1/1	1/2.	1/2.	1/1	2/2	1/2.	2/2	2/2	1/1
22	9	39,003,048	6	1/2.	1/2.	1/1	1/2.	1/1	1/2.	1/1	2/2	1/2.	2/2	1/1	1/2.	1/1	1/1	2/2	1/1
23	9	38,999,843	5	0/0	0/1	0/0	0/0	0/1	0/0	0/0	0/0	0/1	0/0	0/0	0/1	1/1	0/1	0/0	
24	9	38,999,981	5	0/1	1/1	1/1	0/1	1/1	0/0	1/1	1/1	0/0	0/1		0/1	1/1	0/1	0/0	1/1
25	9	39,003,120	4	1/2.	1/2.	1/1	1/2.	1/1	2/2	1/1	2/2	1/2.	2/2	1/1	2/2	1/1		2/2	1/1
26	9	39,000,665	2	0/0	0/0		0/0	0/0	0/1		0/0	0/1	0/0		0/0	0/0	0/0	0/0	
27	9	39,004,669	2	0/0	0/0	0/0	0/0	0/0	0/0	0/0	0/0	0/0	0/0	0/0	0/1	0/1	1/1	0/0	0/0
28	9	39,000,300	1	0/0	0/0	0/1	0/0	0/0	0/0	0/0	0/0	0/0	0/0	0/0	0/0	0/0	0/0	0/0	0/0
		Representation	*All*	*11*	*8*	*4*	*11*	*6*	*6*	*3*	*4*	*11*	*8*		*11*	*7*	*9*		
		heterozygotes	*Selected*	*4*	*3*	*3*	*4*	*4*	*3*	*3*	*4*	*4*	*4*		*4*	*4*	*4*		

Figure 2. SNP selection within a single focal point (FP) of chromosomes (Chrs) 1 and 9. Genotype calls are presented for the discovery panel members: 0/1 and 1/2 indicates heterozygosity, all other codes indicate absence of polymorphism within an individual. Green, blue and orange filled cells relate to selected SNPs, having 5–8, 2–4, or just 1 heterozygous panel member/s, respectively. The physical positions at the extremities of the FP are in bold and italics (most proximal) or underlined (most distal). Within each FP, SNPs are sorted firstly by being selected or not, and secondly according to the number of heterozygous members (HET). In the Chr1-FP example, SNP-6 was not selected, as its heterozygosity pattern is identical to that of SNP-2. The single-member-heterozygote SNP-5 was included because no other more polymorphic SNP markers were available to reach the target of 11 SNPs and no nearby FPs had a specific SNP for the same panel member. In the Chr9-FP example all highly specific SNPs were ignored, as there were sufficient markers of higher heterozygosity available. The selected SNPs show a homogeneous representation of the diploid panel members.

Illumina re-sequencing. Genotyping data were exported from GenomeStudio genotyping report with the option "Forward Strand" for consistency with the re-sequencing strand. Regarding re-sequencing data, only calls with a support of at least 10 reads were extracted from the .vcf file and used for the comparison.

Results

SNP Detection and Filtering

Sequencing resulted in 3,376 million reads from the 14 accessions (13 *M.* × *domestica* apple cultivars and one accession of *M. micromalus*), and an additional 809 million reads from the two DH accessions (http://bioinformatics.tecnoparco.org/fruitbreedomics/node/2). Table 1 details the depth of sequencing and the estimated genome coverage obtained for each of the 14 genotypes of the discovery panel. Variant calling resulted in a variant call format (.vcf) file containing 16,614,171 variants including indels, derived from all the re-sequenced accessions (Figure 1). A total of 662,559 variants were found to be heterozygous in one or both of the two DH accessions re-sequenced and were removed from further analysis. Thus 15,951,612 variants were retained after this filtering step. Following quality filtering described in the paragraph "Read Alignment, Variant Detection and Quality Filtering" of the Methods section, 13,373,299 variants were removed from further analysis, leaving 2,578,313 variants.

SNP Selection

When quality filtered variants were processed to meet the Illumina Infinium II array requirements, a further 1,719,293 variants were removed, leaving 859,020 SNPs that were submitted to the Illumina Assay Design Tool pipeline. Following the Illumina recommendations and the parameters used by [3], 108,257 SNPs with SNP_Score below 0.7 were discarded leaving 750,763 high quality SNPs from the discovery panel for the selection of targets to be included on the array.

Using the focal point approach detailed in the methods, 16,330 SNPs were identified from the re-sequencing of 14 genotypes of the discovery panel, as well as 3,670 validated markers from a previous IRSC array [3] and submitted to Illumina for array production. The Infinium array manufacturing produced a total of 18,019 SNP probes (14,714 newly identified SNPs and 3,305 from the IRSC array) to be incorporated in the final array. A total of 15,669 SNPs were located in 2,019 focal points (Figure 3). This included 955 SNPs from the IRSC array, which meant that some FPs contained more than 11 SNPs. The remaining 2,350 SNPs from the IRSC array did not fall in a FP and their physical position has been highlighted in Figure 3. The number of SNPs per FP ranged from 1 to 15, with a mean of 7.7. The mean distance between focal points was 311 Kbps, with two regions larger than 1 Mbps at the distal end of chromosomes 9 and 13. Overall, most regions that were not effectively covered by the IRSC array are now well represented by newly designed markers.

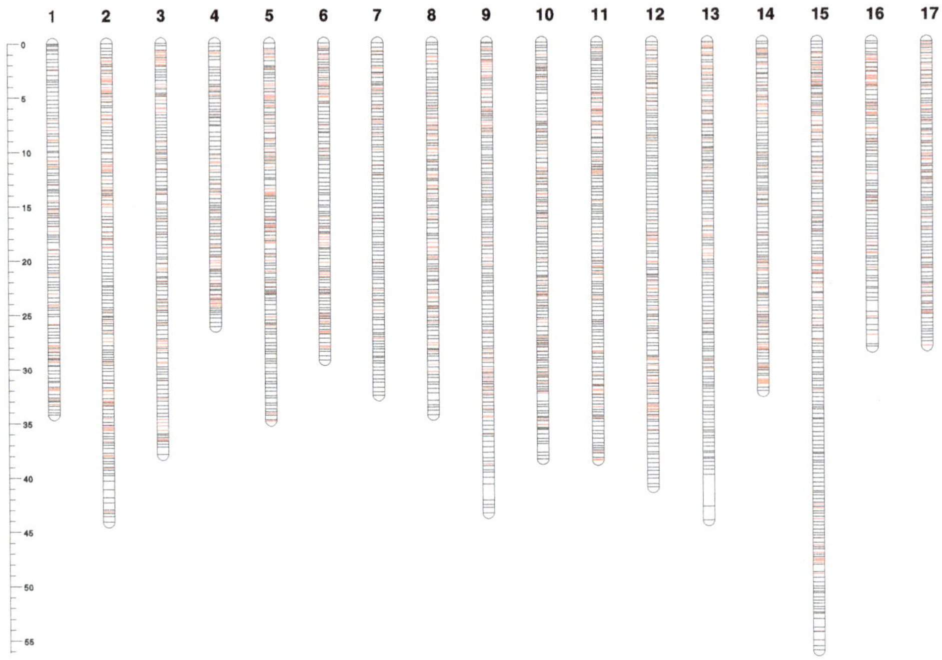

Figure 3. Distribution of focal points (FPs) (black lines) and IRSC SNPs (red lines) in the apple physical map v2. All positions are in Mbps. The average distance between FPs is 311 Kbps. Only two regions longer than 1 Mbps located at the distal end of Chrs 9 and 13 are not covered by FPs and SNPs from the 8 K IRSC SNP array. Scale bars = 5 Mbps.

SNP Mapping and Validation

Within the framework of the FruitBreedomics project, 21 full sib families were SNP genotyped, using the 20 K array, and were used for the generation of linkage maps. This resulted in the genetic mapping of 15.8 K SNP markers, and included 12,611 (success rate 86%) newly developed SNPs and 3,160 (success rate 96%) from the IRSC array that were informative in this new array and were mapped in at least one of the full-sib families screened (Table S2, Dataset S1). Of the non-mapped markers, 271 currently true monomorphic SNP markers may be mapped once screened over a wider germplasm set, as they are based on a single polymorphism in one of the more specific discovery panel members (e.g. *M. micromalus* or 'Common Antonovka'). Additionally, 747 SNPs that featured complex cluster patterns (suggesting the probes annealed to paralogous genomic regions or contained additional SNPs within the probe binding sites) were found informative by visual inspection but would have required additional elaboration on the calling pipeline or manual annotation as previously reported in [6].

The number of SNPs successfully mapped for each of the 21 full-sib families varied from ~5.2 K to ~8.5 K. Examples of a GenomeStudio cluster for a robust mapped SNP and a failed/difficult to score SNP are shown in Figure 4.

Genotype Call Comparison

The concordance of genotype calls made through re-sequencing with those obtained from genotyping with the 20 K array was assessed for the set of 15,771 validated SNPs. Excluding missing data, on average 12,347 comparisons were made for each of the 14 accessions (min: 10,402 for *M. micromalus*; max: 12,647 for 'Fuji') and the average concordance was 95%. In the majority of cases, the genotype call discordance identified was for low frequency SNPs observed in just one or two accessions of the discovery panel. This could have been the result of unreliable genotype calls in the re-sequencing analysis, owing to a lower read depth in those specific regions, or it might have been caused by unreliable genotype calls made in the array analysis owing to the presence of additional SNPs in probe binding sequences.

Discussion

The outcome of our present work, a 20 K WGG array for *M. × domestica* using Illumina Infinium technology, comprising a total of 18,019 SNPs, has greatly increased the genotyping and analysis opportunities for apple researchers and breeders. High throughput array-based genotyping has revolutionized the study of genome-wide genetic variation, reducing costs and increasing the reliability and efficiency of data produced, as well as significantly reducing the time spent on genotyping itself. Medium or high density arrays are available for a range of other crop plant species, including cherry [26], grapevine [27], maize [28], peach [22], potato [29], soybean [30], tomato [31], sorghum [32], white spruce [33], alfalfa [34], and rice [35]. Of our 18,019 apple SNPs, 88% were validated in 21 full-sib families, with a further reservoir of a thousand of potentially informative SNPs if a wider germplasm base had been surveyed or if the remaining markers had been called manually. An efficiency rate of 88% corresponds well with efficiencies calculated for arrays developed for other plant species such as peach (84.3%; [22]), and is significantly higher than that of the previous *Malus* IRSC array (72%) [3]. Taking into account the polyploid origin of apple, the efficiency rate of 88% is the highest reached to date for a species that has undergone a recent whole genome duplication (WGD). This increased array efficiency was achieved through the implementation of a novel SNP detection and filtering pipeline in the design of the array, together with a new SNP calling pipeline for the use of the array [23] and the screening of a relatively large number of mapping populations. The new SNP calling pipeline successfully avoided the problem of extensive genome paralogy in apple [2], which caused calling

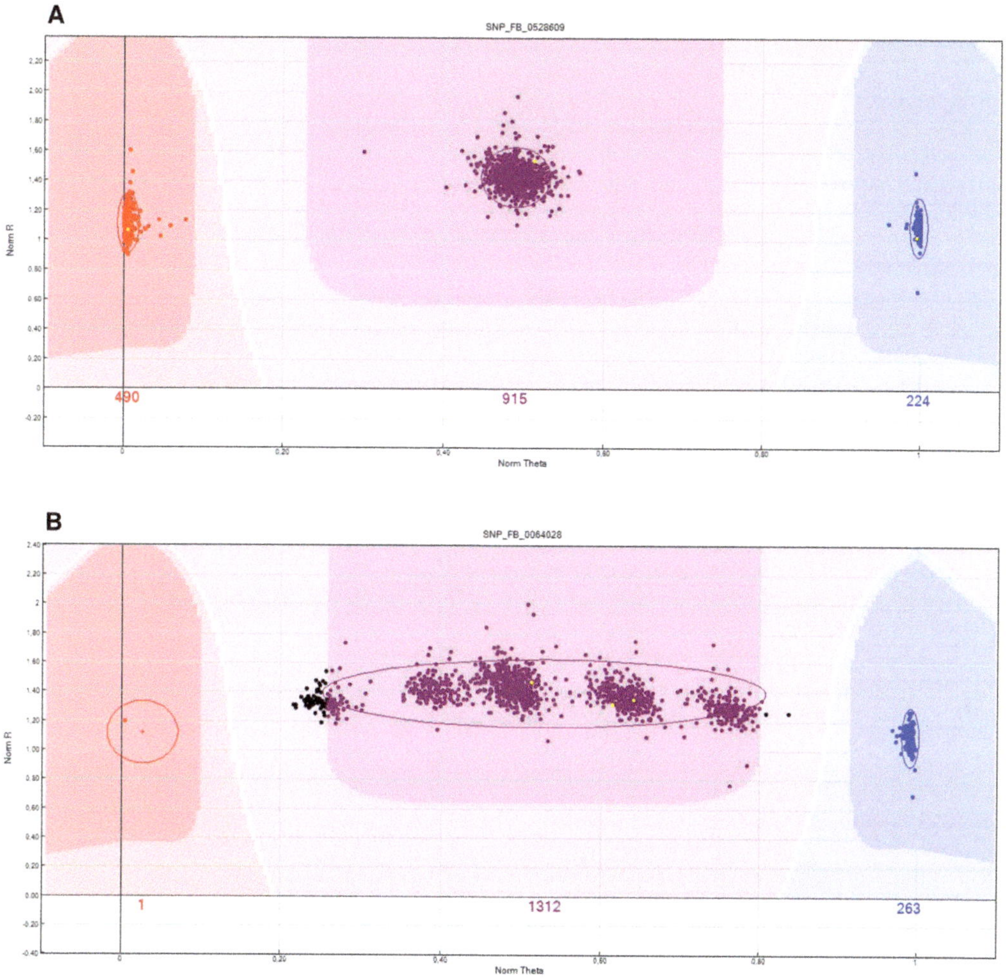

Figure 4. GenomeStudio cluster plot for the 21 F_1 full-sib families used for the construction of genetic maps. Yellow dots indicate panel members that were re-sequenced and used as parents. Genotypes are called for each sample (dot) by their signal intensity (norm R) and Allele Frequency (Norm Theta) relative to canonical cluster positions (dark shading) for a given SNP marker (red = AA, purple = AB, blue = BB) a) GenomeStudio cluster plot of a newly developed robust SNP marker. b) GenomeStudio cluster plot of a failed/difficult to score SNP marker.

difficulties using the IRSC apple array [6]. We employed a more stringent probe design by removing regions that showed multiple matches in the genome of the 24 nucleotides from either side of the SNP. Moreover, more stringent SNP filtering was achieved by including two DH accessions as discovery panel members and removing heterozygous SNPs that were identified in either of those genotypes during the SNP selection process. Finally, the high coverage re-sequencing strategy applied in this work allowed genotypes to be called for most panel members, and a minimum of ten reads per variant was imposed to successfully call a genotype. The comparison of sequenced reads for a potential variant site across multiple samples has the potential to differentiate systematic sequencing errors from real SNPs [36]. Such an approach permitted the detection of variant carriers without any need for complicated bioinformatics algorithms such as the one reported by [37], and reduced the number of false positive SNPs included in the array design. Comparison of read alignments across multiple samples also has the potential to filter out SNPs that are an artifact of inaccurate read alignments [36]. Moreover, it allowed the selection of sets of both more widely and more narrowly polymorphic markers that together represented the diversity of the discovery panel. This combination is expected to be useful for the tracing of markers (SNP haploblocks) along pedigrees for several successive generations. We did not include many SNPs of low heterozygosity (MAF) as they would have low probability of heterozygosity, and thus a low probability of being informative. As a result of these measures, the number of true monomorphic SNPs contained on the final array was very low, approximately 1% of the total. The use of this array to screen the FruitBreedomics families (Table S1) demonstrated the robust performance of the array, as the maps constructed had an average of 6.8 K SNPs uniformly distributed along the genome of each parent.

The final set of 2,578,313 high quality SNPs detected using the 14 re-sequenced accessions of the discovery panel corresponds to an average number of 4.8 SNPs/1,000 bp, in the 530 Mbps assembled and anchored 'Golden Delicious' reference genome sequence. This value was similar to the value reported in the apple genome paper [2] (4.4 SNPs/1,000 bp) but is somewhat higher than that found when surveying the polymorphism rate within a set of *M.* × *domestica* cultivars [38] (3.8 SNPs/1,000 bp), where analyses were limited to genic regions. Since, in contrast to what was done for the IRSC apple array [3], SNPs in this study were not selected exclusively from coding regions of the 'Golden Delicious' reference genome, the SNP heterozygosity rate reported here probably more closely represents the real genome-wide heterozygosity rate of *M.* × *domestica*. However, the actual SNP density may still be higher, as some of the filtered-out SNPs may actually be true polymorphisms.

Other methods have been developed recently for high-throughput genotyping of eukaryotic genomes using short-read sequencing technologies [39–41], and these techniques have been employed in the development of linkage maps and the identification of markers linked to agronomic traits in plants. Such studies include, for instance, genotyping by sequencing (GBS) based genetic maps of *Rubus idaeus* [42], *Hordeum vulgare* [43,44], and restriction-site associated DNA (RAD) sequencing in *Lolium perenne* [41], *Hordeum vulgare* [45] and *Lupinus angustifolius* [46]. These techniques employ reduced genome representation achieved through restriction enzyme digestion and subsequent PCR analysis from adapted linker sequences, and require no *a priori* knowledge of the SNPs being interrogated, making them useful for genetic analysis in species where no reference sequence is currently available. In addition, since up to 96 samples can be multiplexed in a single lane of Illumina HiSeq sequencing, the genotyping cost per sample can be as low as USD12.00 at the time of writing (http://www.igd.cornell.edu/index.cfm/page/GBS/GBSpricing.htm). However, despite clear advantages to the use of GBS under certain experimental conditions for under-resourced species, GBS datasets contain a large proportion of missing values and false homozygote calls, due to low, and uneven genome coverage among individuals (36% in the case of *Rubus*; [42]), and hence data imputation strategies are required for effective data analysis. Moreover sampling chromosomes based on restriction digestion may introduce a bias in allele frequency estimation due to polymorphisms in restriction sites [47]. Whilst the cost of consumables for genotyping per individual is higher employing WGG arrays than for GBS, and only previously characterized SNPs that are present on the array can be interrogated, the data produced are robust and reliable, typically containing almost no missing values. Moreover when a robust and efficient SNP calling pipeline is available, there is less need for bioinformatic capacity, reducing the time and cost associated with data analysis. Coupled with the cost-effective production of arrays such as the one described in this investigation, containing nearly 16,000 validated SNPs, genetic analysis using WGG arrays provides tangible advantages over GBS, particularly for PBA, where genotyping is employed in related germplasm, and variation in hybridization efficiency and indel related null-alleles can be accounted for.

Concluding remarks

We have developed a high-throughput WGG array for apple containing over 16 K validated SNP markers, spanning the apple genome at over 2,000 focal points evenly distributed throughout the 17 chromosomes of *M.* × *domestica*. Through the use of a novel SNP selection strategy informed through the design of the IRSC WGG array [3] and validation of data generated with that array [5,6,10], we have increased the robustness of the markers contained on the array, and reduced the occurrence of monomorphic SNPs and those that display cluster patterns indicating binding to paralogous loci or binding sites containing additional SNPs. The newly developed array, with its high density of polymorphic, validated SNPs, and its suitability for building multi-allelic SNP haploblocks, is expected to be of great utility for pedigree-based analysis, genomic selection and population genetics studies in *M.* × *domestica*. Moreover, the approach in SNP filtering and array design may be of use for the development of much higher density arrays for genome-wide association studies.

Supporting Information

Table S1 Full-sib families screened with the 20 K SNP array. Parents and number of seedlings are listed. Pedigrees of the X-numbered accessions are reported by [11].

Table S2 List of the 18,019 SNPs included in the 20 K SNP array. For each SNP, the physical location (Chromosome, Position), minor allele frequency (MAF), validation in 21 full-sib families, SNP genotype detected by the Illumina array, Illumina strand, source strand, source sequence, probe sequence, genotype calls (gt) made through re-sequencing and no. of high-quality bases (dp) for each discovery panel member, are reported.

Acknowledgments

We wish to thank: INRA-Angers for providing us with the DNA of the two double haploids, Yolanda Noordijk, Andreas Peil (JKI), Anna Pikunova (VNIIISPK), Gennaro Fazio (GPRU), Mark Davey (KUL), and Stefano Tartarini (UNIBO) for providing leaf material for the 14 genotypes of the discovery panel, Yolanda Noordijk (DLO) for preparing the DNA samples, Diego Micheletti, Paolo Fontana, Pietro Franceschi and Ross Crowhurst for the useful discussions on data analysis, Sue Gardiner for manuscript editing, and the International Doctoral Programme in the Genomics and Molecular Physiology of Fruit Plants (GMPF) and the Graduate School of Experimental Plant Sciences of Wageningen University (The Netherlands) for a fellowship awarded to Mario di Guardo.

Author Contributions

Conceived and designed the experiments: R. Velasco SS MT LB EVDW FL. Performed the experiments: SD IG EB. Analyzed the data: AC LB MT EVDW MDG DJS JJ. Contributed reagents/materials/analysis tools: IG SD. Contributed to the writing of the manuscript: MT LB DJS EVDW DC R. Viola.

References

1. Shulaev V, Korban SS, Sosinski B, Abbott AG, Aldwinckle HS, et al. (2008) Multiple models for Rosaceae genomics. Plant Physiol 47(3): 985–1003.
2. Velasco R, Zharkikh A, Affourtit J, Dhingra A, Cestaro A, et al. (2010) The genome of the domesticated apple (*Malus x domestica* Borkh.). Nat Genet 42(10): 833–839.
3. Chagne D, Crowhurst RN, Troggio M, Davey MW, Gilmore B, et al. (2012) Genome-Wide SNP Detection, Validation, and Development of an 8K SNP Array for Apple. PLoS One 7(2): e31745.
4. Montanari S, Saeed M, Knaebel M, Kim Y, Troggio M, et al (2013) Identification of *Pyrus* Single Nucleotide Polymorphisms (SNPs) and Evaluation for Genetic Mapping in European Pear and Interspecific Pyrus Hybrids. PLoS One 8(10): e77022.
5. Antanaviciute L, Fernandez-Fernandez F, Jansen J, Banchi E, Evans KM, et al. (2012): Development of a dense SNP-based linkage map of an apple rootstock progeny using the *Malus* Infinium whole genome genotyping array. BMC Genomics 13: 203.
6. Troggio M, Surbanovski N, Bianco L, Moretto M, Giongo L, et al. (2013) Evaluation of SNP Data from the Malus Infinium Array Identifies Challenges for Genetic Analysis of Complex Genomes of Polyploid Origin. PLoS One 8(6): e67407.
7. Clark MD, Schmitz CA, Rosyara UR, Luby JJ, Bradeen JM (2014) A consensus'Honeycrisp' apple (*Malus* × *domestica*) genetic linkage map from three full-sib progeny populations. Tree Genet Genomes DOI 10.1007/s11295-014-0709-1.

8. Kumar S, Garrick DJ, Bink MCAM, Whitworth C, Chagne D, Volz RK (2013) Novel genomic approaches unravel genetic architecture of complex traits in apple. BMC Genomics 14: 393.
9. Kumar S, Chagne D, Bink MCAM, Volz RK, Whitworth C, et al. (2012) Genomic Selection for Fruit Quality Traits in Apple (*Malus x domestica* Borkh.). PLoS One 7(5): e36674.
10. Pikunova A, Madduri M, Sedov E, Noordijk Y, Peil A, et al. (2014) 'Schmidt's Antonovka' is identical to 'Common Antonovka', an apple cultivar widely used in Russia in the breeding for biotic and abiotic stresses. Tree Genet Genomes 10: 261–271.
11. Bink M, J J, Madduri M, Voorrips R, Durel C, Kouassi A, et al. (2014) Bayesian QTL analyses using pedigreed families of an outcrossing species, with application to fruit firmness in apple. Theor Appl Genet 127: 1073–1090.
12. Khan MA, Korban SS (2012) Association mapping in forest trees and fruit crops. J Exp Bot 63(11): 4045–4060.
13. Noiton DAM, Alspach PA (1996) Founding clones, inbreeding, coancestry, and status number of modern apple cultivars. J Am Socr Hortic Sci 121(5): 773–782.
14. Durel CE, Laurens F, Fouillet A, Lespinasse Y (1998) Utilization of pedigree information to estimate genetic parameters from large unbalanced data sets in apple. Theor Appl Genet 96(8): 1077–1085.
15. Salvi S, Micheletti D, Magnago P, Fontanari M, Viola R, et al. (2014) One-step reconstruction of multigeneration pedigree networks in apple (*Malus* × *domestica* Borkh.) and the parentage of Golden Delicious. Mol Breed DOI 10.1007/s11032-014-0054-y.
16. Lespinasse Y, Bouvier L, Djulbic M, Chevreau E (1999) Haploidy in apple and pear. Acta Horticult 484: 49–54.
17. Evans KM, Patocchi A, Rezzonico F, Mathis F, Durel CE, et al. (2011) Genotyping of pedigreed apple breeding material with a genome-covering set of SSRs: trueness-to-type of cultivars and their parentages. Mol Breed 28(4): 535–547.
18. Marco-Sola S, Sammeth M, Guigo R, Ribeca P, (2012) The GEM mapper: fast, accurate and versatile alignment by filtration. Nat Meth 9(12): 1185–1188.
19. Homer N, Merriman B, Nelson SF (2009) BFAST: An Alignment Tool for Large Scale Genome Resequencing. PLoS One 4(11): e7767.
20. Li H, Handsaker B, Wysoker A, Fennell T, Ruan J, et al. (2009) Genome Project Data P: The Sequence Alignment/Map format and SAMtools. Bioinformatics 25(16): 2078–2079.
21. Derrien T, Estelle J, Marco Sola S, Knowles DG, Raineri E, et al. (2012). Fast Computation and Applications of Genome Mappability. PLoS One 7(1): e30377.
22. Verde I, Bassil N, Scalabrin S, Gilmore B, Lawley CT, et al. (2012) Development and Evaluation of a 9K SNP Array for Peach by Internationally Coordinated SNP Detection and Validation in Breeding Germplasm. PLoS One 7(4): e35668.
23. Van de Weg W, Di Guardo M, Koehorst-Putten H, Longhi S, Noordijk Y, et al. (2014) A Pipeline for Robust Marker Calling from Infinium SNP Arrays for Diploid Crops. In: International Plant & Animal Genome XXII, San Diego, California, U.S.A.
24. Van Ooijen J (2006) JoinMap® 4, Software for the calculation of genetic linkage maps in experimental populations. In. Wageningen, Netherlands: Kyazma B.V.
25. Van Ooijen JW (2011) Multipoint maximum likelihood mapping in a full-sib family of an outbreeding species. Genet Res 93(5): 343–349.
26. Peace C, Bassil N, Main D, Ficklin S, Rosyara UR, et al. (2012) Development and Evaluation of a Genome-Wide 6K SNP Array for Diploid Sweet Cherry and Tetraploid Sour Cherry. PLoS One 7(12): e48305.
27. Myles S, Chia J-M, Hurwitz B, Simon C, Zhong GY, et al. (2010) Rapid Genomic Characterization of the Genus *Vitis*. PLoS One 5(1): e8219.
28. Ganal MW, Durstewitz G, Polley A, Berard A, Buckler ES, et al. (2011) A Large Maize (Zea mays L.) SNP Genotyping Array: Development and Germplasm Genotyping, and Genetic Mapping to Compare with the B73 Reference Genome. PLoS One 6(12): e28334.
29. Felcher KJ, Coombs JJ, Massa AN, Hansey CN, Hamilton JP, et al. (2012) Integration of Two Diploid Potato Linkage Maps with the Potato Genome Sequence. PLoS One 7(4): e36347.
30. Song Q, Hyten DL, Jia G, Quigley CV, Fickus EW, et al. (2013) Development and Evaluation of SoySNP50K, a High-Density Genotyping Array for Soybean. PLoS One 8(1): e54985.
31. Sim S-C, Durstewitz G, Plieske J, Wieseke R, Ganal MW, et al. (2012) Development of a Large SNP Genotyping Array and Generation of High-Density Genetic Maps in Tomato. PLoS One 7(7): e40563.
32. Bekele WA, Wieckhorst S, Friedt W, Snowdon RJ (2013) High-throughput genomics in sorghum: from whole-genome resequencing to a SNP screening array. Plant Biotech J 11(9): 1112–1125.
33. Pavy N, Gagnon F, Rigault P, Blais S, Deschenes A, et al. (2013) Development of high-density SNP genotyping arrays for white spruce (*Picea glauca*) and transferability to subtropical and nordic congeners. Mol Ecol Resour 13(2): 324–336.
34. Li X, Han Y, Wei Y, Acharya A, Farmer AD, et al. (2014) Development of an Alfalfa SNP Array and Its Use to Evaluate Patterns of Population Structure and Linkage Disequilibrium. PLoS One 9(1): e84329.
35. Chen H, Xie W, He H, Yu H, Chen W, et al. (2014) A high-density SNP genotyping array for rice biology and molecular breeding. Mol Plant 7(3): 541–553.
36. Bansal V, Harismendy O, Tewhey R, Murray SS, Schork NJ, et al. (2010) Accurate detection and genotyping of SNPs utilizing population sequencing data. Genome Res 20(4): 537–545.
37. Wang Z, Cai H, Bai X, Ji W, Li Y, et al. (2012) Isolation of GsGST19 from Glycine soja and analysis of saline-alkaline tolerance for transgenic *Medicago sativa*. Acta Agron Sin 38(6): 971–979.
38. Micheletti D, Troggio M, Zharkikh A, Costa F, Malnoy M, et al. (2011) Genetic diversity of the genus *Malus* and implications for linkage mapping with SNPs. Tree Genet Genomes 7(4): 857–868.
39. Baird NA, Etter PD, Atwood TS, Currey MC, Shiver AL, et al. (2008) Rapid SNP Discovery and Genetic Mapping Using Sequenced RAD Markers. PLoS One 3(10): e3376.
40. Elshire RJ, Glaubitz JC, Sun Q, Poland JA, Kawamoto K, et al. (2011) A Robust, Simple Genotyping-by-Sequencing (GBS) Approach for High Diversity Species. PLoS One 6(5): e19379.
41. Pfender WF, Saha MC, Johnson EA, Slabaugh MB (2011) Mapping with RAD (restriction-site associated DNA) markers to rapidly identify QTL for stem rust resistance in *Lolium perenne*. Theor Appl Genet 122(8): 1467–1480.
42. Ward JA, Bhangoo J, Fernández-Fernández F, Moore P, Swanson JD, et al. (2013) Saturated linkage map construction in *Rubus idaeus* using genotyping by sequencing and genome-independent imputation. BMC Genomics 14: 2.
43. Liu H, Bayer M, Druka A, Russell JR, Hackett CA, et al. (2014) An evaluation of genotyping by sequencing (GBS) to map the Breviaristatum-e (ari-e) locus in cultivated barley. BMC Genomics 15: 104.
44. Poland JA, Brown PJ, Sorrells ME, Jannink J-L (2012) Development of High-Density Genetic Maps for Barley and Wheat Using a Novel Two-Enzyme Genotyping-by-Sequencing Approach. PLoS One 7(2): e32253.
45. Chutimanitsakun Y, Nipper RW, Cuesta-Marcos A, Cistue L, Corey A, et al. (2011) Construction and application for QTL analysis of a Restriction Site Associated DNA (RAD) linkage map in barley. BMC Genomics 12: 4.
46. Yang H, Tao Y, Zheng Z, Zhang Q, Zhou G, et al. (2013) Draft Genome Sequence, and a Sequence-Defined Genetic Linkage Map of the Legume Crop Species *Lupinus angustifolius* L. PLoS One 8(5): e64799.
47. Arnold B, Corbett-Detig RB, Hartl D, Bomblies K (2013) RADseq underestimates diversity and introduces genealogical biases due to nonrandom haplotype sampling. Mol Ecol 22(11): 3179–3190.

8

Genetic Diversity and Evolution of Satellite RNAs Associated with the Bamboo Mosaic Virus

Ing-Nang Wang[1,2], Chung-Chi Hu[3], Ching-Wei Lee[3], Sih-Min Yen[3], Wen-Bing Yeh[4], Yau-Heiu Hsu[3], Na-Sheng Lin[1]*

1 Institute of Plant and Microbial Biology, Academia Sinica, Taipei, Taiwan, Republic of China, **2** Department of Biological Sciences, University at Albany, Albany, New York, United States of America, **3** Graduate Institute of Biotechnology, National Chung Hsing University, Taichung, Taiwan, Republic of China, **4** Department of Entomology, National Chung Hsin University, Taichung, Taiwan, Republic of China

Abstract

Satellite RNAs (satRNAs) are subviral agents that depend on cognate helper viruses for genome replication and encapsidation. Their negative impacts on helper viruses have been exploited to control plant viral diseases. SatBaMV is a commonly found satRNA associated with *Bamboo mosaic virus* (BaMV) that infects diverse bamboo species in the field. To investigate the genetic diversity and evolution of satRNAs, we examined seven satBaMV populations derived from five bamboo species and cultivars from Taiwan, China, and India and one from the greenhouse. We found 3 distinct clades among the seven populations. Clade I is consisted of all satBaMV isolates, except for those from *Dendrocalamus latiflorus* in Taiwan and *Bambusa vulgaris* in India, which belong to Clades II and III, respectively. Interestingly, nucleotide diversity was lower for Clade I than II and III. However, the nucleotide diversity did not seem to depend on bamboo species or geographic location. Our population genetic analyses revealed the presence of excessive low-frequency polymorphic sites, which suggests that the satBaMV population was under purifying selection and/or population expansion. Further analysis of *P20*, the only satBaMV gene that encodes a non-structural protein involved in the long-distance movement of satBaMV, showed evidence of purifying selection. Taken together, our results suggest that purifying selection against defective P20 protein is responsible at least in part for the evolution of the satBaMV genome.

Editor: A.L.N. Rao, University of California, Riverside, United States of America

Funding: This study was supported by grants from the Institute of Plant and Microbial Biology (IPMB), Academia Sinica, Taiwan, to NSL. INW received funding from the National Science Council, Taiwan (NSC 101-2811-B-001-024), IPMB, and Academia Sinica during his sabbatical visit. The funders had no role in study design, data collection and analysis, decision to publish, or preparation of the manuscript.

Competing Interests: The authors have declared that no competing interests exist.

* Email: nslin@sinica.edu.tw

Introduction

One of the most striking and under-appreciated differences between plant and animal viruses is the predominant association of satellite RNAs (satRNAs) with the plant viruses [1–3]. The genomes of satRNAs are small, usually <1,500 nucleotides (nt) [1]. Their sequences do not reveal any recognizable open reading frame (ORF) or show only a single ORF encoding a small protein. Because of their small genome size, satRNAs do not encode their own RNA-dependent RNA polymerases (RdRp) or coat proteins (CPs), so they depend on their cognate helper viruses for genome replication and encapsidation [1–5]. Therefore, satRNAs are commonly seen as molecular parasites [3,6] engaging in a zero-sum game by exploiting vital functions provided by their helper viruses.

One manifestation of this antagonistic interaction between satRNAs and their helper viruses is the attenuation of disease symptoms, presumably as a result of reduced viral titers in the host plants. Engineered satRNAs exploit this interfering capacity as antiviral agents for crop protection [7–10]. However, empirical studies have shown that the effects of satRNAs on their helper viruses can run the full gamut of all possible interactions, from antagonistic to beneficial [2,11], resulting in a wide range of disease symptoms [12]. One contributing factor to such variable interactions may be the fascinating interplay among three parties: the satRNA, the helper virus, and the host defense mechanisms, most notably the RNA silencing pathway [1].

Besides the satRNAs associated with *Cucumber mosaic virus* (CMV) and *Turnip crinkle virus* [13], satRNAs associated with the *Bamboo mosaic virus* (satBaMVs) are another of the intensively studied model systems. BaMV was first discovered in Brazil [14]. However, the most extensively studied isolates were first isolated from infected green bamboos (*Bambusa oldhamii* Munro) in Taiwan [15]. BaMV is a plant potexvirus with a filamentous, flexuous morphology of approximately 490 by 15 nm. The BaMV-O genome is a positive-stranded RNA of 6366 nt, with 5 ORFs [16]. The genome for satBaMV is 836 nt (without the poly-A tail) and encodes a single gene product of 183 amino acid residues [17] with similarity to the CP encoded by the satellite virus of *Panicum mosaic virus* [18]. The encoded 20-kDa protein is not required for satBaMV replication [19], but it does preferentially bind to the satBaMV RNA [20], presumably participating in the regulation of systemic movement of satBaMV in the host plant [19,21,22].

Because the replication of satBaMV depends on the RdRp encoded by BaMV, both the 5′ [23,24] and 3′ [25] untranslated regions (UTRs) of the satBaMV and BaMV RNAs, not surprisingly, share structural similarities. Therefore, competition for the limited intracellular resources (*e.g.*, RdRp) can lead to an antagonistic interaction between these two entities, thus resulting in reduced BaMV titers and consequently attenuated symptoms [26–28].

However, not all satBaMV isolates behave antagonistically toward their helper virus. For example, the satBaMV isolate BSL6, first isolated from Ma bamboo (*Dendrocalamus latiflorus* Munro) [29], strongly interferes with BaMV replication [30], whereas BSF4, first isolated from *B. vulgaris* [17], does not seem to have a significant effect on BaMV [19,30]. Detailed genetic studies showed that various mutations at the 5′ apical hairpin stem loop (AHSL) region affect the ability of satBaMV to interfere with the replication of its helper virus [27,28]. A single nucleotide change of U to C at position 82 within the AHSL region of BSF4 can turn the non-interfering BSF4 into an interfering satRNA [28]. Therefore, the interfering phenotype is quite malleable and, presumably, is easily accessible via mutation.

In contrast to many detailed molecular studies, evolutionary and population studies of satRNAs are not as broad or as in-depth. The best-studied example is the various satRNAs associated with CMV [31–34]. At present, no systematic population study has investigated satBaMV, except for the previous surveys of field samples of bamboo species in Taiwan [23,29].

In this study, we collected, sequenced, and analyzed by phylogenetic and population genetics analyses of satBaMV isolates from five bamboo species and cultivars in three main geographic locations in Asia. We identified three distinct satBaMV clades and inferred the evolutionary forces responsible for the observed variations in our collection of satBaMV sequences.

Materials and Methods

Sample collection and preparation

Leaf samples showing visible mosaic symptoms from various bamboo species were collected from locations in northern (*B, vulgaris* Schrad. *ex* J. C. Wendl. and *D. latiflorus* Munro at the Taipei Botanical Garden; GPS coordinates: 24.91571, 121.67394), central (*B. oldhamii* Munro at Chung Hsing Lake of National Chung Hsing University at Taichung; GPS coordinates: 24.23321, 120.94174) and southern (*D. latiflorus* Munro cv. Mei-nung W. C. Lin at Pingtung; GPS coordinates: 22.67611, 120.49417) Taiwan in 2006; Hainan Island of China (*B. ventricosa* McClure; GPS coordinates: 19.00000, 109.50000) in 2007; and Delhi, India (*B. vulgaris* Schrad. *ex* J. C. Wendl.; GPS coordinates: 28.63531, 77.22496) in 2009. The presence of BaMV and satBaMV was checked by dot or northern blot analysis with BaMV and satBaMV-specific probes, respectively [19,35]. To ensure a fair representation of the satBaMV population, leaves from several individual bamboo plants in a given location were combined for total RNA extraction (see below). Figure 1 shows the locations of leaf samples and name designations for satBaMV populations.

To establish a baseline evolution in the greenhouse condition, *Nicotiana benthamiana* leaves were co-infected with 0.1 μg BaMV-S transcripts [36] and 0.1 μg BSF4 satBaMV RNA transcripts [19] obtained by T7 RNA polymerase (Promega) transcription, then harvested at 7 days post-infection (dpi).

RNA isolation, RT-PCR cloning, and sequencing

Total RNA was extracted from leaves as described [37] and was used as a template for RT-PCR reaction. The polyadenylation primer Tn836 (5′-CGTACCGAATTCT$_{15}$-3′, *Eco*RI site underlined) was used for first-strand cDNA synthesis with SuperScriptIII one-step RT-PCR system (Invitrogen), followed by PCR amplification with the high-fidelity Platinum *Taq* DNA polymerase (Invitrogen) and primer pairs for BS-5′ (5′-GAAAACTCACCG-3′) and Tn836. The PCR product was then *Eco*RI-digested, gel-purified, and cloned into the *Stu*I/*Eco*RI sites of the pCass2 vector [38]. Plasmids were purified and used as templates for DNA sequencing to determine the identity of satBaMV isolates.

Sequence analyses

Clustal Omega [39] from EMBL-EBI (http://www.ebi.ac.uk/Tools/msa/clustalo/) was used to align satBaMV sequences, with manual adjustment. The sequence alignment file is available upon request. Several recombination detection algorithms, such as RDP [40], GENECONV [41], Bootscan [42], MaxChi [43], Chimaera [44], SiScan [45], and 3Seq [46], all implemented in RDP4 beta 4.24 [47], were used to detect recombination among satBaMV sequences. MEGA 5.2.1 [48] was used to estimate nucleotide diversity and conduct selection tests. To determine the standard error associated with the estimated nucleotide diversity, 500 bootstraps were conducted. The online service, Intrapop neutrality tests (http://wwwabi.snv.jussieu.fr/achaz/neutralitytest.html), was used for Tajima's D [49] and Fu and Li's D_2* and F* [50] tests. MrBayes 3.2 [51] was used to reconstruct phylogenetic relationships. Instructions in the command block was generated with the web service by R. Page and S. R. Santos (https://131.204.120.103/srsantos/mrbayes_form/index.html) and modified to fit our dataset. GTR (General Time Reversible) was used for the nucleotide substitution model and gamma-distribution the variable rate model. In this study, 1,000,000 generations were generated and 25% burn-in was used for each run. The resulting consensus trees were visualized by FigTree 1.3.1 [52].

Results

We collected seven leaf samples – six from various bamboo species from locations in Taiwan, China, and India and one from inoculated *Nicotiana benthamiana*, a commonly used tobacco plant in plant virus study, in the greenhouse (Figure 1) – to survey the population structures of satBaMVs. The satBaMV genomic sequences were determined by sequencing cloned RT-PCR products. A total of 728 cDNA clones were sequenced, with about 60 to 120 clones from each sample, although not all clones contained the satBaMV sequences (Table 1). Unexpectedly, more clones from the field samples did not contain the satBaMV sequences (13–34%) as compared with the greenhouse sample (~4%) (Table 1). After removing failed clones, 595 satBaMV-containing sequences were used for further analyses.

Baseline error rate from RT-PCR

To establish the baseline error rate introduced by RT-PCR, we used a plasmid-borne satBaMV cDNA as the template to generate RNA transcripts with the T7 RNA polymerase, followed by reverse transcription and PCR amplification under the same conditions used for field samples. We sequenced 18 clones and found only 4 substitution polymorphic sites, all singletons, in 17 full-length sequences of 836 nt and one length polymorphism of 732 nt (with deletion from nucleotides 189 to 292, based on the BSF4 sequence in GenBank accession no. AY205227). The mean nucleotide diversity [53], a population genetic statistic estimating the average number of nucleotide differences per site between two random sequences in a given population, was estimated to be 0.0006±0.0003.

To detect the presence of recombination in our data, we subjected aligned full-length satBaMV sequences to various recombination-detection algorithms (see Materials and Methods).

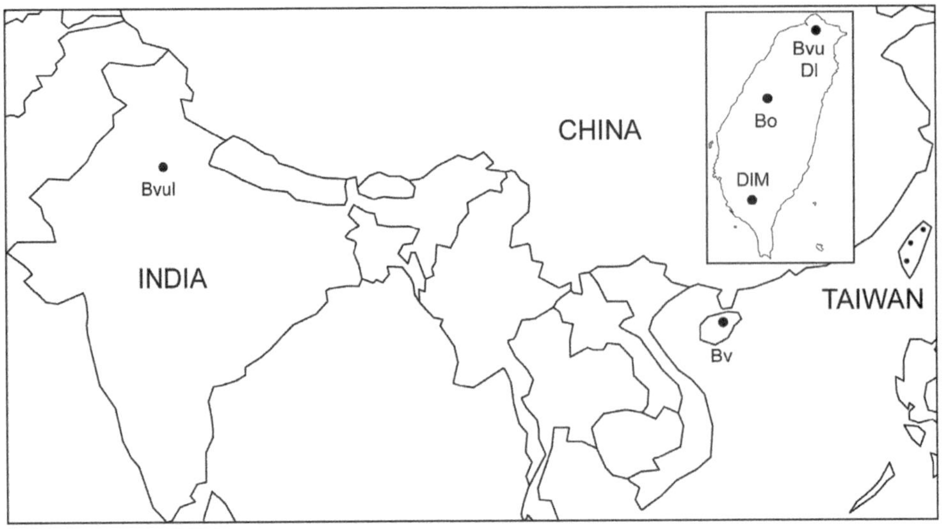

Figure 1. Map showing the geographic locations of collected bamboo samples. Inset shows an enlarged map of Taiwan for clarity. The name designations for each satBaMV populations are: Bvul, from *B. vulgaris* Schrad. *ex* J. C. Wendl. at Delhi, India; Bv, from *B. ventricosa* McClure at Hainan Island, China; Bvu, from *Bambusa vulgaris* Schrad. *ex* J. C. Wendl. at Taipei; Dl, from *Dendrocalamus latiflorus* Munro at Taipei; Bo, from *B. oldhamii* Munro at Taichung, Taiwan; and DlM, from *D. latiflorus* Munro cv. Mei-nung W. C. Lin at Pingtung, Taiwan.

Except for 2 sequences (L-89 and L-104), both from *D. latiflorus* Munro at the Taipei Botanical Garden, we found no evidence of recombination in our samples.

Overall, our results confirmed that RT-PCR introduces a small amount of sequence diversity in a given population study.

Apparent genome length polymorphism

Of the 595 sequences we analyzed, 496 more or less contained the full-length satBaMV sequences, with genome sizes ranging from 799 to 842 nt. Most of the genomes (69%) in the "full-length" category were 836 nt. The remaining 99 sequences all have deletions to various degrees and were classified into 7 categories based on genome length (Table 1). The group of 693–733 nt ("Short 1" category) was the second most common genome size found in all populations, except for populations from *D. latiflorus* Munro at the Taipei Botanical Garden (Dl) and *B. vulgaris* from India (Bvul). Both populations had ~1% genomes in the "Short 1" category, whereas others had 7% to 22% sequences belonging to this category. SatBaMVs isolated from both *D. latiflorus* Munro (Dl and DlM) hosts had a wider range of genome sizes, down to approximately 173–221 nt ("Short 7" category in Table 1).

As shown above, the 732-nt variant is likely a laboratory RT-PCR artifact. For the sequences in the 731–733-nt range, all had exactly the same deletion. Furthermore, individuals from a shorter category (*e.g.*, "Short 3") may be derived from a longer category (*e.g.*, "Short 1") (see Figure S1). Because of this uncertainty, we removed all deletion mutants for subsequent analysis.

Levels of genetic diversity among satBaMV populations

We adopted nucleotide diversity π [53] to measure the genetic diversity of satBaMV populations. The mean estimated nucleotide diversity ranged from 0.0014 ± 0.0006 (the greenhouse-inoculated *N. benthamiana*) to 0.0187 ± 0.0030 (*D. latiflorus* Munro, Dl), approximately a 13-fold difference (Figure 2). Among the field samples, the lowest nucleotide diversity was with *B. vulgaris* Schrader ex Wendland (Bvu), with a value of 0.0020 ± 0.0006.

To test whether these estimated values significantly differed, we used the Tukey-Kramer method for unplanned multiple comparisons [54] (Box 9.7, Part IV). SatBaMV populations from *D. latiflorus* Munro (Dl) and *B. vulgaris* Schrader ex Wendland in Inida (Bvul) had the highest nucleotide diversity, and the rest had a lower value (Figure 2). Among our leaf samples were several contrasting satBaMV populations, derived from the same bamboo species but different locations or from the same location but different species. For example, for *B. vulgaris* Schrader ex Wendland, we had one sample from the Taipei Botanical Garden and one from India (Bvu vs. Bvul). As well, for *D. latiflorus* Munro, we had one sample from the Taipei Botanical Garden and one from Pingtung (Dl vs. DlM). We also had satBaMV populations from *B. vulgaris* or *D. latiflorus* at the Taipei Botanical Garden (Bvu vs. Dl). Interestingly, all pairwise comparisons revealed significant differences in nucleotide diversity (Figure 2).

Our analysis showed that nucleotide diversity among satBaMV populations varied greatly, indicating the presence of different demographic and/or evolutionary histories among these satBaMV populations. Furthermore, neither host species nor geographic location predominantly determined the level of nucleotide diversity within a given satBaMV population.

Phylogeny of the satBaMVs

The mean within-population nucleotide diversity was estimated at 0.0080 ± 0.0012 (or 0.0071 ± 0.0009 with the greenhouse sample), and the mean between-population diversity 0.0430 ± 0.0066 (or 0.0391 ± 0.0056 with the greenhouse sample), approximately a five-fold difference. This result, together with previous findings [23], suggests that each satBaMV population evolves independently, thus accumulating sequence divergence across all populations. This interpretation is further corroborated by the observation that most satBaMV sequences clustered with members from the same population, forming three well-supported clades (Figure 3). Clade I is consisted of all satBaMV populations, except most of those isolated from *D. latiflorus* Munro at the Taipei Botanical Garden (Dl) and all of those from *B. vulgaris*

Table 1. Cloning effort and genome length distribution of satBaMV.

Population*		Bo	Bv	Bvu	DI	DIM	Bvul	B+F4
		cDNA cloning effort						
	Success	69	102	87	77	52	99	109
	Fail	30	16	19	40	9	15	4
	Fail rate	0.303	0.136	0.179	0.342	0.148	0.132	0.035
		Percentage of genome length						
Genome size†	Full-length	78.3	84.3	90.8	66.2	67.3	98.0	86.2
	Short 1	21.7	6.9	9.2	1.3	15.4	1.0	13.8
	Short 2				10.4	1.9	1.0	
	Short 3		1.0			1.9		
	Short 4				5.2			
	Short 5		7.8		9.1	13.5		
	Short 6				3.9			
	Short 7				3.9			

*For population designation, see Figure 1 legend.
†Genome size in nucleotides. Full-length: 799–842; Short 1: 693–733; Short 2: 612–661; Short 3: 556; Short 4: 473–478; Short 5: 397–408; Short 6: 324–357; Short 7: 173–221.

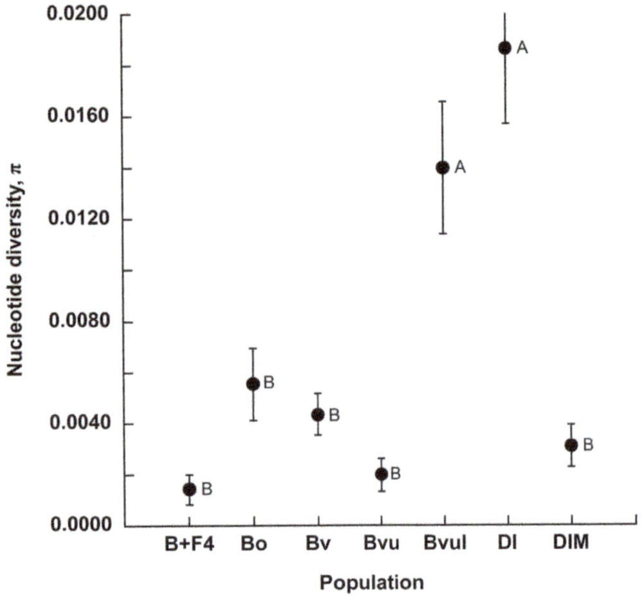

Figure 2. Nucleotide diversity of satBaMV populations. MEGA 5.2.1 [48] was used to estimate the nucleotide diversities and associated standard errors (vertical bars) of various satBaMV populations. Letters denote statistically different groups based on unplanned multiple comparisons with the Tukey-Kramer method [54]. For population designation, see Figure 1. B+F4 denotes the population from greenhouse-inoculated *N. benthamiana*.

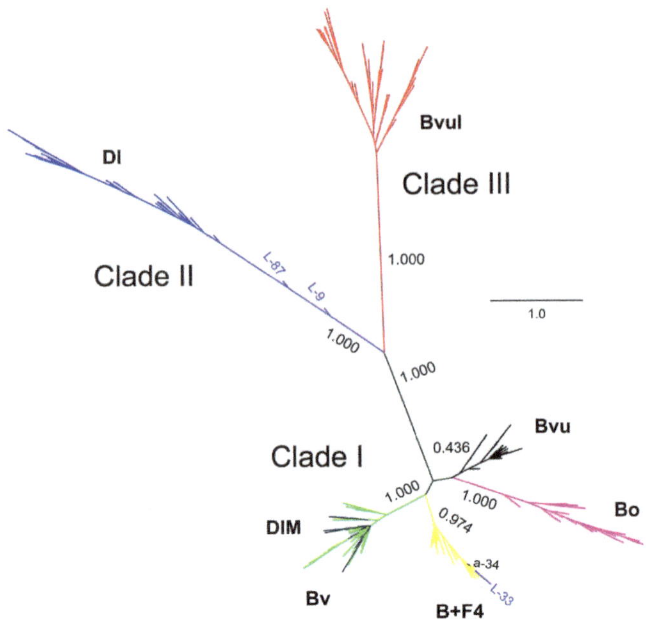

Figure 3. Phylogenetic relationship among different satBaMV populations. Unrooted Bayesian tree was constructed with full-length satBaMV genomic sequences. For clarity of presentation, sequence names are not shown, except for the listed (see text for details). Yellow, B+F4; magenta, Bo; green, Bv; black, Bvu; red, Bvul; blue, Dl; purple, DlM. Posterior probabilities are shown for major branches.

Schrader ex Wendland in India (Bvul). Clade II is consisted of sequences found exclusively in *D. latiflorus* Munro (Dl) at the Taipei Botanical Garden and Clade III sequences exclusively from *B. vulgaris* Schrader ex Wendland in India (Bvul). These three clades also corresponded with the levels of nucleotide diversity shown above, with all populations in Clade I having lower nucleotide diversity and both Clades II and III having significantly higher nucleotide diversity.

Most individual sequences tended to cluster with individuals from the same population (Figure 3), with some exceptions. The most notable exception was the well-supported clustering of sequences from *D. latiflorus* Munro cv. "Mei-nung" (DlM) in southern Taiwan and those from *B. ventricosa* McClure (Bv) in Hainan Island, China (Figure 3). This clustering suggests a recent migration event(s) of satBaMV between these two geographic locations. The second exception is the clustering of a single sequence from *B. vulgaris* (isolate a-34) and from *D. latiflorus* (isolate L-33), both from the Taipei Botanical Garden, with members from the greenhouse *N. benthamiana*. This second exception may not be too surprising, because the BSF4 isolate used to inoculate greenhouse *N. benthamiana* was originally isolated from *B. vulgaris* at the Taipei Botanical Garden in 1994 [17].

Selection of satBaMV P20 function

Among our satBaMV populations, the proportion of single nucleotide polymorphism (nucleotide substitutions that appear only once in the sample population) was relatively high, from 58% (Bo) to 79% (DlM) or to 88% when including the greenhouse (B+F4) population. The finding of the presence of excess rare variants is further corroborated with a more rigorous analysis with the Tajima's *D* test [49], a statistical test designed to distinguish between neutral evolution of nucleotide sequences from non-neutral mechanisms, such as natural selection. All populations had significant negative *D* values, indicating the presence of excess rare variants (Table 2). Similar tests, such as the Fu and Li's D^* and F^* [50] tests, showed the same pattern (data not shown).

Typically, population expansion and purifying selection are the two mechanisms commonly invoked to explain the finding of a negative Tajima's *D* value [49], although these two explanations are not mutually exclusive. In our case, the sole protein-encoding gene *P20* in the satBaMV genome provided us the opportunity to test whether natural selection for P20 function plays a role in the evolution of the satBaMV sequence. One of the commonly used methods for detecting the signatures of natural selection is the dN/dS test [55]. If the ratio (ω) is <1, then it is customarily seen as evidence of purifying selection. However, if $\omega>1$, then it is indicative of positive selection favoring amino acid substitution. If $\omega=1$, then neither purifying selection nor positive selection is responsible for the observed nucleotide sequence pattern. With the commonly used dN/dS test implemented in MEGA [48], we found that 4 of the 7 populations (*i.e.*, Bvu, Dl, DlM, and Bvul) showed significantly smaller dN values than corresponding dS values (Table 2), indicating the presence of purifying selection. Not all populations had significant differences between dS and dN values (*e.g.*, B+F4, Bo, and Bv), but all had a higher dS than dN value (Table 2). Together with the finding of negative Tajima's *D* values, these results suggest a consistent explanation that purifying selection against less-fit P20 variants are at least in part responsible for the evolution of satBaMV.

Discussion

Genome-size polymorphism and RT-PCR artifact

RT-PCR cloning followed by DNA sequencing is commonly used for population genetic studies of RNA viruses. However, the very act of RT-PCR can introduce sequence artifacts in the form

Table 2. Population genetic statistics and tests of satBaMV populations.

Population*	Bo	Bv	Bvu	DI	DIM	Bvul	B+F4
Tajima's *D* (*p*)†	−1.508 (0.042)	−2.629 (<0.001)	−2.230 (<0.001)	−1.530 (0.037)	−1.981 (0.007)	−1.466 (0.041)	−2.268 (<0.001)
Z (*p*)‡	1.278 (0.102)	1.189 (0.118)	2.094 (0.019)	2.315 (0.012)	2.094 (0.019)	2.474 (0.007)	0.284 (0.389)

*For population designation, see Figure 1 legend.
†Tajima's *D* test and associated probability in bracket. For Tajima's *D* test, $H_A = D_{obs} > D_{exp}$, with 100,000 replicates.
‡*Z* test and associated probability in bracket. $Z = dS - dN$ and the $H_A = dN < dS$, with 500 replicates.

of point mutations, deletion/insertion [56,57], and sometimes recombination [58]. To minimize laboratory artifacts, we used a high-fidelity DNA polymerase for our amplifications. Nevertheless, we could not prevent the occurrence of deletion mutations in our samples, specifically the 732-nt variant, and possibly others as well. The deletion is possibly due to a secondary structure in the satBaMV genome encountered by the reverse transcriptase during the first-strand cDNA synthesis [57]. While the 732-nt variant may be an RT-PCR artifact, the phylogenetic relationship within this group still mirrored that of its full-length counterpart. For example, individual sequences from the same population still clustered together, with the same exception for those isolated from *D. latiflorus* Munro cv. "Mei-nung" in southern Taiwan and those from *B. ventricosa* McClure in Hainan Island, China, which formed a well-supported clade (data not shown). Furthermore, within each satBaMV population, the 732-nt sequences were interspersed among the full-length ones, which suggests rather than originating from a single or a few founding mutational events, most of the 732-nt sequences represent independent deletion events (data not shown). One possible explanation is that within a given satBaMV population, a portion of the genomes assumes a stable alternative secondary structure that, when encountered by a reverse transcriptase during RT-PCR, would facilitate the creation of a deletion mutation.

The 732-nt variant was demonstrated to be a laboratory artifact, but the remaining shorter genomes, especially those in both of the *D. latiflorus* Munro hosts (Table 1), may have a more biologically relevant origin. An *in planta* study with a satRNA associated with CMV (satCMV) [59] showed that the deletion mutation rate of the helper viral replicase depended on the plant host. The deletion rate was approximately 7-fold higher in pepper than tobacco. A similar dependency of the BaMV replicase on bamboo species could explain an increased proportion of the satBaMV genomes deleted when in *D. latiflorus* Munro.

Phylogenetic relationships among satBaMV populations

Our previous surveys of satBaMV in Taiwan revealed two readily discernible clades: one with only satBaMVs isolated from *D. latiflorus* Munro (but see below) and the other including all other samples, as well as some from *D. latiflorus* Munro [23,29]. We revealed a third exclusive clade composed of only individuals from *B. vulgaris* in India (BvuI) (Figure 3). When previous and current full-length satBaMV sequences were used in phylogenetic reconstruction, the same three main clades were retrieved (Figure 4). Because collection of these samples spanned more than a decade, the existence of these distinct clades suggests that they are quite stable in the satBaMV populations.

Clade II was composed of almost entirely individuals isolated from *D. latiflorus* Munro from various locations in Taiwan. Only two other bamboo species, *B. dolichoclada cv.* Stripe [29] and *G. levis* [23], were found to carry satBaMVs in this clade. However, at least eight different bamboo species were found infected with Clade I in Taiwan (Figure 5). Whether the paucity of host species for Clade II is due to biased sampling or genuine restricted host range is not known. Nevertheless, Clade II seems to have a narrower host range but approximately 7-fold higher genetic diversity (Figure 2), than Clade I. Clade I may have initially derived from the high-diversity Clade II, with subsequent host range expansion. Alternatively, Clade II may have derived from Clade I, accompanied by an increased mutation rate of the helper BaMV replicase in a different host, as occurred with the satCMV [59], thus maintaining a high steady-state nucleotide diversity in the population. So far, we cannot differentiate between these 2 scenarios. As for Clade III, which is only found in India with *B.*

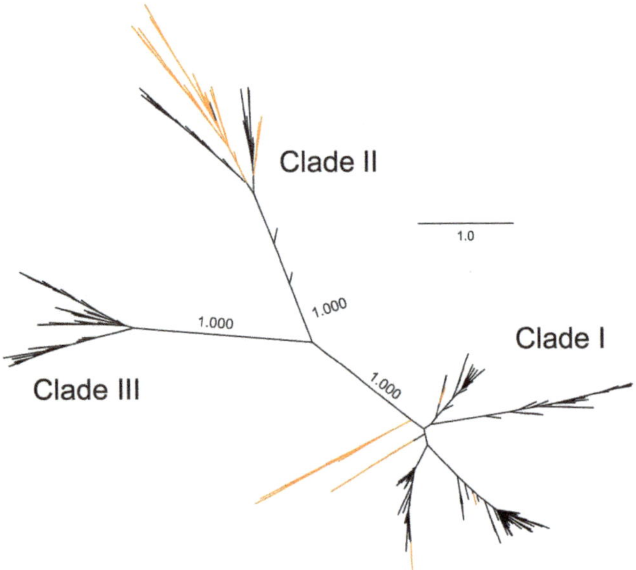

Figure 4. Phylogenetic relationship among all known full-length satBaMV. Unrooted Bayesian tree was constructed using full-length satBaMV genomic sequences from the current (black) and previous (orange) studies. Previous sequences used are from GenBank accession numbers AY205159–AY205231, except AY205176, AY205199, and AY205209. Posterior probabilities are shown for major branches.

vulgaris, our current sampling is too limited to determine whether this clade would have a narrow host range or not.

The existence of distinct satBaMV clades prompted us to examine the phylogenetic relationship of the helper BaMV isolates. Our preliminary analysis showed no immediately discernible clustering among the BaMV isolates (Figure 6), as compared with what is found for satBaMV. This preliminary result suggests that satRNAs, while obligatorily requiring the helper viral functions for replication, encapsidation, and efficient movement, could have very different evolutionary histories from their cognate helper viruses.

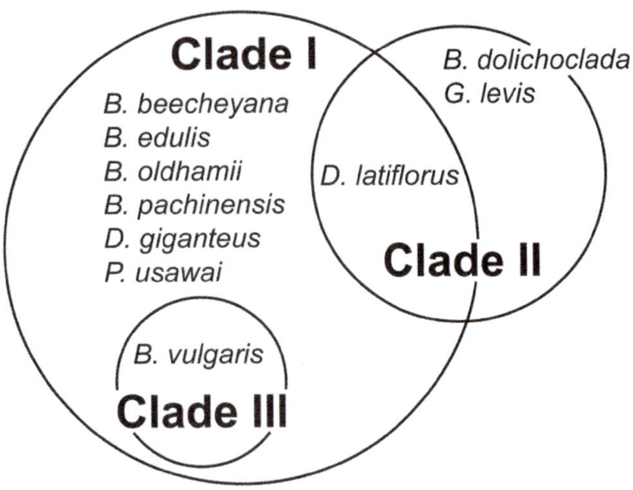

Figure 5. Distribution of bamboo species infected with different satBaMV clades. Information from both current and previous studies [23,29] were used to group bamboo species according to the satBaMV clade(s) found infecting them.

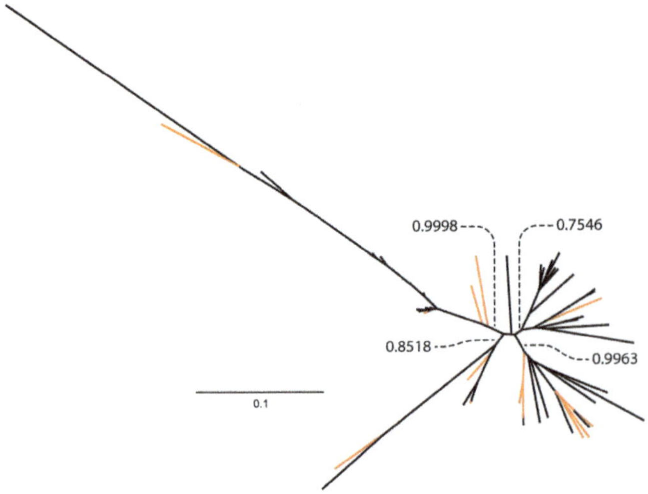

Figure 6. Phylogenetic relationship among BaMV isolates. Unrooted Bayesian tree was constructed using sequences encoding the RNA-dependent RNA polymerase from BaMV isolated from various bamboo species. Highlighted isolates (orange) denote BaMVs from the Ma bamboo (*D. latiflorus* Munro). Posterior probabilities (dashed lines) are shown for major branches.

Evolution of interference phenotype

Two prototypical isolates, BSL6 [29] and BSF4 [17], are commonly used in studies of satBaMV. The former interferes greatly with the replication of the helper BaMV [30], thus reducing the severity of disease symptoms, whereas the latter interferes only slightly, thus showing a typical mosaic symptom on co-infection with BaMV [19,30]. Competition for access of the BaMV-encoded replicase is thought to be the basis for replication interference [28,60]. One important determinant for interference is located at the AHSL at the 5′ UTR, of which both the helper BaMV and satBaMV share a similar secondary structure [23,24]. Within the 5′ AHSL, two nucleotide positions, 60 and 82 (based on the BSF4 sequence [17]), are critical for determining the interference phenotype [28]. If either position contains a U instead of C, the resulting satBaMV would not be able to (or only slightly) interfere with BaMV replication [28]. In our satBaMV populations, all sequences have a C at position 60 (or homologous positions), and almost all have a C at position 82 as well. The only exceptions are the satBaMV population from the greenhouse-infected *N. benthamiana*, for which 38.3% have a U in position 82, and one sequence (isolate a-34) from the *B. vulgaris* and three sequences (isolates L-9, L-33, and L-87) from *D. latiflorus* Munro that also have a U at the homologous position 82. The original sequence used for infecting *N. benthamiana* is BSF4, which contains a U at position 82 [17]. Therefore, even within a short time (*i.e.*, 7 dpi), we observed the evolution of U to C at position 82 from the initial 100% to 38.3%. Presumably, selection for accessing the replicase, against both the ancestral BSF4 and the helper BaMV, drives the evolution at this position. Interestingly, both sequences a-34 and L-33 are found clustered with the satBaMV population from the greenhouse-infected *N. benthamiana* (Figure 2), suggesting that satBaMV sequences very similar to BSF4, first reported in 1994 [17], still exist in the field and coexist with other satBaMV clades. Also, all sequences isolated from *D latiflorus* Munro (Dl) (except the L-33) clustered together, forming a well-supported Clade II, but both L-9 and L-87 seem to be distantly related to all the other sequences in this clade (Figure 3).

These two sequences may represent an intermediate stage between Clade I and II satBaMVs.

With our current phylogeny, BSF4 belongs to Clade I and BSL6 to Clade II. However, mapping of the nucleotide state at positions 60 and 82 showed that, in the field, both clades are dominated by sequences that may greatly interfere with BaMV replication. The same pattern was also found with individuals in Clade III. Most of the satBaMV sequences found in the field probably also have a C in positions that are homologous to position 82. Other features of the genome, such as the AHSL structure and the surrounding sequence in the 5′ UTR region, could influence the interference phenotype, but the evolution of these satBaMV clades is unlikely to be driven by the evolution of the interference phenotype.

Selection of the P20 protein

The satBaMV-encoded P20 protein is important in long-distance movement through the phloem [17,21,22]. P20 plays a supporting role in satBaMV replication as well. For example, none of the truncated P20 mutants could reach the same accumulation level as that of the wild-type P20 in the protoplast assay [19]. Furthermore, other satRNA-encoded proteins are required for satRNA replication [61–63]. Therefore, P20 likely facilitates its replication as well as long-distance movement within the infected host.

Our analysis suggests that the *P20* gene is under purifying selection against less-fit P20 proteins, which, presumably, are not efficient in facilitating satBaMV movement within the host plant and/or its own replication. However, for purifying selection to be the main driving force for the evolution of satBaMV, the interaction between individual satBaMV genomes and the P20 proteins inside an infected cell needs to be predominantly *cis* in nature; that is, the genome that sustained a deleterious mutation in the *P20* gene should, on average, "pay" a fitness cost for expressing a less fit P20 protein, either as a result of less-efficient long-distance and/or less-efficient replication of the satBaMV genome. However, if P20 is a diffusible protein that interacts with any satBaMV genomes for movement and/or replication (*i.e.*, a mainly *trans*-acting interaction), then other fitter P20 proteins can still facilitate the movement and/or replication of the satBaMV genome that carries a less-fit *P20* mutation. The within-cell complementation of fit P20 proteins with mutant satBaMV genomes could greatly reduce or completely abolish the effect of purifying selection.

The replication of the potexviruses, of which BaMV is a member, may provide clues to whether the prerequisite *cis* interaction is possible. Replication of many positive-strand plant RNA viruses is accompanied by extensive intracellular membrane reorganization, usually involving the formation of viral-induced membrane vesicles or invagination [64]. One consequence of RNA genome compartmentalization is the almost exclusive interaction between the protein products and the genomes expressing them, effectively a *cis* interaction. Whether BaMV replication, and by extension satBaMV replication, also involves rearranged membrane structures is unclear. Although other potexviruses [65] adopt such a strategy for genome replication. In the case of satBaMV, it would be interesting to know if co-infection of a wild-type satBaMV with a P20 mutant, such as the one with the P20 gene replaced by *cat* [19], may result in the translocation of the mutant satBaMV, thus establishing the *cis*-acting nature between P20 and the genome expressing it.

Supporting Information

Figure S1 Genomic deletions in satBaMV isolates. The prototypical satBaMV genome of the BSF4 isolate (GenBank accession no. AY205227) was used to illustrate region(s) of deletion in the genomes of the most common isolate in each "short" category listed in Table 1. Open bar shows the encoded *P20* gene and hatched bars the regions of deleted genome. Numbers with vertical bars indicate the nucleotide positions (of BSF4) for the start and end of the *P20* gene or the deleted regions. Numbers at the end of each genome show the specific genome length. Various single-nucleotide deletions, found in some genomes, are not shown in the illustration.

Acknowledgments

We thank Drs. Andrew O. Jackson (University of California, Berkeley, CA, USA) and Jia-Lin Yu (China Agricultural University, Beijing, China) for their help in collecting BaMV-infected bamboo from Hainan Island, China and Adam Bell for preparing high-resolution figures for publication. This study was supported by grants from the Institute of Plant and Microbial Biology (IPMB), Academia Sinica, Taiwan, to NSL. INW received funding from the National Science Council, Taiwan (NSC 101-2811-B-001-024), IPMB, and Academia Sinica during his sabbatical visit.

Author Contributions

Conceived and designed the experiments: YHH NSL. Performed the experiments: CWL SMY. Analyzed the data: INW CCH WBY. Contributed reagents/materials/analysis tools: CWL WBY YHH NSL. Wrote the paper: INW NSL.

References

1. Hu CC, Hsu YH, Lin NS (2009) Satellite RNAs and Satellite Viruses of Plants. Viruses 1: 1325–1350.
2. Roossinck MJ, Sleat D, Palukaitis P (1992) Satellite RNAs of plant viruses: structures and biological effects. Microbiol Rev 56: 265–279.
3. Simon AE, Roossinck MJ, Havelda Z (2004) Plant virus satellite and defective interfering RNAs: new paradigms for a new century. Annu Rev Phytopathol 42: 415–437.
4. Murant AF, Mayo MA (1982) Satellites of plant viruses. Annual Review of Phytopathology 20: 49–70.
5. Tien P, Wu GS (1991) Satellite RNA for the biocontrol of plant disease. Adv Virus Res 39: 321–339.
6. Nee S (2000) Mutualism, parasitism and competition in the evolution of coviruses. Philos Trans R Soc Lond B Biol Sci 355: 1607–1613.
7. Harrison BD, Mayo MA, Baulcombe DC (1987) Virus resistance in transgenic plants that express cucumber mosaic virus satellite RNA. Nature 328: 799–802.
8. Gerlach WL, Llewellyn D, Haseloff J (1987) Construction of a plant disease resistance gene from the satellite RNA of tobacco ringspot virus. Nature 328: 802–805.
9. Lin KY, Hsu YH, Chen HC, Lin NS (2013) Transgenic resistance to Bamboo mosaic virus by expression of interfering satellite RNA. Mol Plant Pathol 14: 693–707.
10. Kong Q, Wang J, Simon AE (1997) Satellite RNA-mediated resistanceto Turnip crinkle virus in *Arabidopsis* involves a reduction in virus movement. Plant Cell 9: 2051–2063.
11. Collmer CW, Howell SH (1992) Role of satellite RNA in the expression of symptoms caused by plant viruses. Annu Rev Phytopathol 30: 419–442.
12. Sleat DE, Palukaitis P (1990) Site-directed mutagenesis of a plant viral satellite RNA changes its phenotype from ameliorative to necrogenic. Proc Natl Acad Sci USA 87: 2946–2950.
13. Roossinck MJ (2001) Cucumber mosaic virus, a model for RNA virus evolution. Mol Plant Pathol 2: 59–63.
14. Lin MT, Kitajima EW, Cupertino FP, Costa CL (1977) Partial purification and some properties of bamboo mosaic virus. Phytopathology 67: 1439–1443.
15. Lin N-S, Lin F-Z, Huang T-Y, Hsu Y-H (1992) Genome properties of bamboo mosaic virus. Phytopathology 82: 731–734.
16. Lin NS, Lin BY, Lo NW, Hu CC, Chow TY, et al. (1994) Nucleotide sequence of the genomic RNA of bamboo mosaic potexvirus. J Gen Virol 75: 2513–2518.

17. Lin NS, Hsu YH (1994) A satellite RNA associated with bamboo mosaic potexvirus. Virology 202: 707–714.
18. Liu JS, Lin NS (1995) Satellite RNA associated with bamboo mosaic potexvirus shares similarity with satellites associated with sobemoviruses. Arch Virol 140: 1511–1514.
19. Lin NS, Lee YS, Lin BY, Lee CW, Hsu YH (1996) The open reading frame of bamboo mosaic potexvirus satellite RNA is not essential for its replication and can be replaced with a bacterial gene. Proc Natl Acad Sci USA 93: 3138–3142.
20. Tsai MS, Hsu YH, Lin NS (1999) Bamboo mosaic potexvirus satellite RNA (satBaMV RNA)-encoded P20 protein preferentially binds to satBaMV RNA. J Virol 73: 3032–3039.
21. Palani PV, Kasiviswanathan V, Chen JC, Chen W, Hsu YH, et al. (2006) The arginine-rich motif of Bamboo mosaic virus satellite RNA-encoded P20 mediates self-interaction, intracellular targeting, and cell-to-cell movement. Mol Plant-Microbe Interact 19: 758–767.
22. Palani PV, Chen JC, Liou MR, Chen HC, Hsu YH, et al. (2012) Phosphorylation of bamboo mosaic virus satellite RNA (satBaMV)-encoded protein P20 downregulates the formation of satBaMV-P20 ribonucleoprotein complex. Nuc Acids Res 40: 638–649.
23. Yeh WB, Hsu YH, Chen HC, Lin NS (2004) A conserved secondary structure in the hypervariable region at the 5′ end of Bamboo mosaic virus satellite RNA is functionally interchangeable. Virology 330: 105–115.
24. Chen SC, Desprez A, Olsthoorn RC (2010) Structural homology between bamboo mosaic virus and its satellite RNAs in the 5′ untranslated region. J Gen Virol 91: 782–787.
25. Huang YW, Hu CC, Lin CA, Liu YP, Tsai CH, et al. (2009) Structural and functional analyses of the 3′ untranslated region of Bamboo mosaic virus satellite RNA. Virology 386: 139–153.
26. Hsu YH, Chen HC, Cheng J, Annamalai P, Lin BY, et al. (2006) Crucial role of the 5′ conserved structure of bamboo mosaic virus satellite RNA in downregulation of helper viral RNA replication. J Virol 80: 2566–2574.
27. Chen HC, Hsu YH, Lin NS (2007) Downregulation of Bamboo mosaic virus replication requires the 5′ apical hairpin stem loop structure and sequence of satellite RNA. Virology 365: 271–284.
28. Chen HC, Kong LR, Yeh TY, Cheng CP, Hsu YH, et al. (2012) The conserved 5′ apical hairpin stem loops of bamboo mosaic virus and its satellite RNA contribute to replication competence. Nuc Acids Res 40: 4641–4652.
29. Liu JS, Hsu YH, Huang TY, Lin NS (1997) Molecular evolution and phylogeny of satellite RNA associated with bamboo mosaic potexvirus. J Mol Evol 44: 207–213.
30. Hsu Y-H, Lee Y-S, Liu J-S, Lin N-S (1998) Differential interactions of bamboo mosaic potexvirus satellite RNAs, helper virus, and host plants. Mol Plant-Microbe Interact 11: 1207–1213.
31. Kurath G, Palukaitis P (1989) RNA sequence heterogeneity in natural populations of three satellite RNAs of cucumber mosaic virus. Virology 173: 231–240.
32. Aranda MA, Fraile A, Garcia-Arenal F (1993) Genetic variability and evolution of the satellite RNA of cucumber mosaic virus during natural epidemics. J Virol 67: 5896–5901.
33. Alonso-Prados JL, Aranda MA, Malpica JM, Garcia-Arenal F, Fraile A (1998) Satellite RNA of cucumber mosaic cucumovirus spreads epidemically in natural populations of its helper virus. Phytopathology 88: 520–524.
34. Garcia-Arenal F, Escriu F, Aranda MA, Alonso-Prados JL, Malpica JM, et al. (2000) Molecular epidemiology of Cucumber mosaic virus and its satellite RNA. Virus Res 71: 1–8.
35. Lin N-S, Chai Y-J, Huang T-Y (1993) Incidence of bamboo mosaic potexvirus in Taiwan. Plant Disease 77: 448–450.
36. Lin MK, Chang BY, Liao JT, Lin NS, Hsu YH (2004) Arg-16 and Arg-21 in the N-terminal region of the triple-gene-block protein 1 of Bamboo mosaic virus are essential for virus movement. J Gen Virol 85: 251–259.
37. Verwoerd TC, Dekker BM, Hoekema A (1989) A small-scale procedure for the rapid isolation of plant RNAs. Nuc Acids Res 17: 2362.
38. Shi BJ, Ding SW, Symons RH (1997) Plasmid vector for cloning infectious cDNAs from plant RNA viruses: high infectivity of cDNA clones of tomato aspermy cucumovirus. J Gen Virol 78: 1181–1185.
39. Sievers F, Wilm A, Dineen D, Gibson TJ, Karplus K, et al. (2011) Fast, scalable generation of high-quality protein multiple sequence alignments using Clustal Omega. Mol Syst Biol 7: 539.
40. Martin D, Rybicki E (2000) RDP: detection of recombination amongst aligned sequences. Bioinformatics 16: 562–563.
41. Padidam M, Sawyer S, Fauquet CM (1999) Possible emergence of new geminiviruses by frequent recombination. Virology 265: 218–225.
42. Martin DP, Posada D, Crandall KA, Williamson C (2005) A modified bootscan algorithm for automated identification of recombinant sequences and recombination breakpoints. AIDS Res Hum Retroviruses 21: 98–102.
43. Maynard Smith J (1992) Analyzing the mosaic structure of genes. J Mol Evol 34: 126–129.
44. Posada D, Crandall KA (2001) Evaluation of methods for detecting recombination from DNA sequences: computer simulations. Proc Natl Acad Sci USA 98: 13757–13762.
45. Gibbs MJ, Armstrong JS, Gibbs AJ (2000) Sister-scanning: a Monte Carlo procedure for assessing signals in recombinant sequences. Bioinformatics 16: 573–582.
46. Boni MF, Posada D, Feldman MW (2007) An exact nonparametric method for inferring mosaic structure in sequence triplets. Genetics 176: 1035–1047.
47. Martin DP, Lemey P, Lott M, Moulton V, Posada D, et al. (2010) RDP3: a flexible and fast computer program for analyzing recombination. Bioinformatics 26: 2462–2463.
48. Tamura K, Peterson D, Peterson N, Stecher G, Nei M, et al. (2011) MEGA5: molecular evolutionary genetics analysis using maximum likelihood, evolutionary distance, and maximum parsimony methods. Mol Biol Evol 28: 2731–2739.
49. Tajima F (1989) Statistical method for testing the neutral mutation hypothesis by DNA polymorphism. Genetics 123: 585–595.
50. Fu YX, Li WH (1993) Statistical tests of neutrality of mutations. Genetics 133: 693–709.
51. Ronquist F, Teslenko M, van der Mark P, Ayres DL, Darling A, et al. (2012) MrBayes 3.2: Efficient Bayesian Phylogenetic Inference and Model Choice across a Large Model Space. Syst Biol 61: 1–4.
52. Rambaut A (2009) FigTree v1.3.1: Tree Figure Drawing Tool.
53. Nei M, Li WH (1979) Mathematical model for studying genetic variation in terms of restriction endonucleases. Proc Natl Acad Sci USA 76: 5269–5273.
54. Sokal RR, Rohlf FJ (2011) Biometry. New York: W. H. Freeman.
55. Hughes AL, Nei M (1988) Pattern of nucleotide substitution at major histocompatibility complex class I loci reveals overdominant selection. Nature 335: 167–170.
56. Bracho MA, Moya A, Barrio E (1998) Contribution of *Taq* polymerase-induced errors to the estimation of RNA virus diversity. J Gen Virol 79: 2921–2928.
57. Domingo-Calap P, Sentandreu V, Bracho MA, Gonzalez-Candelas F, Moya A, et al. (2009) Unequal distribution of RT-PCR artifacts along the E1–E2 region of *Hepatitis C virus*. J Virol Methods 161: 136–140.
58. Mader RM, Schmidt WM, Sedivy R, Rizovski B, Braun J, et al. (2001) Reverse transcriptase template switching during reverse transcriptase-polymerase chain reaction: artificial generation of deletions in ribonucleotide reductase mRNA. J Lab Clin Med 137: 422–428.
59. Pita JS, de Miranda JR, Schneider WL, Roossinck MJ (2007) Environment determines fidelity for an RNA virus replicase. J Virol 81: 9072–9077.
60. Huang YW, Hu CC, Liou MR, Chang BY, Tsai CH, et al. (2012) Hsp90 interacts specifically with viral RNA and differentially regulates replication initiation of Bamboo mosaic virus and associated satellite RNA. PLoS Pathog 8: e1002726.
61. Liu YY, Cooper JI (1993) The multiplication in plants of arabis mosaic virus satellite RNA requires the encoded protein. J Gen Virol 74: 1471–1474.
62. Hemmer O, Oncino C, Fritsch C (1993) Efficient replication of the in vitro transcripts from cloned cDNA of tomato black ring virus satellite RNA requires the 48 K satellite RNA-encoded protein. Virology 194: 800–806.
63. Hans F, Pinck M, Pinck L (1993) Location of the replication determinants of the satellite RNA associated with grapevine fanleaf nepovirus (strain F13). Biochimie 75: 597–603.
64. den Boon JA, Ahlquist P (2010) Organelle-like membrane compartmentalization of positive-strand RNA virus replication factories. Annu Rev Microbiol 64: 241–256.
65. Verchot-Lubicz J, Ye CM, Bamunusinghe D (2007) Molecular biology of potexviruses: recent advances. J Gen Virol 88: 1643–1655.

9

Analysis of Maize (*Zea mays* L.) Seedling Roots with the High-Throughput Image Analysis Tool *ARIA* (Automatic Root Image Analysis)

Jordon Pace[1]*, Nigel Lee[2], Hsiang Sing Naik[2], Baskar Ganapathysubramanian[2], Thomas Lübberstedt[1]

1 Department of Agronomy, Iowa State University, Ames, Iowa, United States of America, **2** Department of Mechanical Engineering, Iowa State University, Ames, Iowa, United States of America

Abstract

The maize root system is crucial for plant establishment as well as water and nutrient uptake. There is substantial genetic and phenotypic variation for root architecture, which gives opportunity for selection. Root traits, however, have not been used as selection criterion mainly due to the difficulty in measuring them, as well as their quantitative mode of inheritance. Seedling root traits offer an opportunity to study multiple individuals and to enable repeated measurements per year as compared to adult root phenotyping. We developed a new software framework to capture various traits from a single image of seedling roots. This framework is based on the mathematical notion of converting images of roots into an equivalent graph. This allows automated querying of multiple traits simply as graph operations. This framework is furthermore extendable to 3D tomography image data. In order to evaluate this tool, a subset of the 384 inbred lines from the Ames panel, for which extensive genotype by sequencing data are available, was investigated. A genome wide association study was applied to this panel for two traits, *Total Root Length* and *Total Surface Area,* captured from seedling root images from WinRhizo Pro 9.0 and the current framework (called *ARIA*) for comparison using 135,311 single nucleotide polymorphism markers. The trait *Total Root Length* was found to have significant SNPs in similar regions of the genome when analyzed by both programs. This high-throughput trait capture software system allows for large phenotyping experiments and can help to establish relationships between developmental stages between seedling and adult traits in the future.

Editor: Pawan L. Kulwal, Mahatma Phule Agricultural University, India

Funding: National Science Foundation Grant Number EPS-1101284 and CMMI-1149365. The funders had no role in study design, data collection and analysis, decision to publish, or preparation of the manuscript.

Competing Interests: The authors have declared that no competing interests exist.

* Email: jmpace1@iastate.edu

Introduction

The maize (*Zea mays* L.) root is designed to provide anchorage as well as to secure uptake of water and nutrients, including nitrogen (N), in an efficient manner [1,2]. Maize roots are formed partly during embryonic and partly during post-embryonic development [3]. There are five main types of roots in maize: crown, seminal, primary, lateral, and brace roots [4]. The major portion of root biomass of mature plants is derived from postembryonic, shoot-borne roots. These postembryonic roots include crown roots, formed below soil surface, and brace roots, formed above soil surface [5]. Their function is important to plant performance as they are responsible for the majority of water and nutrient uptake in maize [5].

Two to three week old seedling root systems are made up of primary roots, lateral roots, seminal roots, and root hairs [4,6]. Lateral roots branch outward from the primary root. These root types are called the axial roots and determine root architecture. Lateral roots increase the surface area of the root system and all root types contribute to water and nutrient uptake [2,7,8]. Moreover, lateral roots contain root initiation points, leading to secondary, tertiary, and higher order root structures, with major influence on the overall root architecture of the root stock [2].

There is extensive genetic variation in root architecture. However, root traits have not been considered by plant breeders to select for improved nutrient uptake efficiency or yield improvement due to the difficulty in measuring root traits and their quantitative mode of inheritance [9]. Studying adult roots using maize "shovelomics", a high-throughput phenotyping technique that measures adult root traits, is time consuming and laborious. This method of phenotyping is also destructive because roots are dug out of the ground. This limits the number of experiments that can be completed in a season [10]. Changes in maize root architecture may strongly affect yield [11]. Seminal roots play a key role in the acquisition of immobile and mobile nutrients such as phosphorus and nitrogen respectively and can determine spatial and temporal domains of its environment and inter-root competition [6]. The relationship between seminal root biomass in hydroponics and root lodging in a field study focusing on root strength and pulling resistance has been explored. Respective correlations were low, but statistically significant.

Correlations found in hydroponic seedling root traits compared to adult field traits were r = 0.44* for shoot weight and adult plant height, and r = 0.22* for lateral root length with brace root development [12,13].

Seedling phenotyping takes less time, is less laborious, and can be repeated many times during the year allowing for quicker turnover of results. Positive but low correlations were found between maize seedling and adult root traits, such as number of seminal roots and weight of seminal roots to root pulling resistance (r = 0.07 and r = 0.36*, respectively) [14]. Expanding the number of seedling root traits and improving respective phenotyping procedures, may increase the chance of capturing strong relationships between different growth stages in maize.

Using digital imaging software to automate phenotypic analysis is an innovative and efficient way of accurately taking measurements of plant physiological traits [15–19]. Roots have been difficult to phenotype in a high throughput manner due to a lack of simple access and their highly plastic nature. With the development of custom root analysis systems, quantitative studies of root systems are now possible [20,21]. There are several software frameworks that extract root morphology traits in two-dimensions in various hierarchies of automation. This ranges from manual root labeling like DART (Le Bot and Serra, 2009), to semi-automated software like WinRhizo (Pro, 2004), a commercial root analysis tool, and EzRhizo [22], a freely available software, all the way to full integrated imaging-analysis platforms like SmartRoot [23] for small root systems and recent platforms, allowing for automated measurements as well as invoking a ranking system for root traits [17].

These software frameworks have substantially enhanced the research community's ability to efficiently analyze and accumulate massive amounts of data. They also pioneered the utilization of graphical user interface (GUI) that enables ease of use. However, most of these software frameworks are either expensive, not expandable to increased (or decreased) dimensions, or cannot be fine-tuned to a specific setup. We developed an open-source, modular, easy-to-use and efficient root system architecture characterization software called *ARIA* (Automatic Root Image Analysis). This is based on a mathematically rigorous approach of converting root images into graphs. We show how extracting a variety of traits becomes a simple process of utilizing various graph algorithms. There are several major advantages to such a graph based approach to extracting root system traits: (a) graph based methods are well-studied and have very fast and efficient algorithms (for example, used in Google, Facebook, most GPS devices etc.) that enable fast, real time data analysis, (b) graph based methods are easily scalable (having almost linear computational complexity) and, hence, can be easily extended to larger problem sizes without compromising on time (with direct implication to large 3D tomography datasets), and (c) a graph-based approach is generic. That is, by making trivial modifications to the definitions of parameters like edges, weights, and labels, a huge variety of traits can be accessed. This makes a graph based framework trivially extendable. Furthermore, graphs are dimension independent, and hence this framework is trivially extendable to 3D root image analysis.

In this study, the utility of *ARIA* has been tested by phenotyping 384 maize inbred lines using scanned images of seedling roots. These data were then applied to a genome wide association study (GWAS) to detect marker-trait associations. Measurements of the trait Total Root Length were analyzed for a comparative GWAS study, as this is the only trait shared between the current platform WinRhizo Pro 9.0 and *ARIA*. The objective of this study is to show that our new and freely accessible root phenotyping software *ARIA* is a fast and accurate platform for automated phenotyping, with the potential of adding additional features when compared to the established software WhinRizo Pro 9.0. For both programs, significant marker trait associations were found using a general linear model. Also, phenotypic measurements with both programs were compared using a 74 maize inbred line panel [24] to further validate utility of *ARIA*. The results of this study show that *ARIA* is an accurate and dependable tool for completing large phenotyping experiments, needed for many quantitative genetic studies. Its flexibility makes *ARIA* a very useful tool to breeders and biologists studying root architecture.

Figure 1. Image of a 14 day root.

Table 1. Traits captured by *ARIA*.

Trait Name	Symbol	Trait Description
Total Root Length	TRL	Cumulative length of all the roots in centimeters
Primary Root Length	PRL	Length of the Primary root in centimeters
Secondary Root Length	SEL	Cumulative length of all secondary roots in centimeters
Center of Mass	COM	Center of gravity of the root.
Center of Point	COP	Absolute center of the root regardless of root length.
Center of Mass (Top)	CMT	Center of gravity of the top 1/3 of the root (Top).
Center of Mass (Mid)	CMM	Center of gravity of the middle 1/3 root (Middle).
Center of Mass (Bottom)	CMB	Center of gravity of the bottom 1/3 root (Bottom).
Center of Point (Top)	CPT	Absolute center of the root regardless of root length (Top).
Center of Point (Mid)	CPM	Absolute center of the root regardless of root length (Middle).
Center of Point (Bottom)	CPB	Absolute center of the root regardless of root length (Bottom).
Maximum Number of Roots	MNR	The 84th percentile value of the sum of every row.
Perimeter	PER	Total number of network pixels connected to a background pixel.
Depth	DEP	The maximum vertical distance reached by the root system.
Width	WID	The maximum horizontal width of the whole RSA.
Width/Depth ratio	WDR	The ratio of the maximum width to depth.
Median	MED	The median number of roots at all Y-location.
Total Number of Roots	TNR	Total number of roots.
Convex Area	CVA	The area of the convex hull that encloses the entire root image
Network Area	NWA	The number of pixels that are connected in the skeletonized image
Solidity	SOL	The fraction equal to the network area divided by the convex area
Bushiness	BSH	The ratio of the maximum to the median number of roots.
Length Distribution	LED	The ratio of TRL in the upper one-third of the root to the TRL.
Diameter	DIA	Diameter of the primary root.
Volume	VOL	Volume of the primary root
Surface Area	SUA	Surface area of the primary root.
SRL	SRL	Total root length divided by root system volume

Results

Root Traits Captured by ARIA

Using *ARIA*, 27 different root traits were extracted from each scanned image of seedling roots (Table 1). Some traits are more suitable for 3D root scan image analysis such as Depth, Width, and the Width/Depth ratio. All simple statistics as well as heritability estimates for all root traits are found in Table S2. This program is free software and can be accessed using the following link: http://www3.me.iastate.edu/bglab/pages/software.html *ARIA* captures more traits than existing programs such as WinRhizo Pro 9.0, which lists eight different traits that can be obtained from a single root scanned image when buying a standard package. *ARIA* is fully automated with the ability to capture up to three separate seedling roots from a single image, and to conduct all analyses with limited user interference. Each image was a high resolution scan (around 4400×6200 pixels) of three seedling roots placed side-by-side (Figure 1). Within each image the bounding boxes were automatically identified for each root. Each of the three roots is then individually analyzed and its 27 traits extracted. Data is then exported into an Excel file. This process takes approximately 20 seconds on a standard desktop (2.8 GHz machine). We used a total of 1059 images, each containing up to three roots per image. *ARIA* ran autonomously and extracted traits within 12 hours, allowing for fast turnaround of phenotypic data. Thus, trait capture is very fast and efficient when analyzing multiple roots of large experiments.

Seedling Trait Correlations

Pearson correlations were calculated using SAS 9.3 for all 27 seedling root traits compared to one another. Correlations between traits (Table S1) ranged from very close ($r = 0.998$) between traits such as secondary root length and PRL to no significant correlations ($r = -0.061$) for TRL and BSH. BSH did not correlate closely with other root traits with the highest r^2 value of 0.166. Similarly, SRL did not show close correlations with other seedling root traits, with its closest correlation of 0.5 with TSA. Conversely, it was found that seedling root trait DEP had close correlations with various other root traits, especially with PRL ($r = 0.95$). A principle component analysis (PCA) was conducted to visualize trait relationships. The first two components explain 45.9% of the variation (with PCA 1 explaining 35.5%). Based on the first two principle components (Figure 2), there are four trait clusters. These clusters are comprised of (1) CMT, WDR, CPT, (2) MNR, and MED, (3) SEL, TRL, NWA, and (4) SCS, WID, PER, CVA, and TSA. All of these traits had close correlations within clusters while traits outside of clusters were not closely correlated (Table S1).

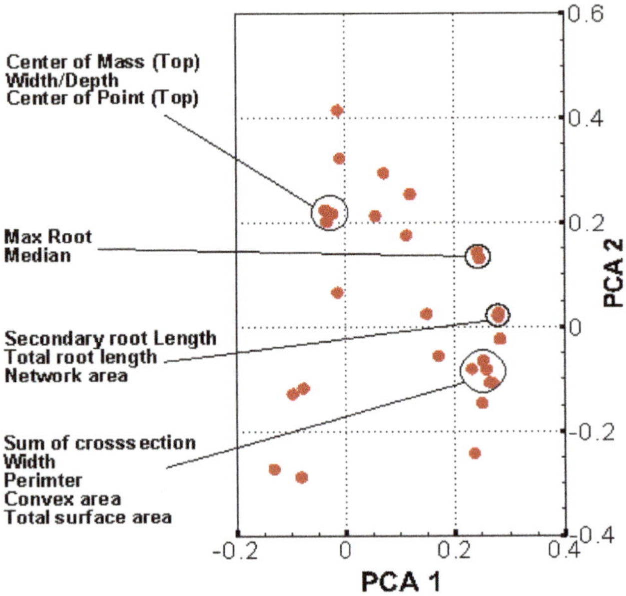

Figure 2. PCA plot of all *ARIA* trait Pearson correlations, clusters of traits have been marked showing which traits are closest related.

Validation of Measurements

In order to validate measurements made by *ARIA* with those obtained by WinRhizo Pro 9.0 (Regent Instruments, Quebec, Canada), the same images of hydroponically grown maize seedling roots were analyzed by both programs and data compared. *Total Root Length* was found to be closely correlated with $r = 0.97$ ($P = 0.0001$) when analyzing data within the Ames Panel. For the ASI panel, total root length was correlated between the two programs at $r = 0.92$ ($P = 0.0001$), and root surface area was closely correlated at $r = 0.90$ ($P = 0.0001$). Broad sense heritabilities (H^2) were calculated for both association mapping populations (Table 2). Heritability estimates were generally higher for measurements extracted using *ARIA* at $H^2 = 0.42$ compared to $H^2 = 0.41$ for total root length measured in the Ames Panel, as well as root surface area in the ASI panel with $H^2 = 0.54$ using *ARIA* compared to $H^2 = 0.50$ using Whin Rhizo Pro 9.0.

Genome Wide Association Study Using ARIA vs WinRhizo

A GWAS experiment was conducted in order to show the utility of this new program and its ability to analyze many root images in a high-throughput manner compared to WinRhizo, the current platform used. Further GWAS analyses will be documented in a future publication. TRL was extracted from a single scan of three roots from each inbred line. This process was repeated three times, once for each replication. Analysis of TRL measured with both *ARIA* and WinRhizo combined with genotypic information on 135,311 single nucleotide polymorphism markers across the entire genome identified significant associations at $p < 5.3 \times 10^{-7}$. Markers found to be significant were located on chromosomes 1, 2, and 4 for *ARIA* (Figure 3) while WinRhizo analysis resulted in additional SNPs on chromosomes 3, 5, 6, and 8 (Figure 4). Both programs identified significant markers in similar regions of the genome specifically on chromosome 2 and chromosome 4. Moreover, significant SNPs on Chromosome 4 were identical for both programs.

Discussion

Quality of ARIA Trait Estimates, Limitations and Prospects

ARIA is a reliable program that results in accurate measurements comparable to established programs such as WinRhizo Pro 9.0. The close correlation and higher heritability estimates of TRL and TSA are encouraging for using *ARIA* to obtain accurate measurements in future quantitative studies. A limitation for using *ARIA* in the current study was that only three roots were analyzed at a time. *ARIA* can be extended to allow a larger number of roots to be analyzed within a single image, depending on the scanning or image capture device. Since *ARIA* can automatically crop pictures for the user, keeping roots separate is important for accurate measurements, as crossed over roots could cause uneven cropping or erroneous paths. When comparing the amount of time needed to extract root measurements with *ARIA* and extracting measurements with WinRhizo, *ARIA* simplifies the process and cuts the time taken measurements to less than half the amount of time needed for WhinRhizo. This is in part due to the automatic cropping system as well as exporting measurement values into an Excel spreadsheet all at once within 20 seconds per seedling root. In WhinRhizo, each root has to be cropped manually; data are extracted into a.txt file, which needs to be edited for data analysis. Exporting data directly into a user friendly format *ARIA* by-passes all of these intermediate steps. The current version of *ARIA* is automated for roots exhibiting a distinguishable primary root. However, *ARIA* should work equally well with multiple equal order roots with minor changes due to the graph based formulation: *ARIA* finds all lengths of roots as distances from kernel to root tips and subsequently picks the single longest root (this can be modified to account for multiple equal order roots, for example in rice). A potential limitation is when a secondary root curl ends exactly at the primary root. This creates circular loops in the graph that impair further analysis. However, none of the 3000 images analyzed exhibited this issue. A way to resolve this minor issue is to consider a quasi-temporal approach to 'growing' the

Table 2. Comparison of repeatability estimates for both WhinRhizo Pro 9.0 and *ARIA*.

Analyzing tool	Trait	Heritability (H^2)
WinRhizo Pro 2009	Total Root Length (Ames Panel)	.41
ARIA	Total Root Length (Ames Panel)	.42
WinRhizo Pro 2009	Total Root Length (ASI Panel)	.42
ARIA	Total Root Length (ASI Panel)	.42
WinRhizo Pro 2009	Root Surface area (ASI Panel)	.50
ARIA	Root Surface area (ASI Panel)	.54

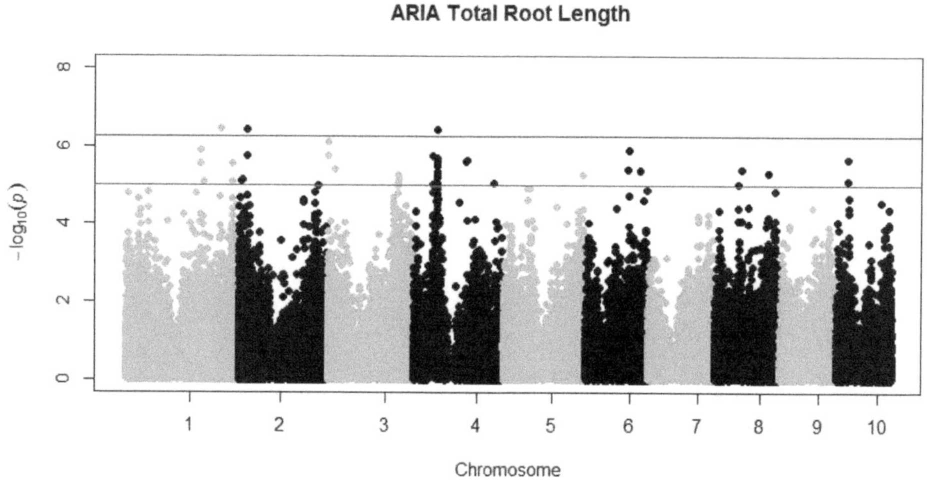

Figure 3. Manhattan plot displaying all 10 maize chromosomes, showing SNP markers significantly associated with trait Total Root Length measured with *ARIA*. Significant SNPs are located on chromosomes 1, 2, and 4.

graph vertex-by-vertex that will distinguish these overlaps (work in progress).

The graph based formulation makes this framework easily extendable to multiple purposes [25,26]. This same framework of trait extraction has been applied in other disciplines including chemistry [25] and materials science [26]. Examples of extensions include 3D phenotyping where magnetic resonance imaging (MRI), X-ray or optical tomography data can be curated and traits extracted. Furthermore, *ARIA* can work with a variety of data formats including photographs, scanned images, microscopy images as well as X-ray based reconstructions.

Significance of ARIA

While current root analysis programs are available to make measurements of root traits, none currently offers the flexibility and functionality as *ARIA*. When comparing WinRhizo 9.0 to *ARIA*, the larger numbers of traits that can be captured, ability to capture 3D image measurements, and shorter time spent to extract trait measurements from images, are key advantages of. *ARIA*, automatically crops root images, after a mouse click defines the starting point for measurements. Furthermore, *ARIA* has the ability to mark a batch of images enabling batch analysis. In *ARIA*, measurements are exported into an Excel spreadsheet, while WinRhizo gives a text file that must be converted. *ARIA's* ability to do this automatically makes this program high-throughput and decreases chances of human error. Another key advantage to this program is the fact that measurement capabilities can easily be added, as additional key architectural attributes of roots are determined.

Using *ARIA*, mapping studies for root traits can be implemented on a larger scale due to the reduced time needed for phenotyping. This software system aids plant scientists by relieving the phenotyping bottleneck for quantitative traits such as root architectural traits by adding to existing technologies in phenomics [27]. Not only is this program fast, its ability to analyze both 2D and 3D images also offers a unique opportunity to look at the same traits, with the same analysis program, but from two different perspectives. Previous programs such as RootReader2D [28] and

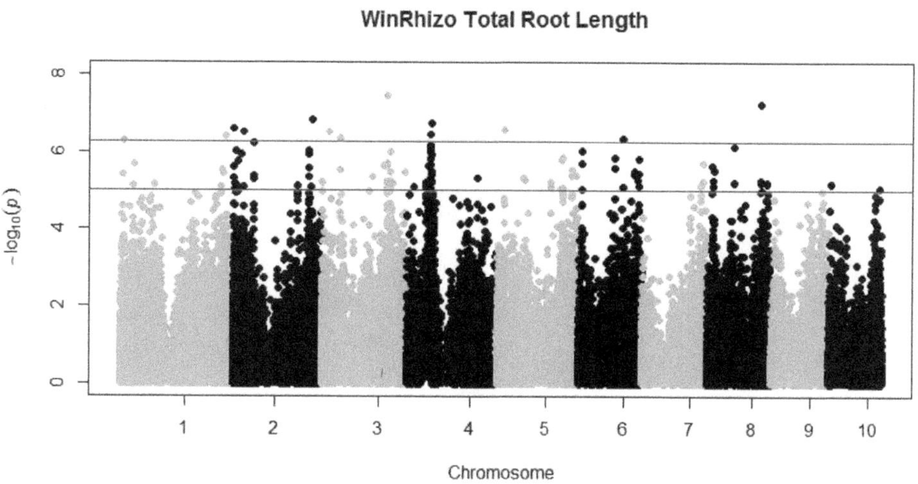

Figure 4. Manhattan plot displaying all 10 maize chromosomes, showing SNP markers significantly associated with trait Total Root Length measured with WinRhizo, significant SNPs are consistent with *ARIA*, with additional SNPs on chromosomes 3, 5, 6, and 8.

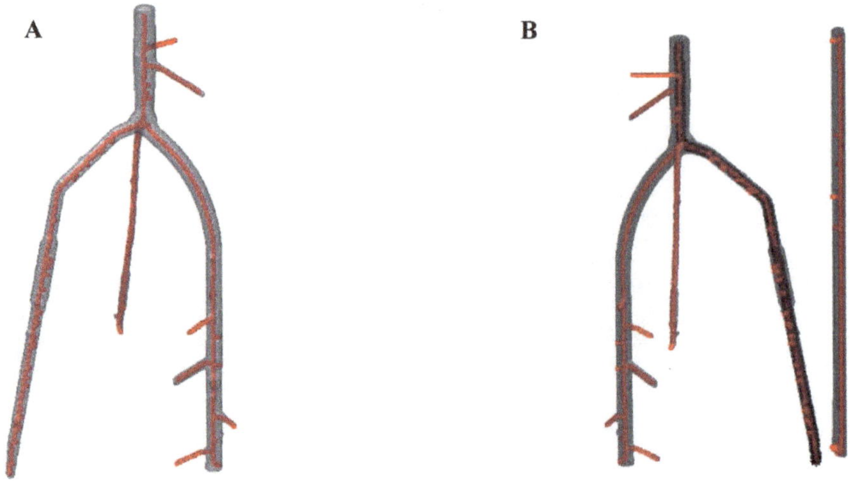

Figure 5. Demonstration of a 3D image analysis. A) Skeletonization of a 3D root image using *ARIA*. B) After skeletonization process is complete, the primary root is found and highlighted in black for accurate measurement of PRL measured at 1479 pixels, TRL measured at 582 pixels, and SEL with a measurement of 897 pixels. Pixel count can be converted to standard measurement notation with the inclusion of a baseline ruler to count pixels per cm or inch.

RootReader3D [29] offer extensive trait collection, but are hindered by the fact that each program is restricted to analyze at either 2D or 3D. *ARIA* in comparison is able to not only analyze 2D flat plane images such as those presented here, but also 3D images of roots. To show this feature, a simple 3D image of a root was analyzed using *ARIA* (Figure 5). Here, we demonstrate that skeletonization and outlining of the primary root can be completed as in 2D. The actual measurements of select traits have also been included in pixels (Figure 5). Based on multiple points of view of the same root system, *ARIA* extracts 27 root traits in a single root analysis. Figure 6 shows how the mathematical foundation (graph based analysis) coupled with the open-source framework can be trivially extended to other trait extraction.

A similar program described by Pascuzzi [17] was used to analyze rice varieties within a gel medium. This program has the ability to capture many of the same traits as *ARIA*. The major advantage of *ARIA* is that it can directly analyze those same gel medium images in both 2D and 3D formats. This adds to the flexibility of this free access program. Existing phenotyping systems can utilize this analysis tool without changing their growth procedures, whereas the other program is not as dimensionally flexible. No changes need to be made in the GUI or procedures to analyze images. This allows for an expanded number of environmental conditions, whether controlled by humans or nature, in which root architecture could be studied and for connections between how root develop in a hydroponic environment compared to soil or other growth medium.

Exploring Roots as a Model for Selection

Large scale mapping studies such as quantitative trait locus (QTL) mapping and GWAS require large mapping populations that must be phenotyped in an accurate manner. Genomic selection (GS) [30] is a method in which a training population is used to collect phenotype information and coupled with extensive genetic information. Then, a model is developed to make predictions for the performance of traits of interest, solely based on genetic information. This requires massive amounts of phenotypic information that are highly accurate, especially for plastic traits such as root architecture [31]. Currently, root architecture is not used for selection, because of the resources needed for extensive phenotyping and the quantitative mode of inheritance of root traits [9]. New phenotyping software such as *ARIA* may facilitate to include root architecture in selection schemes. Comparative GWAS for TRL obtained with both

Figure 6. Analysis of a diseased maple leaf, *ARIA's* flexible mainframe will allow multiple uses of the program beyond root phenotyping.

Figure 7. Thresholding and skeletonization stages.

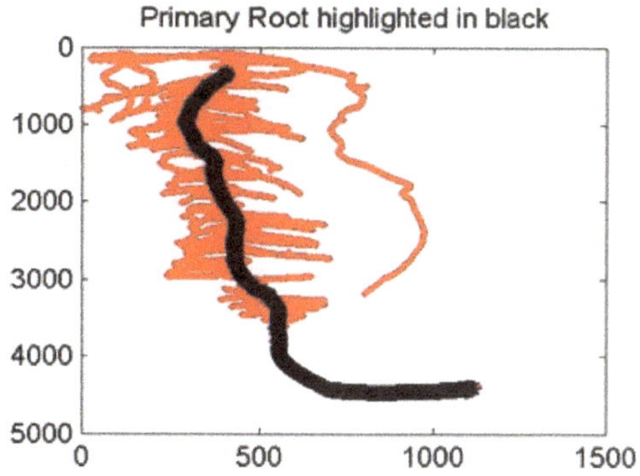

Figure 8. Automated identification of primary and secondary roots.

WinRhizo Pro 9.0 and *ARIA* identified similar or identical regions of the genome associated with this trait. Associations found in only one program may be due to low power of detecting a polymorphism with small genetic effect.

The major goal was to develop an easy to use image software analysis tool for measuring root traits from simple scans or photographs. A free to use software platform with ability to investigate both 2D and 3D root architectural characteristics for plants has been developed to facilitate measuring multiple root traits in a high-throughput, accurate manner. We compared this new program to existing programs. *ARIA* showed close correlations to traits measured with established software, supporting accurate measurements. The 27 root traits measured give an example for the utility of this program and offer an extensive amount of traits to be studied for large scale phenotypic analysis of roots or mapping studies looking at the genetic control of root architecture. Future studies using this program include root characterization for particular maize or other plant species of interest as well as phenotyping for quantitative trait studies such as GWAS, QTL mapping, and GS.

Materials and Methods

ARIA (Automatic Root Image Analysis)

ARIA is custom software written in the programming framework, MATLAB (Natick, Massachusetts, United States). *ARIA* has a user friendly GUI interface to enable easy and rapid data extraction. The operational concept of the software is to convert the root image (after standard image pre-processing) into a graph. The software framework can read in most standard image formats. Each image is loaded (Figure 1), and after a sequence of pre-processing steps, converted into a graph. A graph is a mathematical construct consisting of a set of vertices that are connected by a set of edges. This is done by labeling each pixel of the root image into a vertex, and linking nearest neighbor pixels with edges. The key steps of the software are:

a) Thresholding. The background is first identified (using morphological operations in Matlab) and renormalized to black. This effectively eliminates most of the background signal. Then the image threshold is calculated using Otsu's method. The grey scale image is converted into a black and white image. This is done by comparing the intensity of each pixel with a threshold value. The pixel is marked as black (or white), if it's grey scale value is smaller (or larger) than the threshold (Figure 7).

b) Connected components. Since the root is one large connected system, everything else that is not connected to the root can be removed from the image. This idea is encoded in the graph concept of connected components, which enumerates all the distinct connected components in the image. The largest connected component is the root, all the other connect components are noise or other foreign artifacts. Note that if the image resolves finer root hairs (which our imaging process does not do) these will still be part of the largest connected component.

c) Skeletonization. A 'wire-frame' skeleton of the binary image is constructed by thinning (or eroding). Skeletonization is a fundamental tool with many applications in image processing and visualization. Here, skeletonization is essential to identify and distinguish between the primary and secondary roots (Figure 7).

d) Primary and secondary root identification. The primary root is identified as the graph path that has the longest path length (Figure 8). This is accomplished by Dijkstra's algorithm to estimate shortest paths between two points of the graph [32]. Dijkstra's algorithm is used to compute the shortest paths from each free end of the root to every other free end. The longest "shortest path" is identified as the primary root. Secondary roots are identified easily by subtracting the primary root from the original image and enumerating the remaining distinct connected components.

e) Graph querying and post processing. The graph is queried to construct several traits starting from simple traits like total root length, to more complex measures like bushiness. All data are exported into an Excel sheet for ease of analysis and use. This will allow one to place a series of images for analysis at a time and export it to Excel. The data are also displayed on the GUI. All traits are analyzed automatically and can be viewed when clicking display results (Figure 9).

Plant materials. The first association mapping population or "Ames panel" is comprised of 384 inbred lines obtained from the USDA-ARS North Central Regional Plant Introduction Station (NCRPIS) in Ames, Iowa. All lines used in this study are a subset of a larger collection of lines called the Ames panel [33], consisting of 2815 maize inbred lines conserved at USDA-ARS NCRPIS. The 384 lines were selected based on maturity in view of future

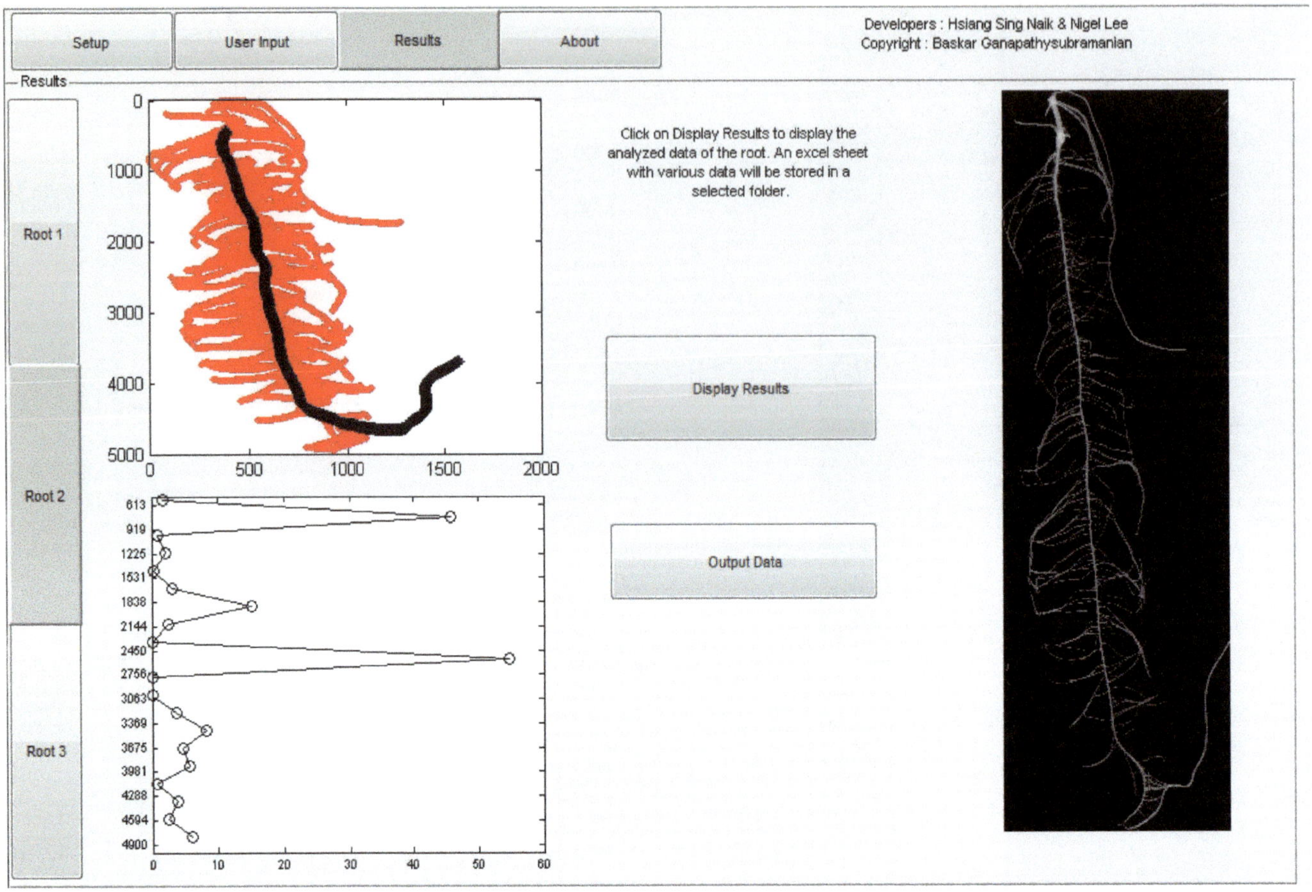

Figure 9. Screen capture of the *ARIA* framework. The picture on the right is the root image. The plot on the top left is automated identification of primary and secondary roots. The graph on the bottom left represents detailed analysis of root architecture, specifically a histogram of secondary roots across each 10% of the primary root.

field trials in central Iowa. The second panel of 74 maize inbred lines called "ASI panel" includes ex-PVPs (Plant Variety Protection) and Germplasm enhancement of Maize (GEM) inbred lines [24].

Root Phenotyping

Cigar Roll Growth Conditions. A paper roll assay described by [24] was used for germination and growth of maize seedlings. Seedlings were grown in 2 L glass beakers filled with 1.4 L of sterilized water. Seedlings were placed in a growth chamber for 14 days at 16/8 hrs light/darkness (25/22°C). Light intensity was 200 μmol photons $m^{-2}s^{-1}$, and a relative humidity maintained at 65%. Each paper roll with four seedlings was considered as experimental unit. After 14 days seedlings were removed from the growth chamber and phenotypic traits measured. If not all traits were measured the same day, plants were preserved in 30% ethanol to prevent aging of roots.

Image Acquisition. Seedling roots were imaged using a high resolution scanner. Three separate seedling roots were imaged at a time using an EPSON Expression 10000 XL scanner system.

Phenotype Data Analysis

Experimental Design. Ames panel lines were grown in three experiments starting June 12, 2012, July 3, 2012, and October 5, 2012. Each experiment was grown in the same growth chamber and at the same growing conditions, as described above. Lines were grown in a completely randomized design (CRD) and trait data were collected per experimental unit: three seedlings out of four within each seed roll were sampled, to eliminate possible outliers within lines, and means taken. The ASI panel of 74 maize inbred lines were grown under the same conditions and replicated twice under one experiment. Analysis of variance of root traits was performed, the additive model for analysis of variance was:

$$yij = \mu + R_i + G_j + E_{ij}$$

where y_{ij} represents the observation from the *ijth* experimental unit, μ is the overall mean, R_i is the *ith* experiment and G_j is the *jth* genotype. The interaction between the fixed effects G_j and the random effect experiment is confounded with the error E_{ij}. The statistics software package SAS 9.3 (Copyright © 2014 SAS Institute Inc.) was used to obtain ANOVA tables, expected mean squares, and least square means for association analyses. Function PROC GLM was implemented and type 3 sums of squares were used to account for missing data. Genotypic (σ_g^2), and phenotypic (σ_p^2) variances as well as broad sense heritability (H^2) were all calculated on an entry mean basis. Heritability on an experimental unit basis was calculated as follows:

$$H^2 = \frac{\sigma_G^2}{\sigma_P^2}, \sigma_G^2 = \left(\frac{MSG - MSE}{rep\#}\right),$$

$$\sigma_P^2 = \left(\frac{MSG - MSE}{rep\#}\right) + MSE,$$

$$H^2 = \frac{\left(\frac{MSG - MSE}{rep\#}\right)}{\left(\frac{MSG - MSE}{rep\#}\right) + MSE}$$

Function PROC GLM was implemented. Pearson correlations were calculated using the SAS function CORR to determine the relationships between seedling traits.

Marker Data

Genotyping-by-sequencing (GBS) [34] was used to genotype the association mapping population with 681,257 single nucleotide polymorphism (SNP) markers across the maize genome. Imputation as described by [33] was employed. In an effort to reduce the number of non-informative markers, all monomorphic SNP markers and those with more than 20% missing data were omitted. SNP markers with a minor allele frequency less than 5% were removed, leaving 135,311 SNP markers spread across all 10 chromosomes of the maize genome to calculate population structure, kinship, and to perform GWAS.

Association Analyses

Population structure was estimated from a reduced number of unimputed SNPs (1,665 SNP markers) using program Structure 2.3.4 [35]. Parameter settings for estimating membership of coefficients of coancestry for lines are a burn-in length of 50,000 with 50,000 iterations for each cluster (K) from 1–15, with each K being run five times. We applied an admixture model with independent allele frequencies. To pick the most probable K value, we used an ad hoc (ΔK) statistic based on the ordering rate of change of $P(X|K)$ [36]. Software program TASSEL 4.0 [37] was used to calculate LD as well as Loiselle kinship coefficients between lines based on 135,311 SNP markers. Population structure (Q matrix) was used in association analyses to decrease the amount of type 1 errors [38]. TASSEL 4.0 was used to conduct genome wide association analyses (GWAS) using a General Linear Model (GLM) and population structure as a fixed factor with model $y = X\beta + U$, where y are the values measured, X is the marker value, β is a matrix of parameters to be estimated, and U uses the Q values as fixed factors. To account for multiple testing during GWAS, statistical package simpleM was implemented in R 3.0 [39]. Based on a α level of $P = 0.05$, the multiple testing threshold level was set to 5.3×10–7 with the equation α/n, where n equals the effective number of independent tests. Only the Ames panel was analyzed, as genomic marker data were not available for the ASI panel.

Author Contributions

Conceived and designed the experiments: JP TL. Performed the experiments: JP. Analyzed the data: JP. Contributed reagents/materials/analysis tools: NL HN BG. Contributed to the writing of the manuscript: JP TL BG.

References

1. Aiken RM, Smucker AJM (1996) Root System Regulation Of Whole Plant Growth1. Annual Review of Phytopathology 34: 325–346.
2. Lynch J (1995) Root Architecture and Plant Productivity. Plant Physiology 109: 7–13.
3. Feldman L (1994) The Maize Root. In: Freeling M, Walbot V, editors. The Maize Handbook: Springer New York. 29–37.
4. Hochholdinger F, editor (2009) Handbook of Maize: Its Biology. 145 p.
5. Hoppe DC, McCully ME, Wenzel CL (1986) The nodal roots of Zea: their development in relation to structural features of the stem. Canadian Journal of Botany 64: 2524–2537.
6. Zhu J, Mickelson SM, Kaeppler SM, Lynch JP (2006) Detection of quantitative trait loci for seminal root traits in maize (Zea mays L.) seedlings grown under differential phosphorus levels. Theor Appl Genet 113: 1–10.
7. Liu J, Li J, Chen F, Zhang F, Ren T, et al. (2008) Mapping QTLs for root traits under different nitrate levels at the seedling stage in maize (Zea mays L.). Plant and Soil 305: 253–265.
8. Jordan WR, Douglas WA, Shouse PJ (1983) Strategies for Crop Improvement for Drought-Prone Regions. Agriculture Water Management 7: 281–299.
9. Salvi RTaS (2007) From QTLs to Genes controlling Root Traits in Maize. 13–22.
10. Trachsel S, Kaeppler SM, Brown KM, Lynch JP (2010) Shovelomics: high throughput phenotyping of maize (Zea mays L.) root architecture in the field. Plant and Soil 341: 75–87.
11. Hammer GL, Dong Z, McLean G, Doherty A, Messina C, et al. (2009) Can Changes in Canopy and/or Root System Architecture Explain Historical Maize Yield Trends in the U.S. Corn Belt? Crop Science 49: 299.
12. Landi P (1998) Seedling characteristics in hydroponic culture and field performance of maize genotypes with different resistance to root lodging. Maydica 43: 111.
13. Landi P, Giuliani M, Darrah L, Tuberosa R, Conti S, et al. (2001) Variability for root and shoot traits in a maize population grown in hydroponics and in the field and their relationships with vertical root pulling resistance. Maydica 46: 177–182.
14. Nass HG, Zuber MS (1971) Correlation of Corn (Zea mays L.) Roots Early in Development to Mature Root Development1. Crop Sci 11: 655–658.
15. Brewer MT, Lang L, Fujimura K, Dujmovic N, Gray S, et al. (2006) Development of a Controlled Vocabulary and Software Application to Analyze Fruit Shape Variation in Tomato and Other Plant Species. Plant Physiology 141: 15–25.
16. Chavarria-Krauser A, Nagel KA, Palme K, Schurr U, Walter A, et al. (2008) Spatio-temporal quantification of differential growth processes in root growth zones based on a novel combination of image sequence processing and refined concepts describing curvature production. New Phytol 177: 811–821.
17. Iyer-Pascuzzi AS, Symonova O, Mileyko Y, Hao Y, Belcher H, et al. (2010) Imaging and analysis platform for automatic phenotyping and trait ranking of plant root systems. Plant Physiol 152: 1148–1157.
18. Nagel KA, Putz A, Gilmer F, Heinz K, Fischbach A, et al. (2012) GROWSCREEN-Rhizo is a novel phenotyping robot enabling simultaneous measurements of root and shoot growth for plants grown in soil-filled rhizotrons. Functional Plant Biology 39: 891.
19. Wang L, Uilecan IV, Assadi AH, Kozmik CA, Spalding EP (2009) HYPOTrace: Image Analysis Software for Measuring Hypocotyl Growth and Shape Demonstrated on Arabidopsis Seedlings Undergoing Photomorphogenesis. Plant Physiology 149: 1632–1637.
20. Le Bot J, Serra V, Fabre J, Draye X, Adamowicz S, et al. (2009) DART: a software to analyse root system architecture and development from captured images. Plant and Soil 326: 261–273.
21. Zeng G, Birchfield ST, Wells CE (2008) Automatic discrimination of fine roots in minirhizotron images. New Phytol 177: 549–557.
22. Armengaud P, Zambaux K, Hills A, Sulpice R, Pattison RJ, et al. (2009) EZ-Rhizo: integrated software for the fast and accurate measurement of root system architecture. The Plant Journal 57: 945–956.
23. Lobet G, Pagès L, Draye X (2011) A novel image-analysis toolbox enabling quantitative analysis of root system architecture. Plant physiology 157: 29–39.
24. Abdel-Ghani AH, Kumar B, Reyes-Matamoros J, Gonzalez-Portilla PJ, Jansen C, et al. (2012) Genotypic variation and relationships between seedling and adult

plant traits in maize (Zea mays L.) inbred lines grown under contrasting nitrogen levels. Euphytica 189: 123–133.
25. Wodo O, Roehling JD, Moulé AJ, Ganapathysubramanian B (2013) Quantifying organic solar cell morphology: a computational study of three-dimensional maps. Energy & Environmental Science 6: 3060.
26. Samudrala S, Wodo O, Suram SK, Broderick S, Rajan K, et al. (2013) A graph-theoretic approach for characterization of precipitates from atom probe tomography data. Computational Materials Science 77: 335–342.
27. Furbank RT, Tester M (2011) Phenomics–technologies to relieve the phenotyping bottleneck. Trends Plant Sci 16: 635–644.
28. Clark RT, Famoso AN, Zhao K, Shaff JE, Craft EJ, et al. (2013) High-throughput two-dimensional root system phenotyping platform facilitates genetic analysis of root growth and development. Plant Cell Environ 36: 454–466.
29. Clark RT, MacCurdy RB, Jung JK, Shaff JE, McCouch SR, et al. (2011) Three-dimensional root phenotyping with a novel imaging and software platform. Plant Physiol 156: 455–465.
30. Meuwissen THE, Hayes BJ, Goddard ME (2001) Prediction of Total Genetic Value Using Genome-Wide Dense Maker Maps. Genetics 157: 1819–1829.
31. Gruber BD, Giehl RF, Friedel S, von Wiren N (2013) Plasticity of the Arabidopsis root system under nutrient deficiencies. Plant Physiol 163: 161–179.
32. Knuth DE (1977) A generalization of Dijkstra's algorithm. Information Processing Letters 6: 1–5.
33. Romay MC, Millard MJ, Glaubitz JC, Peiffer JA, Swarts KL, et al. (2013) Comprehensive genotyping of the USA national maize inbred seed bank. Genome Biol 14: R55.
34. Elshire RJ, Glaubitz JC, Sun Q, Poland JA, Kawamoto K, et al. (2011) A robust, simple genotyping-by-sequencing (GBS) approach for high diversity species. PLoS One 6: e19379.
35. Pritchard JK, Stephens M, Donnelly P (2000) Inference of Population Structure Using Multilocus Genotype Data. Genetics 945–959.
36. Evanno G, Regnaut S, Goudet J (2005) Detecting the number of clusters of individuals using the software STRUCTURE: a simulation study. Mol Ecol 14: 2611–2620.
37. Bradbury PJ, Zhang Z, Kroon DE, Casstevens TM, Ramdoss Y, et al. (2007) TASSEL: software for association mapping of complex traits in diverse samples. Bioinformatics 23: 2633–2635.
38. Yu J, Pressoir G, Briggs WH, Vroh Bi I, Yamasaki M, et al. (2006) A unified mixed-model method for association mapping that accounts for multiple levels of relatedness. Nat Genet 38: 203–208.
39. Gao X, Starmer J, Martin ER (2008) A multiple testing correction method for genetic association studies using correlated single nucleotide polymorphisms. Genet Epidemiol 32: 361–369.

Genome Features of the Endophytic Actinobacterium *Micromonospora lupini* Strain Lupac 08: On the Process of Adaptation to an Endophytic Life Style?

Martha E. Trujillo[1]*, Rodrigo Bacigalupe[1], Petar Pujic[2], Yasuhiro Igarashi[3], Patricia Benito[1], Raúl Riesco[1], Claudine Médigue[4], Philippe Normand[2]

1 Departamento de Microbiología y Genética, Edificio Departamental, Campus Miguel de Unamuno, Universidad de Salamanca, Salamanca, Spain, 2 Université Lyon 1, Université de Lyon, CNRS-UMR5557 Ecologie Microbienne, Villeurbanne, France, 3 Biotechnology Research Center, Toyama Prefectural University, Kurokawa, Imizu, Toyama, Japan, 4 Genoscope, CNRS-UMR 8030, Atelier de Génomique Comparative, Evry, France

Abstract

Endophytic microorganisms live inside plants for at least part of their life cycle. According to their life strategies, bacterial endophytes can be classified as "obligate" or "facultative". Reports that members of the genus *Micromonospora,* Gram-positive Actinobacteria, are normal occupants of nitrogen-fixing nodules has opened up a question as to what is the ecological role of these bacteria in interactions with nitrogen-fixing plants and whether it is in a process of adaptation from a terrestrial to a facultative endophytic life. The aim of this work was to analyse the genome sequence of *Micromonospora lupini* Lupac 08 isolated from a nitrogen fixing nodule of the legume *Lupinus angustifolius* and to identify genomic traits that provide information on this new plant-microbe interaction. The genome of *M. lupini* contains a diverse array of genes that may help its survival in soil or in plant tissues, while the high number of putative plant degrading enzyme genes identified is quite surprising since this bacterium is not considered a plant-pathogen. Functionality of several of these genes was demonstrated *in vitro,* showing that Lupac 08 degraded carboxymethylcellulose, starch and xylan. In addition, the production of chitinases detected *in vitro,* indicates that strain Lupac 08 may also confer protection to the plant. *Micromonospora* species appears as new candidates in plant-microbe interactions with an important potential in agriculture and biotechnology. The current data strongly suggests that a beneficial effect is produced on the host-plant.

Editor: Holger Brüggemann, Aarhus University, Denmark

Funding: MET received financial support from the Spanish Ministerio de Economía y Competitividad under project CGL2009-07287. PN acknowledges financial support from the ANR (Sesam). The funders had no role in study design, data collection and analysis, decisión to publish, or preparation of the manuscript.

Competing Interests: The authors have declared that no competing interests exist.

* Email: mett@usal.es

Background

For a long time, it was considered that a healthy plant was a plant without microbes within its tissues. However, this view has started to change with new approaches to allow strains to grow for a longer time upon isolation as well as the use of NGS, which has permitted the identification of several strains present in the tissues of healthy plants, in particular several actinobacteria [1,2].

Endophytic microorganisms live inside plants for at least part of their life cycle. According to their life strategies, bacterial endophytes can be classified as "obligate" or "facultative". Obligate endophytes are strictly dependent on the host plant for their growth and survival while facultative endophytes have a stage in their life cycle during which they exist outside host plants [3]. These endophytes originate from soil, initially infecting the host plant by colonizing, for instance, the cracks formed at points of emergence of lateral roots from where they quickly spread to the intercellular spaces in the root [4]. Thus, a series of environmental and genetic factors is presumed to have a role in enabling a specific bacterium to become endophytic [5]. Conversely, Marchetti and co-workers [6] recently showed how a pathogen can evolve in a few generations to become a symbiotic endophyte by losing specific transporters and regulators linked to pathogenesis.

Micromonospora is a genus of Gram-positive Actinobacteria that was first isolated from soil [7]. This bacterium has received a lot of attention during natural product screening programs, given its ability to produce a very rich array of secondary metabolites [8,9,10]. The distribution of members of *Micromonospora* is wide-ranging since these bacteria have been isolated from different geographical zones. In addition, its habitats are also diverse and include: soil, freshwater and marine sediments, mangrove soils, rocks, and nitrogen fixing nodules of both leguminous and actinorhizal plants [11,12,13]. The recent report [13] that *Micromonospora* inhabits nitrogen-fixing nodules in a systematic way, has opened up a question as to what is the potential ecological role of this bacterium in the plant and whether this bacterium is in a process of adaptation from a terrestrial to a facultative endophytic life style.

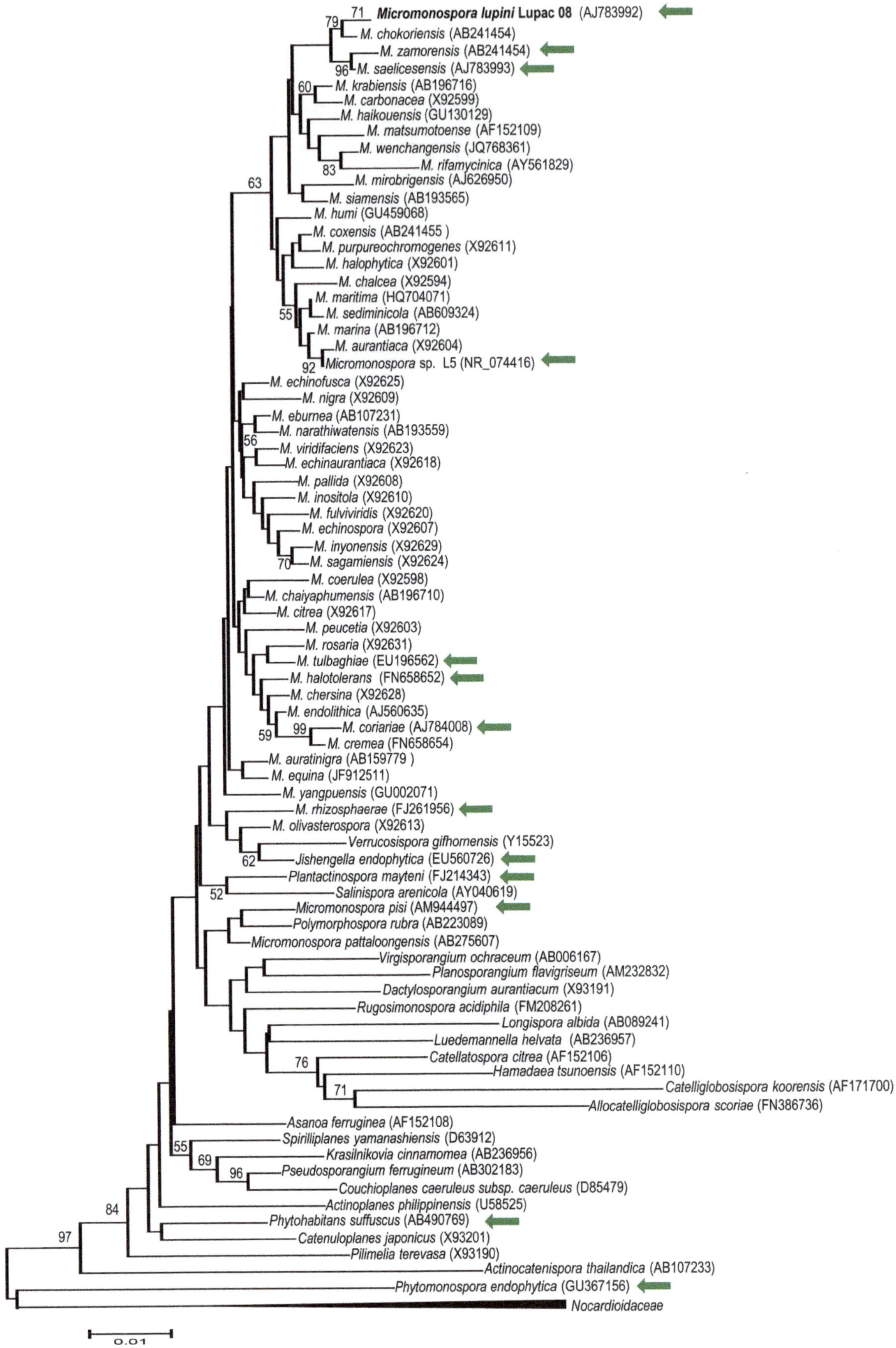

Micromonospora lupini Lupac 08 (AJ783992)
M. chokoriensis (AB241454)
M. zamorensis (AB241454)
M. saelicesensis (AJ783993)
M. krabiensis (AB196716)
M. carbonacea (X92599)
M. haikouensis (GU130129)
M. matsumotoense (AF152109)
M. wenchangensis (JQ768361)
M. rifamycinica (AY561829)
M. mirobrigensis (AJ626950)
M. siamensis (AB193565)
M. humi (GU459068)
M. coxensis (AB241455)
M. purpureochromogenes (X92611)
M. halophytica (X92601)
M. chalcea (X92594)
M. maritima (HQ704071)
M. sediminicola (AB609324)
M. marina (AB196712)
M. aurantiaca (X92604)
Micromonospora sp. L5 (NR_074416)
M. echinofusca (X92625)
M. nigra (X92609)
M. eburnea (AB107231)
M. narathiwatensis (AB193559)
M. viridifaciens (X92623)
M. echinaurantiaca (X92618)
M. pallida (X92608)
M. inositola (X92610)
M. fulviviridis (X92620)
M. echinospora (X92607)
M. inyonensis (X92629)
M. sagamiensis (X92624)
M. coerulea (X92598)
M. chaiyaphumensis (AB196710)
M. citrea (X92617)
M. peucetia (X92603)
M. rosaria (X92631)
M. tulbaghiae (EU196562)
M. halotolerans (FN658652)
M. chersina (X92628)
M. endolithica (AJ560635)
M. coriariae (AJ784008)
M. cremea (FN658654)
M. auratinigra (AB159779)
M. equina (JF912511)
M. yangpuensis (GU002071)
M. rhizosphaerae (FJ261956)
M. olivasterospora (X92613)
Verrucosispora gifhornensis (Y15523)
Jishengella endophytica (EU560726)
Plantactinospora mayteni (FJ214343)
Salinispora arenicola (AY040619)
Micromonospora pisi (AM944497)
Polymorphospora rubra (AB223089)
Micromonospora pattaloongensis (AB275607)
Virgisporangium ochraceum (AB006167)
Planosporangium flavigriseum (AM232832)
Dactylosporangium aurantiacum (X93191)
Rugosimonospora acidiphila (FM208261)
Longispora albida (AB089241)
Luedemannella helvata (AB236957)
Catellatospora citrea (AF152106)
Hamadaea tsunoensis (AF152110)
Catelliglobosispora koorensis (AF171700)
Allocatelliglobosispora scoriae (FN386736)
Asanoa ferruginea (AF152108)
Spirilliplanes yamanashiensis (D63912)
Krasilnikovia cinnamomea (AB236956)
Pseudosporangium ferrugineum (AB302183)
Couchioplanes caeruleus subsp. caeruleus (D85479)
Actinoplanes philippinensis (U58525)
Phytohabitans suffuscus (AB490769)
Catenuloplanes japonicus (X93201)
Pilimelia terevasa (X93190)
Actinocatenispora thailandica (AB107233)
Phytomonospora endophytica (GU367156)
Nocardioidaceae
71
79
96
60
83
63
55
92
56
70
59
99
62
52
76
71
55
69
96
84
97
0.01

Figure 1. Neighbour-joining tree based on 16S rRNA gene sequences showing the relationship of *Micromonospora* species and other members of the family *Micromonosporaceae*. Strains isolated from plant related sources are indicated by a green arrow.

Taxonomically, *Micromonospora* belongs to the family *Micromonosporaceae* which currently contains 27 genera and includes aerobic, non-acid fast and mesophilic microorganisms. Many strains produce mycelial carotenoid pigments giving the colonies an orange to red appearance, but blue-green, brown or purple pigmented strains have also been isolated. The family *Micromonosporaceae* also harbors the genus *Salinispora*, which is widely distributed in tropical and sub-tropical marine sediments. This taxon was described as the first marine actinomycete given its inability to grow in low salinity medium. Indeed, genomic information obtained from the genomes of *Salinispora tropica* and *Salinispora arenicola* provide evidence of marine adaptation of *Salinispora* species [14]. Thus, it appears that *Salinispora* evolved from a terrestrial environment to a marine habitat. In the case of some *Micromonospora* lineages, the question is whether this bacterium has followed a comparable evolution process, changing from a terrestrial to an endophytic lifestyle.

Further examples of closely related actinobacteria with different lifestyles reflected in their genomes include, among others, the genera *Frankia*, *Mycobacterium* and *Streptomyces*. In the case of *Frankia*, comparative genomic analysis of three representative strains, differing by less than 2% in their 16S rRNA genes revealed significant differences in their genome sizes (5.4–9.0 Mb) suggesting that these differences (e.g. gene deletion, acquisition and duplication, etc.) reflect their rapid adaptation to contrasted host plants and to their environments [15]. Similarly, several mycobacterial genomes were analyzed both at the nucleotide and protein levels. One of the most striking features was lipid metabolism genes with marked expansions of the number of genes related to saturated fatty acid metabolism in the pathogenic mycobacteria compared to the soil-dwelling strains [16].

In an effort to identify the genomic traits which make possible adaptation from a soil dwelling way of life to an endophytic habitat, the aim of this work was to present the genome sequence analysis of a representative strain, *Micromonospora lupini* Lupac 08, isolated from a nitrogen fixing nodule of the legume *Lupinus angustifolius*. This strain is part of a collection of more than 2000 strains isolated from nitrogen fixing root nodules of diverse legume [17,18] and actinorhizal species [13]. Strain Lupac 08 was selected as it showed good plant growth promotion, was used previously for in situ localization studies *in planta* [11] and produced several new secondary metabolites [9,10]. The results presented here show that the genome of *M. lupini* Lupac 08 contains a diverse array of genes that may help its survival in soils or in plant tissues, while the high number of putative plant degrading enzyme genes identified in its genome is quite surprising since this bacterium is not considered a plant-pathogen and may instead reflect their ability to bind to plant structural compounds.

Results

Phylogenetic position of *M. lupini* Lupac 08 and general genome features

The phylogenetic position based on 16S rRNA gene sequence analysis of strain Lupac 08 with respect to currently described *Micromonospora* species and other members of the family *Micromonosporaceae* is presented in Figure 1. Those strains associated with plant/rhizosphere sources are highlighted. Strain Lupac 08 was clearly positioned within the genus *Micromonospora* and forms a subgroup together with the species *Micromonospora saelicesensis, Micromonospora zamorensis* and *Micromonospora chokoriensis*. These strains were isolated from a nitrogen fixing nodule, the rhizosphere of a *Pisum sativum* plant and a sandy soil, respectively. Nevertheless, a clear picture based on the habitat cannot emerge from this analysis.

M. lupini Lupac 08 was shown to have a circular chromosome of 7,327,024 bp with a GC content of 71.96% and no plasmid. A total of 7158 genomic objects were identified: 7,054 protein-coding, 10 rRNAs, 77 tRNAs, and 12 miscRNAs genes. The average gene length was 964 bp with an average intergenic distance of 126 bp. After manual validation of the automatic annotation, 61.5% (4338 CDSs) of the genes were assigned a biological function while 38.5% were registered as open reading frames (ORFs) with an unknown function. Based on the G+C skew analysis and position of *dnaA*, the probable origin of replication (*oriC*), was mapped close to the ribosomal protein *rpmH*. A circular representation of the *M. lupini* chromosome is provided in Figure 2 indicating some of the features described above.

The genomic characteristics of strain Lupac 08 and three additional *Micromonospora* genomes deposited in the public databases including *Micromonospora* sp. strain L5 isolated from root nodules of *Casuarina equisetifolia* [19]; *M. aurantiaca* ATCC 27029^{T} and *Micromonospora* sp. ATCC 39149 isolated from soil (Table 1) were compared. An important difference between the four strains was the number of tRNAs identified. *M. lupini* 08 contained by far the highest number with 77 tRNAs while the other strains had between 51 and 53. At present, *M. lupini* Lupac 08 contains one of the largest numbers of tRNAs reported among the actinobacteria sequenced. The number of rRNA and tRNA genes in a genome can be seen as an indication of positive selection. A high number of rRNA genes increases ribosome synthesis, which in turn increases the protein synthesis rate [20] and growth rate [21].

Comparative genome analysis

COG distribution. Nearly 70% of the CDS were classified into clusters of orthologous groups (COGs, Table S1). Thus, 4873 out of 7054 CDS were assigned to 24 different categories, including those for amino acid transport and metabolism (E, 12.7%), transcription (K, 10.8%), carbohydrate transport and metabolism (G, 9.7%), inorganic ion transport and metabolism (P, 8.7%), energy production and conversion (C, 5.5%), and signal transduction mechanisms (5.5%).

The COG distribution of *M. lupini* was similar to that observed in other bacteria in the family *Micromonosporaceae*, however various differences were detected such as the abundance of genes related to carbohydrate transport and metabolism. Among the *Micromonospora* genomes currently available, *M. lupini* Lupac 08 contained the highest percentage of genes (9.7%, 685) related to this category, followed by *Micromonospora* sp. L5 (8.9%, 598) and *M. aurantiaca* ATCC 27029 (8.5%, 576). The gene contents (in the same COG category) of other bacterial genomes classified in the family *Micromonosporaceae* were lower as in the case of *S. tropica* CNS-205 (7.4%, 391) and *S. arenicola* CNH-643 (6.4%, 374) two obligate marine actinomycetes. On the other hand, the overall COG profiles of *Verrucosispora maris* AB-18-032^{T} (genome size 6.7 Mb) and *M. lupini* Lupac 08 were very similar and no clear differences were found. Although *V. maris* was isolated from a sea sediment, it does not require sea salts for growth and it is not considered an obligate marine microorganism

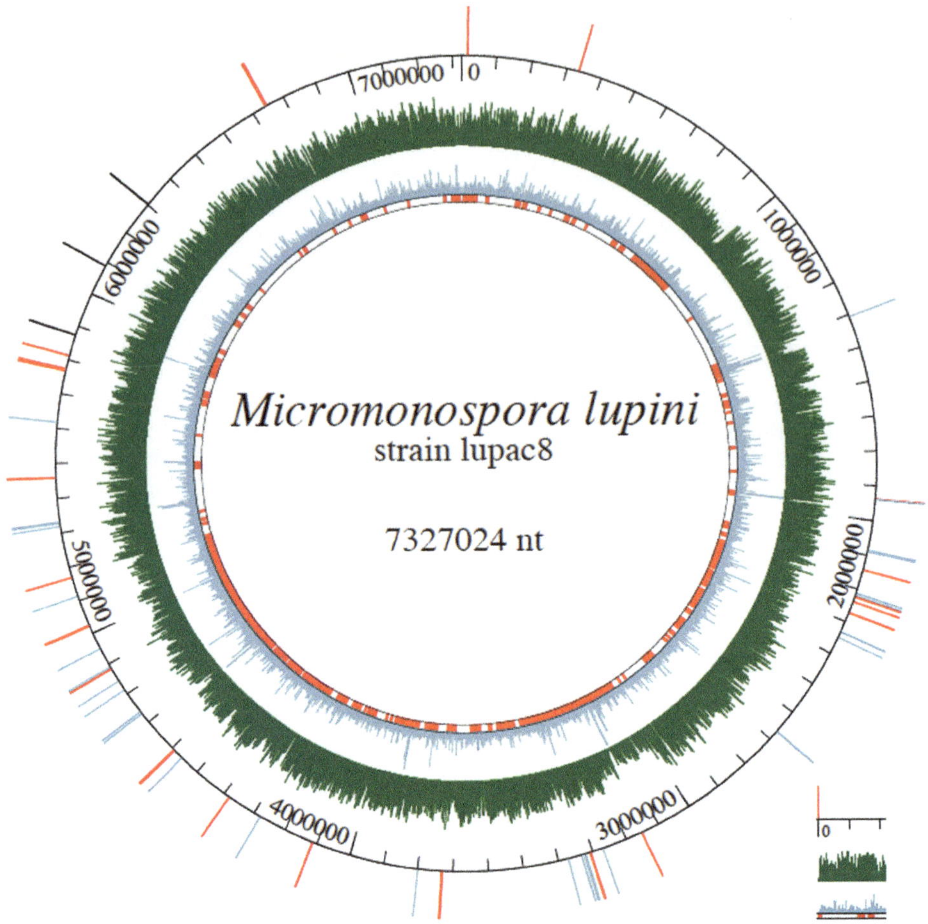

Figure 2. Circular representation of *Micromonospora lupini* Lupac 08. Circles displayed from the outside in: 1. Cellulose-binding genes in black, chitin-binding genes in red, lectin genes in lavender blue; 2. Genome coordinates; 3. MW; 4. GC% (linear range between 65 and 80%); 5. Regions of genome plasticity according to the RGP_Finder method (Mage platform) based on synteny breaks between the query genome (Lupac 08) and close genomes (*Micromonospora aurantiaca* ATCC 27029^T, *Micromonospora* sp. L5 and *Verrucosispora maris* AB-18-032^T) correlated with HGT features (tRNA hotspot, DNA repeats, mobility genes), and compositional bias and GC deviation computation. C1 to C15 indicate the position of the 15 clusters of genes coding for secondary metabolites of Table 4.

unlike *S. tropica* and *S. arenicola*. Thus, its metabolism suggests a terrestrial life style. Micromonosporae are well known for their ability to degrade complex polysaccharides such as cellulose, chitin and lignin [22,23]. In particular, cellulose is frequently utilized as a carbon source [24,25]. Therefore the abundance of these genes in the genome of strain Lupac 08, at first glance may not seem surprising, however, the value of 9.7% is comparable to that of highly active cellulolytic microorganisms such as *Cellulomonas*

Table 1. Comparative genomic characteristics of *M. lupini* Lupac 08 and three *Micromonospora* genomes publicly available.

Feature	*M. lupini* Lupac 08	*M. aurantiaca* ATCC 27029^T	*Micromonospora* sp. L5	*Micromonospora* sp. ATCC 39149
Size (Mb)	7.3	7.0	6.9	6.8
GC%	72	73	73	72
rRNA Operon	10	9	9	6
tRNA	77	52	53	51
CDS number	7054	6676	6617	5633
Average gene size (kb)	946	964	969	975
Protein-coding density (%)	90.1	90.4	90.4	89.9
Genes in COGs (%)	70.2%	68.3%	69%	nd

nd, not determined.

flavigena 134^{T} (9.5%) and *Thermobifida fusca* XY (7.9%), which are abundant in cellulose enriched environments such as soil, or plant tissues.

Synteny. The genome sequence of strain Lupac 08 was aligned with those of *Micromonospora* sp. L5, *M. aurantiaca* ATCC 27029^{T} and *Micromonospora* sp. ATCC 39149^{T} (Fig. 3). Although the four genomes share a significant amount of genetic characteristics, they have undergone various inversions and translocations and *M. lupini* Lupac 08 contains the highest number of non-conserved regions. In addition, this alignment shows a high homology between strains *Micromonospora* sp. L5 and *M. aurantiaca* ATCC 27029^{T} confirming their close phylogenetic relationship as suggested by 16S rRNA gene phylogeny (Fig. 1); nevertheless, strain L5 shows a large inversion event. Thus, although the four *Micromonospora* genomes share many common features, it is also evident that *M. lupini* contains unique genomic regions as compared to *M. aurantiaca* ATCC 27029^{T} or *Micromonospora* sp. L5.

Diversity of *Micromonosporae*: core vs. flexible gene pool

Using the *Micromonospora* genomes of strains M. *lupini* Lupac 08, *M. aurantiaca* ATCC 27029^{T} and that of *Micromonospora* sp. L5 available in the NCBI [19], the core genome was calculated using the SiLix software [26]. The core genome was composed of 2294 CDSs, which correspond to approximately 32% of the predicted proteome. In addition, *M. lupini* Lupac 08 contained the highest number of strain specific CDSs, 4702 (66.6%), which is a very high value when compared to *Micromonospora* sp. L5 and *M. aurantiaca* ATCC 27029^{T} (13–14%, Figure 4), which both share a high gene similarity (85–86%).

Horizontal gene transfer is universally recognized as an efficient mechanism for microorganisms to acquire functions that enable them to adapt to environments with different selective pressures. Therefore insertion elements, transposases, integrated phages, and plasmids can be related to the plasticity of a genome. Strain Lupac 08 contained 49 CDSs (0.7%, of total CDSs) related to gene exchange including eight integrases and eleven recombinases. Except for seven CDSs, most of these genes were grouped into 20 clusters. Interestingly, eight of these mobile element clusters were found near genes related to carbohydrate transport and metabolism.

Metabolic Features. A metabolic pathway reconstruction was performed between the genome of strain Lupac 08 and 20 additional strains among which plant pathogens, symbiotic and saprophytic bacteria were included. The distribution and grouping of the microorganisms analyzed using 798 metabolic routes are presented in Figure 5. A good correlation was obtained between the microorganisms, their life style and phylogeny. Two main groups were obtained, the proteobacteria and actinobacteria. Within the actinobacteria, three clusters were clearly identified: the first one contained strains that belonged to the family *Micromonosporaceae*, the second cluster corresponded to various streptomycetes and the third cluster included the three *Frankia* genomes. Surprisingly, *Micromonospora lupini* Lupac 08 showed a closer metabolic relationship with the three *Frankia* strains (ACN14a, CcI3 and EAN1pec) than with the other two *Micromonospora* genomes.

Plant/Soil-associated life style

Transport systems. Organisms living in endophytic associations need to share resources with their host. Membrane transport systems play essential roles in cellular metabolism and activities. Current data suggest a correlation of transporter profiles to both evolutionary history and the overall physiology and lifestyles of organisms [27].

A total of 631 CDSs were located in the genome of *M. lupini* coding for a large diversity of transporters, representing approximately 8.9% of the genome. The majority of CDSs were related

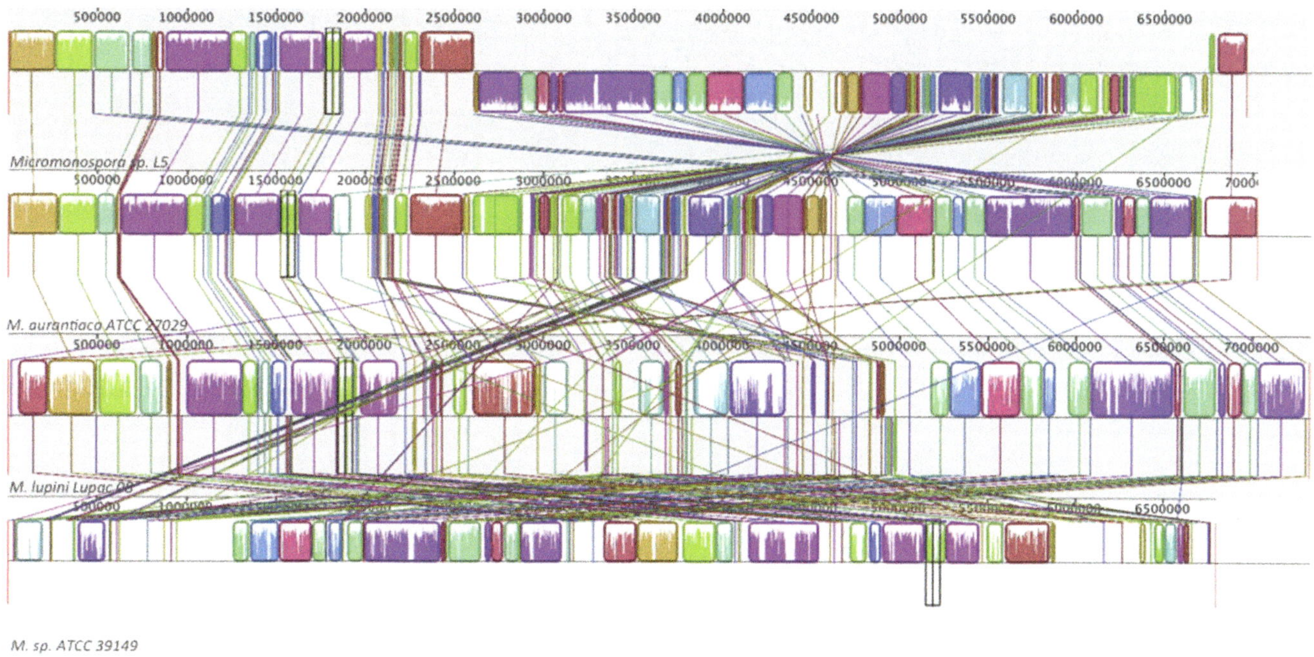

Figure 3. MAUVE alignment of the genome sequences of *Micromonospora lupini* Lupac 08, *Micromonospora* sp. L5, *Micromonospora aurantiaca* ATCC 27029^{T} and *Micromonospora* sp. ATCC 39149. When boxes have the same colour, this indicates syntenic regions. Boxes below the horizontal line indicate inverted regions. Rearrangements are shown by coloured lines. Scale is in nucleotides.

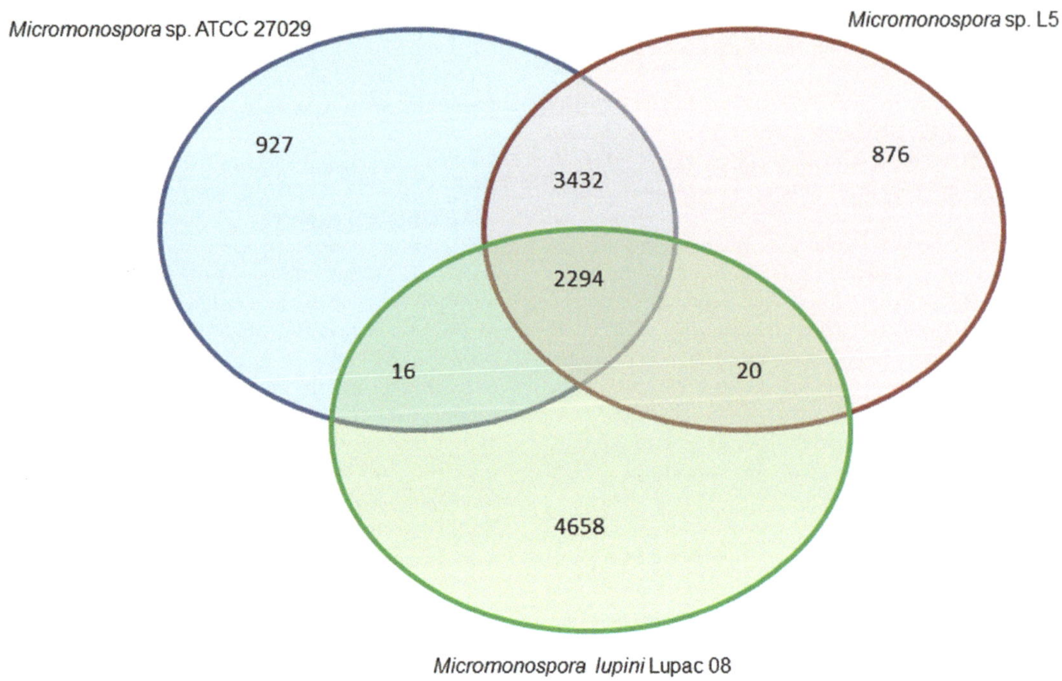

Figure 4. Venn diagram showing the number of clusters of orthologous genes, shared and unique, between *M. lupini* Lupac 08, *Micromonospora* sp. L5 and *M. aurantiaca* ATCC 27029^T.

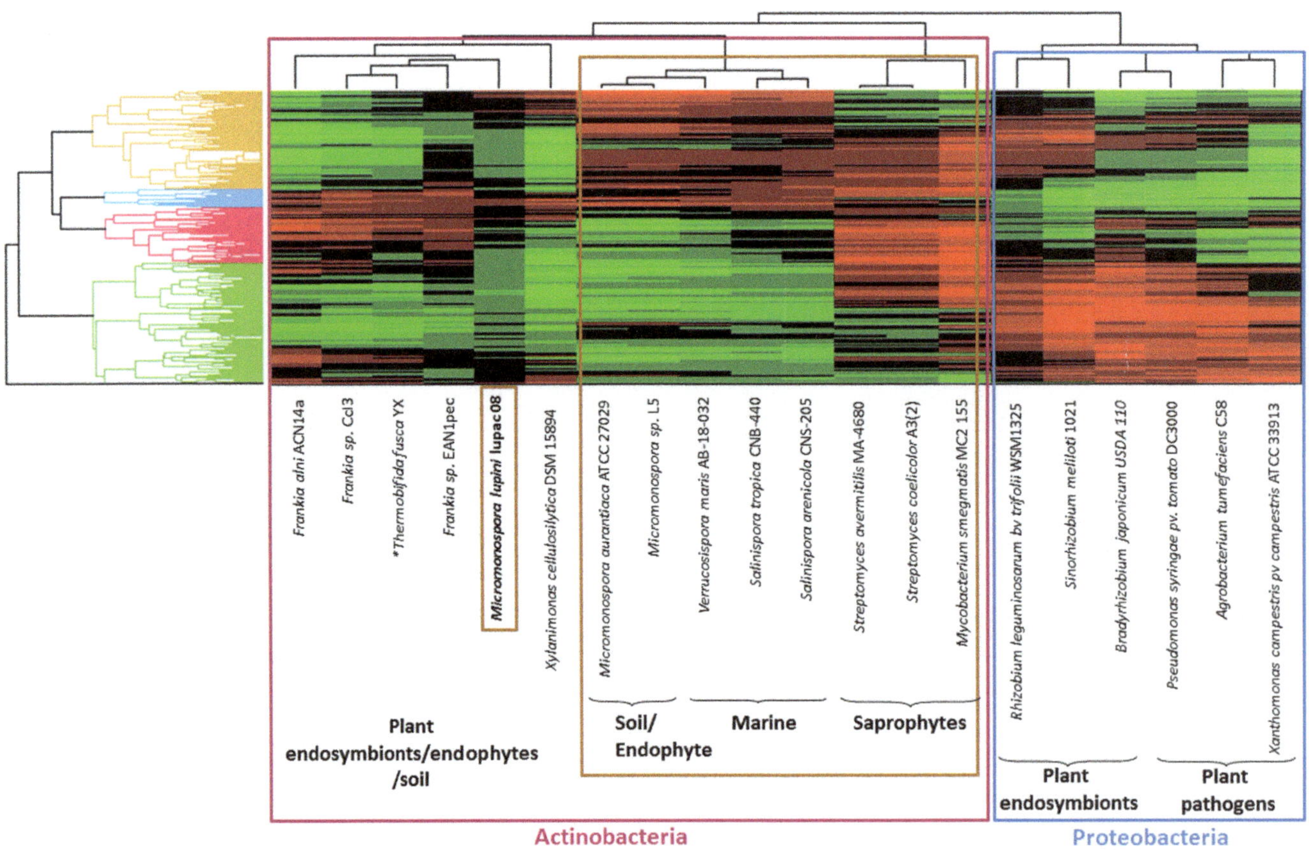

Figure 5. Bicluster plot of the metabolic profiles of *M lupini* Lupac 08 and 20 other bacterial genomes.

to ATP-binding (dependent) transporters of which 362 corresponded to ABC transporters; the next most abundant (215 CDSs) coded for secondary transporters, with 105 classified in the Major Facilitator Superfamily -MFS-); 17 transporters belonged to ion channels and 20 were unclassified. The number of transporters determined in *M. aurantiaca* ATCC 27029^T and *Micromonospora* sp. L5 were lower with 575 and 587, respectively (Table 2).

The number of transporters identified in the genome of *M. lupini* Lupac 08 is correlated with those in other bacteria with a plant/soil associated life style, which requires an efficient nutrient uptake system to obtain nutrients produced by the host plant, in addition to those found in the rhizosphere and the soil (Table 2) [27,28]. However, the number of transporters identified in strain Lupac 08 was lower than those present in other bacteria such as *Bradyrhizobium japonicum* USDA6^T (1138, 11.8%), *Mesorhizobium loti* MAF303099 (968, 14.2%), *Sinorhizobium meliloti* Sm1021 (1024, 16.4%) and *Rhizobium leguminosarium* bv. *trifolii* WSM 3125 (1087, 15.5%), which form a very close interaction with legumes. Nevertheless, the overall distribution (types) and the percentages of these values were similar. An additional difference was the absence of phosphotransferase system transporters (PTS) in *M. lupini* as compared to the strains mentioned above and other soil/plant bacteria included in Table 2. On the other hand, the overall profile of *M. lupini* Lupac 08 was very similar to those of *Frankia* sp. ACN14a and *Frankia* sp. CcI3 which also lack a PTS system.

Secretion systems. Secreted proteins play a number of essential roles in bacteria, including the colonization of niches and host–pathogen interactions. In Gram-positive bacteria, the majority of proteins are exported out of the cytosol by the conserved Sec translocase system or, alternately, by the twin-arginine translocation system. In addition, a unique protein export system, the type VII or ESX secretion system also exists in some Gram positive bacteria [29].

The genome of *M. lupini* Lupac 08 encodes for 537 (7.6%) secreted proteins including several protein secretion systems (Table 3). All genes related to the Sec-dependent pathway were located and included the SecY and SecE proteins which form the membrane channel and interact with the cytoplasmic membrane protein SecG; the auxiliary proteins SecD, YajC and the ATPase SecA. In addition, the heterodimer Ffh-FtsY (MiLup08_41486 and Milup08_41460) was also present. As in other Gram-positive bacteria, *M. lupini* Lupac 08 lacks homologs of SecB, the chaperone that targets proteins to the Sec translocon for passage through the cytoplasmic membrane [30].

Genes related to the Sec-independent twin-arginine translocation pathyway (TAT), which exports prefolded proteins across the cytoplasmic membrane using the transmembrane proton gradient as the main driving force for translocation were also located in strain Lupac 08 (Table 3). Homologs of TatA and TatC were identified, however no homolog for TatB was found. Similar to other actinobacteria (e.g. *Frankia* sp. ACN14a) the *tatA* gene was found next to *tatC*. Only an ORF encoding TatC was located in the genomes of *Micromonospora* sp. L5 and *M. aurantiaca* ATCC 27029^T while no copies of *tatA* or *tatB* were found.

A set of fifteen genes identified as part of the type VII secretion system were located in *M. lupini* Lupac 08 (Table 3). These are arranged in three different clusters and included the essential proteins for secretion EccC, EccD, EsxA and EsxB [31]. The first cluster contains eight genes: *eccC, esxA, esxB, eccD, eccB, eccE* and two copies of *mycP*, a subtilisin-like serine protease which also appears essential but the function of which is not yet known [32]. The second cluster includes a copy of *esxA* (MiLup08_40381), *esxB* (MiLup08_40380) and *mycP*, annotated as S8 S53 subtilin kexin sedolisin (MiLup08_40382). Finally a third cluster contains the genes *eccC* (MiLup08_46744), *eccD* (MiLup08_46743) and *mycP* (MiLup08_46745).

Gram-negative bacteria use the type II secretion system to transport a large number of secreted proteins from the periplasmic space into the extracellular environment. Many of the secreted proteins are major virulence factors in plants and animals [33]. Type II secretion systems have been found in all completely sequenced plant pathogenic bacterial genomes, except in *Agrobacterium tumefaciens*. In addition, other bacteria have been shown to use secretion systems for the delivery of toxins, proteases, cellulases and lipases [34–37]. Genes coding for this system have also been reported for the three symbiotic strains *Frankia* [38].

Fifteen genes in *M. lupini* were annotated as components of the Type II secretion system, grouped into clusters of three to five genes (Table 3). Nine of these genes were annotated as Type II secretion system proteins including protein E and protein F; four were recorded as TadE family proteins and Milup08_40403 was annotated as an uncharacterized protein closest to one found in the *Frankia* symbiont of *Datisca glomerata*.

The secretion systems III and IV which are commonly related to plant-associated bacteria transport a wide variety of effector proteins into the extracellular medium or into the cytoplasm of eukaryotic host cells thus affecting the interaction [39]. In addition, a functional type IV system has been described in the plant symbiont *M. loti* strain R7A [40]. A gene annotated as *virB4* and related to secretion system IV was located in Lupac 08 (MiLup08_42651), this ORF is surrounded by proteins with unknown function related to those present in the genomes of *Micromonospora* sp. L5 and *M. aurantiaca* ATCC 27029^T.

Survival against plant defenses. Reactive oxygen species (ROS) play a major role in plant defense against pathogens. In response to attempted invasion, plants mount a broad range of defense responses, including the synthesis of ROS. *M. lupini* needs to survive under an oxidative environment in the rhizosphere before it can colonize plant roots and its genome revealed several genes encoding proteins to neutralize oxidative stress. The following genes were identified: three *sod* genes (MiLup08_45788, MiLup08_46012 and MiLup08_46604) that code for superoxide dismutases; a catalase HPII *katE* (MiLup08_44247); a catalase-peroxidase (*katG*, MiLup08_44435) and a catalase hydroperoxidase (*katA*, MiLup08_45857); four hydroperoxide reductases (MiLup08_40110, MiLup08_40293, MiLup08_41393, MiLup08_45407); a chloroperoxidase (MiLup08_44157) and a thiol peroxidase (MiLup08_43629).

In addition, a putative organic hydroperoxide resistance protein (Ohr, MiLup08_45098); a 4-hydroxyphenylpyruvate dioxygenase (Hpd, MiLup08_46664) and a homogentisate 1,2-dioxygenase (MiLup08_46677) were identified. Other enzymes include a glutathione peroxidase (MiLup08_45173); two glutathione transferases (MiLup08_46358 and MiLup08_41529) and four glutathione-S-transferases (*fdh*, MiLup08_42270, MiLup08_42834, MiLup08_44416 and 45648). Experimental data indicated that *M. lupini* indeed yields a catalase positive reaction [17] confirming the functionality of some of these genes. Therefore, to successfully reach the internal plant tissues, these genes may defend the bacterium against a ROS release by the plant.

Regulation as a means of adaptation

Lifestyle can be viewed as the set of biotopes an organism can thrive into and the relationships that it establishes with other species and its abiotic components. It is one of the driving forces that contribute to the overall characteristics of bacterial genomes [41].

Table 2. Transporters identified in the genome of *M. lupini* Lupac 08 and comparison with other bacteria with a plant/soil associated life styles.

	M. lupini **Lupac 08**	*M. aurantiaca* **ATCC 27029^T**	*Micromono-spora* **sp. L5**	*S. arenicola* **CNS 205**	*S. tropica* **CNB 440**	*F. alni* **ACN14**	*Frankia* **sp. Ccl3**	*F. symbiont Dastica glomerata*	*B. japo-nicum* **USDA6^T**	*R. leguminosarum Bv trifolii WSM 3125*	*Enterobacter* **sp. 368**	*S. coelicolor A32*	*S. scabiei* **8722**	*Pseudomonas syringae pv phaseolicola 1448A*
Genome size (Mb)	7.3	7.0	6.9	5.7	5.2	7.5	5.4	5.3	9.6	7.4	4.6	9.1	10	5.9
Total transport proteins	631	575	587	405	413	433	253	300	1138	1087	662	798	775	670
Tran-sporters (%)	8.9	8.5	8.7	7.1	7.8	6.4	4.5	7.1	11.8	15.5	15.6	9.7	8.2	12.5
No. Transporters/ Mb genome	0.08	0.08	0.08	0.07	0.08	0.06	0.05	0.06	0.12	0.15	0.14	0.09	0.08	0.11
ATP dependent (% of total)	**379 (60.1%)**	**362 (63.0%)**	**366 (62.4%)**	**244 (60.2%)**	**247 (59.8%)**	**281 (64.9%)**	**146 (57.7%)**	**210 (70%)**	**684 (60.1%)**	**800 (73.6%)**	**317 (49.7%)**	**461 (57.8%)**	**480 (61.9%)**	**392 (58.5%)**
ABC family*	362 (95.5%)	341 (94%)	342 (93.4%)	225 (92.2%)	228 (92.3%)	262 (93.2%)	127 (87%)	189 (90%)	645 (94.3%)	769 (96.1%)	287 (90.5%)	433 (93.9%)	455 (94.8%)	346 (88.3%)
Ion channels (% of total)	**17 (2.7%)**	**14 (2.4%)**	**15 (2.6%)**	**9 (2.2%)**	**11 (2.7%)**	**12 (2.8%)**	**7 (2.8%)**	**7 (2.3%)**	**24 (2.1%)**	**26 (2.4%)**	**23 (3.5%)**	**19 (2.4%)**	**22 (2.8%)**	**32 (4.8%)**
Phospho-transferase system (PTS)	–	–	–	**4 (1%)**	–	–	–	–	**4 (0.4%)**	**6 (0.6%)**	**47 (7.1%)**	**10 (1.3%)**	**6 (0.8%)**	**5 (0.7%)**
Secondary transporter	**215 (34.1%)**	**185 (32.3%)**	**192 (32.7%)**	**9 (2.2%)**	**145 (35.1%)**	**130 (30%)**	**89 (35.2%)**	**75 (25%)**	**408 (35.9%)**	**234 (21.5%)**	**256 (38.7%)**	**286 (35.8%)**	**252 (32.5%)**	**226 (33.7%)**
MFS family*	105 (48.8%)	68 (36.8%	70 (36.5%)	65 (46.8%)	74 (51%)	64 (49.2%)	36 (40.4%)	33 (44%)	114 (27.9%)	69 (29.5%)	84 (32.8%)	120 (42%)	111 (44%)	72 (31.9%)
RND family		6 (3.2%)	7 (3.6%)	7 (5%)	8 (5.5%)	9 (6.9%)	7 (7.9%)	7 (9.3%)	31 (7.6%)	13 (5.6%)	18 (7%)	15 (5.2%)	18 (7.1%)	16 (7.1%)
Unclassified	**20 (3%)**	**13 (2.3%)**	**13 (2.2%)**	**8 (2%)**	**9 (2.2%)**	**10 (2.3%)**	**11 (4.3%)**	**8 (2.7%)**	**11 (1%)**	**16 (1.5%)**	**14 (2.1%)**	**21 (2.6%)**	**14 (1.8%)**	**11 (1.6%)**

*Number and percentage in relation to the total number of ATP dependent and secondary transporters respectively.

Table 3. Secretion system genes present in the genome of *M. lupini* Lupac 08.

Secretion System	Gene (Milup08_X)	Product
Sec-dependent	*secY* (*prlA*)(46297)	Preprotein translocase, membrane component
	secE (46336)	Preprotein translocase subunit secE
	secG (44961)	Preprotein translocase SecG subunit
	secD (42464)	Protein-export membrane protein secD
	secF (42465)	Protein-export membrane protein secF
	yajC (42463)	Preprotein translocase, YajC subunit
	secA (41087)	Protein translocase subunit secA
	ffh (41468)	Signal recognition particle protein
	scRNA (misc_RNA-12)	SRP, Ribosome-nascent chain complex (RNC)
	yidC (30220)	Cytoplasmic insertase into membrane protein
	yidC-like (43138)	Membrane protein insertase, YidC/Oxa1 family
	yidC (45964)	Inner membrane protein translocase component YidC
	Milup_08_41485	Signal peptidase I
	Milup_08_41486	Signal peptidase I
	Milup_08_42560	Conserved protein of unknown fuction (probable signal peptidase I)
	lspA (45113)	Lipoprotein signal peptidase
	lgt (45071)	Prolipoprotein diacylglyceryltransferase
TAT-	*tatA* (43424)	Sec-independent protein translocase protein tatA/E homolog
	tatC (43425)	Sec-independent protein translocase protein tatC homolog
Type II- (T2SS)	Milup_08_40403	Similar to uncharacterized protein from *Frankia* symbiont of *Diastica glomerata*
	Milup_08_40405	Putative helicase/secretion neighbourhood TadE-like protein
	tadE (40223)	TadE Family protein
	tadE (40224)	TadE Family protein
	tadE (42690)	Similar to TadE family protein
	tadE (42691)	Similar to TadE family protein
	Milup_08_40226	Type II secretion system protein
	Milup_08_40227	Type II secretion system protein
	Milup_08_40228	Type II secretion system protein E
	Milup_08_40398	Type II secretion system protein E
	Milup_08_40399	Similar to Type II secretion system protein E
	Milup_08_40401	Similar to Type II secretion system protein
	Milup_08_42693	Type II secretion system protein F
	Milup_08_42694	Type II secretion system protein F
	Milup_08_42695	Type II secretion system protein
Type IV- (T4SS)	Milup_08_42651	VirB4 protein-like protein
Type VII/WXG100-	*eccB* (40554)	ESX-4 secretion system protein eccB4
	eccC (40438)	FtsK/SpoIIIE family protein
	eccC (40557)	ESX-4 secretion system protein/cell division protein ftsK/spoIIIE
	eccC (46744)	FtsK/SpoIIIE-like transmembrane protein
	eccD (40556)	ESX-4 secretion system protein eccD4/Putative secretion protein snm4
	eccD (46743)	FtsK/SpoIIIE family protein
	eccE (40555)	Putative uncharacterized protein
	esxA (40381)	Putative uncharacterized protein
	esxA (40559)	Putative uncharacterized protein
	esxB (40380)	Putative uncharacterized protein
	esxB (40558)	Putative uncharacterized protein
	mycP (40382)	Peptidase S8 and S53 subtilisin kexin sedolisin
	mycP (40560)	Peptidase S8 and S53 subtilisin kexin sedolisin
	mycP (40564)	Peptidase S8 and S53 subtilisin kexin sedolisin
	mycP (46745)	Peptidase S8 and S53 subtilisin kexin sedolisin

TAT, twin-arginine translocation; X, corresponds to the annotation gene numbers given in parenthesis.

The *M. lupini* genome shows a strong emphasis on regulation, with 643 proteins (~10%) predicted to have a regulatory function. This value is lower than that reported for the saprophytic strain *Streptomyces coelicolor* A3(2) with an exclusively terrestrial lifestyle (965 proteins; 12.3%) [42], but higher than the endosymbiotic strains *M. loti* MAFF303099 (542 proteins, 7.7%) [43]; *Frankia alni* ACN14a (515 proteins, 7.6%); *Frankia* sp. EAN1pec (555, 6.1%) and *Frankia* sp. CcI3 (244 proteins, 4.3%).

The genome codes for various regulator families such as TetR, AraC, LacI, ArsR, MerR, AsnC, MarR, DeoR, GntR and Crp. In addition, thirty-three ECF (extra-cytoplasmic function) sigma factors were located. Furthermore, 147 genes were related to two-component regulatory systems of which 34 were LuxR proteins. These two-component systems appear to play a crucial role in quorum sensing of Gram-positive bacteria and a positive correlation between plant-microbe interactions and the number of LuxR proteins has been suggested [44,45].

Many regulatory genes (~18%) were located near polysaccharide related loci including those involved in plant cell wall degradation. Specifically, 63% of cellulose degradation or cellulose binding genes had a nearby regulator (proximity ranged from 2–4 genes up or downstream). In the case of xylan metabolism, regulators were identified for 50% of the genes, while 43% of pectin metabolism genes also had a regulator nearby. An extended overview of the regulators and their associated carbohydrate genes is presented in Table S2.

An endophytic bacterium highly equipped with an array of plant cell wall degrading enzymes

The ability of *M. lupini* lupac 08 to assimilate a wide range of sugars was previously reported [19] and this is clearly reflected in its genome. The range of simple and complex saccharides assimilated by this strain include cellobiose, cellulose, glucose, mannitol, starch, sucrose, trehalose, xylan and xylose among others. Genomic analyses confirmed the presence of a large number of genes devoted to the metabolism of carbohydrates, including many compounds of plant origin. Plant-polymer degrading enzymes such as cellulases, xylanases and pectinases have been suspected to play a role in internal plant colonization [46]. In the case of plant pathogenic bacteria and fungi, these gain access by actively degrading plant cell wall compounds using glycoside hydrolases including cellulases and endoglucanases. However, genomic analyses show that non-pathogenic endophytic microorganisms such as *Enterobacter* sp. 638 [47], *Azoarcus* BH72 [39] or the symbiotic actinobacterium *Frankia sp.* [48] have only a reduced set of cell-wall degrading enzymes.

The genome of *M. lupini* Lupac 08 revealed a significant number of genes encoding enzymes potentially involved in plant-polymer degradation but also an important number of cellulose-binding related genes. Overall, about 10% of the genome coded for genes related to carbohydrate metabolism of which 192 had a hydrolytic function. At least 79 genes putatively involved in interactions with plants and with the potential to hydrolyze plant polymers were identified (Table S1). These genes were placed into the glycosyl hydrolase families GH5, GH6, GH9, GH10, GH11, GH18, GH20, GH43, GH44 and GH62, or into the carbohydrate binding modules CBM2, CBM13, CBM33, CBM3, CBM46, CBM42, CBM5, CBM4, CBM6 and CBM32. The CBM2 family was the most abundant appearing in 46 of the 79 genes identified.

Fourteen genes were further identified as lectins or proteins with lectin binding domains, which presumably bind to and interact with carbohydrates. Some of these loci (e.g. Milup_42969, Milup_42975, Milup_44484, and Milup_44962) appear to be related to cellulases and xylanases, respectively. These proteins are important as they serve as a means of attachment between a bacterium and its host (animal or plant) and are produced by either of the two interacting organisms [48].

Compared to the 45 enzymes predicted to act on oligo- and/or polysaccharides reported for *T. fusca* XY [49], the number of these enzymes present in the genome of *M. lupini* is significantly higher.

Cellulose metabolism. Aerobic cellulolytic actinobacteria have been shown to use a system for cellulose degradation consisting of sets of soluble cellulases and hemicellulases. Most of these independent cellulolytic enzymes contain one or more carbohydrate binding domains [50].

A total of 46 genes were found to present a hydrolytic or binding fuction towards cellulose (Table S1). Several endoglucanases were detected in strain Lupac 08 (e.g. C1, C2 C10 and C14), these enzymes hydrolyze internal bonds at random positions of amorphous regions of cellulose and generate chain ends for the processive action of cellobiohydrolases (exoglucanases). A copy of the exoglucanase gene *cbhA* (C16) was also located in the genome. Exoglucanases act on the ends of cellulose polysaccharide chains, liberating cellobiose as the major product. β-D-glucosidases such as M108 and M109 which would further hydrolyze cellobiose were also identified. In addition, several extracellular cellulase coding genes were identified including *celA* (C3 and C6), *celB* (C5) and *celD* (C13). These results strongly suggest that strain Lupac 08 is potentially capable of completely degrading cellulose.

Strain Lupac 08 was tested for *in vitro* production of cellulases. Very high cellulase activity was detected in minimal agar supplemented with carboxymethylcellulose (CMC, 0.5%) (Fig. 6A). When the culture medium was supplemented with glucose (1%) similar results were obtained indicating that this sugar did not repress nor derepress the expression of the genes responsible for the production of cellulases.

Hemicellulosic substrates. Genome analysis also revealed the ability of *M. lupini* to convert various hemicellulosic substrates to sugars. Twelve putative genes related to the metabolism of xylan included several copies of extracellular xylanases (X1, X3, X4, X5, X6, X7, X9, X10 and X12; see Table S1); an extracellular bifunctional xylanase/deacetylase (X8); and an arabinofuranosidase (X2) which work synergistically with xylanases to degrade xylan to its component sugars. Genes for several α-arabinofuranosidases were also identified (C17, M33, and M39); these are exo-acting enzymes which hydrolyze nonreducing arabinofuranose residues from arabinoxylan, pectins, and shorter oligosaccharides.

In vitro xylanase activity was detected in strain Lupac 08 when tested in a minimal medium supplemented with xylan (1%). Production of xylanases was detected after incubation for 4 days increasing significantly after 14 days (Fig. 6D). The substrate was assayed with and without glucose with similar results.

Starch degradation. Starch is a ubiquitous and easily accessible source of energy. In plant cells it is usually deposited as large granules in the cytoplasm. Several genes coding for amylo-α-1,6-glucosidases (e.g. M26, M32, M44, M63, M111 and M121; Table S1) were located in addition to two *amyE* homologs that code for an extracellular α-amylase. Furthermore, strain Lupac 08 was able to degrade this polymer under laboratory conditions (Fig. 6C) and it was previously shown that Lupac 08 can utilize this substrate as a carbon source [17].

Pectin degradation. Pectinolytic enzymes can degrade pectic substances either through hydrolysis (hydrolases) or trans-elimination (lyases) [51] and are important virulence mechanisms in many soft-rotting and macerating pathogens [52]. Six pectate lyases (P1, P3, P4, P5 and P6; Table S1) were located in the

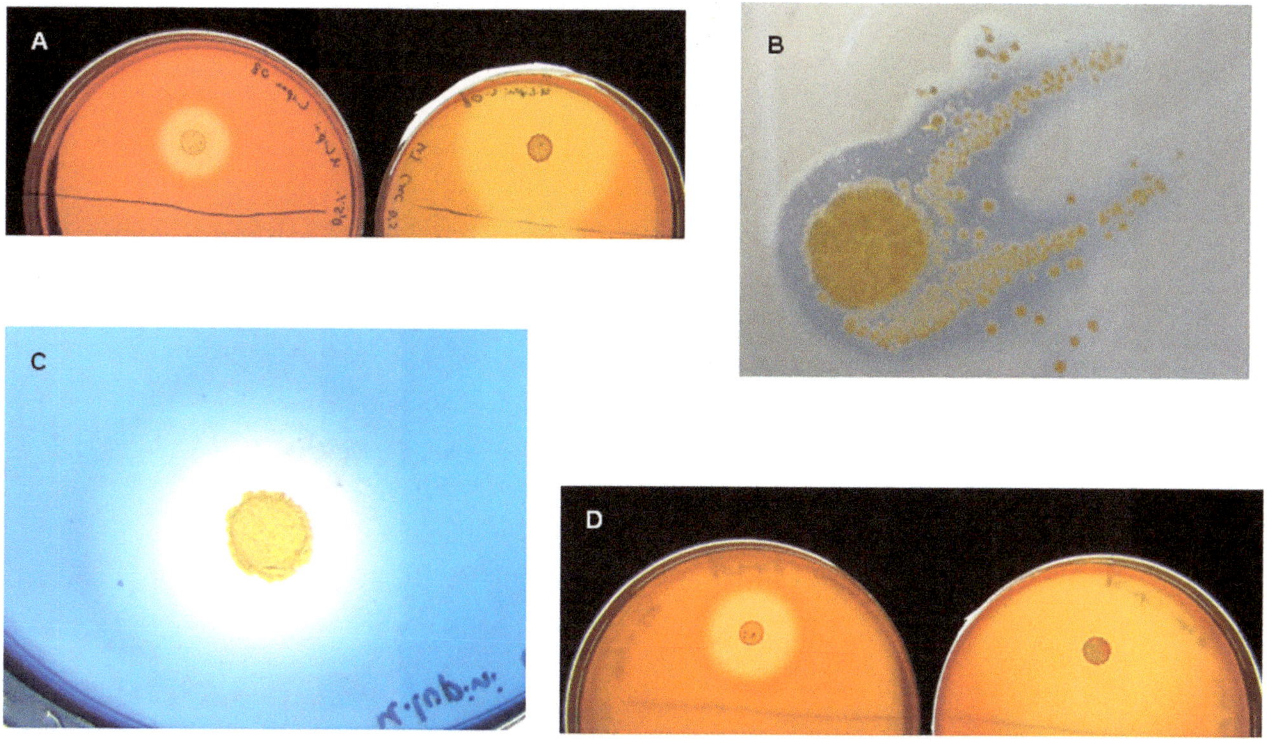

Figure 6. Expression of cellulose, starch, xylan and chitin degrading genes in *Micromonospora lupini* Lupac 08. (A) carboxymetheylcellulose hydrolysis at 4 (left) and 14 (right) days after inoculation. (B), starch hydrolysis at 4 days after inoculation. (C), chitin degradation at 7 days after inoculation. (D), xylan degradation at 4 (left) and 14 (right) days after inoculation.

genome of *M. lupini*, two of which were annotated as virulence factors (P5 and P6). In addition, an extracellular pectin methylesterase gene, *pmeA*, and a gene coding for a pectate lyase involved in D-galacturonic acid hydrolysis (P7) were also identified. Interestingly, *T. fusca* XY contains two pectin lyase homologs but does not appear to possess a pectin methylesterase or a pectin acetylesterase gene. Pectinase-encoding genes are reported to be absent in other endophytic microorganisms such as *Azoarcus* sp. BH72 or *Enterobacter* sp. 638 [39,47]. Production of pectinases was observed under laboratory conditions and activity was visualized after 8 days of incubation (Fig. 7D).

Expansin-like proteins. Expansins are proteins that were first described from plants [53]. These molecules function as cell wall loosening proteins by disrupting the noncovalent binding of matrix polysaccharides to cellulose [54], resulting in physical effects, such as polymer creep and stress relaxation of extended cell walls [55,56]. Many plant-associated microorganisms including several pathogenic actinobacterial species have been shown to contain proteins with expansin-like domains [57].

Two genes (MiLup_41274 and MiLup_45306) were identified in the genome of strain Lupac 08 that encode for a secreted protein showing 42% and 48% sequence similarity to the corresponding *celA* genes of *Clavibacter michiganensis* subsp. *michiganensis* and *Clavibacter michiganensis* subsp. *sepedonicus*, respectively. This gene corresponds to a secreted β-1,4-endoglucanase (CelA) that is required for virulence and contains a C-terminal α-expansin like domain [58,59]. In the case of *C. michiganesis* subsp. *michiganensis* CelA, this expansin-like domain is essential for development of wilting symptoms [58]. It is suggested that microbial expansins function to promote microbe-plant interactions, both harmful and beneficial ones [60].

Plant growth promotion traits of *Micromonospora* lupini Lupac 08

Our current knowledge of plant-microbe interactions indicates that populations inhabiting a host plant are not restricted to a single microbial species but comprise several genera and species. Few reports are available regarding the presence of other microorganisms (associated or endophytic) in nitrogen fixing nodules, in spite of the fact that nodules are much richer in nutrients as compared to roots [61]. The recent reports on the isolation of large *Micromonospora* populations from nitrogen fixing nodules clearly suggest that this bacterium plays an important role which has yet to be defined.

Effect of *M. lupini* Lupac 08 on *Trifolium*. *Micromonospora lupini* Lupac 08 clearly produced a plant growth enhancing effect when it was co-inoculated with *Rhizobium* sp. E11 under laboratory conditions on clover plantlets. In general, the number of nodules was higher in those plants co-inoculated (18–24 nodules) with both bacteria as compared to the plants inoculated only with *Rhizobium* sp. E11 (11–15 nodules). Overall, the co-inoculated plants showed better growth and were larger in size as compared to the other two treatments (Fig. 7C). Similar results were previously observed when strain Lupac 08 was inoculated in its original host, *Lupinus* [62].

Nitrogen fixing capacity. Indirect evidence of nitrogen fixing genes was obtained by partial amplification of *nifH*-like gene fragments in strains *Micromonospora* sp. L5 [12] and *M. lupini* Lupac 08 [11]. In the present work the genomes of *Micromonospora lupini* Lupac 08 and *Micromonospora sp.* L5 were screened for the presence of nitrogen fixing genes to confirm this earlier finding. After thorough analysis of the complete genome, no sequences related to this biological process were detected, supporting the results reported for strain

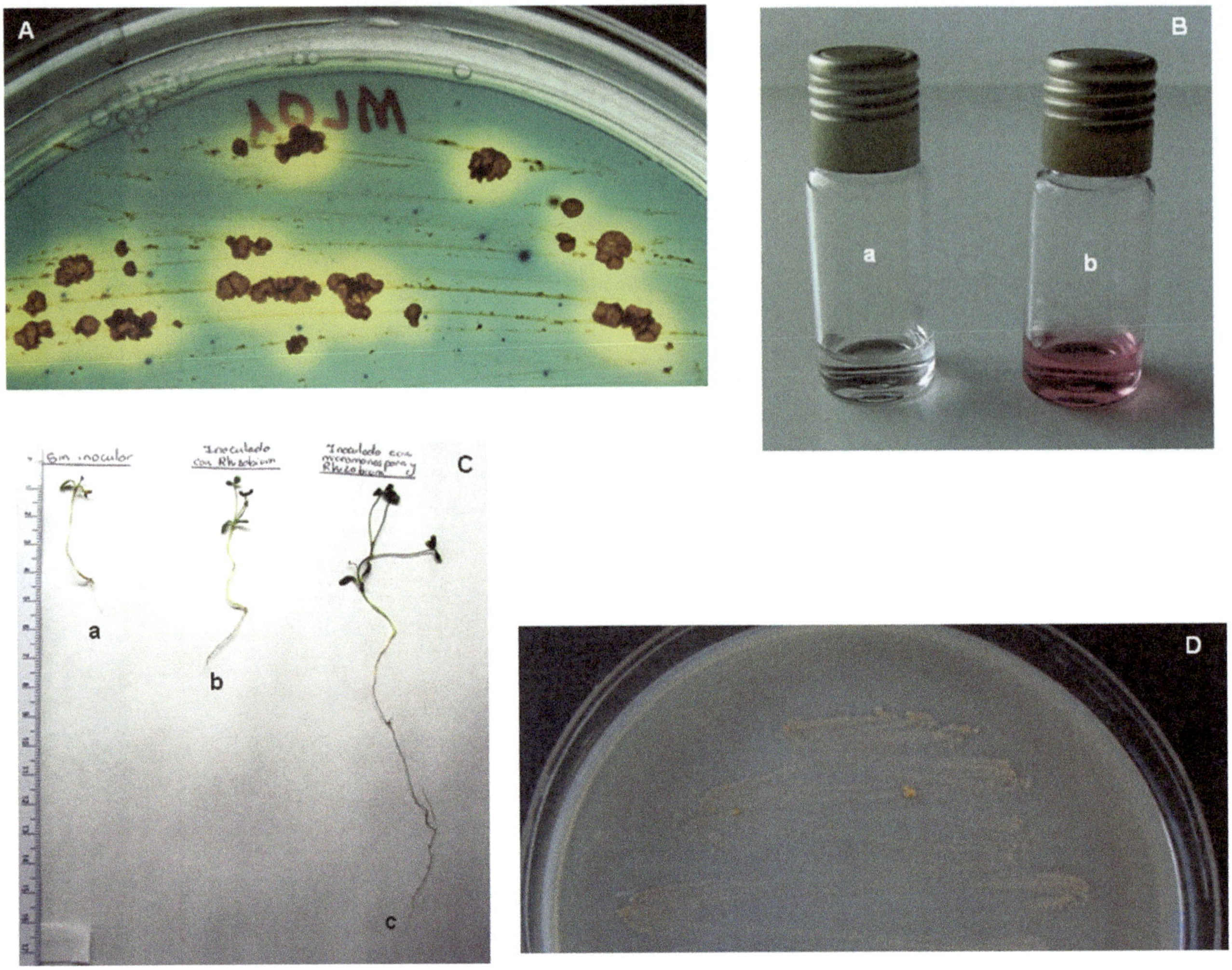

Figure 7. Plant growth promotion and biological control features of *M. lupini* Lupac 08. (A) Siderophore, (B) indole-3-acetic acid [a, negative control *E. coli* DH5α; b, Lupac 08] and pectinase production (D) by *M. lupini* strain Lupac 08;. (C) Plant growth promoting effect of *M. lupini* Lupac 08 on clover plantlets. a) control; b) inoculated with *Rhizobium* sp. E11; c) co-inoculated with *Rhizobium* sp. E11 and *M. lupini* Lupac 08.

Micromonospora sp. L5 [19]. Nitrogenase activity detection by acetylene reduction assays carried out with strain Lupac 08 over a period of two weeks were negative. A positive result was reported for strain L5 [12.].

Trehalose and its role in nodulation and bacteroid survival. Trehalose is a common reserve disaccharide in the root nodules of legumes, present at high concentrations in bacteroids at the onset of nitrogen fixation [63]. It has been reported that in the interaction between *Phaseolus vulgaris* and *Rhizobium*, enhanced germination, quality and grain yield have been correlated with trehalose content, and a higher tolerance to abiotic stress [64,65]. On the other hand, the trehalose content appears to be regulated by trehalase, a nodule stimulated plant enzyme [66,67]. Although trehalose metabolism in leguminous plants is still poorly understood, it has been shown that in senescent nodules, trehalose becomes the most abundant non-structural carbohydrate [68] and it is proposed that trehalose, a stress protectant accumulated in bacteria, could offset membrane injuries and/or serve as an intermediate energy reserve. Indeed, Müller *et al.* [68] showed that during terminal senescence of nodules an appreciable part of the bacteria maintained their trehalose pools and survived.

Eight genes related to the metabolism of trehalose were detected in the genome of Lupac 08; seven genes were related with trehalose synthesis (Mlup08_40949, Mlup08_43225, Mlup08_43226, Mlup_45189, Mlup_45758, Mlup08_45759 and Mlup08_45961) and one (*treA*, Mlup08_45961) with the enzyme trehalase. Barraza *et al.* [67] proposed that modification of the trehalose content in the nodules could trigger physiological alterations that would enhace carbon and nitrogen metabolism, as well as bacteroid fitness (greater survival) and nitrogen fixation, which in turn would positively impact on symbiotic interactions. *Micromonospora* may contribute to the survival of rhizobia by helping to maintain high levels of trehalose.

Chitin degradation and protection against pathogens. Plant β-1,3-glucanases are directly involved in defense by hydrolyzing the cell walls of fungal pathogens most commonly in combination with chitinases. Nine chitin-related ORFs were identified in *M. lupini*. Specifically, six code for a chitooligosaccharide deacetylase, several extracellular endo- and exo-chitinases and a β-N-acetyl-hexosaminidase (MiLup08_41789, MiLup08_41912, MiLup08_43481, MiLup08_44343, MiLup08_45172, MiLup08_45568), while three CDS code for putative chitin-binding domain proteins (MiLup08_41110, MiLup08_41724, MiLup08_41729).

Chitinases often work synergistically with chitin-binding proteins (CBPs). The biological roles of bacterial chitinases and carbon binding proteins are easily understood in an environmental context, especially in soil (that harbour fungi and insects) and marine (shellfish) habitats and their impact on chitin cycling. However, there is an increasing amount of direct or indirect evidence suggesting that some chitinases and CBPs additionally serve as virulence factors for bacterial pathogens during infection of non-chitinous substrates [69].

Experimental data confirmed *in vitro* chitin degradation of strain Lupac 08 (Figure 6B). As with other endophytic bacteria *Micromonospora* may produce chitinases to inhibit fungal pathogens, or may produce these molecules to elicit the plant defense mechanism. Either way, it seems that *Micromonospora* would provide a benefit to its host.

Siderophores (Iron-transport) and other secondary metabolites

Iron is an element essential for every living organism, as a cofactor of numerous proteins. Siderophores produced by plant growth promoting bacteria may reduce the growth of phytophathogens by depriving them of iron. Thus, an efficient iron uptake system can contribute to protect the host plant against pathogens. Interestingly, siderophores can also act as important virulence determinants for both plant and animal pathogens [70].

The genome of strain Lupac 08 revealed several siderophore related genes including specific iron uptake transporters, secretion of different siderophores and synthesis of siderophore receptors. Namely, a zinc/iron permease (MiLup_40258), a ferrous iron permease FTR1 (efeU, MiLup_41076) and eight iron ABC transporters (MiLup_42281-MiLup_42285). The number of the latter transporters is similar to the number of those found in the genome of the endophytic bacterium *Enterobacter* sp. 638 [47] while the plant pathogen *Erwinia amylovora* CFBP 1430 presents only three such transporters [71].

Several gene clusters related to the biosynthesis and transport of the siderophores enterobactin (MiLup_44069-MiUp_44071), aerobactin (iucA/iucC family protein, MiLup_44063 and MiLup_44064; MiLup_40326) and alcaligin (MiLup_44065) were also located. In the case of aerobactin, the gene *iucA* is highly correlated with virulence in avian pathogenic *E. coli* strains [72].

Two siderophore-interacting proteins (MiLup_40648 and MiLup_45559) were also found. One of these genes (MiLup_40648) was located next to a siderophore transporter of the RhtX/FptX family; RhtX from *S. meliloti* 2011 and FptX from *Pseudomonas aeruginosa* appear to be single polypeptide transporters from the major facilitator family for import of siderophores as a means to import iron [73]. In addition, a thiazolinyl imide reductase involved in siderophore biosynthesis was also identified (MiLup_43551).

The genome of Lupac 08 also contained several regulators including an iron-dependent repressor (IdeR, MiLup_41668), two ferric uptake regulation proteins (MiLup_40794 and MiLup_44436) and a putative iron-regulated membrane protein which suggests that these systems are highly regulated. Production of siderophores was detected experimentally (Fig. 7A).

Actinobacteria are well known to be capable of producing a vast diversity of natural secondary metabolite compounds with applications in medicine, agriculture, and other biotechnological areas [74]. Endophytic bacteria are currently of significant interest as an untapped resource of novel bioactive small molecules because their metabolites are speculated to affect the physiological conditions of host plants including growth and disease resistance. *Micromonospora* strains are well known for their capacity to produce many secondary metabolites and *M. lupini* Lupac 08 was previously screened for the production of novel compounds with antitumoral activity and the results obtained confirmed the production of a new family of molecules named Lupinacidins A, B and C [9,10].

Fifteen clusters involved in the biosynthesis of secondary metabolites were identified in the genome of *M. lupini* Lupac 08. These included siderophores (see above), terpenes, butyrolactones, polyketides (PKS), nonribosomal peptides (NRPS), chalcone synthases and bacteriocins (Table 4). A DNA stretch of 544 kb was estimated to code predominantly for secondary metabolites, accounting for about 7.4% of the genome. This percentage is lower than that reported for the marine actinobacterium *S. tropica* (9.9%, [75]) but it is within the range of other actinobacteria e.g. *S. coelicolor* (8.2%, [43]). Interestingly, *Frankia* strains ACN14a and EAN1pec dedicate about 5% of their genomes to natural product assembly while the potential of CcI3, which has the smallest genome of the three *Frankia* strains, has a much reduced host range and is absent from most soils is significantly smaller (~3%) [76].

Several clusters identified in the genome of *M. lupini* were also located in other genomes of phylogenetically related bacteria, especially in *S. tropica* CNB-440, *S. arenicola* CNS-205 and *V. maris* AB-18-032. Nevertheless other clusters were unique to *M lupini* (Table 4). Eight of the 15 clusters identified were located in the region between coordinates 4,000 kb and 5,000 kb of the genome, close to the terminus of replication. This area of the genome also contains a high density of genes coding for the biosynthesis of various plant cell wall degrading enzymes and several transposases.

Terpene related enzymes present in the genome of *M. lupini* are involved in the synthesis of carotenoids, sugar-binding lipids and the production of pentalenolactone type antibiotics. Similar molecules have also been predicted from the genomes of the three sequenced *Frankia* strains [77]. Various polyketide biosynthetic and non-ribosomal peptide synthase pathways were also identified specifically as PKI, PKII (2 clusters), PKIII types, NRPS (2 clusters) and hybrid PKS/NRPS clusters (2 clusters). The presence of these gene clusters suggests that *M. lupini* is capable of producing a vast diversity of secondary metabolites such as the antitumor anthraquinone derivative lupinacidins reported earlier [9,10]. Some of these metabolites may perform specialized functions in ecological niches and recent studies have reported on the importance of PKS and NRPS molecules and their potential role in communication during root colonization [78,79]. In addition, cluster 10 contains genes that putatively code for the production of granaticin. Granaticins are antibiotics of the benzoisochromanequinone class of aromatic polyketides, the best known member of which is actinorhodin produced by *S. coelicolor* A3(2). Production of granaticins has mainly been reported from *Streptomyces* strains [80]. NRPS cluster 11 (see Table 4), appeared to be unique to strain Lupac 08 as this group of genes was not detected in any of the other genomes compared except in *S. tropica* CNB-440 where it seems to be only partially conserved.

The PKS type III cluster corresponds to several genes that code for the production of naringenin, a central precursor of many flavonoids. It has recently been proposed that flavonoids play an important role in the establishment of plant root endosymbioses. In the case of legume-*Rhizobium* interactions, flavonoids released by plant roots induce genes involved in nodulation [81]. In a similar way it has also been suggested that these molecules play an important role during the early stages of the symbiotic association between *Frankia* and actinorhizal plants [82]. *Micromonospora* flavonoids may contribute to support communication between the nitrogen fixing bacteria and their host plants.

Table 4. Comparison of secondary metabolite clusters found in the genome of *M. lupini* Lupac 08 and other related microorganisms.

Cluster	Type	*M. lupini* **Lupac 08 (Milup08_X)**	***Micromonospora* sp. L5**	***M. aurantiaca* ATCC 27029^T**	***Verrucosispora maris* AB-18-032**	***Salinispora tropica* CNB-440**	***Salinispora arenicola* CNS-205**	***Streptomyces coelicolor* A3(2)**
1	Terpene	40204–40210	Conserved	Conserved	Conserved	Conserved	Conserved	Conserved
2	Terpene	40306–40320	Conserved	Conserved	Conserved	Absent	Absent	Present
3	Butyrolactone	40602–40668	Absent	Absent	Absent	Conserved	Conserved	Conserved
4	Type I PKS	41995–42009	Conserved	Conserved	Conserved	Conserved	Conserved	Absent
5	Terpene	43134–43144	Conserved	Conserved	Conserved	Conserved	Conserved	Conserved
6	NRPS+PKS	43546–43581	Absent	Absent	Conserved	Conserved	Absent	Partially conserved
7	Type II PKS	43804–43844	Conserved	Conserved	Conserved	Conserved	Conserved	Partially conserved
8	Siderophore	44063–44071	Conserved	Conserved	Conserved	Conserved	Conserved	Conserved
9	NRPS-Type I PKS	44386–44405	Conserved	NRPS Absent	NRPS Absent	NRPS Absent	NRPS Absent	NRPS Absent
10	Type II PKS	44613–44624	Partially conserved	Partially conserved	Partially conserved	Partially conserved	Partially conserved	Partially conserved
11	NRPS	44684–44691	Absent	Absent	Absent	Partially conserved	Absent	Absent
12	Bacteriocin	44929–44933	Conserved	Conserved	Conserved	Conserved	Conserved	Partially conserved
13	Terpene	45087–45093	Conserved	Conserved	Conserved	Absent	Absent	Conserved
14	NRPS	45439–45446	Conserved	Conserved	Conserved	Conserved	Conserved	Conserved
15	Type III PKS	46684–46700	Conserved	Conserved	Conserved	Conserved	Conserved	Absent

PKS, polyketide synthases; NRPS, non-ribosomal peptide synthases.

Genomic information also revealed that *M. lupini* has the potential to produce bacteriocins (cluster 12) as suggested by the presence of a putative short-chain dehydrogenase/reductase.

Phytohormones

Indole-3 acetic acid. Diverse bacterial species have the ability to produce auxinic phytohormones such as indole-3-acetic acid (IAA) and a few can also produce phenyl-acetate (PAA) such as *Frankia alni* [83,84]. Different biosynthesis pathways have been identified and redundancy for IAA biosynthesis is widespread among plant-associated bacteria [83]. Interactions between IAA-producing bacteria and plants may lead to several outcomes, from pathogenesis to phytostimulation [85]. The genome of *M. lupini* Lupac 08 contains a gene (Milup_45687) potentially involved in the biosynthesis of IAA via the indole-3-acetonitrile pathway. This gene corresponds to the conversion of indole-3-acetonitrile to indole-3-acetic acid. Nitrilases with specificity for indole-3-acetonitrile have been reported in *Alcaligenes faecalis* [86]. In *A. tumefaciens* and *Rhizobium* spp. nitrile hydratase and amidase activity could be identified, indicating the conversion of indole-3-acetonitrile to indole-3-acetic acid via indole-3-acetamide [87]. Analysis of IAA production by strain Lupac 08 was carried out, yielding a positive result (Fig. 7B).

Acetoin and 2,3-butanediol. The volatile compounds acetoin and 2,3-butanediol produced by bacteria such as *Bacillus subtilis* GB03 and *Bacillus amyloliquefaciens* IN937a have been reported as plant growth promoting hormones [88]. Several genes were located in the genome of strain Lupac 08 which could be involved in the production of these compounds. Two copies of the gene *pdhB* (MiLup_40114 and MiLup_43782) that encode the enzyme pyruvate dehydrogenase were identified. This enzyme transforms pyruvate to acetaldehyde and in this process a small fraction of pyruvate is converted to acetoin as a by-product. In addition, three acetolactate synthases involved in the synthesis of acetolactate from pyruvate are present (MiLup_41336, MiLup_41383 and MiLup_41384). Under aerobic conditions, acetolactate is converted to acetoin by the enzyme acetoin dehydrogenase (MiLup_41670).

Discussion

The most extensively studied bacteria interacting with plants are Gram-negative proteobacteria because they are readily isolated from plant tissues and genetically handled for interaction studies. However, the impact of Gram-positive bacteria on plants should not be underestimated as has been done for many years mainly due to their slow growth. *M. lupini* Lupac 08, a Gram-positive actinobacterium was isolated from the internal root nodule tissues of *Lupinus angustifolius* but it is only a representative of a large collection of more than 2000 *Micromonospora* strains isolated from diverse legumes and actinorhizals from different geographical locations. So far, several genomes of root symbionts and soil saprophytes have been studied; therefore we decided to focus on an intermediary category, that of facultative endophytes.

Lupac 08 was isolated from lupine nodules, and shown to produce the anticancer agents Lupinadicin A, B and C. The genome of strain Lupac 08 was sequenced to obtain information about the potential ecological role of *Micromonospora* in interaction with legumes and actinorhizal plants. Genomic analysis revealed several strategies which are probably necessary to lead a successful lifestyle as a saprophyte in the rhizosphere, a competitive and harsh environment, and as an endophyte capable of colonizing the internal plant tissues. *Micromonospora* species have less than 3% distance in their 16S rRNA genes, which can be roughly translated to a time of 150MY according to the equivalence proposed by Ochman and Wilson [89]. In the current phylogeny (Fig. 1), *M. lupini* has as closest neighbours *M. chokoriensis M. saelicesensis and M. zamorensis*, isolated from sandy soil, root nodules of *L. angustifolius* and the rhizosphere of *P. sativum* respectively. *Micromonospora* sp. L5 and *M. aurantiaca* are located further away, with a distance of 1.2% that would translate to 60 MY for the emergence of a group of species that interact with plants, a date that would be close to the postulated time of emergence of *Fabaceae* and that of many actinorhizal plant families [90]. The separation from the *Salinispora* and *Verrucosispora* lineages would constitute two independent events that would have occurred slightly earlier at 160MY and 170MY, while the emergence of the *Actinoplanes* and that of the *Dactylosporangium* would have occurred 250MY ago, a time when dicotyledons had not yet appeared but when continents and thus soils had appeared that did permit the growth of primitive plants such as the gymnosperms.

The size of the *Micromonospora* genomes analyzed is quite uniform, with that of Lupac 08 slightly larger. The chromosome size of *M. lupini* Lupac 08 appears to reflect a wealth of genes allowing for adaptation to a complex saprophytic/endophytic lifestyle, which means adapting to a wider range of environmental conditions with the ability to metabolize a large variety of nutrient sources. Considering that *Micromonospora* sp. L5 and *M. lupini* Lupac 08 were both isolated from nitrogen fixing nodules (actinorhizal and legume plants, respectively), it would be expected that the genomes of these strains be more similar to each other than to *M. aurantiaca* ATCC 27029^T which was originally isolated from soil. Surprisingly this was not the case as confirmed by the high number of strain specific genes identified in the genome of Lupac 08, suggesting a high capacity of adaptation to a fluctuating environment by this microorganism. On the other hand, *Micromonospora* sp. L5 and *M. aurantiaca* ATCC 27029^T share a high number of orthologous genes (86%) suggesting that the niche of origin is not crucial.

An interesting result was the distribution of the metabolic profiles of 20 bacteria representing different living environments (Fig. 5). There was a clear proximity between *M. lupini* Lupac 08 and the three *Frankia* genomes. This result suggests that strain Lupac 08 contains metabolic functions similar to those found in *Frankia* strains that are probably useful for its interaction with plants. This metabolic versatility combined with a diverse transport system make Lupac 08 an organism fit to adapt to a soil/plant environment.

The emergence and evolution of nitrogen fixation ability among the domains *Bacteria* and *Archaea* is complex and has not yet been fully elucidated. Although it was previously reported that *Micromonospora* strains isolated from legume and actinorhizal root nodules contained *nifH*-like gene fragments [11,12], we could not confirm these results. In a similar approach based on PCR-amplification, other authors reported the presence of *nif*-H like sequences for bacterial isolates obtained from legumes collected in arid zones including *Microbacterium*, *Agromyces*, *Mycobacterium* and *Ornithinicoccus* [91]. One recurrent problem with the use of a PCR-based approach is that it is limited to a single gene amplified billions of times, which may provide false-positive results [92] and for this reason must always be confirmed by an independent approach.

Plant-polymer degrading enzymes such as cellulases and pectinases have been suspected to play a role in internal colonization. Most plant pathogens secrete cellulases, pectinases, xylanases, or other enzymes to hydrolyze plant cell wall polymers, while a lack of secreted hydrolases has been proposed to be favourable for microorganisms that form beneficial association with plants. Examples of endophytic plant growth promoting bacteria that lack large amounts of cell wall degrading enzymes include *Frankia* [38], *Enterobacter* sp. 638 [47], *Azoarcus* sp. BH72 [60] and *Herbaspirillum seropedicae* [93]. An *Azospirillum* sp. that does not colonize root tissues proper, but only the rhizosphere, has a genome containing a large number of putative cellulases similar to soil cellulolytic bacteria with 26–34 glycosyl hydrolases [94], as compared to the 37 present in *T. fusca*, a highly cellulolytic actinobacterium isolated from soil.

The genome of *M. lupini* revealed a high number of putative genes that encode for hydrolytic enzymes and specifically cellulolytic, xylanolytic, chitinolytic and pectinolytic activities were confirmed in the laboratory, indicating the capacity of *Micromonospora* to degrade plant polymers in a way similar to that of plant pathogen microorganisms. In the case of *Micromonospora*, there seems to be a paradox since strain *M. lupini* Lupac 08 shows a very high *in vitro* activity for cellulases and xylanases, however, preliminary inoculation experiments in our laboratory indicate that the microorganism does not behave as a pathogen, on the contrary, *Micromonospora* appears to interact in a tripartite relationship stimulating nodulation and plant growth (Fig. 7c). Therefore the question arises as to what is the function of these enzymes when *Micromonospora* interacts with the host plant. Alternatively some of the genes present, especially those related to the metabolism of cellulose may not necessarily imply that the bacterium is involved in plant cell wall degradation but have a different role, yet to be defined [95]. In addition many of the cellulose-related genes contain binding-domains suggesting that these may be related to the adhesion of the bacterium to the plant. These genes could also help *Micromonospora* digest plant cell walls upon senescence of the nodules.

M. lupini Lupac 08 contains several secondary metabolite gene clusters, many of which appear to be involved in the synthesis of siderophores and also of antibiotics. These would also in all likelihood be involved in the synthesis of the antitumor anthraquinone molecules described previously [9,10]. *Micromonospora*, like many other endophytic bacteria is a facultative plant colonizer that must compete with other microorganisms in the rhizosphere before entering the plant. In this sense, the NRPS and PKS gene clusters identified in the genome of *M. lupini* Lupac 08 may be involved in defense as well as in interaction and communication with its host plant. Thus, it will be necessary to identify these compounds and their functional attributes to further expand our knowledge of this plant-microbe interaction.

Conclusions

We have provided experimental data which supports the hypothesis that *M. lupini* Lupac 08 is a plant growth promoting bacterium. *Micromonospora lupini* Lupac 08 clearly produces a plant growth enhancing effect as observed in laboratory experiments. The localization of several genes involved in plant growth promotion traits such as the production of siderophores, phytohormones, the degradation of chitin (biocontrol) and the biosynthesis of trehalose may all contribute to the welfare of the host plant. *Micromonospora* appears as a new candidate in plant-microbe interactions with an important potential in agriculture and other biotechnological applications. The current data is promising but it is still too early to determine which specific roles are played by this microorganism in interactions with nitrogen fixing plants.

Methods

Genome sequencing, annotation and analysis

The genome sequence of *M. lupini* Lupac 08 was determined using the 454 FLX system and Titanium platform (454 Life Sciences) as previously reported [96]. Sequences were assembled into 50 contigs and four scaffolds ranging from 583 to 7,083,659 nucelotides using the MaGe (Magnifying Genomes) interface [97]. This Whole Genome Shotgun project has been deposited at European Nucleotide Archive under accession number NZ_CAIE00000000.01.

16S rRNA gene phylogeny

Sequences obtained from public databases (Genbank/EMBL) were manually aligned using clustal X software [98]. Phylogenetic distances were calculated with the Kimura 2-parameter model [99] and the tree topologies were inferred using the maximum-likelihood method [100]. All analyses were carried out using the MEGA5 program [101].

Comparative genome analysis

Genome rearrangement of the *Micromonospora* strains *M. lupini* Lupac 08, *Micromonospora* sp. L5, *M. aurantiaca* ATCC 27029^T and *Micromonospora* sp. ATCC 39149 were carried out using MAUVE software [102]. The number of shared and unique genes present in the respective genomes were calculated and represented by a Venn-diagram using the EDGAR software [103]. Potential horizontally transferred genes were predicted using the "Region of Genomic Plasticity Finder" method implemented on the MicroScope platform. First we selected genomes included in the PkGDB and NCBI RefSeq databases that presented high synteny percentages with the Lupac 08 strain. Automatic results were manually curated according to several features such as base composition, DNA repetitions, presence of near mobile elements and information provided by SIGI and IVON programs [104].

Comparative analysis of metabolic profiles

A bicluster plot of the metabolic profiles for *M lupini* Lupac 08 and 20 other bacterial genomes was performed with Multibiplot [105]. A comparison of 798 MicroCyc metabolic pathways was made using MaGe. This comparison is based on the calculation of 'pathway completion' values, scaled from 0 to 1, where 0 means that a particular organism does not contain any enzyme for a given pathway and 1 that it contains all the reactions of the pathway. These values were transformed applying row standardization and a JK-Biplot was constructed after performing a PCA (Singular Value Decomposition estimation method). The heatmap was then obtained with the expected values computing the Euclidian distance and average linkage for rows and columns.

Transport proteins identification and classification

Information about transport proteins of genomes was obtained from the TransportDB relational database when available (http://www.membranetransport.org/). The identification and classification of the transporters of strain Lupac 08 was performed using the TransAAP tool based on TransportDB [106] followed by manual validation.

Cellulose, starch and xylan degradation

Strain Lupac 08 was cultivated on yeast-malt agar for 5 days and subsequently transferred to M3 agar [107] with and without glucose and supplemented with one of the substrates in the following way: carboxymethylcellulose (CMC, 0.5%), starch (1%)

and xylan (1%). A bacterial suspension of 10^6 per ml was prepared in saline solution (0.85%) and 200 µl were inoculated on the different plates which were then incubated at 28°C and results were recorded at 4, 7 and 14 days after inoculation. Xylan and CMC plates were stained with Congo red [108] while starch plates were flooded with iodine solution [109].

Pectin degradation

Pectinolytic acitivity was determined as described in Williams et al. [109]. Agar plates supplemented with pectin (0.5%) were streaked with strain Lupac 08 and incubated at 28°C. Hydrolysis zones were detected after 14 days incubation by flooding plates with an aqueous solution of cetyltrimethyl ammonium bromide (CTAB, 1%) and examining them after 30 min.

Chitin degradation

Chitinolytic acitivity was determined as described in Murthy et al. [110]. Agar plates supplemented with colloidal chitin at 0.5% (Gift of France-Chitine, Orange, http://france-chitine.com/), partly hydrolysed by stirring in 0.5 M HCl for 2h, were inoculated with strain Lupac 08 and incubated at 28°C. Hydrolysis zones were detected as cleared zones after 14 days incubation.

Siderophore production

Siderophore production was assessed using a modified chrome azurol S (CAS) assay [111]. Strain Lupac 08 was cultured on yeast-malt agar and incubated for 7 days, subsequently it was streaked onto CAS agar plates and incubated at 28°C for 7–10 days. A positive result was indicated by an orange halo around bacterial colonies.

Indole-3-acetic acid production

Production of indole acetic acid was assayed following the method of Glickmann and Dessaux [112]. Strain Lupac 08 was inoculated in 5 ml of yeast-malt medium supplemented with L-tryptophan (0.2%) and incubated at 28°C at 150 rpm during 7 days. The culture was then centrifuged at 12,000 x g for 10 min and 1 ml of the supernatant was mixed with 2 ml of Salkowski's reagent [113] and incubated at room temperature for 30 min. IAA production was measured spectrophotometrically at 530 nm to assess the development of a pink colour.

Plant growth

Surface-sterilized seeds of *Trifolium* sp. were germinated axenically in Petri dishes on 1.4% w/v agar. Seedlings were transferred to sterile square plastic plates that contained a nitrogen-free nutrient solution [114]. Fifteen plantlets were inoculated in the following manner (5 per treatment): 500 µl (each) of bacterial suspensions (10^6 cells per ml) of *M. lupini* Lupac 08 and *Rhizobium* sp. E11 for coinoculation treatment; inoculation with *Rhizobium* sp. E11; and uninoculated plants as negative controls.

Acetylene reduction activity

Nitrogenase activity was measured using acetylene reduction [115] in sterile 150 ml plasma flasks with a rubber stopper. Cells of Lupac 08 were cultured in liquid minimal glucose medium without nitrogen at 28°C with shaking. The air in flasks was replaced with mixture of air and acetylene (ration 90:10 v/v). One mililiter of mixture was sampled for each measure using gas chromatography with a flame ionization detector (Girdel 30, France). *Mesorhizobium melitolti* Sm1021 was used as positive control.

Author Contributions

Conceived and designed the experiments: MET PN. Performed the experiments: MET PP PB RR. Analyzed the data: MET RB PP PN CM. Contributed reagents/materials/analysis tools: MET PN PP YI CM. Contributed to the writing of the manuscript: MET PN RB PP.

References

1. Conn VM, Franco CM (2004) Analysis of the endophytic actinobacterial population in the roots of wheat (*Triticum aestivum* L.) by terminal restriction fragment legth polymorphism and sequencing of 16S rRNA clones. Appl Environ Microbiol 70: 1787–94.
2. Kaewkla O, Franco CM (2013) Rational approaches to improving the isolation of endophytic actinobacteria from Australian native trees. Microb Ecol 65: 384–393.
3. Hardoim PR, van Overbeek LS, Elsas DJ (2008) Properties of bacterial endophytes and their proposed role in plant growth. Trends Microbiol 16: 463–471.
4. Chi F, Shen SH, Cheng HP, Jing YX, Yanni YG, et al. (2005) Ascending migration of endophytic rhizobia, from roots to leaves, inside rice plants and assessment of benefits to rice growth physiology. Appl Environ Microbiol 71: 7271–7278.
5. Reinhold-Hurek B, Hurek T (1998) Life in grasses: diazotrophic endophytes. Trends Microbiol 6: 139–144.
6. Marchetti M, Capela D, Glew M, Cruvellier S, Chane-Woon-Ming B, et al. (2010) Experimental evolution of a plant pathogen into a legume symbiont Plos Biol 8: e1000280. doi: 10.1371/journal.pbio.1000280.
7. Ørskov J (1923) Investigations into the morphology of the ray fungi. Copenhagen.
8. Igarashi Y, Ogura H, Furihata K, Oku N, Indananda C, et al. (2011a) Maklamicin, an antibacterial polyketide from an endophytic *Micromonospora* sp. J Nat Prod 74: 670–674.
9. Igarashi Y, Trujillo ME, Martinez-Molina E, Yanase S, Miyanaga S, et al. (2007) Antitumor anthraquinones from an endophytic actinomycete *Micromonospora lupini* sp. nov. Bioorg Med Chem Lett 17: 3702–3705.
10. Igarashi Y, Yanase S, Sugimoto K, Enomoto M, Miyanaga S, et al. (2011b) Lupinacidin C, an inhibitor of tumor cell invasion from *Micromonospora lupini*. J Nat Prod 74: 862–865.
11. Trujillo ME, Alonso-Vega P, Rodriguez R, Carro L, Cerda E, et al. (2010) The genus *Micromonospora* is widespread in legume root nodules: the example of *Lupinus angustifolius*. ISME J 4: 1265–1281.
12. Valdés M, Perez NO, Estrada-de Los Santos P, Caballero-Mellado J, Pena-Cabriales JJ, et al. (2005) Non-*Frankia* actinomycetes isolated from surface-sterilized roots of *Casuarina equisetifolia* fix nitrogen. Appl Environ Microbiol 71: 460–466.
13. Carro L, Pujic P, Trujillo ME, Normand P (2013) *Micromonospora* is a normal inhabitant of actinorhizal nodules. J Biosci 38: 685–693.
14. Penn K, Jensen PR (2012) Comparative genomics reveals evidence of marine adaptation in *Salinispora* species. BMC Genomics 13: 86.
15. Normand P, Lapierre P, Tisa LS, Gogarten JP, Alloisio N, et al. (2007) Genome characteristics of facultatively symbiotic *Frankia* sp. strains reflect host range and host plant biogeography. Genome Res 17: 7–15.
16. Smith SE, Corneli-Showers P, Dardenne CN, Harpending HH, Martin DP, et al. (2012) Comparative genomic and phylogenetic approaches to characterize the role of genetic recombination in mycobacterial evolution. PLoS ONE 7: e50070.
17. Trujillo ME, Kroppenstedt RM, Fernandez-Molinero C, Schumann P, Martinez-Molina E (2007) *Micromonospora lupini* sp. nov. and *Micromonospora saelicesensis* sp. nov., isolated from root nodules of *Lupinus angustifolius*. Int J Syst Evol Microbiol 57: 2799–2804.
18. Carro L, Spröer C, Alonso P, Trujillo ME (2012) Diversity of *Micromonospora* strains isolated from nitrogen fixing nodules and rhizosphere of *Pisum sativum* analyzed by multilocus sequence analysis. Syst Appl Microbiol 35: 73–80.

19. Hirsch AM, Alvarado J, Bruce D, Chertkov O, De Hoff PL, et al. (2013) Complete genome sequence of *Micromonospora* strain L5, a potential plant-growth regulating actinomycete, originally isolated from *Casuarina equisetifolia* root nodules. Genome Announc 1: 2-00759-13.
20. Lethlefsen L, Schmidt TM (2007) Performance of the translational apparatus varies with the ecological strategies of bacteria J Bacteriol 189: 3237–3245.
21. Yano K, Wada T, Suzuki S, Tagami K, Matsumoto T, et al. (2013) Multiple rRNA operons are essential for efficient cell growth and sporulation as well as outgrowth in *Bacillus subtilis*. Microbiol 159: 2225–2236.
22. McCarthy AJ, Broda P (1984) Screening for lignin degrading actinomycetes and characterization of their activity against ^{14}C-lignin labelled wheat lignocellulose. J Gen Microbiol 130: 905–2913.
23. Jendrossek D, Tomasi G, Kroppenstedt R (1997) Bacterial degradation of natural rubber: a privilege of actinomycetes? FEMS Microbiol Lett 150: 179–188.
24. Sandrak NA (1977) Degradation of cellulose by micromonospores. Mikrobiologiia 46 478–481.
25. de Menezes AB, Lockhart RJ, Cox MJ, Allison HE, McCarthy AJ (2008) Cellulose degradation by micromonosporas recovered from freshwater lakes and classification of these actinomycetes by DNA gyrase B gene sequencing. Appl Environ Microbiol 74: 7080–7084.
26. Miele V, Penel S, Duret L (2011) Ultra-fast sequence clustering from similarity networks with SiLiX. BMC Bioinformatics 12: 116 doi: 10.1186/1471-2105-12-116.
27. Ren Q, Paulsen IT (2007) Large-scale comparative genomic analyses of cytoplasmic membrane transport systems in prokaryotes. J Mol Microbiol Biotechnol 12: 165–179.
28. Ren Q, Paulsen IT (2005) Comparative analyses of fundamental differences in membrane transport capabilities in prokaryotes and eukaryotes. PLoS Comp Biol 1: e27.
29. Sutcliffe I (2011) New insights into the distribution of WXG100 protein secretion systems. Antonie van Leeuwenhoek 99: 127–131.
30. Scott JR, Barnett TC (2006) Surface proteins of Gram-positive bacteria and how they get there. Annu Rev Microbiol 60: 397–423.
31. Fyans JK, Bignell D, Loria R, Toth IT, Palmer T (2013) The ESX7type VII secretion system modulates development, but not virulence, of the plant pathogen *Streptomyces scabies* Mol Plant Pathol 14: 119–130.
32. Abdallah AM, Gey van Pittius NC, DiGiuseppe Champion PA, Cox J, Luirink J, et al. (2007) Type VII secretion – mycobacteria show the way. Nat Rev 5: 883–891.
33. Johnson TL, Abendroth J, Hol WGJ, Sandkvist M (2006) Type II secretion: from structure to function. FEMS Microbiol Lett 255: 175–186.
34. Dow JM, Daniels MJ, Dums F, Turner PC, Gough C (1989) Genetic and biochemical analysis of protein export from *Xanthomonas campestris*. J Cell Sci Suppl 11: 59–72.
35. Filloux A, Bally M, Ball G, Akrim M, Tommassen J, et al. (1990) Protein secretion in gram-negative bacteria: transport across the outer membrane involves common mechanisms in different bacteria. EMBO J 9: 4323–4329.
36. Reeves PJ, Whitcombe D, Wharam S, Gibson M, Allison G, et al. (1993) Molecular cloning and characterization of 13 out genes from *Erwinia carotovora* subspecies *carotovora*: genes encoding members of a general secretion pathway (GSP) widespread in Gram-negative bacteria. Mol Microbiol 8: 443–456.
37. DeShazer D, Brett PJ, Burtnick MN, Woods DE (1999) Molecular characterization of genetic loci required for secretion of exoproducts in *Burkholderia pseudomallei*. J Bacteriol 181: 4661–4664.
38. Mastronunzio JE, Tisa LS, Normand P, Benson DR (2008) Comparative secretome analysis suggests low plant cell wall degrading capacity in *Frankia* symbionts. BMC Genomics 9: 47 doi: 10.1186/1471-2164-9-47.
39. Krause A, Ramakumar A, Bartels D, Battistoni F, Bekel T, et al. (2007) Complete genome of the mutualistic, N_2-fixing grass endophyte *Azoarcus* sp. strain BH72. Nat Biotechnol 24: 1385–1391.
40. Hubber A, Vergunst AC, Sullivan JT, Hooykaas PJJ, Ronson CW (2004) Symbiotic phenotypes and translocated effector proteins of the *Mesorhizobium loti* strain R7A VirB/D4 type IV secretion system. Mol Microbiol 54: 561–574.
41. Cases I, de Lorenzo V, Ouzounis CA (2003) Transcription regulation and environmental adaptation in bacteria. Trends Microbiol 11: 248–253.
42. Bentley SD, Chater KF, Cerdeño-Tárraga AM, Challis GL, Thomson NR, et al. (2002) Complete genome sequence of the model actinomycete *Streptomyces coelicolor* A3(2). Nature 417: 141–147.
43. Kaneko T, Nakamura Y, Sato S, Asamizu E, Kato T et al. (2000) Complete genome structure of the nitrogen-fixing symbiotic bacterium *Mesorhizobium loti*. DNA Res 7: 331–338.
44. Lopes-Santos C, Correia-Neves M, Moradas-Ferreira P, Vaz-Mendes M (2009) A walk into the LuxR regulators of actinobacteria: phylogenomic distribution and functional diversity. Plos One 7: e46758.
45. Patankar AV, Gonzalez JE (2009) Orphan LuxR regulators of quorum sensing. FEMS Microbiol Rev 33: 739–756.
46. Compant S, Duffy B, Nowak J, Clément C, Barka EA (2005) Use of plant-growth-promoting bacteria for biocontrol of plant diseases: principles, mechanisms of action, and future prospects. Appl Environ Microbiol 71: 4951–4959.
47. Taghavi S, van der Lelie D, Hoffman A, Zhang YB, Walla MD, et al. (2010) Genome sequence of the plant growth promoting endophytic bacterium *Enterobacter* sp. 638. Plos Genet 6 e1000943. doi:10.1371/jounal.pgen.1000943.
48. Pujic P, Fournier P, Alloisio N, Hay AE, Maréchal J, et al. (2012) Lectin genes in the *Frankia alni* genome. Arch Microbiol 194: 47–56.
49. Lykidis A, Mavromatis K, Ivanova N, Anderson I, Land M, et al. (2007) Genome sequence and analysis of the soil cellulolytic actinomycete *Thermobifida fusca* YX. J Bacteriol 189: 2477–2486.
50. Anderson I, Abt B, Lykidis A, Klenk HP, Kyrpides N, et al. (2012) Genomics of aerobic cellulose utilization sytems in actinobacteria. PlosOne 7 e39331.
51. Jayani RS, Saxena S, Gupta R (2005) Microbial pectinolytic enzymes: a review. Process Biochem 40: 2931–2944.
52. Jakob K, Kniskern J, Bergelson MJ (2007) The role of pectate lyase and the jasmonic acid defense response in *Pseudomonas viridiflava* virulence. Mol Plant Microb Interact 20: 146–158.
53. McQueen-Mason S, Durachko DM, Cosgrove DJ (1992) Two endogenous proteins that induce cell wall extension in plants. Plant Cell 4: 1425–1433.
54. Georgelis N, Tabuchi A, Nikolaidis N, Cosgrove DJ (2011) Structure-function analysis of the bacterial expansin EXLX1. J Biol Chem 286: 16814–16823.
55. McQueen-Mason SJ, Cosgrove DJ (1995) Expansin Mode of Action on Cell Walls - Analysis of wall hydrolysis, stress relaxation, and binding. Plant Physiol107: 87–100.
56. Cosgrove DJ (2000) Loosening of plant cell walls by expansins. Nature 407: 321–326.
57. Bignell DR, Huguet-Tapia JC, Joshi MV, Pettis GS, Loria R (2010) What does it take to be a pathogen: genomic insights from *Streptomyces* species. Antonie van Leeuwenhoek 98: 179–194.
58. Jahr H, Dreider J, Meletzus D, Bahro R, Eichenbalub R (2000) The endo-beta-1,4-glucanase CelA of *Clavibacter michiganensis* subsp. *michiganensis* is a pathogenicity determinant required for induction of bacterial wilt of tomato. Mol Plant Microb Interact 13: 703–14.
59. Laine MJ, Haapalainen M, Wahlroos T, Kankare K, Nissinen R, et al. (2000) The cellulase encoded by the native plasmid of *Clavibacter michiganensis* ssp. *sepedonicus* plays a role in virulence and contains an expansin-like domain. Physiol Mol Plant Pathol 57: 221–233.
60. Kerff F, Amoroso A, Herman R, Sauvage E, Petrella S, et al. (2008) Crystal structure and activity of *Bacillus subtilis* YoaJ (EXLX1), a bacterial expansin that promotes root colonization. Proc natl Acad Sci USA 105: 16876–16881.
61. Dudeja SS, Giri R, Saini R, Suneja-Madan P, Kothe E (2012) Interaction of endophytic microbes with legumes. J Basic Microbiol 52: 248–60.
62. Cerda ME (2008) Aislamiento de *Micromonospora* de nódulos de leguminosas tropicales y análisis de su interés como promotor del crecimiento vegetal. Ph.D. Thesis. Universidad de Salamanca, Spain.
63. Streeter JG (1985) Accumulation of alpha, alpha-trehalose by *Rhizobium* bacteria and bacteroids. J Bacteriol 164: 78–84.
64. Farías-Rodriguez R, Mellor RB, Arias C, Peña-Cabriales JJ (1998) The accumulation of trehalose in nodules of several cultivars of common bean (*Phaseolus vulgaris*) and its correlation with resistance to drought stress. Physiol Plant 102 353–359.
65. Altamirano-Hernández J, López MG, Acosta-Gallegos JA, Farías-Rodríguez R, Peña-Cabriales JJ (2007) Influence of soluble sugars on seed quality in nodulated common bean (*Phaseolus vulgaris* L.): the case of trehalose. Crop Sci 47: 1193–1205.
66. Aeschbacher RA, Müller J, Boller T, Wiemken A (1999) Purification of the trehalase GMTRE1 from soybean nodules and cloning of its cDNA. GMTRE1 is expressed at a low level in multiple tissues. Plant Physiol 119: 489–495.
67. Barraza AG, Estrada-Navarrete G, Rodriguez-Alegria ME, Lopez-Munguia A, Merino E, et al. (2013) Down-regulation of PvTRE1 enhances nodule biomass and bacteroid number in the common bean. New Phytol 197: 194–206.
68. Müller J, Boller T, Wiemken A (2001) Trehalose becomes the most abundant non-structural carbohydrate during senescence of soybean nodules. J Experimental Bot 52 943–947.
69. Frederiksen RF, Paspaliari DK, Larsen T, Storgaard BG, Larsen MH, et al. (2013) Bacterial chitinases and chitin-binding proteins as virulence factors. Microbiology 159: 833–847.
70. Taguchi F, Suzuki T, Inagaki Y, Toyoda K, Shiraishi T, et al. (2010) The siderophore pyoverdine of *Pseudomonas syringae* pv. tabaci 6605 is an intrinsic virulence factor in host tobacco infection. J Bacteriol 192: 117–126.
71. Smits THM, Jaenicke S, Rezzonico F, Kamber T, Blom J, et al. (2010) Complete genome sequence of the fire blight pathogen *Erwinia amylovora* CFBP 1430 and comparsion to other *Erwinia* spp. Mol Plant Microbe Interact 23: 384–393.
72. Tivendale KA, Allen JL, Ginns CA, Crabb BS, Browning GF (2004) Association of *iss* and *iucA*, but not *tsh*, with plasmid-mediated virulence of avian pathogenic *Escherichia coli*. Infect Immun 72: 6554–6560.
73. Cuív PO, Clarke P, Lynch D, O'Connell M (2004) Identification of *rhtX* and *fptX*, novel genes encoding proteins that show homology and function in the utilization of the siderophores rhizobactin 1021 by *Sinorhizobium meliloti* and pyochelin by *Pseudomonas aeruginosa*, respectively. J Bacteriol 186: 2996–3005.
74. Genilloud O (2014) The re-emerging role of microbial natural products in antibiotic discovery. Antonie van Leeuwenhoek 106: 173–178.
75. Udwary DW, Zeigler L, Asolkar RN, Vasanth S, Alla L, et al. (2007) Genome sequencing reveals complex secondary metabolome in the marine actinomycete *Salinispora tropica*. Pro Natl Acad Sci 104: 10376–10381.

76. Nett M, Ikeda H, Moore BS (2009) Genomic basis for natural product biosynthetic diversity in the actinomycetes. Nat Prod Rep 26: 1362–1384.
77. Udwary DW, Gontang EA, Jones AC, Jones CS, Schultz AW, et al. (2011) Significant natural product biosynthetic potential of actinorhizal symbionts of the genus *Frankia*, as revealed by comparative genomic and proteomic analyses. Appl Environ Microbiol 77: 3617–3625.
78. Velázquez-Robeldo R, Contreras-Cornejo H, Macías-Rodriguez LI, Hernández-Morales A, Aguirre J, et al. (2011) Role of the 4-phosphopantetheinyl transferase of *Trichoderma virens* in secondary metabolism, and induction of plant defense responses. Mol Plant Microb Interact 24: 1459–1471.
79. Mukherjee PK, Buensanteai N, Moran-Diez ME, Druzhinina IS, Kenerley CM (2012) Functional analysis of non-ribosomal peptide synthetases (NRPs) in *Trichoderma virens* reveals a polyketide synthase (PKS)/NRPS hybrid enzyme involved in the induced systemic resistance response in maize. Microbiol 158: 155–165.
80. Tornus D, Floss HG (2001) Identification of four genes from the granaticin biosynthetic gene cluster of Stretpomyces violaceoruber Tü22 involved in the biosynthesis of L-rhodisone. J Antibiot 54: 91–101.
81. Tadra-Sfeir MZ, Souza EM, Faoro H, Müller-Santos M, Baura VA, et al. (2011) Naringenin regulates expression of genes involved in cell wall synthesis in *Herbaspirillum seropedicae* Appl Environ Microbiol 77: 2180–2183.
82. Abdel-Lateif K, Boguz D, Hocher V (2012) The role of flavonoids in the establishment of plant roots endosymbioses with arbuscular mycorrhiza fungi, rhizobia and *Frankia* bacteria. Plant Signal Behav 7: 636–641.
83. Hammad Y, Nalin R, Marechal J, Fiasson K, Pepin R, et al. (2003) A possible role for phenyl acetic acid (PAA) on *Alnus glutinosa* nodulation by *Frankia* bacteria. Plant and Soil 254: 193–206.
84. Duca D, Lorv J, Patten CL, Rose D, Glick BR (2014) Indole-3-acetic acid in plant-microbe interactions. Antonie van Leeuwenhoek 106: 85–125.
85. Spaepen S, Vanderleyden J, Remans R (2007) Indole-3-acetic acid in microbial and microorganism-plant signaling. FEMS Microbiol Rev 31: 425–448.
86. Kobayashi M, Izui H, Nagasawa T, Yamada H (1993) Nitrilase in biosynthesis of the plant hormone indole-3-acetic acid from indole-3-acetonitrile: cloning of the *Alcaligenes* gene and site-directed mutagenesis of cysteine residues. Proc Natl Acad Sci 90: 247–251.
87. Kobayashi M, Suzuki T, Fujita T, Masuda M, Shimizu S (1995) Occurrence of enzymes involved in biosynthesis of indole-3-acetic acid from indole-3-acetonitrile in plant-associated bacteria, *Agrobacterium* and *Rhizobium*. Proc Natl Acad Sci 92: 714–718.
88. Ryu CM, Farag MA, Hu CH, Reddy MS, Wei HX, et al. (2003) Bacterial volatiles promote growth in Arabidopsis. Proc Natl Acad Sci 100: 4927–4933.
89. Ochman H, Wilson AC (1987) Evolution in bacteria: evidence for a universal substitution rate in cellular genomes. J Mol Evol 26: 74–86.
90. Bell CD, Soltis DE, Soltis PS (2010) The age and diversification of the angiosperms re-revisited. Amer J Bot 97: 1–8.
91. Zakhia F, Jeder H, Willems A, Gillis M, Dreyfus B, et al. (2006) Diverse bacteria associated with root nodules of spontaneous legumes in Tunisia and first report for nifH-like gene within the genera Microbacterium and Starkeya. Microb Ecol 51: 375–393.
92. Gtari M, Ghodhbane-Gtari F, Nouioui I, Beauchemin N, Tisa LS (2012) Phylogenetic perspectives of nitrogen-fixing actinobacteria. Arch Microbiol 194: 3–11.
93. Pedrosa FO, Monteiro RA, Wassem R, Cruz LM, Ayub RA, et al. (2011) Genome of *Herbaspirillum seropedicae* strain SmR1, a specialized diazotrophic endophyte of tropical grasses. PLoS Genet 7: e1002064. doi:10.1371/journal.pgen.1002064.
94. Wisniewski-Dyé F, Borziak K, Khalsa-Moyers G, Alexandre G, Sukharnikov LO, et al. (2011) *Azospirillum* genomes reveal transition of bacteria from aquatic to terrestrial environments. Plos Genetics 7: e1002430. doi: 10.1371/journal.pgen.1002430.
95. Medie FM, Davies GJ, Drancourt M, Henrissat B (2012) Genome analyses highlight the different biological roles of cellulases. Nat Rev Microbiol 10: 227–234.
96. Alonso-Vega P, Normand P, Bacigalupe R, Pujic P, Lajus A, et al. (2012) Genome sequence of *Micromonospora lupini* Lupac 08, isolated from root nodules of *Lupinus angustifolius*. J Bacteriol 194: 4135.
97. Vallenet D, Labarre L, Rouy Z, Barbe C, Bocs S, et al. (2006) MaGe–a microbial genome annotation system supported by synteny results. Nucleic Acids Res 34: 53–65.
98. Thompson JD, Gibson TJ, Plewniak F, Jeanmougin F, Higgins DG (1997) The CLUSTAL_X windows interface: flexible strategies for multiple sequence alignment aided by quality analysis tools. Nucleic Acids Res 25: 4876–4882.
99. Kimura M (1980) A simple method for estimating evolutionary rates of base substitutions through comparative studies of nucleotide sequences. J Mol Evol 16: 111–120.
100. Felsenstein J (1981) Evolutionary trees from DNA sequences: a maximum likelihood approach. J Mol Evol 17: 368–376.
101. Tamura K, Peterson D, Peterson N, Stecher G, Nei M, et al. (2011) MEGA5: molecular evolutionary genetics analysis using maximum likelihood, evolutionary distance, and maximum parsimony methods. Mol Biol Evol 10: 2731–2739.
102. Darling AE, Mau B, Perna NT (2010) progressiveMauve: multiple genome alignment with gene gain, loss and rearrangement. PLoS ONE 5: e11147. doi: 10.1371/journal.pone.0011147.
103. Blom J, Albaum S, Doppmeier D, Pühler A, Vorhölter FJ, et al. (2009) EDGAR: a software framework for the comparative analysis of prokaryotic genomes. BMC Bioinformatics 10: 154. doi: 10.1186/1471–2105–10–154.
104. Vernikos GS, Parkhill J (2006) Interpolated variable order motifs for identification of horizontally acquired DNA: revisiting the *Salmonella* pathogenicity islands. Bioinformatics 22: 2196–2203.
105. Vicente-Villardón JL (2010) MULTBIPLOT: A package for Multivariate Analysis using Biplots. Departamento de Estadística. Universidad de Salamanca. Available: http://biplot.usal.es/ClassicalBiplot/index.html.
106. Ren Q, Chen K, Paulsen IT (2007) TransportDB: a comprehensive database resource for cytoplasmic membrane transport systems and outer membrane channels. Nucleic Acids Res35: 274–279.
107. Rowbotham TJ, Cross T (1977) Ecology of *Rhodococcus coprophilus* and associated actnomycetes in fresh water and agricultural habitats. J Gen Microbiol 100: 231–240.
108. Mateos PF, Jimenez-Zurdo JI, Chen J, Squartini AS, Haack SK, et al. (1992) Cell-associated pectinolytic and cellulolytic enzymes in *Rhizobium leguminosarum* biovar trifolii. Appl Environ Microbiol 58: 1816–1822.
109. Williams ST, Goodfellow M, Alderson G, Wellington EMH, Sneath PHA, et al. (1983) Numerical classification of *Streptomyces* and related genera J Gen Microbiol 129: 1743–1813.
110. Murthy N, Bleakley B (2012) *Simplified method of preparing colloidal chitin used for screening of chitinase-producing microorganisms*. Internet J Microbiol 10: DOI:10.5580/2bc3.
111. Milagres AMF, Machuca A, Napoleão D (1999) Detection of siderophore production from several fungi and bacteria by a modification of chrome azurol S (CAS) agar plate assay. J Microbiol Meth 37: 1–6.
112. Glickmann E, Dessaux Y (1995) A critical examination of the specificity of the salkowski reagent for indolic compounds produced by phytopathogenic bacteria. Appl Environ Microbiol 61: 793–796.
113. Gordon SA, Weber RP (1951) Colorimetric estimation of indole acetic acid. Plant Physiol 26: 192–195.
114. Rigaud J, Puppo A (1975) Indole-3-acetic acid catabolism by soybean bacteroids J Gen Microbiol 88: 223–228.
115. Hardy RWF, Hoilsten RD, Jackson EK, Burns RC (1968) The acetylene-ethylene assay for N_2 fixation: laboratory and field evaluation. Plant Physiol 43: 1185–1207.

The Complete Mitochondrial Genomes of Six Species of *Tetranychus* Provide Insights into the Phylogeny and Evolution of Spider Mites

Da-Song Chen[1], Peng-Yu Jin[1], Kai-Jun Zhang[2], Xiu-Lei Ding[1], Si-Xia Yang[1], Jia-Fei Ju[1], Jing-Yu Zhao[1], Xiao-Yue Hong[1]*

1 Department of Entomology, College of Plant Protection, Nanjing Agricultural University, Nanjing, Jiangsu, China, **2** Department of Entomology, College of Plant Protection, Southwest University, Chongqing, China

Abstract

Many spider mites belonging to the genus *Tetranychus* are of agronomical importance. With limited morphological characters, *Tetranychus* mites are usually identified by a combination of morphological characteristics and molecular diagnostics. To clarify their molecular evolution and phylogeny, the mitochondrial genomes of the green and red forms of *Tetranychus urticae* as well as *T. kanzawai*, *T. ludeni*, *T. malaysiensis*, *T. phaselus*, *T. pueraricola* were sequenced and compared. The seven mitochondrial genomes are typical circular molecules of about 13,000 bp encoding and they are composed of the complete set of 37 genes that are usually found in metazoans. The order of the mitochondrial (mt) genes is the same as that in the mt genomes of *Panonychus citri* and *P. ulmi*, but very different from that in other Acari. The J-strands of the mitochondrial genomes have high (~84%) A+T contents, negative GC-skews and positive AT-skews. The nucleotide sequence of the *cox1* gene, which is commonly used as a taxon barcode and molecular marker, is more highly conserved than the nucleotide sequences of other mitochondrial genes in these seven species. Most tRNA genes in the seven genomes lose the D-arm and/or the T-arm. The functions of these tRNAs need to be evaluated. The mitochondrial genome of *T. malaysiensis* differs from the other six genomes in having a slightly smaller genome size, a slight difference in codon usage, and a variable loop in place of the T-arm of some tRNAs by a variable loop. A phylogenic analysis shows that *T. malaysiensis* first split from other *Tetranychus* species and that the clade of the family Tetranychoidea occupies a basal position in the Trombidiformes. The mt genomes of the green and red forms of *T. urticae* have limited divergence and short evolutionary distance.

Editor: Hector Escriva, Laboratoire Arago, France

Funding: This study was supported in part by a grant-in-aid from the Science and Technology Program of the National Public Welfare Professional Fund (no. 201103020) from the Ministry of Agriculture of China (www.moa.gov.cn), and a grant-in-aid for Scientific Research (nos. 31172131 and 31371944) from the National Natural Science Foundation of China (www.nsfc.gov.cn). The funders had no role in study design, data collection and analysis, decision to publish, or preparation of the manuscript.

Competing Interests: The authors have declared that no competing interests exist.

* Email: xyhong@njau.edu.cn

Background

The spider mite genus *Tetranychus* includes 149 species [1], some of which are of cosmopolitan agronomical pests, such as *Tetranychus urticae*. With expanded gene families for ABC genes [2], detoxification and digestion [3] in the genome, *T. urticae* represents one of the most polyphagous arthropod herbivores [1]. The genes of detoxification families have also been reported to transcriptionally respond to the defense chemistry of plants after *T. urticae* adapts to a challenging host (tomato) [4]. *T. urticae* and other tetranychid species feed on apples, citrus, cottons, cucumbers, cucurbits, eggplants, grapes, maize, papayas, peppers, soy, strawberries and tomatoes both in the field and in greenhouses. However, it is difficult to identify the tetranychid species by morphological characters because the potential diagnostic morphological characters are limited and often exhibit great phenotypic flexibility. Classification of the green and red forms of *T. urticae* is also a source of debate. These problems demonstrate the need for combining morphological and molecular approaches to identifying the species [5].

The second internal transcribed spacer (ITS2) of ribosomal DNA can be used as a barcode for distinguishing tetranychid species. For example, the ITS2 sequences of *T. kanzawai* and *T. hydrangea* indicate that they are synonymous species, which was confirmed by cross-breeding experiments [6]. One molecular diagnostic tool that has been developed to overcome the difficulty of morphological identification by the ITS2 sequence is restriction fragment length polymorphism (RFLP) [7,8]. The PCR-RFLP approach has been used to identify tens of *Tetranychus* species and has the ability to distinguish more species. The 5′ end of the mitochondrial COI gene is extensively used as a barcode to identify *Tetranychus* species and to analyze their phylogenetic evolution. Its higher divergence makes COI suitable for investigating intraspecific variation, but its usefulness for resolving phylogenetic species relationships remains limited [9,10]. The lack and occasional unreliability of sequences in public databases restricts their use as molecular diagnostic tools [5].

Thus, to clarify the molecular phylogeny of *Tetranychus* species, as well as to provide new DNA barcodes, we decided to compare the whole mitochondrial genomes of the green and red forms of *T. urticae* as well as five other major spider mite pests in China. Most metazoan mitochondrial genomes are circular, have a length of approximately 16 kb and encode 37 genes including 13 protein-coding genes (PCGs), two rRNA genes (rRNAs), and 22 tRNA genes (tRNAs) [11]. The databases presently include the mitochondrial genomes of 37 acarids, including 12 of the superorder Acariformes and 25 of the superorder Parasitiformes (Table S1). Several aspects of the mt-genomes of Acari have been examined, including gene rearrangement [12–16], tRNA gene loss [17] and atypical short tRNA [13,14].

Sequencing these genomes will have other benefits. For example it should provide insights into the molecular evolution of acaricide-resistance genes. The rapid development of acaricide resistance of spider mites is a long-standing problem [18,19]. Several acaricides have been identified as mitochondrial respiration inhibitors [19–21]. Resistance to the acaricide bifenazate has been correlated with mutations in the mitochondrial cytochrome b (*cob*) gene [22,23]. The genomes will also provide information on gene rearrangements [24–30], evolutionary pattern and structure of the control region [31,32], strand asymmetry in nucleotide composition [33] and RNA secondary structure [34].

Materials and Methods

Sample origin and identification

Ethics Statement: No specific permits were required for the collection of spider mites because the spider mite is a pest in agriculture and the location is not privately-owned in any way. The field study did not involve endangered or protected species.

Strains of *T. kanzawai*, *T. ludeni*, *T. malaysiensis*, *T. phaselus*, *T. pueraricola* and the green and red forms of *T. urticae* were collected in the field, separately (Table S2). Mites were reared on a leaf of the common bean (*Phaseolus vulgaris* L.) at 25±1°C, 60% r.h. and under a 16/8 h (light/dark) photoperiod. All the species were classified by morphological characteristics [35] and RFLP analyses [7,8]. For the RFLP analyses, the ITS2 fragment was amplified by PCR using the *Tetranychus* universal primers, rD02 and HC2 (Table S3) and genomic DNA as template, and digested by five restriction endonucleases (DraI, RsaI, MboII, DdeI and HinfI) (Fig. S1).

DNA processing

DNA was extracted from individual mites with a Wizard Genomic DNA Purification Kit (Promega) according to the manufacturer's protocol. A fragment of the COI gene was amplified by standard PCR using the *Tetranychus* universal primers, T-CO1-F and T-CO2-R (Table S3). PCR fragments were ligated into pEASY-T3 cloning vector (Beijing TransGen Biotech) and the resulting plasmid DNAs were transformed into competent *Escherichia coli* Trans1-T1 cells provided in the cloning kit. The inserted fragments were sequenced with M13f and M13r primers. Long PCR primers for each species were designed according to the COI fragment sequences (Table S3). The mitochondrial genome was amplified by long PCR in one single fragment according to the manufacturer's rapid PCR protocol. The reaction mixture contained 2 µl PrimeSTAR GXL DNA Polymerase (Takara), 10 µl buffer, 4 µl dNTP mixture (2.5 mM each), 1 µl of each primer (10 mM), 1 µl of DNA and water added to total volume of 50 µl. The cycling conditions were 30 cycles of 98°C for 10 s and 68°C for 5 min. Sequencing libraries for the long PCR fragments were prepared by using a TruSeq DNA Sample Prep Kit (Illumina) following the manufacturer's instructions. Each individual library was tagged for a different multiplexing identifier (MID). The libraries were purified with Certified Low Range Ultra Agarose (Bio-Rad), quantified with a TBS380 fluorometer (Invitrogen), pooled and sequenced using a Miseq V2 Reagent kit to generate pair-end reads (read length: 250 bp). The reads for each sample were sorted by tag sequences and assembled into one contig after trimming the tag adapter sequences. The contigs were assembled with COI sequences respectively to obtain integral genome sequences (NCBI GenBank accession numbers: KJ729017 – KJ729023).

Annotation and analysis

Protein-coding genes (PCGs) were identified by ORF Finder implemented at the NCBI website using the invertebrate mitochondrial genetic code. The sequences of PCGs that matched *T. urticae* mitochondrial genes submitted previously to NCBI [22] were accepted. The tRNA genes were identified using ARWEN with default parameters [36] and the tRNAscan-SE [37] with a cove cutoff score of 0.1, the tRNA-model set to "EufindtRNA-Cove" and source set to "Mixed". Other tRNA genes and two rRNA genes (*rrnL* and *rrnS*) were determined by sequence similarity to genes in *Panonychus citri* [13,23], *P. ulmi* and *T. urticae*. The secondary structure models for tRNA and rRNA genes were constructed by comparison with the published secondary structures for *Dermatophagoides pteronyssinus* [16], *Leptotrombidium pallidum* [38], *Panonychus citri* [13], *Steganacarus magnus* [17] and *Palmaria palmata* [39]. The Map, GC content and GC skew of the mitochondrial genome were drawn with the CGView [40]. The base composition, codon usage, Relative Synonymous Codon Usage (RSCU) values and nucleotide substitution were analyzed with Mega ver. 6 [41]. The secondary structure of the A+T-rich regions (putative control region) were constructed with Mfold Server [42] and RnaViz ver. 2 software [43]. The evolutionary pairwise divergence was estimated with Mega ver. 6 with Kimura 2-parameter model [44]. Multiple alignments of 13 PCGs' amino acid sequences were performed with ClustalW [45] as implemented in Mega ver. 6 and then corrected by eye to ensure that alignment was in agreement with protein coding genes and to minimize the number of uninformative gaps (Appendix S1). The order of the PCGs is *atp6, atp8, cox1-3, cob, nad1-4, nad4l and nad5–6*. The length of the alignment was 2974 amino acids in the final dataset. ProtTest (http://darwin.uvigo.es/software/prottest2_server.html) [46] was used to select the best model of protein evolution. It selected mtREV+G+I+F as the most appropriate model for the combined amino acid dataset under the Bayesian Information Criterion (BIC) [47]. Maximum-likelihood (ML) analysis of the amino-acid dataset was performed with PhyML ver. 3.1 [48] under the mtREV+G+I+F model, and Maximum-parsimony (MP) analysis was performed with Mega ver. 6. Bootstrap percentages (BPs) were calculated with 1000 replications. MrBayes ver. 3.2.2 was used for Bayesian analysis under the mtREV model of amino acid substitution with the nst = 6 rates = invgamma command. Two runs were performed simultaneously, each with four Markov chains (one cold, three heated). The analyses were run for 1,000,000 generations and the trees were sampled every 100 generations. The Markov chain stationarity and run parameter convergence were evaluated with TRACER ver. 1.6. The first 25 percent of the trees were discarded as burn-in with the relburnin = yes burninfrac = 0.25 command. Multiple alignments

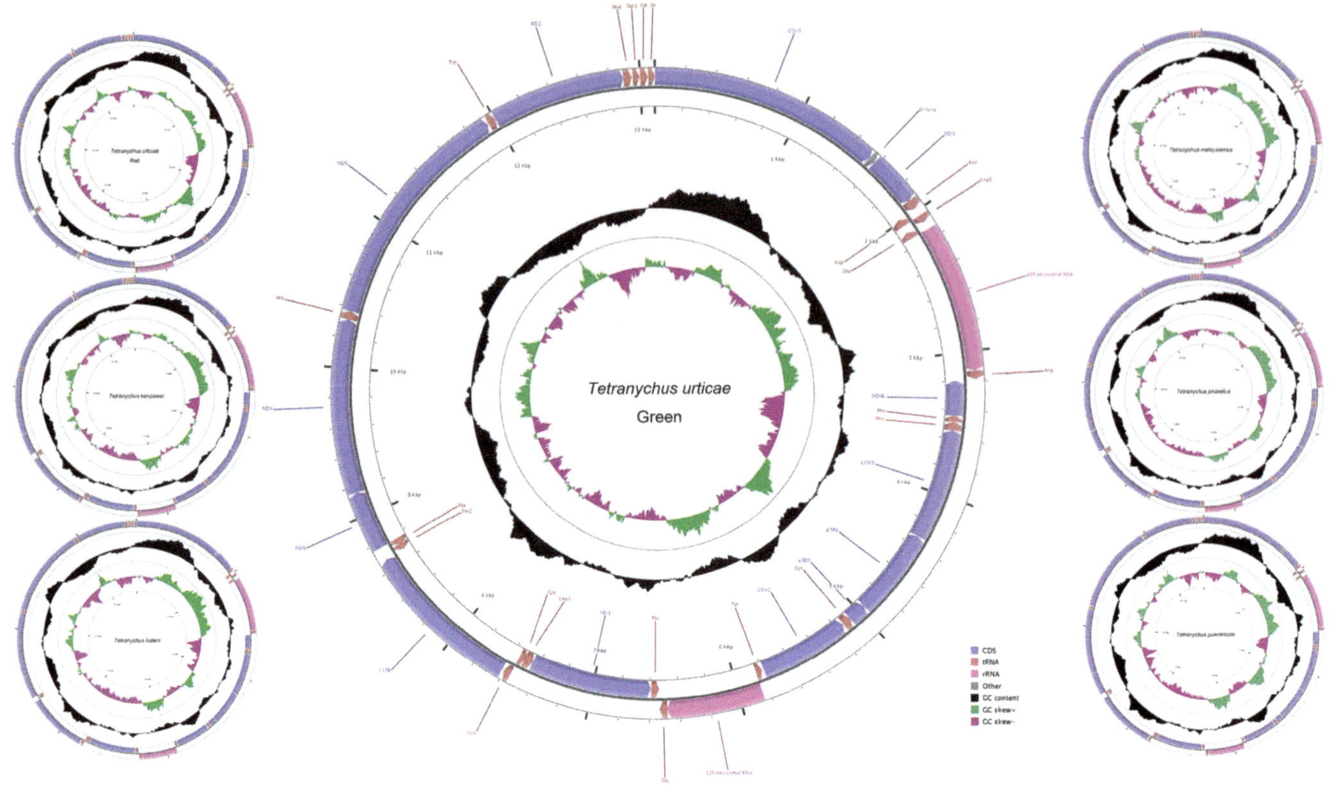

Figure 1. Mitochondrial genome maps of *T. urticae* (green and red forms), *T. kanzawai*, *T. ludeni*, *T. malaysiensis*, *T. phaselus* and *T. puerariccola*. From outer to inner, the 1st circle shows the gene map and tRNA genes are abbreviated by triple letter, with Leu1 = CUN, Leu2 = UUR, Ser1 = AGN and Ser2 = UCN. The 2nd circle shows the GC content and the 3rd shows GC skew calculated as (G−C)/(G+C). GC content and GC skew are plotted as the deviation from the average value of the entire sequence.

of whole genomic DNA sequences of the Tetranychoidea were performed with ClustalW as implemented in Mega ver. 6 (Appendix S1). jModelTest Ver. 2.1.4 [49] selected the GTR+I+G model as the best model for the nucleotide sequences of the mt genome. The ML analysis was performed and bootstrapped with 1000 replications with PhyML v. 3.1 with the GTR+I+G model, and MP analysis was performed by using Mega ver. 6. The Markov chains were run for 10,000,000 generations with sampling every 1000 generations for the Bayesian analysis. The first 25 percent of the trees were discarded and the remaining trees were used to calculate Bayesian posterior probabilities.

Results and Discussion

General features of mitochondrial genome organization

The mitochondrial genomes of the seven common tetranychid mites in China are typical circular DNAs (Fig. 1) with lengths of about 13,000 bp (Table S2). To our knowledge, the mitochondrial genome of *T. malaysiensis* is the smallest within all Acari genomes accessible in the GenBank (status March 17, 2014). The mitochondrial genome sizes of the red and green forms of *T. urticae* differ by 4 bp. Thirty-seven genes (13 PCGs, two rRNA genes, and 22 tRNA genes) were identified in each genome (Table S4), which is typical of presently available in metazoan mitochondrial genomes [11]. Twenty genes are encoded on the majority strand (J-strand), whereas the others are encoded on the minority strand (N-strand). The gene order is the same in all seven genomes and the same as that in *Panonychus citri* and *P. ulmi* which are in the same family as *Tetranychus*. However, this gene order is very different from that in other Acari and chelicerates as Yuan's [13] and Van Leeuwen's [23] reports. This suggests that the mitochondrial genome rearrangement event occurred before the divergence of *Panonychus* and *Tetranychus*.

Base content

The J-strands of the seven mitochondrial DNAs have high A+T contents (83.4–84.5%) which are higher than those for Acariformes (about 74% without the Tetranychidae family) and Parasitiformes (about 77%) (Fig. 2). In Acari, the AT-skew of the mitochondrial genome (average 0.009±0.013) ranges from 0.279 in *Unionicola parkeri* to −0.253 in *Dermatophagoides farina*. Within the genus *Tetranychus*, the species with the highest AT skews in the mitochondrial DNAs are *T. malaysiensis* (0.056), *T. phaselus* (0.050) and *T. ludeni* (0.041). The average GC-skew of Acari mitochondrial genomes is −0.130±0.024, ranging from −0.379 in *Ornithodoros moubata* to 0.231 in *Dermatophagoides farina*. Most Acari have negative mitochondrial genome GC-skews. The exceptions are *Varroa destructor* (0.178), *Dermatophagoides pteronyssinus* (0.194), *D. farinae* (0.231), *Phytoseiulus persimilis* (0.222), *P. citri* (0.033) and *P. ulmi* (0.005). The *Tetranychus* species with the lowest mitochondrial DNA skews are *T. ludeni* (−0.072), *T. phaselus* (−0.072) and *T. malaysiensis* (−0.070). The AT- and GC-skews are quite similar in the mitochondrial genomes of the two *T. urticae* forms. Most metazoan species present a clear strand asymmetry, in which the J-strand is biased in favor of A and C and the N-strand is biased in favor of T and G [50]. The J-strands of the seven mitochondrial DNAs exhibit typical GC-skews, but two completed mitochondrial

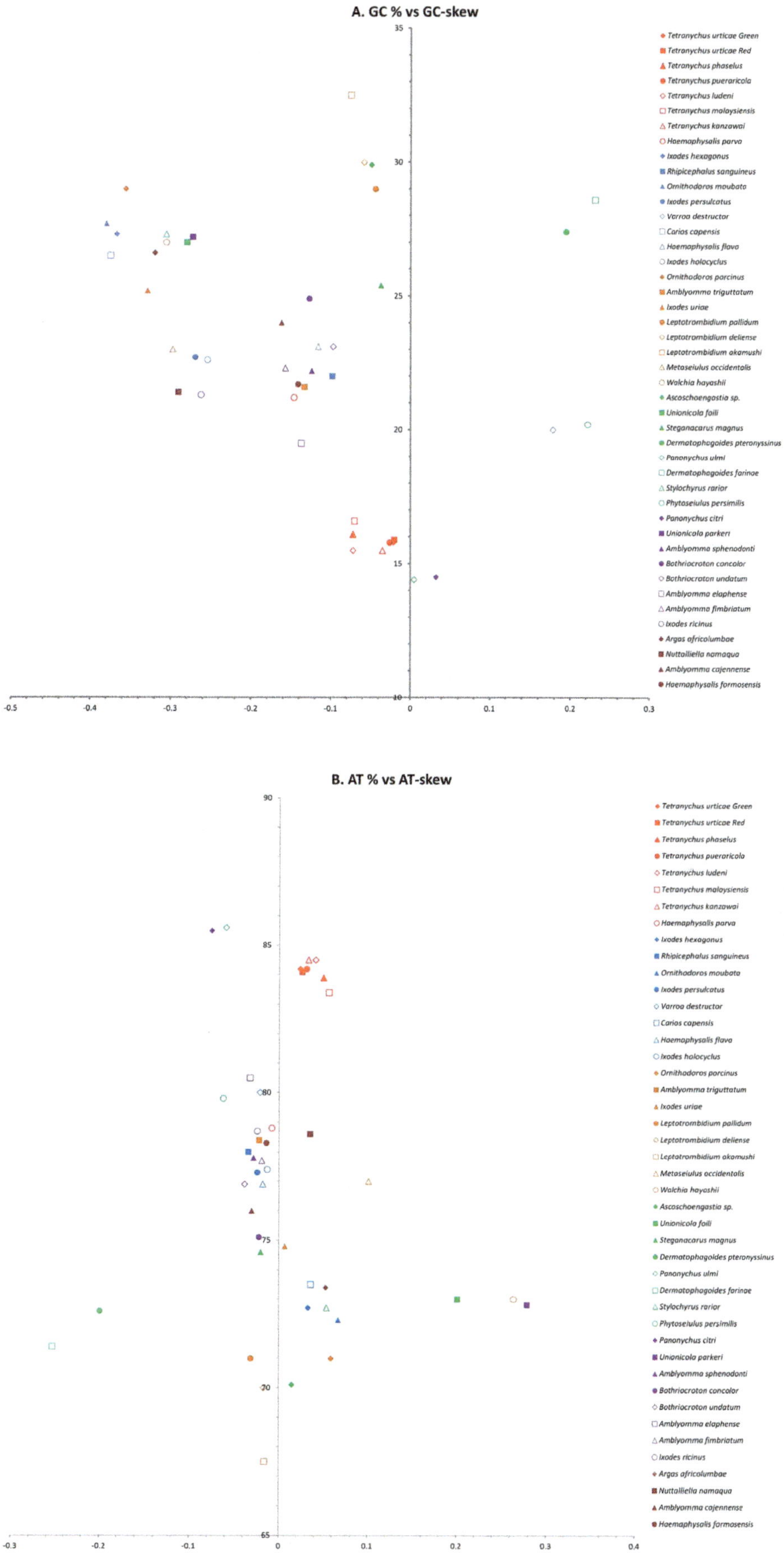
A. GC % vs GC-skew
35
30
25
20
15
10
-0.5
-0.4
-0.3
-0.2
-0.1
0
0.1
0.2
0.3
Tetranychus urticae Green
Tetranychus urticae Red
Tetranychus phaselus
Tetranychus pueraricola
Tetranychus ludeni
Tetranychus malaysiensis
Tetranychus kanzawai
Haemaphysalis parva
Ixodes hexagonus
Rhipicephalus sanguineus
Ornithodoros moubata
Ixodes persulcatus
Varroa destructor
Carios capensis
Haemaphysalis flava
Ixodes holocyclus
Ornithodoros porcinus
Amblyomma triguttatum
Ixodes uriae
Leptotrombidium pallidum
Leptotrombidium deliense
Leptotrombidium akamushi
Metaseiulus occidentalis
Walchia hayashii
Ascoschoengastia sp.
Unionicola foili
Steganacarus magnus
Dermatophagoides pteronyssinus
Panonychus ulmi
Dermatophagoides farinae
Stylochyrus rarior
Phytoseiulus persimilis
Panonychus citri
Unionicola parkeri
Amblyomma sphenodonti
Bothriocroton concolor
Bothriocroton undatum
Amblyomma elaphense
Amblyomma fimbriatum
Ixodes ricinus
Argas africolumbae
Nuttalliella namaqua
Amblyomma cajennense
Haemaphysalis formosensis
B. AT % vs AT-skew
90
85
80
75
70
65
-0.3
-0.2
-0.1
0
0.1
0.2
0.3
0.4
Tetranychus urticae Green
Tetranychus urticae Red
Tetranychus phaselus
Tetranychus pueraricola
Tetranychus ludeni
Tetranychus malaysiensis
Tetranychus kanzawai
Haemaphysalis parva
Ixodes hexagonus
Rhipicephalus sanguineus
Ornithodoros moubata
Ixodes persulcatus
Varroa destructor
Carios capensis
Haemaphysalis flava
Ixodes holocyclus
Ornithodoros porcinus
Amblyomma triguttatum
Ixodes uriae
Leptotrombidium pallidum
Leptotrombidium deliense
Leptotrombidium akamushi
Metaseiulus occidentalis
Walchia hayashii
Ascoschoengastia sp.
Unionicola foili
Steganacarus magnus
Dermatophagoides pteronyssinus
Panonychus ulmi
Dermatophagoides farinae
Stylochyrus rarior
Phytoseiulus persimilis
Panonychus citri
Unionicola parkeri
Amblyomma sphenodonti
Bothriocroton concolor
Bothriocroton undatum
Amblyomma elaphense
Amblyomma fimbriatum
Ixodes ricinus
Argas africolumbae
Nuttalliella namaqua
Amblyomma cajennense
Haemaphysalis formosensis

Figure 2. GC% *vs* GC-skew and AT% *vs* AT-skew in the 44 Acari mitochondrial genomes. Values are calculated for mitochondrial genomes of J-strands. The X-axis provides the nucleotide skew values and the Y-axis shows the nucleotide percentages.

genomes from the genus *Panonychus* have positive GC-skews. It has been suggested that such reversals are caused by inversions of the A+T-rich regions and replication origin [33].

Putative control region

The longest non-coding region, which presumably functions as the mitochondrial control region, is 41–44 bp in length and is flanked by the *cox1* and *nad3* genes. Among Acari, the genus *Tetranychus* has the smallest mitochondrial control region. The A+T-rich region is believed to be characterized by a poly-T stretch at the 5′ end, a poly$[TA(A)]_n$ stretch close to the poly-T stretch, a stem and loop structure flanked by a TATA motif and a G $(A)_n$ T motif [31,32]. The A+T-rich region in the seven *Tetranychus* genomes can be folded into one stem-loop secondary structure and an apparent poly $[TA(A)]_n$ stretch located near the 5′ end of A+T-rich region (Fig. 3). However,none of the hairpin structures is associated with a poly-T stretch or flanked by a TATA or G $(A)_n$ T motif. The loops in the secondary structures of *T. ludeni* and *T. malaysiensis* are smaller than they are in other *Tetranychus* species, and the stem-loop secondary structure in *T. urticae* and *T. pueraricola* are located closer to the 5′ end of the A+T-rich region. Although similar poly $[TA(A)]_n$ stretches are present near the 5′ end of A+T-rich regions in *T. urticae* (green form: TAAAA; red form: TAAAA) and *T. pueraricola* (TAAA), they were folded as stems in the hairpin structures. The A+T-rich regions of *P. citri* and *P. ulmi* are not only longer (57 and 55 bp, respectively), but they also can be folded into two hairpin structures. Short poly-T stretches were identified in the A+T-rich regions of *P. citri* and *P. ulmi*, but the poly $[TA(A)]_n$ stretch, TATA or G $(A)_n$ T motif are not present.

Protein-coding genes and codon usage patterns

The total length of the 13 PCGs in *T. malaysiensis* (10210 bp) is shorter than it is in the other *Tetranychus* mitochondrial genomes (10224–10227 bp). All the PCGs start with typical ATN codons (where N is any nucleotide), as is the case in other metazoan mitochondria (Table S4). Only three of the mitochondrial genes in *T. malaysiensis* (*cox1*, *nad6* and *nad4*) and only one gene in *T. phaselus* (*nad4l*) have ATC start codons. In the other genes, ATA, ATT and ATG start codons are more common. In most cases, a given gene has different start codons in the different mitochondrial genomes. The exceptions are *cox3*, *atp6* and *cox2*, which start with ATG in each genome and *nad2*, which starts with ATA in each genome. Although most PCGs have TAA or TAG stop codons, many of them have incomplete stop codons, such as T or TA. The two forms of *T. urticae* have different start codons for *nad3* and different stop codons for *cox1*. In the codons of the 13 PCGs of the seven genomes, the three codon positions have different nucleotide biases (Fig. S2). The A+T content of the third codon position (about 91%) is higher than the A+T contents of the first and second codon positions (about 82% and 78%, respectively). But the third codon position in the *T. malaysiensis* genome has a lower thymine content (47%) and a higher cytosine content (7%) than do the third codon positions in the others six genomes (in which the average T and C contents are 49% and 5%, respectively).

The amino acid frequencies without start and stop codon are similar between the different *Tetranychus* mitochondrial genomes (Fig. S3). The utmost frequently used amino acids are Phe (14.61–15.26%), Leu (13.14–13.85%), Met (11.25–12.83%), Ile (10.38–11.57%), Ser (9.06–9.65%) and Asn (8.25–8.59%). The mitochondrial genome of *T. malaysiensis* has a slight variety in the amino acid proportion of Met (12.83%, average: 11.77%) and Ile (10.38%, average: 11.29%). The seven AT-rich codons TTT-Phe (13.31–14.35%), ATT-Ile (9.52–11.02%), TTA-Leu (10.30–10.87%), ATA-Met (10.63–12.00%), AAT-Asn (6.71–7.23%), AAA-Lys (4.52–4.84%), and TAT-Tyr (3.48–3.78%) are the most frequently used codons in the *Tetranychus* PCGs. The slightly low ATT codon (9.52%, average: 10.54%) and high ATA codon (12.00%, average: 11.03%) usage in the *T. malaysiensis* mitochondria leads to the variety in the amino acid proportion. A considerable transversion likely occurred between the codon ATT and ATA in the evolution of *Tetranychus* mitochondria. The Relative Synonymous Codon Usage (RSCU) in *Tetranychus* PCGs exhibits a similar pattern and an over-usage of A and T at the third codon positions (Fig. 4). Six codons could not be identified in the *T. malaysiensis* mitochondrial PCGs whereas other *Tetranychus* PCGs abandon only 1–4 codons. Two codons (CTG and CGC) could not be identified in the PCGs of green form of *T. urticae*, while only one codon (AGG) is not present in the PCGs of red form of *T. urticae*. Several arthropods have been reported to translate the codon AGG as lysine instead of serine in mitochondrial genetic codon [51]. However, only one AGG codon was identified in the PSGs of *T. urticae* green form. The other mitochondrial genomes of spider mites including *P. citri* and *P. ulmi* [13] do not use AGG codon. And it is still uncertain which amino acid does the AGG codon in spider mite's mitochondrion translate.

The evolutionary patterns of the PCGs are different (Fig. 5). The *cox1* gene is commonly used as a taxon barcode because of its high rates of interspecific sequence change and constraints on intraspecific divergence [52,53]. But this gene exhibits the lowest nucleotide substitution rate per site (0.208 ± 0.014) and lowest value of nonsynonymous substitutions per nonsynonymous site (Ka) (0.04 ± 0.004) compared to the other genes. The nucleotide substitution rate per site (0.555 ± 0.097) and the number of synonymous substitutions per synonymous site (Ks) (1.098 ± 0.143) of *nad3* gene are highest, while the Ka number of *atp8* is highest (0.315 ± 0.062). In addition, the GC content was found to be negatively correlated with both the nucleotide substitution rate per site ($R = -0.637$, $P = 0.019$) and Ka ($R = -0.715$, $P = 0.006$), while the nucleotide substitution rate per site was found to be positively correlated with Ka ($R = 0.825$, $P = 0.001$). Ks was not found to be associated with Ka, GC content or nucleotide substitution rate per site.

Ribosomal and transfer RNAs

Among the seven *Tetranychus* mitochondrial genomes, the large subunits of rRNA (*rrnL*) are from 982 bp (*T. ludeni*) to 998 bp (*T. urticae* and *T. kanzawai*) in length. The first nucleotide downstream of *trnE* was annotated as the 5′ end of *rrnL* and the first nucleotide upstream of *trnR* was annotated as the 3′ end of *rrnL*. The genes for the small subunit of the rRNA (*rrnS*) are 629 bp (*T. ludeni* and *T. phaselus*) to 645 bp (*T. pueraricola*) in length and are located between *trnY* and *trnG*. Both ribosomal subunits are encoded on the J-strand as is the case in *Metaseiulus occidentalis* [54], *Dermatophagoides farinae* [14], *D. pteronyssinus* [16], *Leptotrombidium pallidum* [55], *P. citri* [13] and *P. ulmi*,

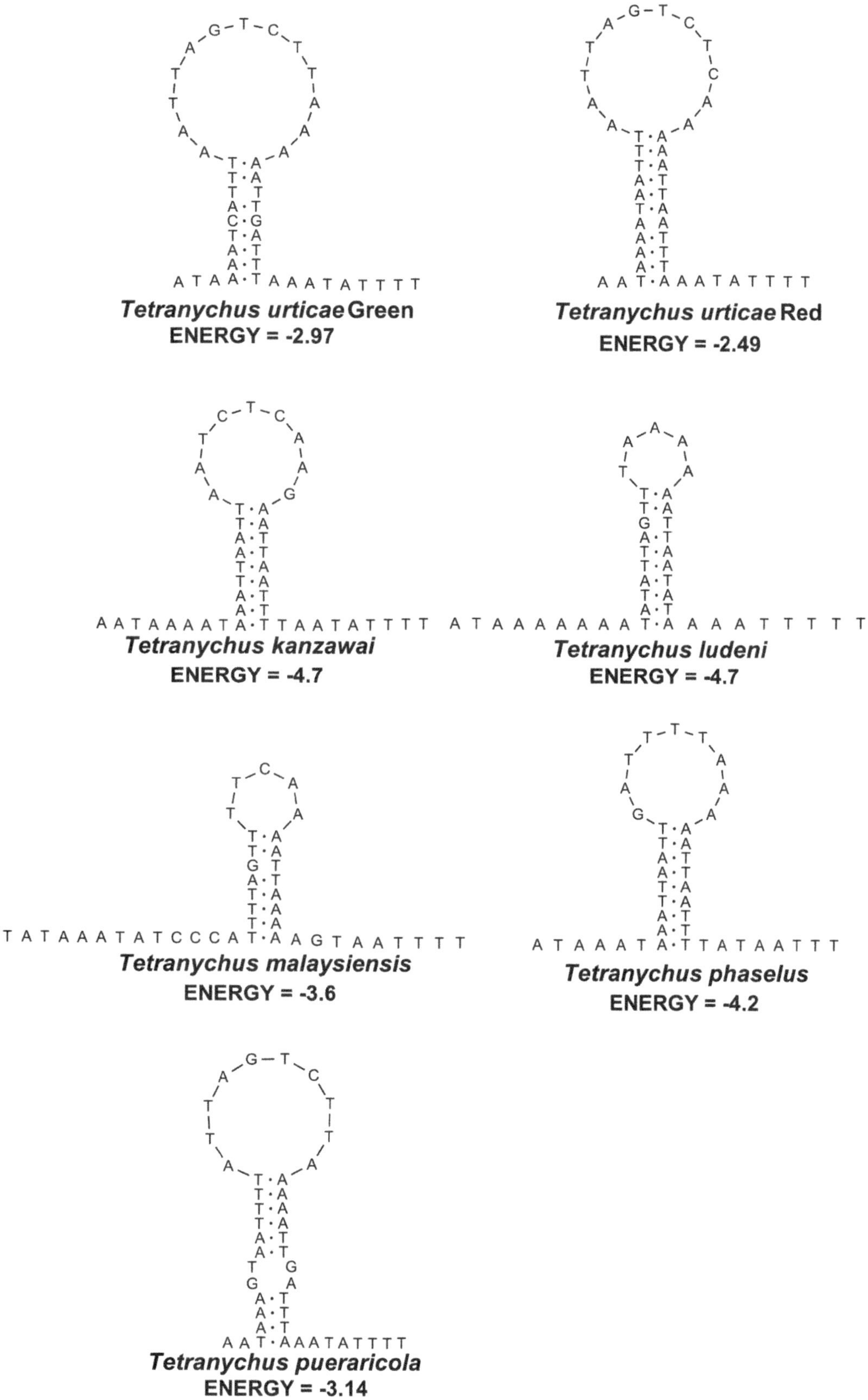

Figure 3. Putative stem-loop secondary structure of A+T-rich regions. The structures were constructed by using Mfold and Watson-Crick bonds are illustrated by black dots. The free energy values (kcal/mol) and the species names are shown below each structure.

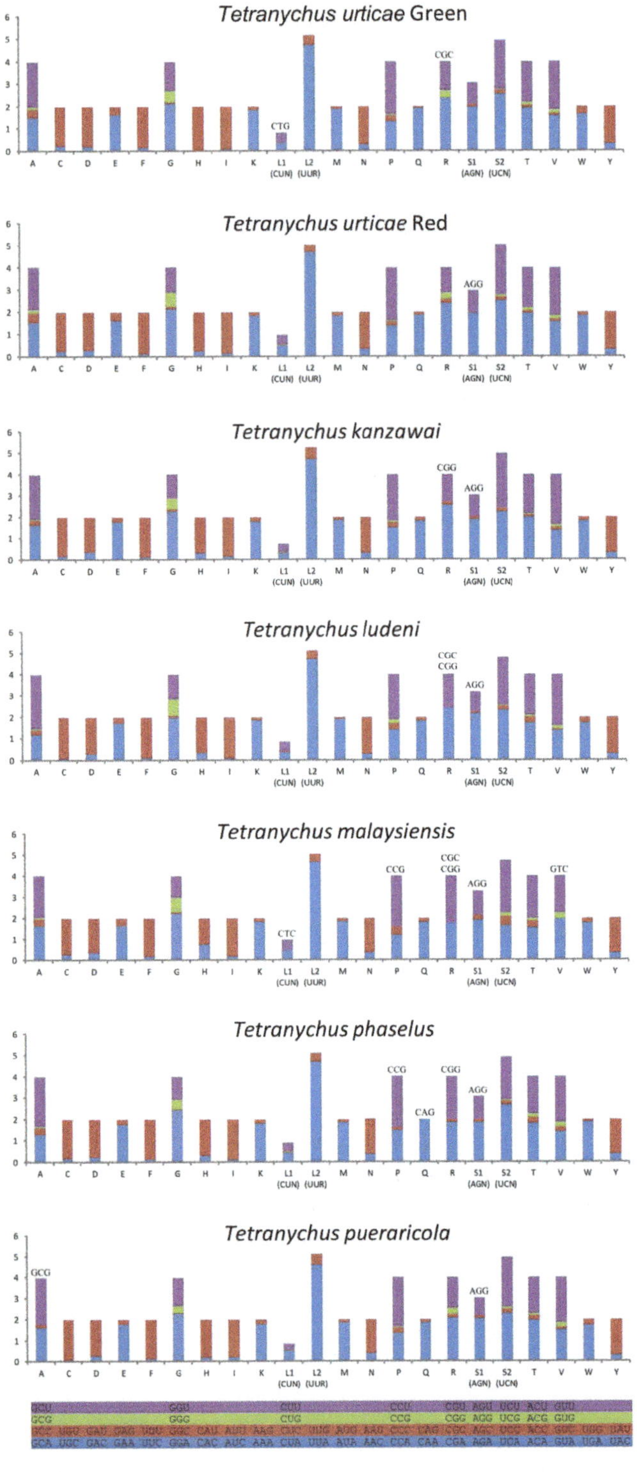

Figure 4. Relative synonymous codon usage (RSCU) for the mitochondrial proteins. The X-axis shows the codon families, while the Y-axis shows RSCU values. Absent codons are shown at the top of the columns.

whereas the rRNA genes in most species of Acari, arthropods and chelicerates are located on the N-strand [12,14,15,17,56,57].

We predicted the *rrnL* and *rrnS* secondary structures of *T. urticae* (Fig S4 and S5). Nucleotides that are identical among the seven *Tetranychus* mitochondrial DNAs are shown in bold in Figures S4 and S5. In contrast, most of the helices found in Acari *rrnL* and *rrnS* genes are present in *T. urticae*. However, four helices (D10, D14, G9 and H3) of *rrnL* and one helix (33) of *rrnS* found in *P. citri* and *L. pallidum* are not present, while *T. urticae* has one more helix (D8) in *rrnL* and three more helices (4, 18 and 22) in *rrnS*. Most ribosomal subunit sequences, especially the 3′-end of both *rrnL* and *rrnS*, are conserved among the seven mitochondrial genomes. The helices D2, D4, D7, D9, and D17 could be identified in all the mitochondrial *rrnL* genes with slight variations, whereas their sequences are not conserved. The sequences of helices 24/25/26 as depicted in *rrnS* are weakly conserved among the genus *Tetranychus*, and the stem-loops in the deduced secondary structures vary in size.

Eleven of the 22 tRNA genes are encoded on the J-strand and the other 11 are encoded on the N-strand. The secondary structures for all tRNAs of the seven mitochondrial genomes were predicted (Table S5). The aminoacyl acceptor stem (AA-arm) and anticodon arm (AC-arm) in the tRNAs are highly conserved among the seven genomes. With a few exceptions among the seven genomes, only seven of the tRNAs (*trnH*, *trnK*, *trnL2*, *trnM*, *trnN*, *trnR* and *trnW*) can be potentially folded into a classical cloverleaf structure, whereas four tRNAs (*trnD*, *trnE*, *trnS1* and *trnS2*) lose the DHU stem (D-arm), eight (*trnA*, *trnC*, *trnF*, *trnG*, *trnL1*, *trnT*, *trnV* and *trnY*) lack the TψC stem (T-arm) and 3 (*trnI*, *trnP* and *trnQ*) appear to lack both the D- and the T-arm. *T. ludeni* differs from the other species in that the D-arm of *trnA* was substituted by a D-replacement loop, and *T. malaysiensis* differs from the others in that the T-arms of *trnM*, *trnR* and *trnS1* were replaced by a variable loop. *trnP* of *T. malaysiensis* and *trnQ* of *T. kanzawai* still have T-arms while the *trnP*s and *trnQ*s in the other species have lost them. The *trnD* has a only 4 bp well-paired anticodon stem. And the *trnP* of *T. ludeni*, *trnQ* of *T. kanzawai* and *T. phaselus* and *trnV* of *T. malaysiensis* also have one mismatch. Commonly, the 3 bp anticodon of the tRNAs was flanked by 2 bp up-stream and 2 bp down-stream, but *trnI* of *Tetranychus* mitochondria has 3 bp down-stream. These aberrant anticodon loops have been reported for the same tRNA of *P. citri* [13] and *P. ulmi*, which implies that this anticodon loop pattern is universal among *trnI* genes of the Tetranychidae mitochondria. Noncanonical anticodon loop structures are also present in *trnL2* (3 bp down-stream) of *D. pteronyssinus* [16], *trnS2* (3 bp up-stream) of *Camelus bactrianus ferus* [58] and *trnH* (3 bp up-stream) and *trnN* (6 bp down-stream) of *Mesobuthus gibbosus* [59].

Phylogenetic analysis

Maximum-parsimony (MP), Maximum-Likelihood (ML) and Bayesian inference (BI) phylogenetic trees were constructed based on the concatenated amino acid sequences of the 13 PCGs of the seven *Tetranychus* species and other Acariformes. The three trees had identical topologies (Fig. 6). Almost all nodes were well supported, whereas the split between *T. phaselus* and the other *Tetranychus* species was not supported by the MP tree. The three species of Sarcoptiformes were monophyletic and clustered as a sister group of Trombidiformes, in agreement with a previous analysis [16]. The superfamily Tetranychoidea, which is considered as a member of Trombidiformes, split as a single clade and formed a sister group to other species of Trombidiformes. Another study [16] obtained a similar topology, except that the bootstrap values and BI posterior probabilities for the Tetranychoidea were low. Mitochondrial genomes evolve at high rates. It has been suggested that the three most important factors are rearrangements in the mitochondrial genome, a parasitic lifestyle and small body size [60]. Most species of Acariformes exhibit all three factors and the great variability in branch lengths suggest a high

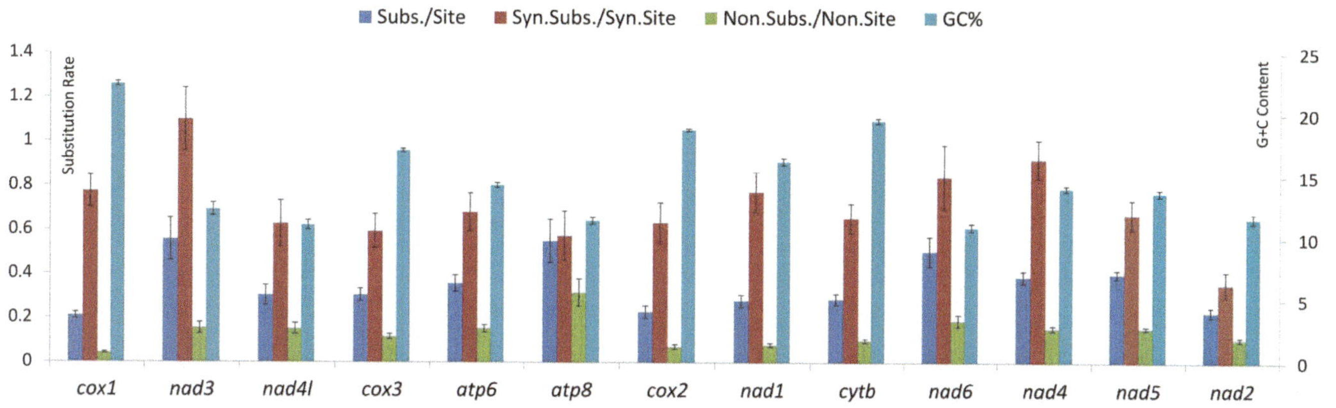

Figure 5. Different evolutionary patterns among protein coding genes of *T. urticae* green form, *T. kanzawai*, *T. ludeni*, *T. malaysiensis*, *T. phaselus* and *T. pueraricola*. Sub./Site, the number of nucleotide substitutions per site from averaging over all sequence pairs, was calculated with JC model. Syn.Subs./Syn.Site, the number of synonymous substitutions per synonymous site (Ks); Non.Subs./Non.Site, the number of nonsynonymous substitutions per nonsynonymous site (Ka); Ks-Ka, the value of Ks minus Ka; the analysis were estimated with Kumar model [68]. The rate variation among sites was modelled with a gamma distribution calculated by Mega ver. 6 and the standard error estimations were obtained by bootstrapping 1000 replicates. Substitution rate is shown on the left Y-axis and G+C content is shown on the right Y-axis.

heterogeneous substitution rate among the mt genomes. These problems led to markedly long branches, and the superabundant phylogenetic signals probably led to the distant phylogenetic position of the Tetranychoidea. Identical topologies were constructed by the three approaches with high bootstrap values and high Bayesian posterior probabilities for the position of Tetranychoidea. However, additional mt genome data or trees based on more conserved sequences are needed to improve the phylogeny of Acariformes.

The seven *Tetranychus* species clustered monophyleticly. *T. malaysiensis* was allocated to the basal position of the genus *Tetranychus*. The position of *T. phaselus* was supported by the BI and ML trees, whereas the bootstrapping for MP questioned this node. A phylogenetic tree was constructed based on the mitochondrial genomic sequences of Tetranychoidea (Fig. S6). The topologies of the trees constructed with the three approaches were consistent. Furthermore, a phylogenetic analysis based on nuclear gene sequences of rRNA and ITS supported this analysis (data not shown). However, the position of *T. phaselus* was not supported well by three trees. Some other mitochondrial genomic sequences are needed to improve the phylogenetic analysis for the genus *Tetranychus*. Especially the species has close relationship with *T. phaselus*.

The green and red forms of *T. urticae*

The green and red forms of *T. urticae* clustered together in the phylogenetic trees and the evolutionary divergence between their mitochondrial genomes was low (0.027±0.002) (Table S6).

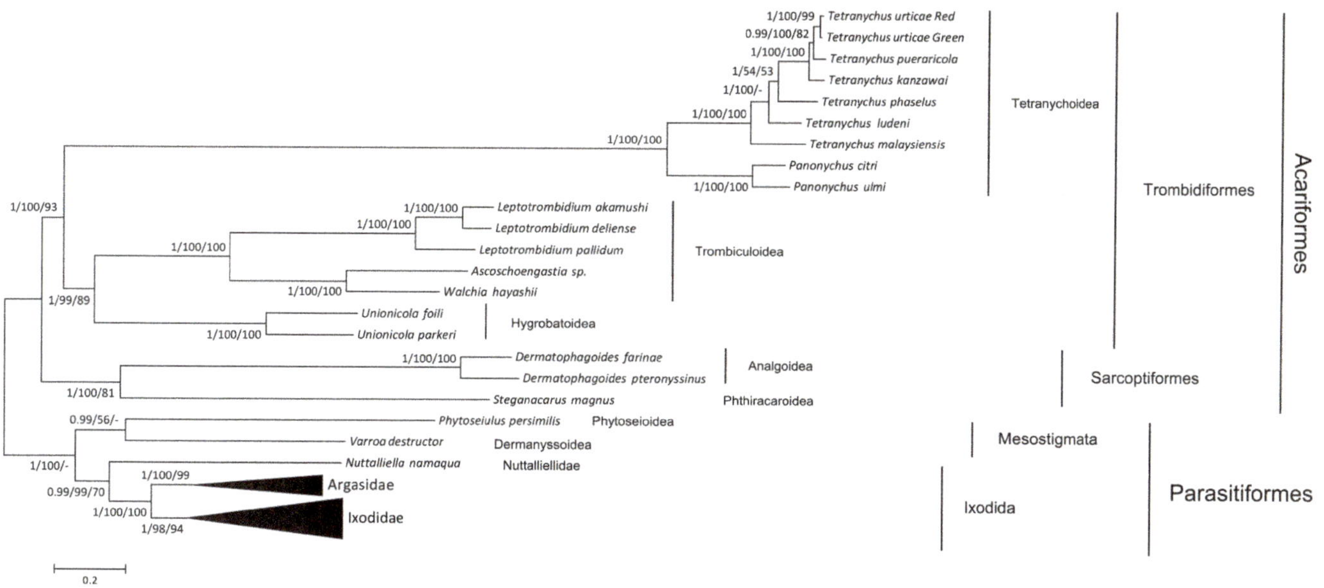

Figure 6. Phylogenetic tree of Acariformes relationships. The tree was inferred from amino acid datasets and rooted with Parasitiformes taxa. Numbers at nodes are percentages from Bayesian posterior probabilities (left), ML bootstrapping (middle) and MP bootstrap support values (right). The nodes that did not support Bayesian inference or had low bootstrap support values were marked as midlines.

Pairwise distances among the *Tetranychus* mt genomes except for the value between the two forms of *T. urticae* were 0.089 to 0.211 which is much higher than the distance between the two forms of *T. urticae*. Some slight divergences between two forms are found in the length of 4 bp difference, an AGG codon found in the PCGs of green form, a different start codon in *nad3* gene and a different stop codon in *cox1* gene. However, these limited differences can not be classified as interspecies divergences. Consequently, the close evolutionary distance and limited mitochondrial genome divergence do not support the red forms of *T. urticae* as a new species or subspecies. Between the two forms of *T. urticae*, partial hybrid infertility was discovered [61] and hybrid affinity strongly restricted the gene flow [62]. However, the harboring of *Wolbachia* and *Cardinium* by the two forms [63–65] confounds the origin of reproductive abnormality in the hybrid analyses. *Wolbachia* is considered to induce cytoplasmic incompatibility (CI) in *T. urticae*, whereas the levels of CI in different populations varied greatly [66,67]. Although *Cardinium* did not appear to distort reproduction in *T. urticae* [63], it is still necessary to investigate the CI level between *Wolbachia* and *Cardinium* in *T. urticae*. In conclusion, intracellular bacteria with the ability to manipulate reproduction complicate investigations of the ability of the two forms to interbreed. Further studies are needed to analyze the hybridization in *T. urticae* without the influence of *Wolbachia*, *Cardinium* and other bacteria that can manipulate reproduction.

Supporting Information

Appendix S1 Multiple alignments of Acari 13 PCGs' amino acid sequences (13PCGs.meg). The order of the PCGs is *atp6, atp8, cox1-3, cob, nad1-4, nad4l and nad5–6*. The alignments were performed with ClustalW as implemented in Mega and require the Mega software version 6 to examine it (http://www.megasoftware.net/).

Appendix S2 Multiple alignments of Tetranychidae mitochondrial genomic sequence (Tetranychidae.meg). The alignments were performed with ClustalW as implemented in Mega and require the Mega software version 6 to examine it (http://www.megasoftware.net/).

Figure S1 Species identification by PCR-restriction fragment-length polymorphism. PCR products were digested by 5 restriction endonucleases (MboII, HinfI, RsaI, DraI, and DdeI). The white arrowheads indicate interspecific variation. M, 100-bp ladder DNA size marker.

Figure S2 Base composition at each codon position of the 13 PCGs. Y-axis shows the percentage of each nucleotide.

Figure S3 Codon usage pattern of each mitochondrial genome. Numbers to the left refer to the percentage of each codon. Codon families are shown on the X-axis.

Figure S4 Putative secondary structure of the large-subunit ribosomal RNA of *T. urticae*. Inferred Watson-Crick bonds are illustrated by black dots, whereas GU bonds are illustrated by grey dots. The nucleotides with bold text show 100% identity among the seven mitochondrial genomes. The numbering of stem-loops is after [69].

Figure S5 Putative secondary structure of the small-subunit ribosomal RNA of *T. urticae*. Inferred Watson-Crick bonds are illustrated by black dots, whereas GU bonds are illustrated by grey dots. The nucleotides with bold text show 100% identity among the seven mitochondrial genomes. The numbering of stem-loops is after [70].

Figure S6 Phylogenetic tree of Tetranychoidea relationships. The tree was inferred from mitochondrial genomic sequences. Numbers at nodes are percentages from Bayesian posterior probabilities (left), ML bootstrapping (middle) and MP bootstrap support values (right).

Acknowledgments

We thank Hao-Sen Li for technical assistance for phylogenetic analysis in the lab. Chao Guo and Xiu-Ting Zhao provided important DNA samples of red form of *T. urticae*.

Author Contributions

Conceived and designed the experiments: DSC PYJ KJZ XLD SXY JFJ JYZ XYH. Performed the experiments: DSC PYJ XLD SXY JFJ. Analyzed the data: DSC PYJ XLD JFJ. Contributed reagents/materials/analysis tools: DSC PYJ KJZ XLD SXY JFJ JYZ. Wrote the paper: DSC PYJ KJZ XLD XYH.

References

1. Migeon A, Nouguier E, Dorkeld F (2010) Spider Mites Web: a comprehensive database for the Tetranychidae. Trends in Acarology: 557–560 http://www.montpellier.inra.fr/CBGP/spmweb/.
2. Dermauw W, Osborne EJ, Clark RM, Grbic M, Tirry L, et al. (2013) A burst of ABC genes in the genome of the polyphagous spider mite *Tetranychus urticae*. BMC Genomics 14: 317.
3. Grbic M, Van Leeuwen T, Clark RM, Rombauts S, Rouze P, et al. (2011) The genome of *Tetranychus urticae* reveals herbivorous pest adaptations. Nature 479: 487–492.
4. Dermauw W, Wybouw N, Rombauts S, Menten B, Vontas J, et al. (2013) A link between host plant adaptation and pesticide resistance in the polyphagous spider mite *Tetranychus urticae*. Proc Natl Acad Sci USA 110: E113–E122.
5. de Mendonça RS, Navia D, Diniz IR, Auger P, Navajas M (2011) A critical review on some closely related species of *Tetranychus* sensu stricto (Acari:

Tetranychidae) in the public DNA sequences databases. Exp Appl Acarol 55: 1–23.
6. Navajas M, Gutierrez J, Williams M, Gotoh T (2001) Synonymy between two spider mite species, *Tetranychus kanzawai* and *T hydrangeae* (Acari: Tetranychidae), shown by ribosomal ITS2 sequences and cross-breeding experiments. B Entomol Res 91: 117–123.
7. Arimoto M, Satoh M, Uesugi R, Osakabe M (2013) PCR-RFLP analysis for identification of *Tetranychus* spider mite species (Acari: Tetranychidae). J Econ Entomol 106: 661–668.
8. Osakabe M, Kotsubo Y, Tajima R, Hinomoto N (2008) Restriction fragment length polymorphism catalog for molecular identification of Japanese *Tetranychus* spider mites (Acari: Tetranychidae). J Econ Entomol 101: 1167–1175.
9. Navajas M, Lagnel J, Gutierrez J, Boursot P (1998) Species-wide homogeneity of nuclear ribosomal *ITS2* sequences in the spider mite *Tetranychus urticae* contrasts with extensive mitochondrial *COI* polymorphism. Heredity 80: 742–752.
10. Ros VI, Breeuwer JA (2007) Spider mite (Acari: Tetranychidae) mitochondrial COI phylogeny reviewed: host plant relationships, phylogeography, reproductive parasites and barcoding. Exp Appl Acarol 42: 239–262.
11. Boore JL (1999) Animal mitochondrial genomes. Nucleic Acids Res 27: 1767–1780.
12. Edwards DD, Jackson LE, Johnson AJ, Ernsting BR (2011) Mitochondrial genome sequence of *Unionicola parkeri* (Acari: Trombidiformes: Unionicolidae): molecular synapomorphies between closely-related *Unionicola* gill mites. Exp Appl Acarol 54: 105–117.
13. Yuan ML, Wei DD, Wang BJ, Dou W, Wang JJ (2010) The complete mitochondrial genome of the citrus red mite *Panonychus citri* (Acari: Tetranychidae): high genome rearrangement and extremely truncated tRNAs. BMC Genomics 11: 597.
14. Klimov PB, Oconnor BM (2009) Improved tRNA prediction in the American house dust mite reveals widespread occurrence of extremely short minimal tRNAs in acariform mites. BMC Genomics 10: 598.
15. Ernsting BR, Edwards DD, Aldred KJ, Fites JS, Neff CR (2009) Mitochondrial genome sequence of *Unionicola foili* (Acari: Unionicolidae): a unique gene order with implications for phylogenetic inference. Exp Appl Acarol 49: 305–316.
16. Dermauw W, Van Leeuwen T, Vanholme B, Tirry L (2009) The complete mitochondrial genome of the house dust mite *Dermatophagoides pteronyssinus* (Trouessart): a novel gene arrangement among arthropods. BMC Genomics 10: 107.
17. Domes K, Maraun M, Scheu S, Cameron SL (2008) The complete mitochondrial genome of the sexual oribatid mite *Steganacarus magnus*: genome rearrangements and loss of tRNAs. BMC Genomics 9: 532.
18. Osakabe M, Uesugi R, Goka K (2010) Evolutionary aspects of acaricide-resistance development in spider mites. Psyche 2009: 1–11.
19. Van Leeuwen T, Vontas J, Tsagkarakou A, Dermauw W, Tirry L (2010) Acaricide resistance mechanisms in the two-spotted spider mite *Tetranychus urticae* and other important Acari: a review. Insect Biochem Mol Biol 40: 563–572.
20. Hayashi N, Sasama Y, Takahashi N, Ikemi N (2013) Cyflumetofen, a novel acaricide - its mode of action and selectivity. Pest Manag Sci 69: 1080–1084.
21. Dekeyser MA (2005) Acaricide mode of action. Pest Manag Sci 61: 103–110.
22. Van Leeuwen T, Vanholme B, Van Pottelberge S, Van Nieuwenhuyse P, Nauen R, et al. (2008) Mitochondrial heteroplasmy and the evolution of insecticide resistance: Non-Mendelian inheritance in action. Proc Natl Acad Sci USA 105: 5980–5985.
23. Van Leeuwen T, Van Nieuwenhuyse P, Vanholme B, Dermauw W, Nauen R, et al. (2011) Parallel evolution of cytochrome *b* mediated bifenazate resistance in the citrus red mite *Panonychus citri*. Insect Mol Biol 20: 135–140.
24. Boore JL, Fuerstenberg SI (2008) Beyond linear sequence comparisons: the use of genome-level characters for phylogenetic reconstruction. Philos T R Soc B 363: 1445–1451.
25. Boore JL (2006) The use of genome-level characters for phylogenetic reconstruction. Trends Ecol Evol 21: 439–446.
26. Dowton M, Castro LR, Austin AD (2002) Mitochondrial gene rearrangements as phylogenetic characters in the invertebrates: the examination of genome 'morphology'. Invertebr Syst 16: 345–356.
27. Dowton M, Austin AD (1999) Evolutionary dynamics of a mitochondrial rearrangement "hot spot" in the Hymenoptera. Mol Biol Evol 16: 298–309.
28. Boore JL, Lavrov DV, Brown WM (1998) Gene translocation links insects and crustaceans. Nature 392: 667–668.
29. Boore JL, Brown WM (1998) Big trees from little genomes: mitochondrial gene order as a phylogenetic tool. Curr Opin Genet Dev 8: 668–674.
30. Dermauw W, Vanholme B, Tirry L, Van Leeuwen T (2010) Mitochondrial genome analysis of the predatory mite *Phytoseiulus persimilis* and a revisit of the *Metaseiulus occidentalis* mitochondrial genome. Genome 53: 285–301.
31. Zhang DX, Hewitt GM (1997) Insect mitochondrial control region: a review of its structure, evolution and usefulness in evolutionary studies. Biochem Syst Ecol 25: 99–120.
32. Zhang DX, Szymura JM, Hewitt GM (1995) Evolution and structural conservation of the control region of insect mitochondrial DNA. J Mol Evol 40: 382–391.
33. Wei SJ, Shi M, Chen XX, Sharkey MJ, van Achterberg C, et al. (2010) New views on strand asymmetry in insect mitochondrial genomes. PLoS One 5: e12708.
34. Masta SE (2010) Mitochondrial rRNA secondary structures and genome arrangements distinguish chelicerates: comparisons with a harvestman (Arachnida: Opiliones: *Phalangium opilio*). Gene 449: 9–21.
35. Ehara S (1999) Revision of the spider mite family Tetranychidae of Japan (Acari, Prostigmata). Species diversity 4: 63–141.
36. Laslett D, Canback B (2008) ARWEN: a program to detect tRNA genes in metazoan mitochondrial nucleotide sequences. Bioinformatics 24: 172–175.
37. Lowe TM, Eddy SR (1997) tRNAscan-SE: a program for improved detection of transfer RNA genes in genomic sequence. Nucleic Acids Res 25: 955–964.
38. Shao R, Barker SC, Mitani H, Takahashi M, Fukunaga M (2006) Molecular mechanisms for the variation of mitochondrial gene content and gene arrangement among chigger mites of the genus *Leptotrombidium* (Acari: Acariformes). J Mol Evol 63: 251–261.
39. Wuyts J, De Rijk P, Van de Peer Y, Pison G, Rousseeuw P, et al. (2000) Comparative analysis of more than 3000 sequences reveals the existence of two pseudoknots in area V4 of eukaryotic small subunit ribosomal RNA. Nucleic Acids Res 28: 4698–4708.
40. Grant JR, Stothard P (2008) The CGView Server: a comparative genomics tool for circular genomes. Nucleic Acids Res 36: W181–W184.
41. Tamura K, Stecher G, Peterson D, Filipski A, Kumar S (2013) MEGA6: Molecular Evolutionary Genetics Analysis Version 6.0. Mol Biol Evol 30: 2725–2729.
42. Zuker M (2003) Mfold web server for nucleic acid folding and hybridization prediction. Nucleic Acids Res 31: 3406–3415.
43. De Rijk P, Wuyts J, De Wachter R (2003) RnaViz 2: an improved representation of RNA secondary structure. Bioinformatics 19: 299–300.
44. Kimura M (1980) A simple method for estimating evolutionary rates of base substitutions through comparative studies of nucleotide sequences. J Mol Evol 16: 111–120.
45. Thompson JD, Higgins DG, Gibson TJ (1994) CLUSTAL W: improving the sensitivity of progressive multiple sequence alignment through sequence weighting, position-specific gap penalties and weight matrix choice. Nucleic Acids Res 22: 4673–4680.
46. Abascal F, Zardoya R, Posada D (2005) ProtTest: selection of best-fit models of protein evolution. Bioinformatics 21: 2104–2105.
47. Adachi J, Hasegawa M (1996) Model of amino acid substitution in proteins encoded by mitochondrial DNA. J Mol Evol 42: 459–468.
48. Guindon S, Dufayard JF, Lefort V, Anisimova M, Hordijk W, et al. (2010) New algorithms and methods to estimate maximum-likelihood phylogenies: assessing the performance of PhyML 3.0. Syst Biol 59: 307–321.
49. Darriba D, Taboada GL, Doallo R, Posada D (2012) jModelTest 2: more models, new heuristics and parallel computing. Nat Methods 9: 772.
50. Hassanin A, Leger N, Deutsch J (2005) Evidence for multiple reversals of asymmetric mutational constraints during the evolution of the mitochondrial genome of metazoa, and consequences for phylogenetic inferences. Syst Biol 54: 277–298.
51. Abascal F, Posada D, Knight RD, Zardoya R (2006) Parallel evolution of the genetic code in arthropod mitochondrial genomes. PLoS Biol 4: e127.
52. Hebert PD, Ratnasingham S, de Waard JR (2003) Barcoding animal life: cytochrome c oxidase subunit 1 divergences among closely related species. Proc R Soc B 270: S96–S99.
53. Hebert PD, Cywinska A, Ball SL (2003) Biological identifications through DNA barcodes. Proc R Soc B 270: 313–321.
54. Jeyaprakash A, Hoy MA (2007) The mitochondrial genome of the predatory mite *Metaseiulus occidentalis* (Arthropoda: Chelicerata: Acari: Phytoseiidae) is unexpectedly large and contains several novel features. Gene 391: 264–274.
55. Shao R, Mitani H, Barker SC, Takahashi M, Fukunaga M (2005) Novel mitochondrial gene content and gene arrangement indicate illegitimate inter-mtDNA recombination in the chigger mite, *Leptotrombidium pallidum*. J Mol Evol 60: 764–773.
56. Black W, Roehrdanz RL (1998) Mitochondrial gene order is not conserved in arthropods: Prostriate and metastriate tick mitochondrial genomes. Mol Biol Evol 15: 1772–1785.
57. Mans BJ, de Klerk D, Pienaar R, de Castro MH, Latif AA (2012) The mitochondrial genomes of *Nuttalliella namaqua* (Ixodoidea: Nuttalliellidae) and *Argas africolumbae* (Ixodoidae: Argasidae): estimation of divergence dates for the major tick lineages and reconstruction of ancestral blood-feeding characters. PLoS One 7: e49461.
58. Cui P, Ji R, Ding F, Qi D, Gao H, et al. (2007) A complete mitochondrial genome sequence of the wild two-humped camel (*Camelus bactrianus ferus*): an evolutionary history of camelidae. BMC Genomics 8: 241.
59. Davila S, Pinero D, Bustos P, Cevallos MA, Davila G (2005) The mitochondrial genome sequence of the scorpion *Centruroides limpidus* (Karsch 1879) (Chelicerata; Arachnida). Gene 360: 92–102.
60. Hassanin A (2006) Phylogeny of Arthropoda inferred from mitochondrial sequences: strategies for limiting the misleading effects of multiple changes in pattern and rates of substitution. Mol Phylogenet Evol 38: 100–116.
61. De Boer R (1982) Partial hybrid sterility between strains of the arrhenotokous spider mite, *Tetranychus urticae* complex (Acari, Tetranychidae). Genetica 58: 23–33.
62. Sugasawa J, Kitashima Y, Gotoh T (2002) Hybrid affinities between the green and the red forms of the two-spotted spider mite *Tetranychus urticae* (Acari: Tetranychidae) under laboratory and semi-natural conditions. Appl Entomol Zool 37: 127–139.

63. Gotoh T, Noda H, Ito S (2007) *Cardinium* symbionts cause cytoplasmic incompatibility in spider mites. Heredity (Edinb) 98: 13–20.
64. Vala F, Breeuwer JA, Sabelis MW (2000) *Wolbachia*-induced 'hybrid breakdown' in the two-spotted spider mite *Tetranychus urticae* Koch. Proc Biol Sci 267: 1931–1937.
65. Breeuwer JAJ (1997) *Wolbachia* and cytoplasmic incompatibility in the spider mites *Tetranychus urticae* and *T. turkestani*. Heredity 79: 41–47.
66. Xie RR, Chen XL, Hong XY (2011) Variable fitness and reproductive effects of *Wolbachia* infection in populations of the two-spotted spider mite *Tetranychus urticae* Koch in China. Appl Entomol Zool 46: 95–102.
67. Gotoh T, Sugasawa J, Noda H, Kitashima Y (2007) *Wolbachia*-induced cytoplasmic incompatibility in Japanese populations of *Tetranychus urticae* (Acari: Tetranychidae). Exp Appl Acarol 42: 1–16.
68. Nei M, Kumar S (2000) Molecular evolution and phylogenetics. New York: Oxford University Press.
69. De Rijk P, Robbrecht E, de Hoog S, Caers A, Van de Peer Y, et al. (1999) Database on the structure of large subunit ribosomal RNA. Nucleic Acids Res 27: 174–178.
70. Van de Peer Y, Robbrecht E, de Hoog S, Caers A, De Rijk P, et al. (1999) Database on the structure of small subunit ribosomal RNA. Nucleic Acids Res 27: 179–183.

12

Whole Plastome Sequences from Five Ginger Species Facilitate Marker Development and Define Limits to Barcode Methodology

Justin N. Vaughn[1], Srinivasa R. Chaluvadi[1], Tushar[2], Latha Rangan[2], Jeffrey L. Bennetzen[1]*

1 Department of Genetics, University of Georgia, Athens, Georgia, United States of America, **2** Department of Biotechnology, Indian Institute of Technology Guwahati, Assam, India

Abstract

Plants from the Zingiberaceae family are a key source of spices and herbal medicines. Species identification within this group is critical in the search for known and possibly novel bioactive compounds. To facilitate precise characterization of this group, we have sequenced chloroplast genomes from species representing five major groups within Zingiberaceae. Generally, the structure of these genomes is similar to the basal angiosperm excepting an expansion of 3 kb associated with the inverted repeat A region. Portions of this expansion appear to be shared across the entire Zingiberales order, which includes gingers and bananas. We used whole plastome alignment information to develop DNA barcodes that would maximize the ability to differentiate species within the Zingiberaceae. Our computation pipeline identified regions of high variability that were flanked by highly conserved regions used for primer design. This approach yielded hitherto unexploited regions of variability. These theoretically optimal barcodes were tested on a range of species throughout the family and were found to amplify and differentiate genera and, in some cases, species. Still, though these barcodes were specifically optimized for the Zingiberaceae, our data support the emerging consensus that whole plastome sequences are needed for robust species identification and phylogenetics within this family.

Editor: Shilin Chen, Chinese Academy of Medical Sciences, Peking Union Medical College, China

Funding: This research was funded by endowment funds from the University of Georgia Giles Professorship, the Georgia Research Alliance and Department of Information Technology, and Ministry of Information Technology, Government of India (DIT Grant No. DIT No: 0526/T/IITG/014/0809/38). The funders had no role in study design, data collection and analysis, decision to publish, or preparation of the manuscript.

Competing Interests: The authors have declared that no competing interests exist.

* Email: maize@uga.edu

Introduction

The Zingiberaceae family includes ginger, turmeric, and cardamom spices, as well as many species with medicinal applications [1]. Although these plants have received little characterization at the genome level, there are many justifications for initiating such studies. Divergence in metabolite production in the Zingiberaceae can occur quite rapidly, exhibiting substantial variation even within the same species [2]. In contrast, morphological evolution is slower and often convergent [3]. Hence, DNA polymorphism analysis may be the most effective way to identify new germplasms that are sources of novel bioactive compounds. Previous work on the phylogenetic characterization of Zingiberaceae has focused on a small number of shared loci. While these loci can be used to differentiate genera, the approach fails to resolve genera into species [3]. Moreover, even for deeper phylogentic relationships, the Zingiberaceae contain many discrepancies between morphological and molecular classification of species; genera such as the *Alpinia* [4] and, to a lesser extent, *Curcuma* [3] are apparently paraphyletic.

Due to their high levels of conservation and maternal inheritance, chloroplasts can be used to develop population markers for phylogenetic and phylogeographic studies. The chloroplast DNA of most flowering plants are circular molecules that are highly conserved in terms of gene content, size and structural organization [5]. Because the size of the chloroplast genome is fairly constrained, highly conserved genes are found close to one another. These genes serve as optimal primer targets for amplifying short stretches of divergent intergenic DNA. In addition, chloroplasts outnumber nuclei in the plant cells. Thus, genes within the chloroplast are present in much greater concentrations than nuclear genes. While this may not be critical for analysis of fresh tissue, it can be a major consideration for dried specimens, gut samples, herbal stocks or other degraded DNA sources [6].

In addition to their pharmacological significance, gingers have become a model system for studying herbivore networks in tropical ecosystems through the use of DNA barcodes [7]. When the correct sequence region is chosen, DNA barcoding is an efficient way to assign an unknown specimen to a taxonomic group [6,8]. In effect, finding that a specimen matches a previously characterized barcode allows a researcher to assign that specimen to the species associated with that barcode. Difficulties arise in taxonomic groups without variation across a chosen barcode. For example,

Minami, *et al.* [9] used thirteen loci in an attempt to differentiate species within the paraphyletic *Curcuma* genus. Only one of the thirteen had haplotypes that were polymorphic across all accessions. Though these polymorphisms could be used to differentiate species, the intra-specific variation was similiar to the inter-specific variation, limiting the ability to assign unknown samples to higher taxonomic levels.

Both phylogenetic analysis and barcode classification are limited by the number of informative sites on which to base a conclusion. The entire chloroplast sequence dramatically increases the number of these sites, and, because of technological advances in sequencing, this increased information does not substantially increase total expenditures. Here, we present representative chloroplast genome sequences, or "plastomes", for *Zingiber officinale* (ginger), *Amomum cardamomum* (cardamom), *Alpinia zerumbet* (shell ginger), *Hedychium coronarium* (white ginger), and *Curcuma longa* (turmeric). These data should serve as a foundation for further characterization of this pharmacologically significant lineage. Indeed, using these sequences, we developed a set of highly variable markers. We demonstrated the utility and limits of a subset of these novel markers on a range of Zingiberaceae species.

Results and Discussion

Assembly and annotation

We performed *de novo* assembly on the *Z. officinale* chloroplast genome using DNA sequences from a shotgun library of total cell DNA. Assembly resulted in two major contigs of 87,626 and 45,356. These were aligned to the *Typha latifolia* (gi|289065068|ref|NC_013823.1) plastome and scaffolded with smaller contigs. Borders between the inverted repeat regions (IRa and IRb) and the long and short single copy regions (LSC and SSC, respectively) were manually assembled and curated by iterative mapping of reads. A final plastome with 162,598 bp was produced (Figure 1A). Excepting a gap at the IRa/SSC border, the *de novo* assembly shows perfect colinearity with the *Musa acuminata* (banana) plastome [10] (Figure 1B). Sequencing reads are fairly even in their coverage of the entire plastome (Figure 1C) - as the IR regions are effectively identical, reads were randomly placed in either region.

Annotation was performed using CpGAVAS [11]. In general, there are no major discrepancies in the gene content of the *Z. officinale* plastome compared to known plastomes, excepting an expansion of the IRa region as described below. The expansion creates a duplication of the first 3,912 bp of the *ycf1* gene. YCF1 and YCF2 are the longest proteins encoded by known plastomes and appear to be indispensable to plant survival [12]. Though the *Zingiber* plastome encodes a full-length YCF1, it is unknown if the large fragment associated with the IRa/SSC boundary is also expressed.

Repeat expansion and structural variation

The borders of inverted repeats and single copy regions are known to be hypermorphic for major structural variation. A major expansion of IRa has been noted in the *Musa* lineage [10]. As seen in Figure 1, the *Z. officinale* plastome does not completely share this expansion, although it does have a 3 kb expansion at this site relative to the otherwise highly colinear *Amborella trichopoda* (gi|34500893|ref|NC_005086.1) and *T. latifolia* plastomes (Figure S1 and S2). (Because of their deep ancestry, the colinear relationship between *Amborella* and *Typha*, among others, suggests that this is the ancestral plastome structure.) Thus, it appears that an initial 3 kb IRa expansion occurred in the *Musa/Zingiber* ancestor and was further expanded in the *Musa* lineage by another 4.7 kb. Alternatively, a 7.7 kb expansion could have occurred in the *Musa/Zingiber* ancestor followed by a deletion at the IRa/SSC boundary in the *Zingiber* lineage after divergence from the *Musa* lineage. The structure of the recently sequenced *Zingiber spectabile* plastome [13] suggests that these were separate events and that the short expansion has only occurred in *Z. officinale*. By mapping our sequencing reads from the *C. longa* plastome to the *Z. spectabile* scaffold, it is apparent that the scaffold has collapsed the 3 kb Zingiberale IRa expansion into the SSC. We see no such discrepancy in coverage when reads from *C. longa* or the more distantly related *A. cardamomum* are mapped to the *Z. officinale* plastome. Since this small expansion appears to have occurred at the root of the Zingiberaceae, the large expansion in *Musa* may have occurred in multiple steps. Whether this sequence is predisposed to expansion or such expansions are preferentially retained by natural selection remain open questions.

pan-Zingiberaceae variation

We then used the *Z. officinale* plastome as a reference on which to assemble data from four other libraries of lower coverage or shorter read length (Table 1). The final sequence reported for each plastome is the consensus sequence of these mapped reads (see Materials and Methods).

Generally, the IRs, which encode ribosomal RNAs, are highly conserved relative to both the LSC and SSC regions. Within the LSC and SSC, highly variable regions correlate with regions that either encode tRNAs or short genes and have a low gene density. Interestingly, regions of the IR with similarly low gene density exhibit much higher levels of conservation. This may be related to regulatory constraint outside of the protein coding regions or to a homology-dependent DNA repair mechanism specific to the IR region [14,15].

The tree generated from alignment of these plastomes (Figure 2) agrees with previous trees based on combined data from the nuclear ITS locus and the chloroplast *matK* gene [3]. Both trees indicate that *Hedychium* and *Zingiber* share a more recent common ancestor than *Zingiber* and *Curcuma*, although the divergence of all three lineages appears to have been nearly simultaneous on an evolutionary time scale.

Microsatellite identification and barcode development

Short sequence repeats (SSRs) or "microsatellites" are hypermorphic due to errors during DNA replication. Such sequences often serve as useful loci for differentiating closely related individuals. In contrast, DNA barcoding exploits point mutations and short indels that occur in rapidly evolving sequences flanked by more slowly evolving sequences. The lower rate of mutation in these barcodes relative to SSRs is somewhat compensated for by the presence of multiple informative sites in a single barcode sequence. Which of these strategies is most appropriate depends on the application, the taxonomic group under study, and the available instrumentation. Indeed, a very short SSR may serve as one among many informative sites within a single barcode. Using *M. accuminata* plastome sequence, researchers were able to identify 15 chloroplastic SSR markers that were polymorphic within the species [10]. By using both the *Z. officinale* sequence and the additional Zingiberaceae data produced in this study and elsewhere [13], we attempted to generate reliable chloroplastic SSR markers using a comparative approach. Briefly, alignment information from a whole plastome comparison between *Z. officinale* and *Z. spectabile* was used to find di- and tri-nucleotide repeats that were polymorphic within the *Zingiber* genus. We avoided homopolymer repeats because, generally, they are

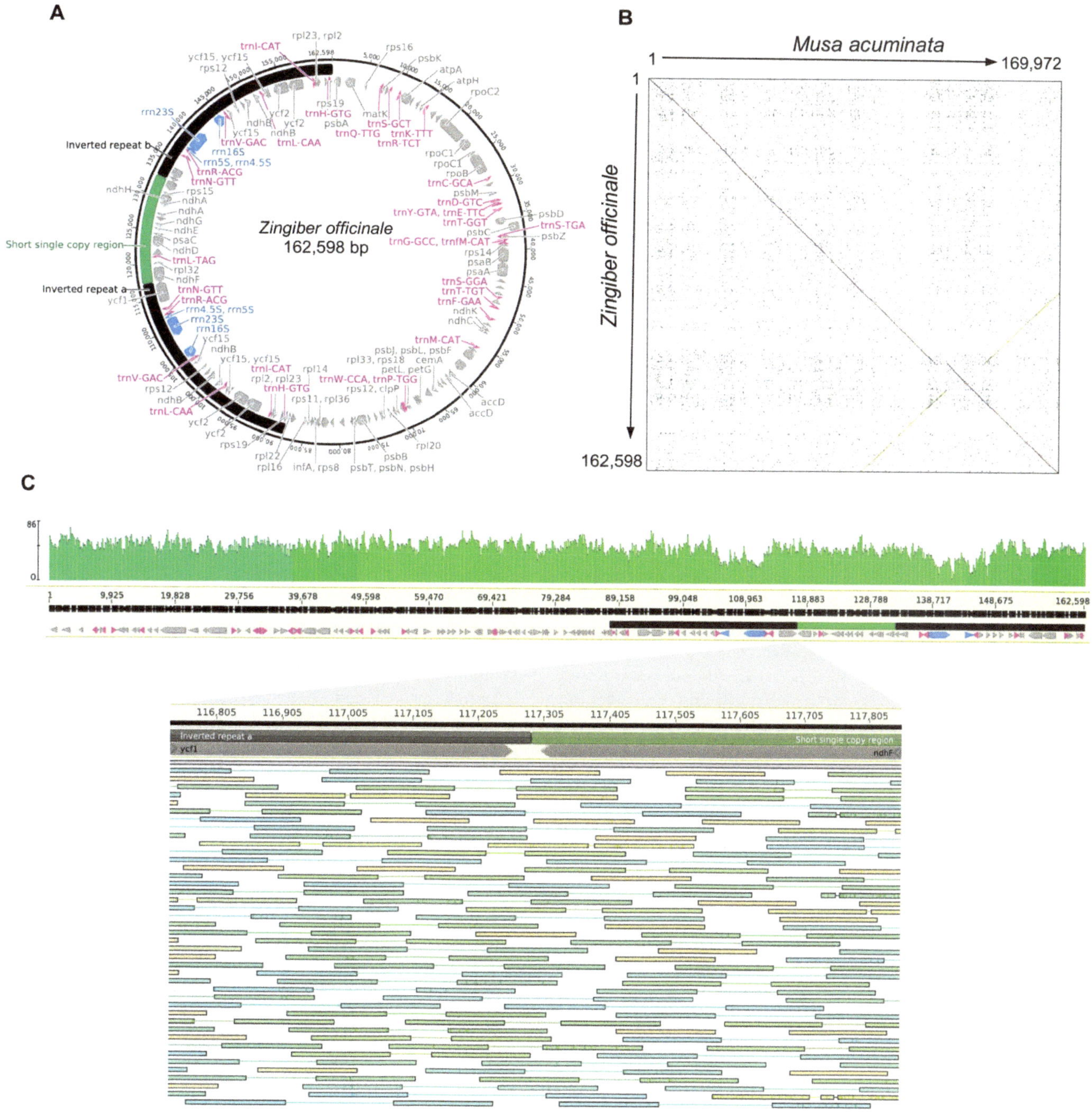

Figure 1. ***Zingiber officinale* assembly quality and annotation. A**, The *Z. officinale* plastome is shown in circular form with IRa and IRb indicated. The short single copy region between the two inverted repeats is shown in green. Protein coding genes are shown in grey, tRNA genes in pink, and rRNA genes in blue. **B**, Dotplot comparison of the *M. acuminata* and *Z. officinale* plastome sequences. The inverted repeat dotplot is transposed in gold. **C**, Total coverage across the *Z. officinale* plastome. The annotation color scheme is as depicted in A. The lower panel magnifies the border between the IRa and the SSC, where the large indel that differentiates *Musa* and *Zingiber* is seen to occur. Reads are shown individually; green is very near the expected insert size, while blue is slightly smaller and gold is slightly longer, respectively, than the expected size. The entire stack of reads is clipped due to space limitations.

technically difficult to use as markers. We also exploited alignment information in order to generate reliable primers for the amplification of these polymorphic regions. Using a similar pipeline for the alignment of all plastomes in this study (Figure 2A, excluding *Musa*), we also identified SSR markers that are likely to be most effective across the Zingiberaceae.

"Universal" plant barcodes, such as *matK* and *trnH-psdA*, have been tested across a range of Zingiberaceae. Shi *et al.* [16] found that the nuclear ITS2 region, because of its high interspecific divergence, was superior to universal choloroplastic barcodes. In contrast, Vinitha *et al.* [17] showed that the utility of both ITS and ITS2 regions is limited by low direct sequencing efficency caused

Table 1. General features of the material used for sequencing and the sequencing method.

Species	Site of Collection	Sequencing Method	Insert Size (bp)	Total Reads (10^6)	Chloroplast Reads (10^3)	Coverage
Zingiber officinale (TARS18100)	USDA Tropical Agriculture Research Station, Puerto Rico	MiSeq-PE150	400–600	3.0	52.9	50×
Curcuma longa (ZSD07)	Assam, India	MiSeq-PE150	400–600	1.3	58.1	54×
Alpinia zerumbet (TARS1718)	USDA Tropical Agriculture Research Station, Puerto Rico	HiSeq- PE100	400–600	3.0	35.4	22×
Amomum cardamomum (ZSB03)	Assam, India	HiSeq- PE100	400–600	10.8	112.4	70×
Hedychium coronarium (UGA-Hyd)	UGA Greenhouse, Athens, GA	HiSeq- PE100	400–600	20.2	337.1	211×

by intragenomic heterogeneity. These researchers determined that *matK* remained the superior barcode for gingers, but *matK* failed to differentiate species in 25% of cases. Similiar findings in other taxonomic groups [18,19] have led researchers to develop group-specific barcodes [20]. One aim of this study was to compare the discriminatory power of "universal" barcodes tested in the referenced works with barcodes that are designed specifically for gingers based on whole plastome sequences spanning the family.

To find suitable barcodes, we searched the alignments for regions that have at least 2% divergence. The regions were also filtered based on the number of single nucleotide polymorphisms (SNPs) that were bordered by 5 conserved bases on both sides of the SNP. Similar to SSR discovery, we identified markers using the intra-genus comparison for *Zingiber* as well as the comparison across all plastomes from this study. For the intra-genus and inter-genus comparison, we required at least 5 and 8 of such SNPs, respectively, across a 300 bp region. As with the SSR marker discovery, primers were only chosen from perfectly conserved sequence blocks.

The "Universal barcode" primers in Figure 3 are from previously proposed sets designed to work across all flowering plants [21]. Additionally, the "LR barcode" track plots another set of commonly used primer sets within the Zingiberaceae (Latha *et al.*, unpublished). Figure 3 also indicates the four sets of primer

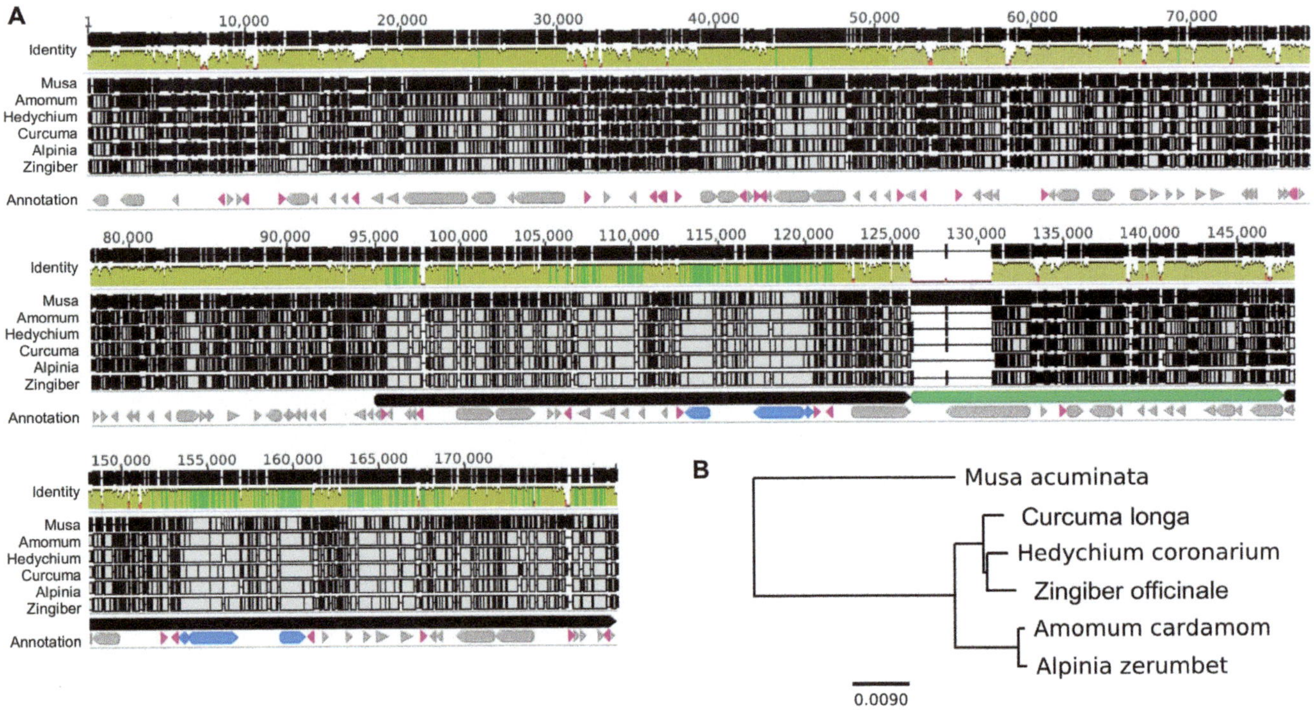

Figure 2. Plastome sequence variation across five Zingiberaceae. A, Alignment of whole plastome alignment between *M. acuminata* and Zingiberaceae species sequenced in this study. The annotation color scheme is as depicted in Figure 1A. In the coverage plot, color indicates percent identity: green, 100%; gold, >30%; red, <30%. Black bars within each species row indicate polymorphisms relative to the consensus. **B**, A phylogenetic tree based on the plastome alignments. *Musa acuminata* was used as the outgroup. Each branch point has 100% bootstrap support. Scale bar indicates the number of substitutions per site estimated to have occurred for such a length along the tree.

pairs resulting from our marker discovery pipeline. Looking across all four sets, we chose a subset of primers that exhibit a high number of polymorphisms at both the intra-genus and intra-family levels ("JNV barcode"). As shown, there is very little overlap between the "JNV barcode" markers and those used in previous studies. Though not shown in detail, only one of the "Universal barcode" primer sets perfectly matches the *Z. officinale* plastome, while the rest that are shown contain one to many mismatches. (Those with more than two mismatches are not shown.) While these mismatched primers may still produce a clear PCR product, they are not optimal with regard to heterogeneous starting material. Again, we aimed to make our primers as unbiased as possible by selecting from only highly conserved sites. Thus, in theory, "JNV barcodes" are optimal markers for the Zingiberaceae.

Marker testing on multiple Zingiberaceae lineages

To demonstrate the utility of JNV barcodes (Table S1), we amplified and Sanger sequenced these regions from a range of previously characterized and uncharacterized Zingiberaceae specimens (Table S2). Generally, all pairs amplified across the Zingiberaceae phylogeny, but JNV-A and JNV-C exhibited superior performance under the conditions tested (Table S2, Figures S4, S5). These pairs were followed in efficacy by JNV-E, JNV-D, and JNV-B. Pairs JNV-G, JNV-H, and JNV-I failed to amplify in any of the tested samples. Importantly, all pairs were tested in a single experiment as a proof-of-concept; we expect that PCR optimization on a pair-specific basis would greatly improve their amplification efficiency and phylogenetic distribution. Pairs JNV-A and JNV-C were able to efficiently amplify sequences for all genera in this study (Table S2, Figures S4, S5). They resolve a major branch point between *Alpinia/Amomum* samples and *Zingiber/Hedychium/Curcuma* samples (Figure 3B–C). The latter group could generally be differentiated by one or more polymorphic sites into three subgroups coinciding with the three different genera. As predicted by the whole plastome alignment, the branch between the *Zingiber/Hedychium* and *Curcuma* split was very short and thus difficult to resolve (Figure 2B). Still, there are clearly polymorphisms in both data sets that cluster *Zingiber* and *Hedychium* to the exclusion of *Curcuma* (Figure 3B–C). In addition, these barcodes clearly discriminate species mis-identified as belonging to the *Alpinia* genus (Figure 3B–C).

In terms of species resolution, there are cases where two species can be resolved, such as within the *Hedychium* and *Zingiber* groups, as well as within the *Alpinia* group (Figure 3B–C and File S1 and S2). *Curcuma* species are more difficult to resolve using these primers. Either this is a result of broad similarity across the whole plastomes of this group, or the markers we have chosen do not correspond to the most variable loci within this group. Additional plastome sequences within the *Curcuma* should resolve this question. As observed in other studies [3,4], the paraphyletic nature of genera such as *Alpinia* and *Curcuma* is also evident from our data, while genera such as *Zingiber*, *Kaempferia*, and *Hedychium* show monophyletic distributions (Figure 3B–C).

Conclusions

There is growing consensus that the entire chloroplast genome sequence is an optimal 'barcode' in plant identification and phlyogenetics [22–25]. Our study supports this consensus: even markers optimized to focus on the most variable (but still alignable) regions of these plastomes failed to produce enough informative sites for differentiating the most closely related species and for generating a tree with comparable bootstrap support of the same nodes shown in Figure 2B. Moreover, given the low cost of next-generation sequencing, the price of sequencing an entire plastome versus amplifying and Sanger sequencing a single barcode are approaching equivalence. This is particularly true when a reference plastome is available to facilitate assembly using lower read coverage.

Still, even a comparison of entire plastome sequences will reach resolution limits as samples are drawn from very closely related individuals [23]. In terms of the *Zingiber officinale/spectabile* comparison, plastomes exhibit hundreds of genuine polymorphisms, while universal barcodes *matK* and *rbcL* possess only 4 and 2, respectively (Figure S3). Though the ginger-specific barcodes reported in this study, JNV-C and JNV-A, have at least 5 polymorphisms between these two species, they too are ineffective with regard to *Curcuma* species (Figure 3B–C), suggesting far fewer polymorphisms exist across plastomes of that genus. As discussed above, whole-plastome sequences from the *Curcuma* genus will be an interesting test case for the differentiating power of organellar sequences.

Short-read sequencing and assembly strategies will also be ineffective when faced with a heterogeneous pool of samples from similar species, as might be associated with an ecological study or with herbal/spice contamination detection. Both the former [7] and latter applications are of particular interest within the Zingiberaceae. Therefore, exploiting the range of chloroplast sequences presented in this study, we have also developed a new set of barcode markers optimized for the Zingiberaceae. Two of those tested primer pairs, individually resolve the major and minor branches within the family. Though closely related species could be resolved in some cases, a combination of these markers is still likely to be required for confident identification of the constituent species in a mixed pool.

Methods

Sampling of Zingiberaceae accessions

Zingiberaceae accessions used in the current study were collected from US and India (Table 1). They are being maintained in greenhouses at the University of Georgia, Athens, GA and the Indian Institute of Technology in Guwahati, Assam, India.

DNA extraction and Illumina Sequencing

DNA extractions were carried out using PlantDNeasy mini kits (Qiagen, USA) following the manufacturer's instructions. The TruSeq sample preparation kit (Illumina, USA) was used to prepare DNA libraries with inserts from 400–600 bp for paired-end multiplexed sequencing. Library preparations and sequencing on an Illumina HiSeq2000 instrument were carried out at the Interdisciplinary Center for Biotechnology Research, University of Florida. Library preparations and sequencing on an Illumina MiSeq instrument were performed at the Georgia Genomics Facility, University of Georgia. Library-specific indexed adapter sequences were used to separate reads from individual samples from pooled libraries. All sequencing reads used in the assemblies have been made available in the Short Read Archive (http://www.ncbi.nlm.nih.gov/sra) under the BioProject accession PRJNA253694.

Assembly and annotation

For the *Z. officinale* plastome, shotgun reads were assembled as follows. Total reads were first assembled using *edena* with default parameters. Reads where then filtered to remove read pairs with polynucleotide runs of >30 bp and consecutive Ns of >75. Contigs resulting from *edena* assembly were used as long pseudo-

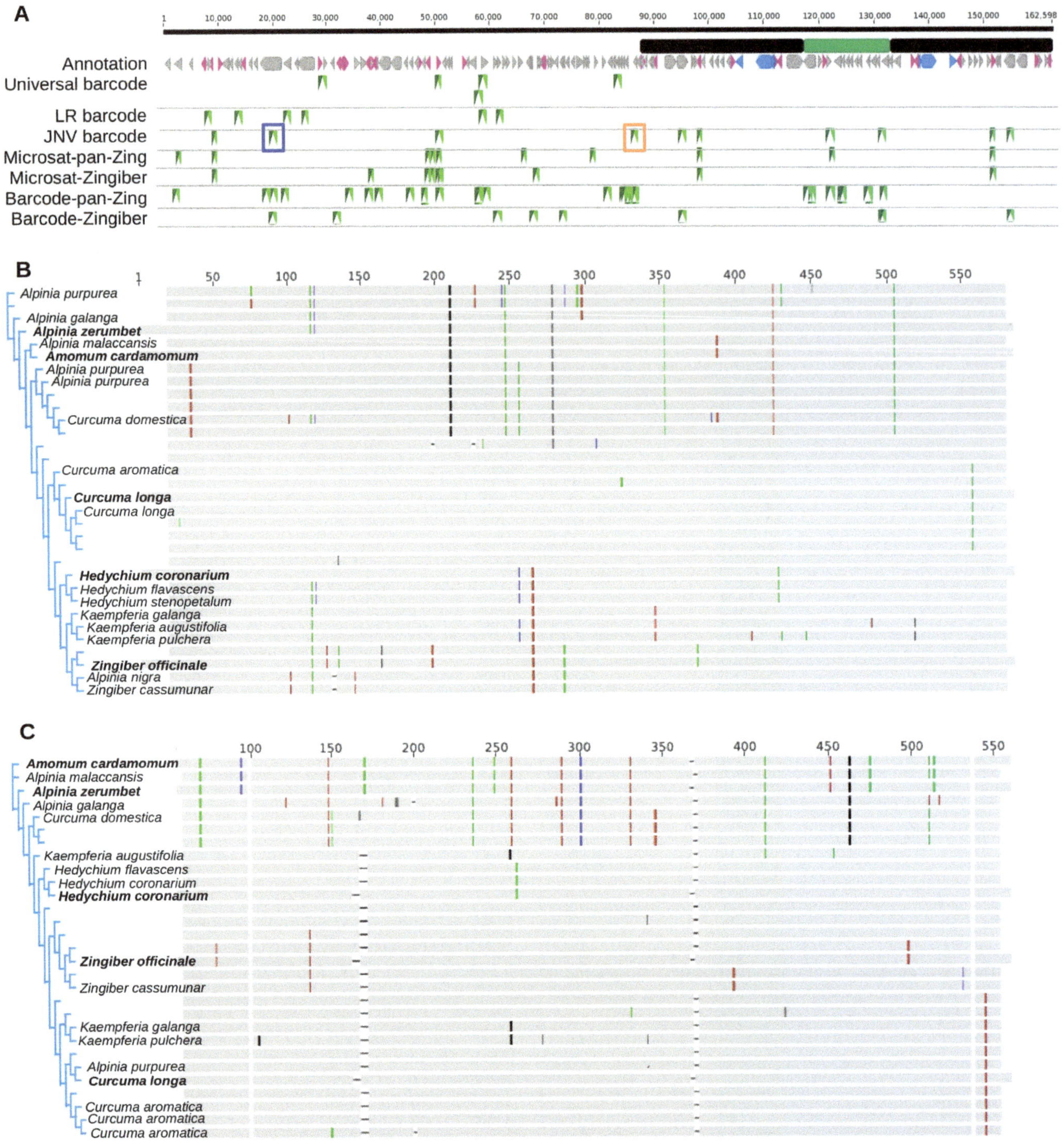

Figure 3. Assorted primer combinations from this and prior studies. A, "Universal barcode" primers are those described in [21]. "LR barcode" primers are from a recent unpublished study. "JNV barcode" primers were developed in this study and were generated based on the microsatellite and barcode primers identified by the computational pipeline described in the text. These are indicated in the figure as the last four tracts, where "Zingiber" indicates the intraspecific *Z. officinale* and *Z. spectabile* comparison and "pan-Zing" indicates a comparison across all sequences described in this study. Underlined markers indicate multiple overlapping markers at that site. The blue and orange boxes demarcate the markers used to generate sequences in Figure 3B and 3C, respectively. **B** and **C**, Sequences from a range of Zingiberaceae species were amplified using the primer pairs indicated in Figure 3A. **B** is associated with the blue box and shows products amplified by barcode set JNV-C (Table S1). **C** is associated with the orange box and shows products amplified by barcode set JNV-A (Table S1). Grey color indicates identity with the consensus of the alignment. Colored sites indicate a polymorphism, where different colors represent 4 different bases and light grey represents an ambiguous base. Appropriate regions from the fully sequenced plastomes were extracted and included in the alignment. These sequences are indicated by bold font. Names are not shown for samples that were not keyed to species-level.

reads ('-long' parameter) in a second round of assembly by *velvet* in paired-end mode with *k*-mer size set to 51 bp and the minimum k-mer coverage set to 2 ('-cov_cutoff 2'). Additionally, we required only 1 long read to scaffold two contigs ('-long_mult_cutoff 1'). Masked reads were then mapped back to the final assemblies using *bowtie-2* in local mode (–local), and scaffolded gaps were manually curated. We arrived at this hybrid strategy - a combination of overlap-consensus-layout and *k*-mer graph approaches – by observing that, while *velvet* was more robust to sequencing errors and could extend contigs further, *edena* was less prone to make errors when resolving the ends of duplicated sequences. The CpGAVAS web service [11], based on the MAKER pipline [26], was used to annotate the *Z. officinale* plastome. The sequence with annotation has been deposited in GenBank (http://www.ncbi.nlm.nih.gov/genbank/) under the accession KM213122.

For the assembly of other species in this study, total shotgun reads from a given individual were mapped to the *Z. officinale* plastome using Geneious (6.0.6) with 'Medium Sensitivity' and 5 iterations. Matches with equal scores were randomly assigned to an appropriate position. A consensus sequence was generated such that 50% of the bases across all mapped reads were required to be identical. An 'N' was called if the quality score was <20. Total shotgun reads were remapped to these consensus sequences and manually curated. All sequence assemblies are available as supporting information (File S3).

The MAUVE aligner ('mauveAligner' algorithm) [27] was used to create whole plastome alignments between the *Musa acuminata* sequence (gi|525312436|embl|HF677508.1) and all Zingiberaceae sequences generated in this study. A seed weight of 8 was used, and plastomes were assumed to be colinear. Resultant alignments and PhyML were used to estimate the topology, branch length, and substitution rate of a tree relating these species. A Kimura (K80) substitution model was used with 50 bootstrap trials. *M. acuminata* was chosen as the outgroup.

Identifying optimal markers

The method we employed to identify both microsatellite and barcode markers use an alignment to maximize both the variability of the marker and the consistency of the primer sites used to amplify the marker. To that end, two consensus sequences were generated from whole plastome alignments: the "best-base" consensus, for which the majority base in every column was used and gaps were ignored, and the "ambiguous" consensus, for which 100% of bases were required to match and gaps were considered. Thus, explicit bases (i.e., A, C, G, and T) in the ambiguous consensus are indicative of a perfectly conserved column.

For microsatellite marker development, we used the best-base consensus in conjunction with MISA software [28] to identify di- and tri-nucleotide repeats of 4 units or more. Using a custom Perl program, the resultant coordinates were then screened against the ambiguous consensus to determine if they were in fact polymorphic across the species analyzed, such that an ambiguous base was required to be present within the microsatellite or within 4 bp to the left or right of it. Primer3 [29] was then used to identify primer sites within the ambiguous consensus that amplify polymorphic positions, with the following parameters: PRIMER_PRODUCT_-SIZE_RANGE = 100–280, PRIMER_MAX_END_STABILITY = 250, and PRIMER_LIBERAL_BASE = 1. Primers were not picked from ambiguous sequence, so, by using the ambiguous consensus, only highly conserved regions were considered.

A similar approach was used for barcode development, except that, instead of MISA software, a custom Perl program was written to find 300 bp blocks that 1) had >2% divergence, 2) did not contain more that 7 consecutive ambiguous bases, 3) did not have >10% or 15% divergence for genera or family comparisons, respectively, and 4) had >4 or >8 single-nucleotide polymorphisms with 5 conserved flanking bases for genera or family comparisons, respectively. Primer sequences were sought for within 300 bp upstream and 300 bp downstream of a 300 bp block passing the above criteria. The same parameters were used as for microsatellite primer identification (excepting the different size range). All programs not from external sources are available upon request.

PCR analysis with novel barcode primers designed in the current study

Genomic DNA was isolated from 26 Zingiberaceae accessions collected from Assam, India and the USA. These accessions included 20 identified species and 6 samples whose species had not been determined. Table S2 lists details of the various accessions used in this study. We tested 11 primer pairs designed in the hypervariable regions of the chloroplast genome. All the primers used are listed in Table S1. Six primer pairs showed consistent amplification across most of the species. PCR products amplified with two primer pairs, A and C (Table S1), were used for Sanger sequencing. PCR reactions were performed in 50-µL reactions: 1× Phusion High Fidelity Buffer (Finzyme, Finland), (NEB, USA), 0.25 µM of each primer, 0.5 µM each dNTP, 1 U Phusion High-Fidelity DNA Polymerase (NEB, USA), and 100 ng DNA. PCR conditions were as follows: 98°C for 2 min; 26 cycles of 98°C for 10 s, 58°C for 20 s, 72°C for 30 s; with a final extension at 72°C for 10 min. Approximately 5 uL of the PCR reaction was tested on the gel to verify the presence of amplification product. PCR reactions were purified with the AGENCOURT AMpure kit (Beckman Coulter, USA) as per manufacturer's instructions. PCR product purification and direct sequencing were performed by standard procedures at the Georgia Genomics Facility (University of Georgia, Athens, GA) on an ABI 3730 instrument.

Supporting Information

Figure S1 Dot-plot alignment of *Zingiber officinale* and *Amborella trichopoda* plastomes.

Figure S2 Dot-plot alignment of *Zingiber officinale* and *Typha latifolia* plastomes.

Figure S3 Alignments for matK and rbcL barcode sequences from *Zingiber officinale*, *Zingiber spectabile*, and *Hedychium coronarium*. Colored letters indicate polymorphic sites.

Figure S4 Gel documenting amplification efficiency of JNV-C barcode sequence among Zingiberaceae accessions. Lanes are coded based on the matrix given below the gel. Plant accessions are as they appear in Table S2.

Figure S5 Gel documenting amplification efficiency of JNV-A barcode sequence among Zingiberaceae accessions. Lanes are coded based on the matrix given below the gel. Plant accessions are as they appear in Table S2.

Table S1 List of primers identified as optimal barcodes. Shown as 'JNV barcodes' in Figure 3.

Table S2 Species list and results for barcode A and C amplification.

File S1 Alignments of sequenced products from barcode A PCR in fasta format.

File S2 Alignments of sequenced products from barcode C PCR in fasta format.

File S3 Fasta file containing each assembled plastome sequence.

Acknowledgments

We offer our sincere thanks to the Botany Department, Gauhati University, for identification and maintenance of plant material. Tushar thanks the Department of Information Technology (DIT) and India-US Fulbright program for his student Fellowship. We thank Brian M. Irish, USDA-ARS Tropical Agriculture Research Station, Mayaguez, PR for providing Zingiberaceae accessions. Thanks to Dr. Jim Leebens-Mack at UGA for useful discussions about the phylogeny of the Zingiberales. We also thank Roger Nelson at the Georgia Genomics Facility and Savita Shankar at the University of Florida for their assistance with Illumina sequencing.

Author Contributions

Conceived and designed the experiments: JNV LR JLB. Performed the experiments: T SRC JNV. Analyzed the data: JNV SRC T. Contributed reagents/materials/analysis tools: LR T. Wrote the paper: JNV LR JLB.

References

1. Tushar, Basak S, Sarma GC, Rangan L (2010) Ethnomedical uses of Zingiberaceous plants of Northeast India. J Ethnopharmacol 132: 286–296.
2. Xiao W-L, Motley TJ, Unachukwu UJ, Lau CBS, Jiang B, et al. (2011) Chemical and genetic assessment of variability in commercial *Radix astragali* (*Astragalus* spp.) by ion trap LC-MS and nuclear ribosomal DNA barcoding sequence analyses. J Agric Food Chem 59: 1548–1556.
3. Kress WJ, Prince LM, Williams KJ (2002) The phylogeny and a new classification of the gingers (Zingiberaceae): evidence from molecular data. Am J Bot 89: 1682–1696.
4. Kress WJ, Liu A-Z, Newman M, Li Q-J (2005) The molecular phylogeny of Alpinia (Zingiberaceae): a complex and polyphyletic genus of gingers. Am J Bot 92: 167–178.
5. Wicke S, Schneeweiss GM, dePamphilis null CW, Müller KF, Quandt D (2011) The evolution of the plastid chromosome in land plants: gene content, gene order, gene function. Plant Mol Biol 76: 273–297.
6. Group CBOLPW (2009) A DNA barcode for land plants. Proc Natl Acad Sci U S A 106: 12794–12797.
7. García-Robledo C, Erickson DL, Staines CL, Erwin TL, Kress WJ (2013) Tropical plant-herbivore networks: reconstructing species interactions using DNA barcodes. PLoS One 8: e52967.
8. Fang Y, Wu H, Zhang T, Yang M, Yin Y, et al. (2012) A complete sequence and transcriptomic analyses of date palm (*Phoenix dactylifera* L.) mitochondrial genome. PLoS One 7: e37164.
9. Minami M, Nishio K, Ajioka Y, Kyushima H, Shigeki K, et al. (2009) Identification of Curcuma plants and curcumin content level by DNA polymorphisms in the trnS-trnfM intergenic spacer in chloroplast DNA. J Nat Med 63: 75–79.
10. Martin G, Baurens F-C, Cardi C, Aury J-M, D'Hont A (2013) The complete chloroplast genome of banana (*Musa acuminata*, Zingiberales): insight into plastid monocotyledon evolution. PLoS One 8: e67350.
11. Liu C, Shi L, Zhu Y, Chen H, Zhang J, et al. (2012) CpGAVAS, an integrated web server for the annotation, visualization, analysis, and GenBank submission of completely sequenced chloroplast genome sequences. BMC Genomics 13: 715.
12. Drescher A, Ruf S, Calsa T, Carrer H, Bock R (2000) The two largest chloroplast genome-encoded open reading frames of higher plants are essential genes. Plant J 22: 97–104.
13. Barrett CF, Davis JI, Leebens-Mack J, Conran JG, Stevenson DW (2013) Plastid genomes and deep relationships among the commelinid monocot angiosperms. Cladistics 29: 65–87.
14. Wolfe KH, Li WH, Sharp PM (1987) Rates of nucleotide substitution vary greatly among plant mitochondrial, chloroplast, and nuclear DNAs. Proc Natl Acad Sci U S A 84: 9054–9058.
15. Birky Jr C, Walsh JB (1992) Biased gene conversion, copy number, and apparent mutation rate differences within chloroplast and bacterial genomes. Genetics 130: 677–683.
16. SHI L-C, Zhang J, HAN J-P, SONG J-Y, Yao H, et al. (2011) Testing the potential of proposed DNA barcodes for species identification of Zingiberaceae Journal of Systematics and Evolution 49: 261–266.
17. Vinitha MR, Kumar US, Aishwarya K, Sabu M, Thomas G (2014) Prospects for discriminating Zingiberaceae species in India using DNA barcodes Journal of integrative plant biology.
18. Ran J-H, Wang P-P, Zhao H-J, Wang X-Q (2010) A test of seven candidate barcode regions from the plastome in Picea (Pinaceae) Journal of integrative plant biology 52: 1109–1126.
19. Ashfaq M, Asif M, Anjum ZI, Zafar Y (2013) Evaluating the capacity of plant DNA barcodes to discriminate species of cotton (*Gossypium*: Malvaceae) Molecular ecology resources 13: 573–582.
20. Hollingsworth ML, ANDRA CLARK A, Forrest LL, Richardson J, Pennington R, et al. (2009) Selecting barcoding loci for plants: evaluation of seven candidate loci with species-level sampling in three divergent groups of land plants. Molecular Ecology Resources 9: 439–457.
21. Kress WJ, Wurdack KJ, Zimmer EA, Weigt LA, Janzen DH (2005) Use of DNA barcodes to identify flowering plants. Proc Natl Acad Sci U S A 102: 8369–8374.
22. Parks M, Cronn R, Liston A (2009) Increasing phylogenetic resolution at low taxonomic levels using massively parallel sequencing of chloroplast genomes. BMC Biol 7: 84.
23. Nock CJ, Waters DLE, Edwards MA, Bowen SG, Rice N, et al. (2011) Chloroplast genome sequences from total DNA for plant identification. Plant Biotechnol J 9: 328–333.
24. Zhang Y-J, Ma P-F, Li D-Z (2011) High-throughput sequencing of six bamboo chloroplast genomes: phylogenetic implications for temperate woody bamboos (Poaceae: Bambusoideae). PLoS One 6: e20596.
25. Yang J-B, Tang M, Li H-T, Zhang Z-R, Li D-Z (2013) Complete chloroplast genome of the genus Cymbidium: lights into the species identification, phylogenetic implications and population genetic analyses. BMC Evol Biol 13: 84.
26. Cantarel BL, Korf I, Robb SMC, Parra G, Ross E, et al. (2008) MAKER: an easy-to-use annotation pipeline designed for emerging model organism genomes. Genome Res 18: 188–196.
27. Rissman AI, Mau B, Biehl BS, Darling AE, Glasner JD, et al. (2009) Reordering contigs of draft genomes using the Mauve aligner. Bioinformatics 25: 2071–2073.
28. Thiel T, Michalek W, Varshney RK, Graner A (2003) Exploiting EST databases for the development and characterization of gene-derived SSR-markers in barley (*Hordeum vulgare* L.). Theor Appl Genet 106: 411–422.
29. Untergasser A, Cutcutache I, Koressaar T, Ye J, Faircloth BC, et al. (2012) Primer3–new capabilities and interfaces. Nucleic Acids Res 40: e115.

13

The Diurnal Logic of the Expression of the Chloroplast Genome in *Chlamydomonas reinhardtii*

Adam D. Idoine, Alix Boulouis, Jens Rupprecht, Ralph Bock*

Max-Planck-Institut für Molekulare Pflanzenphysiologie, Potsdam, Golm, Germany

Abstract

Chloroplasts are derived from cyanobacteria and have retained a bacterial-type genome and gene expression machinery. The chloroplast genome encodes many of the core components of the photosynthetic apparatus in the thylakoid membranes. To avoid photooxidative damage and production of harmful reactive oxygen species (ROS) by incompletely assembled thylakoid protein complexes, chloroplast gene expression must be tightly regulated and co-ordinated with gene expression in the nucleus. Little is known about the control of chloroplast gene expression at the genome-wide level in response to internal rhythms and external cues. To obtain a comprehensive picture of organelle transcript levels in the unicellular model alga *Chlamydomonas reinhardtii* in diurnal conditions, a qRT-PCR platform was developed and used to quantify 68 chloroplast, 21 mitochondrial as well as 71 nuclear transcripts in cells grown in highly controlled 12 h light/12 h dark cycles. Interestingly, in anticipation of dusk, chloroplast transcripts from genes involved in transcription reached peak levels first, followed by transcripts from genes involved in translation, and finally photosynthesis gene transcripts. This pattern matches perfectly the theoretical demands of a cell "waking up" from the night. A similar trend was observed in the nuclear transcripts. These results suggest a striking internal logic in the expression of the chloroplast genome and a previously unappreciated complexity in the regulation of chloroplast genes.

Editor: Frederik Börnke, Leibniz-Institute for Vegetable and Ornamental Plants, Germany

Funding: This work was supported by grants from the BMBF (Systems Biology Initiative FORSYS: GoFORSYS) and the Deutsche Forschungsgemeinschaft (FOR2092, BO 1482/17-1) to RB. The funders had no role in study design, data collection and analysis, decision to publish, or preparation of the manuscript.

Competing Interests: The authors have declared that no competing interests exist.

* Email: rbock@mpimp-golm.mpg.de

Introduction

The unicellular green alga *Chlamydomonas reinhardtii* has become a model organism for many research areas, including chloroplast biology and photosynthesis research [1], [2], [3], [4], [5]. Moreover, its capacity to produce molecular hydrogen and to synthesize and accumulate considerable amounts of oil also have made *C. reinhardtii* an attractive target organism of many efforts to develop algae as an efficient and sustainable production platform for biofuels [6], [7], [8], [9]. Since all algal biofuels and renewable green chemical feedstocks are directly or indirectly derived from photosynthesis and many of the core components of the photosynthetic apparatus are encoded in the chloroplast (plastid) DNA, a comprehensive understanding of the mechanisms that govern expression of the chloroplast genome is of great importance to all future attempts to rationally engineer primary metabolism. However, our knowledge about the regulatory mechanisms operating in the control of chloroplast gene expression in response to endogenous cues and environmental stimuli and the signaling pathways co-ordinating expression of the chloroplast genome with that of the nuclear genome is still far from complete.

The chloroplast genome of *C. reinhardtii* is 204 kb in size and contains 99 genes [10]. The vast majority of these genes belong to the two main gene classes found in plastid genomes of algae and embryophyte plants: photosynthesis genes and genetic system genes [11]. The photosynthesis-related genes comprise genes for subunits of the major multiprotein complexes in the thylakoid membrane that participate in photosynthetic electron transfer and ATP generation (photosystem II, cytochrome b_6f complex, photosystem I, ATP synthase). The genetic system genes encode components of the transcriptional apparatus (subunits of the plastid-encoded RNA polymerase, PEP) and the translational machinery (rRNAs, tRNAs, ribosomal proteins).

Genes in the chloroplast genome of *C. reinhardtii* are transcribed by a multisubunit RNA polymerase that resembles bacterial RNA polymerases of the $\alpha_2\beta\beta'$ type. All of its core subunits are encoded in the chloroplast DNA and, for this reason, the enzyme has been dubbed PEP (for plastid-encoded RNA polymerase). In *Chlamydomonas*, there appears to be no nucleus-encoded plastid RNA polymerase that has been shown to be additionally present in plastids of embryophyte plants [12], [13]. The PEP core complex interacts with the single sigma factor found in the nuclear genome of *C. reinhardtii* (RPOD) to confer sequence specificity in promoter recognition [14], [15]. TATA-like -10 box sequences (and, in some cases, the extended -10 box motif TATAATAT) are found upstream of many reading frames in the *C. reinhardtii* plastid DNA and are thought to act as key promoter elements. However, several other types of promoters have been identified, and a consensus is yet to emerge [16], [17], [18].

Natural conditions for *C. reinhardtii* involve several environmental variables which vary reliably over a day. The strongest signal over a day is light, which is often used in laboratory conditions to simulate a diurnal cycle. Many cellular processes respond to light-dark cycles, such as regulation of the cell cycle, as well as various steps of chloroplast gene expression. As examples, the structure of the entire plastid genome is most accessible early in the light period, and the half-life of several mRNAs is much shorter in the light period than in the dark [19], [20].

So far, nine chloroplast transcripts have been quantified over a diurnal cycle in at least one study (Figure S1 in File S1). The majority of these attain their highest levels in the light period. However, the majority of the mRNAs which have been studied so far are involved in photosynthesis. Neither RNA polymerase subunits nor any subunits of the ribosome have been analyzed. Interestingly, when the diurnal regulation of chloroplast transcription in *C. reinhardtii* was studied, it was found to be accompanied by little or no fluctuation in RNA polymerase activity or in the level of RPOD, the only sigma factor in chloroplasts of *Chlamydomonas* [21].

Here, we have undertaken a systematic study of the expression of the chloroplast genome in *Chlamydomonas* under diurnal conditions. Our work has uncovered a surprising internal logic in the expression of the chloroplast genome in that transcripts of the transcriptional apparatus peak first, followed by transcripts of the translational machinery, and finally by photosynthesis gene transcripts.

Results

Development of a qRT-PCR platform for transcript quantification in *Chlamydomonas*

We have previously designed a microarray for profiling of chloroplast transcripts in the model alga *C. reinhardtii* and used it to determine changes in RNA accumulation in response to changing environmental parameters [22]. While the microarray faithfully detected expression changes in genes responding strongly to environmental cues, its sensitivity and reproducibility was not high enough to accurately determine more subtle changes in transcript levels as they occur, for example, over a diurnal cycle. Since PCR-based transcript profiling assays are potentially more sensitive and less prone to technical variation between experiments [23], we sought to establish a qRT-PCR platform for the quantitative analysis of chloroplast transcripts. To this end, primers were designed for all plastid genes and open reading frames (see Material and Methods; Dataset S1). In addition, the mitochondrial genes and a selected set of nuclear genes were included in the platform (Table S1 in File S1; Dataset S1). Test PCRs confirmed that amplicons of the expected sizes could be generated for most genes and cDNAs derived from their transcripts. Where necessary, several oligonucleotides were tested to optimize the PCRs and prevent amplification of non-specific products. Genes which repeatedly failed quality criteria were excluded from the platform. These included the tRNA genes and several genes encoding small polypeptides, owing to difficulty in designing suitable primers for small sequences. The final design of the fully optimized platform comprises 68 chloroplast, 21 mitochondrial and 51 nuclear transcripts (Table S1 in File S1; Dataset S1).

The qRT-PCR platform was then validated against our previously designed microarray for profiling of chloroplast

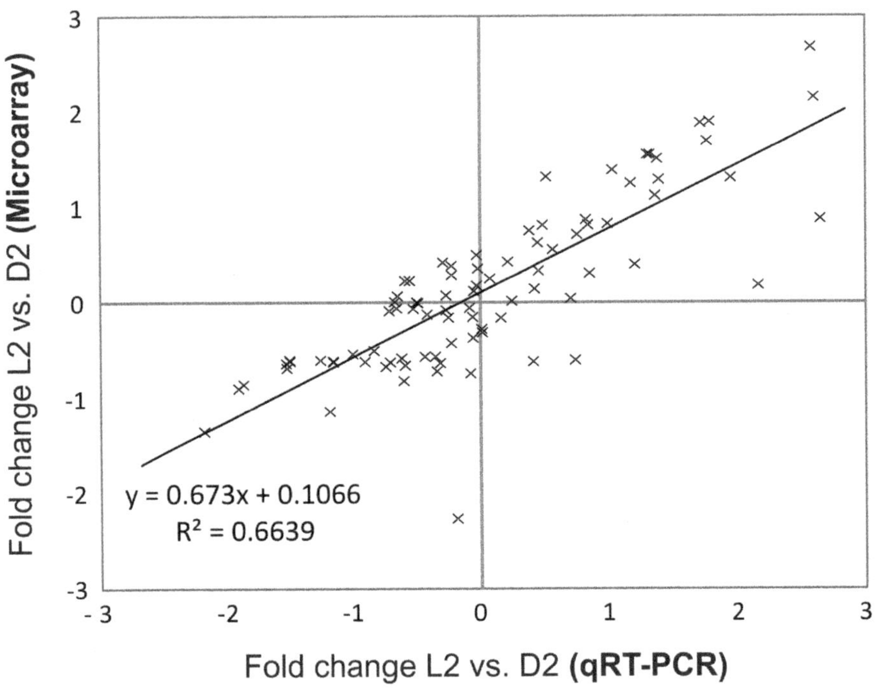

Figure 1. Comparison of microarray and qRT-PCR-based quantification of transcripts in a diurnal rhythm. Changes in organelle (chloroplast and mitochondrial) transcript levels between time points L2 and D2 in a wild-type strain of *C. reinhardtii* (CC-124) were determined by microarray and by qRT-PCR. The resulting fold-changes correlate with an R^2 value of 0.66. It is lower than 1.0, mainly because the microarray produces less accurate results for lowly expressed genes. All data points represent the average of three independent experiments.

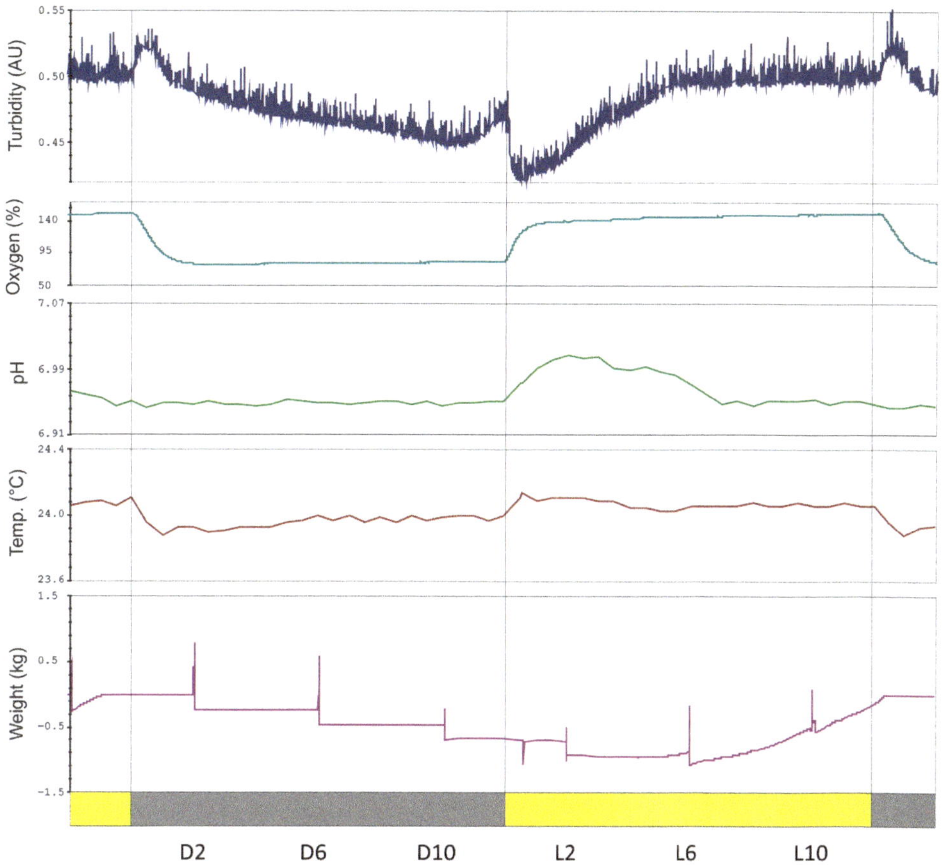

Figure 2. Key bioreactor parameters from a representative bioreactor run (R2) in diurnal conditions. Turbidity (navy blue) slightly decreased during the dark period and increased during the light. The rapid peak after dusk and rapid drop after dawn are technical artefacts of the turbidimeter. The peaks in the weight graph correspond to the sampling times, when the fermenter was disturbed. Dissolved oxygen concentration increases from ca. 80% to ca. 150% very rapidly in the light. The pH of the culture is controlled by titration of acid or base, temperature is controlled by dynamic cooling to maintain reasonably stable values for these parameters. Samples were taken every 4 h, these points are indicated and also visible as peaks in the weight data (purple). Yellow bars represents the 12 h light period (L), flanked by dark periods (D; grey bars).

transcripts in *Chlamydomonas* [22]. For this purpose, total RNA was isolated from algal cells grown in diurnal conditions in batch cultures, converted into cDNA by random priming and analyzed by both methods. The fold changes observed between the time points 2 hours into the dark period (D2) and 2 hours into the light period (L2) for each gene agree excellently between the two methods ($R^2 = 0.66$; Figure 1). However, overall we found the qRT-PCR platform to produce better results for lowly expressed genes (Figure 1).

To validate the qRT-PCR platform to more diverse studies, we compared the absolute expression levels determined in the light with the data generated by three previous *C. reinhardtii* chloroplast transcriptomic studies. The data obtained in this study were compared to data from two independent microarrays [22], [24] and to data obtained from NanoString technology [25]. Excellent correlations were obtained between this work, one of the microarray studies [22] and the NanoString study [25] with correlation coefficients of 0.82 and 0.60, respectively (Figure S1 in File S1). The data produced in the second microarray study [24] showed a lower correlation coefficient ($R^2 = 0.47$), but a similar trend was still visible.

Taken together, these results prove that the qRT-PCR platform developed for this work is of high quality. It, therefore, was used to analyze the accumulation of transcripts in cells grown in fully controlled diurnal conditions.

Diurnal growth

Cells were grown in highly controlled 5 L bioreactors [22] (see Materials and Methods) which allow very precise control and measurement of culture parameters, such as oxygen concentration, temperature, pH and density of the culture (Figure 2). As expected, the dissolved oxygen concentration in the culture medium increased rapidly at the beginning of the light period to approximately 140 % air saturation due to the rapid onset of photosynthetic oxygen evolution. During the dark period, the dissolved oxygen decreased to around 80 % as the cells consumed some of the dissolved oxygen through respiration, and ceased to generate oxygen from photosynthetic water splitting (Figure 2). Constant bubbling with air ensured that the cells did not experience hypoxia. A slight increase in the medium temperature in the light due to heat produced by the light-emitting diode light shells was also observed. Although the temperature of the bioreactor was dynamically controlled with circulating water, a small increase of 0.15–0.20°C was observed in the light. The pH also fluctuated slightly over the diurnal cycle. It increased at the start of the light period from 6.95 to around 7.0 (Figure 2). This is potentially due to the increase in dissolved oxygen and/or the higher amount of CO_2 consumed by photosynthesis. The pH then slowly decreased over the rest of the day until a time point between midday (L6) and dusk (L12), when the pH dropped below 6.95. At

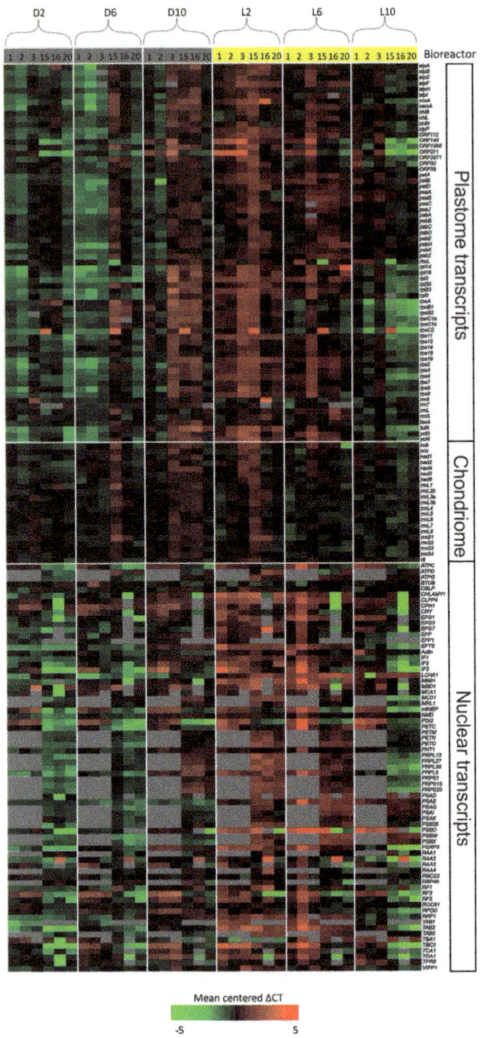

Figure 3. Diurnal transcriptomics on bioreactor cultures of *Chlamydomonas*. The effect of 12 h light/12 h dark cycles on organelle (and selected nuclear) transcripts was analyzed by qRT-PCR. *C. reinhardtii* cells were grown in six independent bioreactor runs (run numbers 1, 2, 3, 15, 16, 20) and harvested at the time points indicated. Data obtained for plastome transcripts is shown in the upper portion of the heatmap, chondriome transcripts in the central region, and nuclear transcripts in the lower region. Within each genome, transcripts are listed alphabetically. Data are normalized to housekeeping transcripts and then to the average across all samples for that gene. Red boxes indicate up-regulation, green boxes: down-regulation, grey boxes: no data. Visualization by Multiexperiment Viewer [38].

this point KOH was titrated into the bioreactor to prevent the pH dropping further. KOH titration continued until shortly before midnight (D6), when the pH stabilized and remained at 6.95 until dawn (D12).

The growth of the cells was monitored by the optical density of the culture. The optical density was set to a target value (0.4 or 0.5), but it still showed a complex diurnal pattern. At dawn, a rapid decrease in the optical density was caused directly by light acting on the turbidimeter. The turbidimeter was positioned in the bioreactor to minimize the magnitude of this effect and no changes greater than 0.06 units were observed. The optical density of the culture increased between approximately L1 and L6 as the cells grew, until the target optical density was attained. The culture was then diluted dynamically with fresh medium to maintain a steady state. After the light was turned off, a small peak was observed (again due to a technical effect on the turbidimeter), then the optical density decreased for most of the dark period (presumably due to degradation of intracellular storage compounds). Shortly before the end of the dark period, the optical density began to increase modestly (Figure 2), most likely due to the release of daughter cells.

Transcript levels over the diurnal cycle determined by qRT-PCR

RNA was extracted from cells which were sampled from the bioreactors every 4 hours in diurnal conditions. There is a general trend for samples harvested in the light to yield greater amounts of RNA than samples harvested during the dark period (data not shown). To determine changes in transcript levels, the RNA was analyzed by qRT-PCR (Figure 3). Most chloroplast and nuclear transcripts show high levels between at the end of the night and first half of the day (D10 to L6), whereas transcripts derived from the mitochondrial genome (chondriome) do not vary greatly over a diurnal cycle (Figure 3). Although there was some variation between the replicates, the altogether six biological replicates performed allow strong conclusions to be drawn.

k-means clustering was used to group transcripts according to the patterns of their responses to the diurnal conditions. Within the chloroplast transcripts, k-means clustering separates the transcripts into three classes: one containing all PEP transcripts (i.e., mRNAs encoding subunits of the plastid-encoded RNA polymerase), a second one strongly enriched in ribosome-related transcripts, and a third one enriched in photosynthesis-related transcripts (Figure 4). These are the three major classes of protein-coding genes encoded in the *C. reinhardtii* plastid genome (plastome), and the degree of separation is remarkable. Interestingly, a similar separation of photosynthesis- and translation-related transcripts was observed in the subset of nuclear-encoded transcripts which were investigated (Figure 4; Dataset S1), but not in the mitochondrial genes.

Figure 4 also shows that most transcripts resemble a sine curve in their diurnal fluctuations. This allowed the quantitation of their phases by comparing each transcript profile to a series of sine curves. The sine curve to which the data for each transcript fitted best, was used to classify the time of the day at which that transcript peaks. This analysis revealed that the defining characteristic of each of the three k-means clusters (Figure 4) is the phase at which the transcripts peak. The PEP cluster peaks first, followed by the ribosome cluster, finally followed by the photosynthesis cluster. Remarkably, in all six experiments, using two different *C. reinhardtii* strains, this order in which the three groups of genes responded remained consistent (Figure 5).

Northern blot analyses confirm the diurnal patterns in chloroplast transcript levels

In order to confirm the observed striking sequence in diurnal expression of the different classes of chloroplast genes, two representative transcripts from each of the three classes (PEP genes: *rpoA* and *rpoC2*; ribosome-related genes: *rps2* and *rps3*; photosynthesis-related genes: *psaB* and *psbB*) were tested by northern blot analysis (Figure 6). In addition, *tufA*, the gene encoding translation elongation factor EF-Tu [26], [27] was investigated. While being a component of the translational apparatus (and thus belonging to the ribosome-related genes), *tufA* is special in that its expression is known to be under stringent circadian control [28], [29].

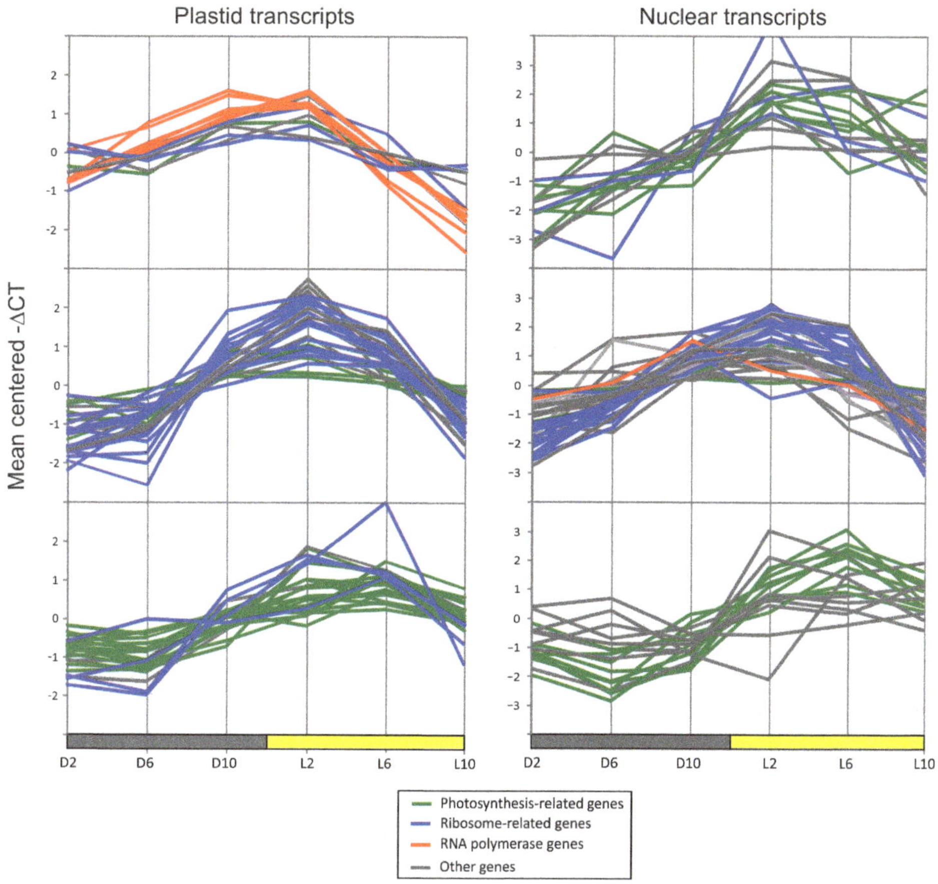

Figure 4. k-means clustering of diurnal qRT-PCR data for chloroplast transcripts (left) and a subset of nuclear transcripts (right). The averaged data for six biological replicates were grouped into three clusters with the Pearson correlation as the distance metric. Gene profiles are colored according to the gene product's function. Green: photosynthesis-related genes, blue: ribosome-related genes, red: plastid-encoded RNA polymerase genes, grey: miscellaneous genes. Clustering and visualization by Multiexperiment Viewer [38].

The expression profiles determined by northern blot analyses for all seven transcripts agreed excellently with the data obtained from qRT-PCR analysis. *rpoC2* and *rpoA* both reached their highest levels at the end of the night, *rps2* and *rps3* reached their highest levels in the first half of the day, and *psaB* and *psbB* reached high levels in the second half of the day. These results confirm that, during the diurnal cycle, chloroplast transcripts are differentially regulated depending on their function. They also confirm the striking internal logic of chloroplast genome expression (Figures 4–6): the genes encoding the components of the transcriptional machinery are induced and peak shortly before dawn, followed by the genes encoding components of the translational machinery, and finally followed by the genes encoding the components of the photosynthetic apparatus.

Discussion

In this work, we have used the model alga *C. reinhardtii* to study the expression of the chloroplast and mitochondrial genomes in comparison to a subset of nuclear genes over the diurnal cycle. The qRT-PCR platform developed for this purpose provides a useful tool for systematic analyses of organellar gene expression in *Chlamydomonas* in response to environmental conditions and developmental cues. With respect to sensitivity and reproducibility, the platform compares favourably to other methods for plastid transcript profiling.

This platform was used to quantify the changes in an unbiased group containing most chloroplast transcripts over the diurnal cycle in *C. reinhardtii*. Strikingly, this analysis revealed that transcripts behave differently depending on the function of their gene product. Unbiased clustering into three groups reliably separated photosynthesis-related transcripts, ribosome-related transcripts and RNA polymerase (PEP) transcripts.

Even more strikingly, these groups of transcripts fluctuate in a temporally ordered manner. PEP transcripts peak first at the end of the night, ribosome-related transcripts peak early in the day, and photosynthesis transcripts attain high levels at the end of the day. This agrees excellently with the hypothetical demands of a cell "waking up" from the night. Assuming that protein complexes decline during the night (which is likely due to the fast growth rate of *C. reinhardtii*, causing "dilution by division"), the cells will have to build up all complexes again for the day. Hence, remaining RNA polymerase can first transcribe its own genes, which can then be translated by remaining ribosomes in the chloroplast. In this way, a feed-forward loop creates more RNA polymerase. After a temporal delay, ribosome transcripts begin to increase, which are in turn translated by remaining ribosomes, creating another feed-forward loop on ribosome content. Finally, after a longer temporal delay, photosynthesis transcripts start to increase, which can then be translated by the large pool of ribosomes already built up.

Although substantial regulation of chloroplast gene expression also occurs at the translational level [30], the observed patterns in

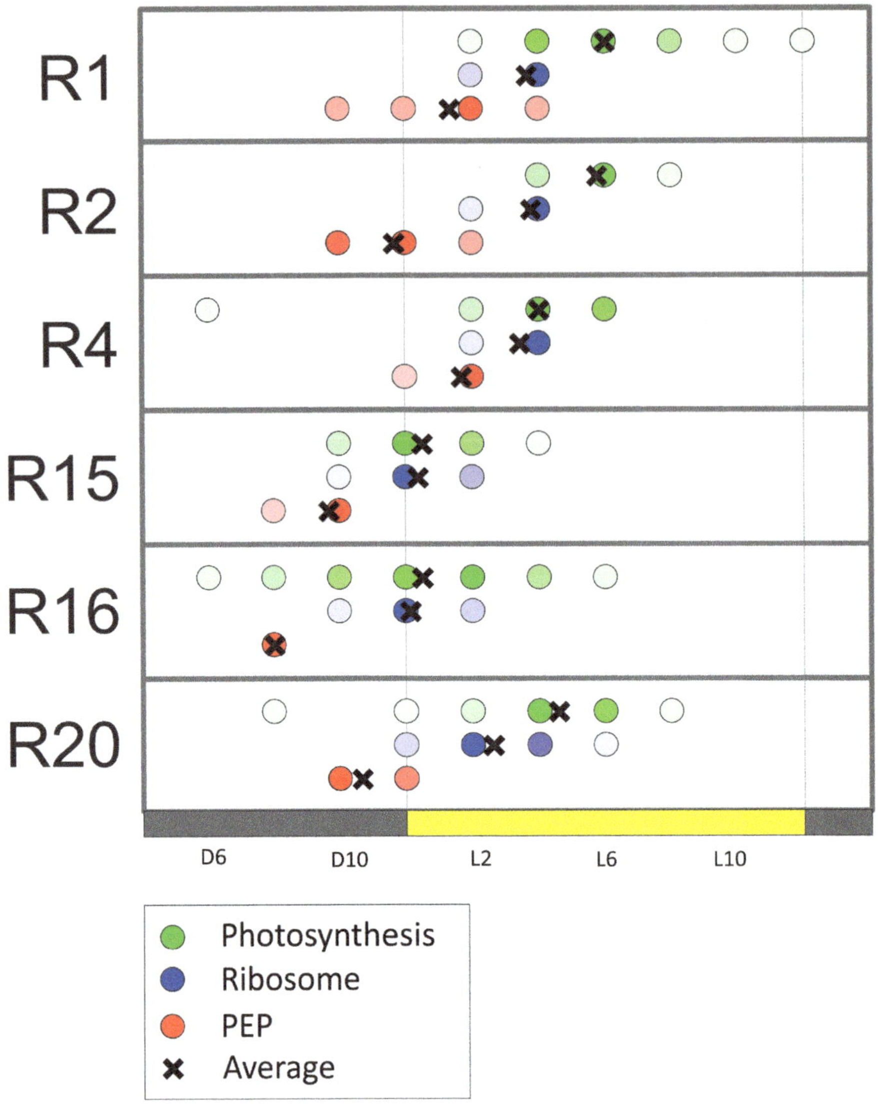

Figure 5. Behavior of functionally related chloroplast transcripts in diurnal conditions. The average peak time of all transcripts belonging to the three functional classes (red: plastid-encoded RNA polymerase genes, PEP; blue: ribosome-related genes; green: photosynthesis-related genes) is shown for each bioreactor experiment. The intensity of the circle represents the number of transcripts peaking at the time, and the black x represents the average of all transcripts. The yellow bar represents the 12 h light period, flanked by dark periods (grey bars).

transcript levels are unlikely to be insignificant. It is also interesting in this respect that the window of peak transcript levels corresponds to the window of increased supercoiling of the chloroplast DNA [19]. Furthermore, our finding that a similar temporal order of gene expression over the diurnal cycle exists in the nucleus illustrates the tight co-regulation of plastid and nuclear gene expression through anterograde and retrograde signalling pathways [31], [32], [33]. In fact, our set of nucleus-encoded genes for chloroplast proteins included several translation factors (Dataset S1), whose expression also cycled diurnally (Figures 3 and 4), raising the possibility that translational regulation reinforces the changes occurring at the transcriptional level. Finally, it is also important to realize that the changes in transcript accumulation measured by microarray hybridization or qRT-PCR represent the result of transcriptional regulation and regulation at the level of RNA turnover. It is noteworthy in this respect that at least some chloroplast genes are also diurnally regulated at the level of mRNA degradation [34], [35], [36]. Consistent with this, several known (nucleus-encoded) proteins that determine the stability of specific chloroplast mRNAs were included in our analyses (Figure 3; Dataset S1) and showed diurnal rhythmicity.

In summary, our data reported here uncover a striking internal logic in the diurnal gene expression program of the model alga *C. reinhardtii*. The peak in the expression of genes related to transcription is followed by the peak in the expression of genes related to translation which in turn is followed by the peak in the expression of genes involved in photosynthesis. It will be interesting to see whether this pattern is conserved also in seed plants. Also, the identification of the *trans*-factors that confer this intriguing pattern of diurnal regulation represents an attractive challenge. In combination with the *cis*-elements they bind, these factors could provide useful tools for controlling gene and transgene expression in both the nuclear and the chloroplast genome.

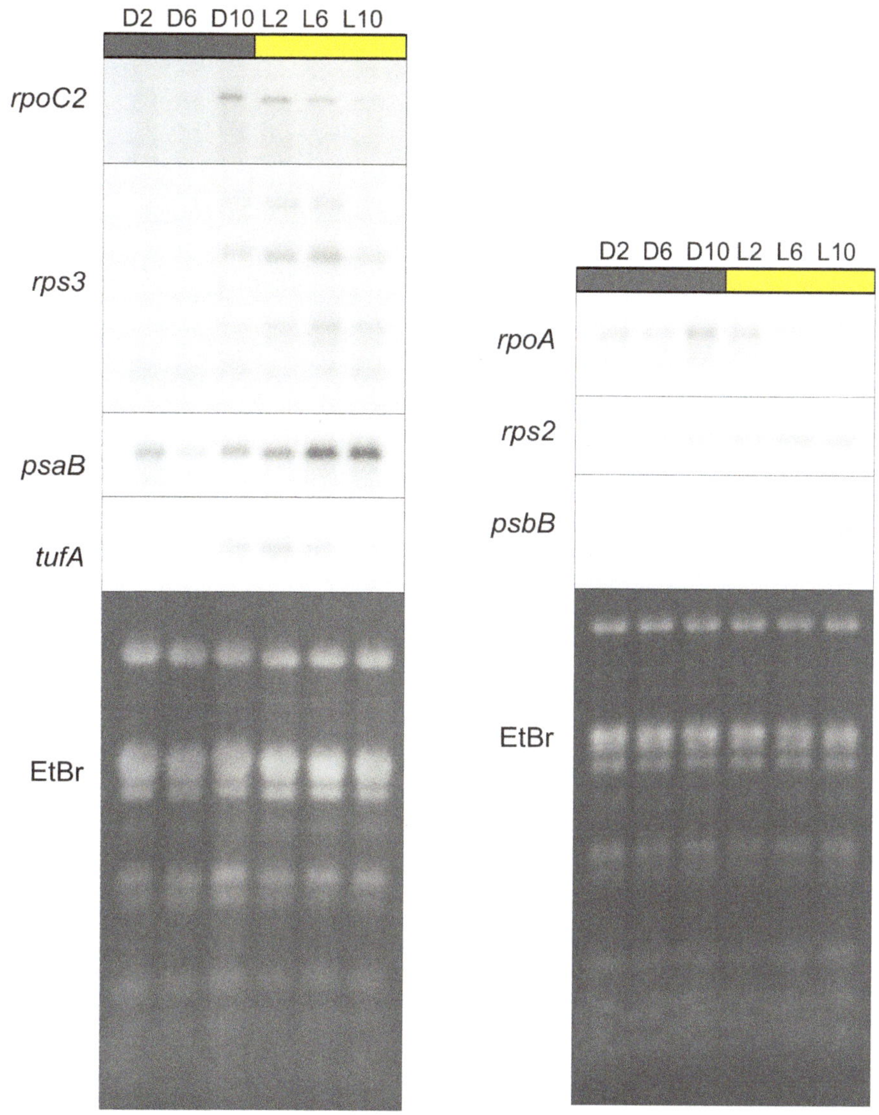

Figure 6. Confirmation of the diurnal expression patterns observed in the pRT-PCR experiments by northern blot analyses. To avoid variation from RNA loading, each membrane was hybridized and stripped multiple times. The order of probing corresponds to the vertical order of the images (and the increasing signal intensity known to be obtained with the different probes). The ethidium bromide fluorescence (EtBr) remaining on the membrane after blotting is shown as a loading control.

Materials and Methods

Algal strains and growth conditions

Chlamydomonas reinhardtii strains CC-124 and SAG73.72 (equivalent to CC-3348 in the Chlamydomonas Resource Center; www.chlamycollection.org) were grown in bioreactors which allow the growth of 5 L of continuous culture in highly controlled conditions [22]. The reactors were bubbled with 200 cm^3 min^{-1} of filtered 5% CO_2 in air and were stirred at 50 rpm with three impellers. Light was supplied from two custom-made light-emitting diode half-shells, and the light was focused with individual lenses to the center of the bioreactor, to minimize light gradients within the bioreactor. Cell density was monitored with a turbidimeter, and when the threshold of 0.4 or 0.5 was exceeded, sterile medium was pumped into the bioreactor to dilute the culture. The addition of medium was sensed as an increase in weight of the bioreactor, and a corresponding volume of culture was removed to maintain a steady volume. When the pH of the culture dropped below 6.95, or above 7.05, sterile 1 M KOH or 1 M HCl was slowly pumped into the bioreactor until pH 7.00 (+/−0.05) was attained. Temperature was maintained at 24.0°C by cooled water circulating within an outer mantle. The dissolved oxygen concentration was also measured. All of these parameters were logged in the BioPAT MFCS/win software (Sartorius; http://www.sartorius.com/en/products/process-control-tools-software/biopat-mfcs/), which also controlled all parameters of the bioreactor, including light in conjunction with a second software program, Quattro Color (ProBioData GmbH, Berlin, Germany).

Cells were grown in H_5P media (Tris-minimal medium with 5 mM HEPES in place of 20 mM Tris; [37]). Cells were precultured in H_5AP (Tris-acetate medium with 5 mM HEPES in place of 20 mM Tris) in continuous light, before bioreactor inoculation. Cultures were grown in constant light until they reached their target optical density, then they were allowed to grow for at least an additional two days in constant light. The cells were then exposed to three days of 12 h light/12 h dark cycles before samples were taken on the fourth day of diurnal conditions. Samples were centrifuged at 4000 ×*g* for 4 min at 4°C, and the cell pellet was frozen in liquid nitrogen until use.

RNA extraction

RNA was extracted from 3–5 × 10^7 cells with the Trifast reagent according to the manufacturer's instructions (Peqlab, Erlangen, Germany). Contaminating DNA was removed by treatment with the TURBO DNA-free kit (Invitrogen, Paisley, UK). RNA quantity was estimated by measuring the absorbance of the RNA solution at 260 nm. RNA quality was determined by denaturing electrophoresis in formaldehyde-containing agarose gels using standard protocols.

Primer design for qRT-PCR

Primers for qRT-PCR were designed with the online software QuantPrime (www.quantprime.de) after the chloroplast and mitochondrial genomes were added to a custom database. Additional primers were designed with Primer3 (http://primer3.ut.ee/). All primers were tested for acceptable melting curves and product sizes. Primer pairs which failed these tests were discarded and redesigned. After testing and redesign, primers for 68 chloroplast, 21 mitochondrial, and 51 nuclear transcripts were retained. These primers are listed in Table S1 in File S1.

qRT-PCR experiments

cDNA was synthesized from 2 μg of total cellular RNA with the SuperScript III First-Strand Synthesis System for RT-PCR (Invitrogen) according to the manufacturer's instructions. Each 5 μL qPCR reaction contained 2.5 μL Power Sybr Green PCR Master Mix (Applied Biosystems), 0.5 μL of diluted cDNA (corresponding to cDNA obtained from approximately 5 ng total RNA), and 2 μL of 5 pmol μL^{-1} forward and reverse primers). Reactions were conducted in opaque 384 well plates in a 7900HT Fast Real-Time PCR System Instrument (Applied Biosystems).

Northern blotting and mRNA detection

Denaturing electrophoresis in formaldehyde-containing agarose gels was performed according to standard protocols. Electrophoretically separated RNAs were blotted onto nylon membranes (GE Healthcare) by capillary transfer. RNA was immobilized on the nylon membrane by UV crosslinking. Specific signals were detected with probes generated by PCR (Table S2 in File S1), purified from excised gel slices and labelled with ^{32}P dCTP using the Megaprime DNA Labeling System (GE Healthcare). Membranes were stripped by washing several times with boiling 1% SDS.

qRT-PCR data analysis

CT values were normalized to the geometric average of three housekeeping genes (*CBLP*, *BTUB* and *RBCS2*) to obtain ΔCT values. ΔCT values were mean centered per gene to the average across all time points.

k-means clustering was performed with the Pearson correlation metric using the Multiexperiment Viewer software package [38], after Figure of Merit analysis was performed to determine that 3 was a suitable value of k.

To classify transcript profiles to a sine curve, each individual transcript profile was compared to twelve sine curves (period 24 h, amplitude 0.5, phase offset by 2 h) using the Pearson correlation metric. The best match was used to classify the time at which that particular transcript peaks.

Supporting Information

File S1 Contains the following files: **Table S1.** List of primers for qPCR used in this study. **Table S2.** List of PCR primers used to generate hybridization probes. **Figure S1.** Average expression levels of plastid transcripts across multiple experiments. The average expression level for each transcript was calculated for each experiment. The expression levels determine by qRT-PCR are shown on the left, and were used to sort the heatmap. This dataset is compared to two microarray experiments [22], [24] and one NanoString experiment [25]. Expression levels are all in log2 and are mean centered. High expression is indicated by red, low expression by green, and grey boxes indicate missing data. All experiments were performed with light-grown algal cultures, but the growth conditions were likely not identical in all details (e.g., light intensity, spectral quality, cell density, media composition), thus potentially explaining part of the variation.

Dataset S1 List of genes, PCR primers, gene functions, gene classifications, and associations with clusters.

Acknowledgments

We are grateful to Dr. Maria Mittag (University of Jena, Germany) and the Chlamydomonas Resource Center for providing algal strains. We thank Mrs. Liliya Yaneva-Roder for excellent technical assistance, and Dr. Michael Schroda (University of Kaiserslautern) and the other members of the *Go*FORSYS consortium for helpful discussion. This work was supported by grants from the BMBF (Systems Biology Initiative FORSYS: GoFORSYS) and the Deutsche Forschungsgemeinschaft (FOR2092, BO 1482/17-1) to RB.

Author Contributions

Conceived and designed the experiments: RB ADI AB JR. Performed the experiments: ADI JR. Analyzed the data: ADI AB JR RB. Contributed to the writing of the manuscript: ADI RB.

References

1. Harris EH (2001) Chlamydomonas as a model organism. Annu Rev Plant Physiol Plant Mol Biol 52: 363–406.
2. Rochaix J-D (2001) Assembly, functions, and dynamics of the photosynthetic machinery in Chlamydomonas reinhardtii. Plant Physiol 127: 1394–1398.
3. Merchant SS, Prochnik SE, Vallon O, Harris EH, Karpowicz SJ, et al. (2007) The Chlamydomonas genome reveals the evolution of key animal and plant functions. Science 318: 245–251.
4. Day A, Goldschmidt-Clermont M (2011) The chloroplast transformation toolbox: selectable markers and marker removal. Plant Biotechnol J 9: 540–553.
5. Terashima M, Specht M, Hippler M (2011) The chloroplast proteome: a survey from the Chlamydomonas reinhardtii perspective with a focus on distinctive features. Curr Genet 57: 151–168.
6. Rupprecht J, Hankamer B, Mussgnug JH, Ananyev G, Dismukes C, et al. (2006) Perspectives and advances of biological H_2production in microorganisms. Appl Microbiol Biotechnol 72: 442–449.
7. Hu Q, Sommerfeld M, Jarvis E, Ghirardi M, Posewitz M, et al. (2008) Microalgal triacylglycerols as feedstocks for biofuel production: perspectives and advances. Plant J 54: 621–639.
8. Beer LL, Boyd ES, Peters JW, Posewitz MC (2009) Engineering algae for biohydrogen and biofuel production. Curr Op Biotechnol 20: 264–271.
9. Esquível MG, Amaro HM, Pinto TS, Fevereiro PS, Malcata FX (2011) Efficient H_2 production via Chlamydomonas reinhardtii. Trends Biotechnol 29: 595–600.
10. Maul JE, Lilly JW, Cui L, dePamphilis CW, Miller W, et al. (2002) The Chlamydomonas reinhardtii plastid chromosome: islands of genes in a sea of repeats. Plant Cell 14: 2659–2679.

11. Shimada H, Sugiura M (1991) Fine structural features of the chloroplast genome: comparison of the sequenced chloroplast genomes. Nucleic Acids Res 19: 983–995.
12. Shiina T, Tsunoyama Y, Nakahira Y, Khan MS (2005) Plastid RNA polymerases, promoters, and transcription regulators in higher plants. Int Rev Cytol 244: 1–68.
13. Liere K, Börner T (2007) Transcription and transcriptional regulation in plastids. Top Curr Genet 19: 121–174.
14. Carter ML, Smith AC, Kobayashi H, Purton S, Herrin DL (2004) Structure, circadian regulation and bioinformatic analysis of the unique sigma factor gene in Chlamydomonas reinhardtii. Photosynth Res 82: 339–349.
15. Bohne A-V, Irihimovitch V, Weihe A, Stern DB (2006) Chlamydomonas reinhardtii encodes a single sigma^{70}-like factor which likely functions in chloroplast transcription. Curr Genet 49: 333–340.
16. Klein U, De Camp JD, Bogorad L (1992) Two types of chloroplast gene promoters in Chlamydomonas reinhardtii. Proc Natl Acad Sci USA 89: 3453–3457.
17. Klein U, Salvador ML, Bogorad L (1994) Activity of the Chlamydomonas chloroplast rbcL gene promoter is enhanced by a remote sequence element. Proc Natl Acad Sci USA 91: 10819–10823.
18. Blowers AD, Ellmore GA, Klein U, Bogorad L (1990) Transcriptional analysis of endogenous and foreign genes in chloroplast transformants of Chlamydomonas. Plant Cell 2: 1059–1070.
19. Salvador ML, Klein U, Bogorad L (1998) Endogenous fluctuations of DNA topology in the chloroplast of Chlamydomonas reinhardtii. Mol Cell Biol 18: 7235–7242.
20. Salvador ML, Klein U (1999) The redox state regulates RNA degradation in the chloroplast of Chlamydomonas reinhardtii. Plant Physiol 121: 1367–1374.
21. Kawazoe R, Mahan KM, Venghaus BE, Carter ML, Herrin DL (2012) Circadian regulation of chloroplast transcription in Chlamydomonas is accompanied by little or no fluctuation in RPOD levels or core RNAP activity. Mol Biol Rep 39: 10565–10571.
22. Mettler T, Mühlhaus T, Hemme D, Schöttler MA, Rupprecht J, et al. (2014) Systems analysis of the response of photosynthesis, metabolism and growth to an increase in irradiance in the photosynthesis model organism Chlamydomonas reinhardtii. Plant Cell 26: 2310–2350.
23. Czechowski T, Bari RP, Stitt M, Scheible W-R, Udvardi MK (2004) Real-time RT-PCR profiling of over 1400 Arabidopsis transcription factors: unprecedented sensitivity reveals novel root- and shoot-specific genes. Plant J 38: 366–379.
24. Erickson B, Stern DB, Higgs DC (2005) Microarray analysis confirms the specificity of a Chlamydomonas reinhardtii chloroplast RNA stability mutant. Plant Physiol 137: 534–544.
25. Ramundo S, Rahire M, Schaad O, Rochaix J-D (2013) Repression of essential chloroplast genes reveals new signaling pathways and regulatory feedback loops in Chlamydomonas. Plant Cell 25: 167–186.
26. Baldauf SL, Manhart JR, Palmer JD (1990) Different fates of the chloroplast tufA gene following its transfer to the nucleus in green algae. Proc Natl Acad Sci USA 87: 5317–5321.
27. Hwang S, Kawazoe R, Herrin DL (1996) Transcription of tufA and other chloroplast-encoded genes is controlled by a circadian clock in Chlamydomonas. Proc Natl Acad Sci USA 93: 996–1000.
28. Matsuo T, Onai K, Okamoto K, Minagawa J, Ishiura M (2006) Real-time monitoring of chloroplast gene expression by a luciferase reporter: Evidence for nuclear regulation of chloroplast circadian period. Mol Cell Biol 26: 863–870.
29. Matsuo T, Okamoto K, Onai K, Niwa Y, Shimogawara K, et al. (2008) A systematic forward genetic analysis identified components of the Chlamydomonas circadian system. Genes & Dev 22: 918–930.
30. Eberhard S, Drapier D, Wollman F-A (2002) Searching limiting steps in the expression of chloroplast-encoded proteins: relations between gene copy number, transcription, transcript abundance and translation rate in the chloroplast of Chlamydomonas reinhardtii. Plant J 31: 149–160.
31. Bräutigam K, Dietzel L, Pfannschmidt T (2007) Plastid-nucleus communication: anterograde and retrograde signalling in the development and function of plastids. Top Curr Genet 19: 409–455.
32. Berry JO, Yerramsetty P, Zielinski AM, Mure CM (2013) Photosynthetic gene expression in higher plants. Photosynth Res 117: 91–120.
33. Jarvis P, López-Juéz E (2013) Biogenesis and homeostasis of chloroplast and other plastids. Nat Rev Mol Cell Biol 14: 787–802.
34. Herrin DL, Michaels AS, Paul AL (1986) Regulation of genes encoding the large subunit of ribulose-1,5-bisphosphate carboxylase and the photosystem II polypeptides D-1 and D-2 during the cell cycle of Chlamydomonas reinhardtii. J Cell Biol 103: 1837–1845.
35. Leu S, White D, Michaels A (1990) Cell cycle-dependent transcriptional and post-transcriptional regulation of chloroplast gene expression in Chlamydomonas reinhardtii. Biochim Biophys Acta 1049: 311–317.
36. Salvador ML, Klein U, Bogorad L (1993) Light-regulated and endogenous fluctuations of chloroplast transcript levels in Chlamydomonas. Regulation by transcription and RNA degradation. Plant J 3: 213–219.
37. Harris EH (1989) The Chlamydomonas Sourcebook. Academic Press, San Diego, CA.
38. Saeed AI, Sharov V, White J, Li J, Liang W, et al. (2003) TM4: a free, open-source system for microarray data management and analysis. BioTechniques 34: 374–378.

14

Phylogenomic Reconstruction Indicates Mitochondrial Ancestor Was an Energy Parasite

Zhang Wang, Martin Wu*

Department of Biology, University of Virginia, Charlottesville, Virginia, United States of America

Abstract

Reconstruction of mitochondrial ancestor has great impact on our understanding of the origin of mitochondria. Previous studies have largely focused on reconstructing the last common ancestor of all contemporary mitochondria (proto-mitochondria), but not on the more informative pre-mitochondria (the last common ancestor of mitochondria and their alphaproteobacterial sister clade). Using a phylogenomic approach and leveraging on the increased taxonomic sampling of alphaproteobacterial and eukaryotic genomes, we reconstructed the metabolisms of both proto-mitochondria and pre-mitochondria. Our reconstruction depicts a more streamlined proto-mitochondrion than these predicted by previous studies, and revealed several novel insights into the mitochondria-derived eukaryotic metabolisms including the lipid metabolism. Most strikingly, pre-mitochondrion was predicted to possess a plastid/parasite type of ATP/ADP translocase that imports ATP from the host, which posits pre-mitochondrion as an energy parasite that directly contrasts with the current role of mitochondria as the cell's energy producer. In addition, pre-mitochondrion was predicted to encode a large number of flagellar genes and several cytochrome oxidases functioning under low oxygen level, strongly supporting the previous finding that the mitochondrial ancestor was likely motile and capable of oxidative phosphorylation under microoxic condition.

Editor: Frank Voncken, University of Hull, United Kingdom

Funding: This research was funded by an award from Thomas F. & Kate Miller Jeffress Memorial Trust to MW. The funder had no role in study design, data collection and analysis, decision to publish, or preparation of the manuscript.

Competing Interests: The authors have declared that no competing interests exist.

* Email: mw4yv@virginia.edu

Introduction

Mitochondria are eukaryotic organelles with a bacterial origin. Known as the endosymbiotic theory, it is now widely accepted that mitochondria originated once from an alphaproteobacterium probably two billion years ago [1]. However, it remains unclear what constituted the initial endosymbiosis between the alphaproteobacterium and its host [2,3,4]. Specifically, what was the role played by the mitochondrial ancestor that initiated the endosymbiosis? Were mitochondria originated under oxic, microoxic, or anoxic condition? Did mitochondria arise at the same time as, or subsequent to, the appearance of nucleus? What was the driving force behind the initial symbiosis [4,5]? Several hypotheses have been proposed to account for circumstances of the founding endosymbiotic events [4,6,7,8]. The "hydrogen hypothesis" proposes the metabolic syntrophy between a H_2-producing alphaproteobacterium and a H_2-dependent archaeon as the driving force behind the endosymbiosis [4]. This hypothesis allows the possibility of a simultaneous origination of mitochondrion and nucleus, with the same alphaproteobacterium also contributing to the rise of nucleus by fusing its genome with the host genome [3]. Therefore, the "hydrogen hypothesis" directly challenges the traditional serial endosymbiosis model in which the host is posited to be a full-fledged, nucleus-containing (but amitochondriate) eukaryote. In contrast, the "oxygen scavenger" hypothesis proposes that the removal of the toxic oxygen by the alphaproteobacterium from the anaerobic host has driven the initial symbiosis [5]. The circumstances under which the founding events occurred remain highly debated.

Reconstructing the gene complement of mitochondrial ancestor can shine light on the origin of mitochondria. It will help us better understand the timing of its signature reductive evolution. More importantly, reconstruction of its metabolism will help elucidate the driving force of the endosymbiosis by testing the alternative hypotheses. For example, whether the initial endosymbiotic condition was aerobic or anaerobic is a key yet debated point among different hypotheses [4,5]. The hydrogen hypothesis supports anaerobic syntrophy whereas the oxygen scavenger hypothesis supports aerobic mutualism. A recent study predicted the presence of both flagella and a cytochrome cbb3 oxidase in the mitochondrial ancestor and suggested that the mitochondrial ancestor was motile and capable of oxidative phosphorylation under microoxic condition [9]. However, this prediction was largely based on the analysis of one bacterial genome and needs to be further evaluated.

Although a key to understanding of the origin of mitochondria, mitochondrial ancestor reconstruction faces a multitude of problems. Firstly, there have been extreme genome reduction and dramatic metabolic turnover during mitochondria's transformation from a bacterium to an organelle. For example, mitochondrion of *Reclinomonas americana*, the least derived mitochondrion recognized so far, encodes only 67 proteins in its genome [10]. Vast majority of genes of the ancestor of mitochondria have been either lost or transferred to the nucleus [3], resulting in the highly reduced genomes of modern mitochondria that only encode proteins functioning in translation and ATP production. To reconstruct the mitochondrial ancestor, these lost genes need to be recovered.

Secondly, a robust phylogenetic framework required for ancestral reconstruction has been lacking. Although mitochondria have been firmly placed within alphaproteobacteria, their exact phylogenetic position is uncertain. Weak phylogenetic signal and serious systematic errors such as long-branch attraction and sequence composition bias have all hampered the effort to pinpoint the origin of mitochondria. Nevertheless, recent phylogenomic studies with increased genomic sampling have started to form a consensus by placing mitochondria in or near the *Rickettsiales* order [2,11,12,13,14].

Previous studies have reconstructed the last common ancestor (LCA) of all mitochondria [15,16]. Although insightful, results from these studies have serious limitations when it comes to understanding the origin of mitochondria. Figure 1 illustrates our point. Assuming the evolution of mitochondria is not reversible, point A represents the LCA of mitochondria and their alphaproteobacterial sister clade (for simplicity, hereafter referred as pre-mitochondria), while point B represents the LCA of all mitochondria (proto-mitochondria). Mitochondria emerged somewhere along the stem between points A and B, i.e., after they split off from alphaproteobacteria but before the divergence of the eukaryotic lineages (point C). All previous studies essentially reconstructed proto-mitochondria by simply pooling mitochondrial genes (including genes that have been transferred to the nucleus). Considering the dramatic transformation after the origin of mitochondria and the massive gene losses associated with this transformation, reconstructing proto-mitochondria would reveal little of what it looked like at the origin of mitochondria and therefore provide limited insights into the initial endosymbiosis event.

In order to understand what was happening at the origin of mitochondria, ideally we should reconstruct the ancestral state at time point C. It is difficult to delineate point C in the tree, however, because the origination event was not associated with any lineage diversification. If point B is an end point for studying the origin of mitochondria, then point A represents a good starting point. To gain better insights into the origin of mitochondria, it would be imperative for us to reconstruct pre-mitochondria at point A.

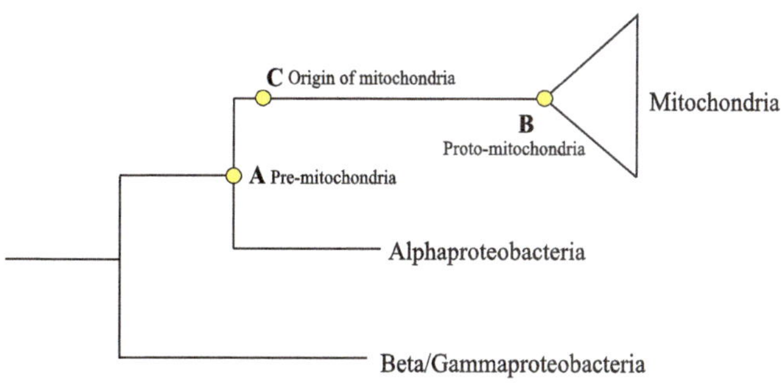

Figure 1. Key time points in mitochondria evolution. Time point A represents pre-mitochondria, the last common ancestor of mitochondria and alphaproteobacteria. Time point B represents proto-mitochondria, the last common ancestor of all contemporary mitochondria. Time point C represents the origin of mitochondria.

Results and Discussion

Identifying mitochondria-derived nuclear genes

Recovering mitochondrial genes that were lost to nucleus (hereafter referred as mitochondria-derived nuclear genes) is a prerequisite for mitochondrial ancestor reconstruction. Previous studies were based on a rather limited availability of bacterial and eukaryotic genomes at the time of their studies [15,16]. Leveraging on a substantially increased representation of eukaryotic and alphaproteobacterial genomes, we performed a phylogenomic analysis to systematically identify mitochondria-derived nuclear genes. Eukaryotic genes with a top mitochondria/alphaproteobacteria BLAST hit were first clustered into gene families. A phylogenetic tree was reconstructed for each family and nuclear genes that clustered with alphaproteobacteria in the trees were identified as mitochondria-derived. Starting with 427,186 genes from 30 eukaryotic genomes representing a broad range of phylogenetic diversity, we identified 4,459 genes belonging to 394 families as mitochondria-derived nuclear genes. To eliminate recent, lineage-specific gene transfers between alphaproteobacteria and eukaryotes (e.g., between *Trichoplax adhaerens* and its rickettsial endosymbiont [17]), gene families were required to be present in at least two alphaproteobacterial and two eukaryotic lineages.

We estimated the specificity and sensitivity of our method following the procedure of [16]. For the false positive rate, we used the phylum of *Deinococcus/Thermus*, which has no close relationship with eukaryotic lineages, as a benchmark. Of all the 427,186 eukaryotic sequences we screened, only 278 sequences were clustered with *Deinococcus/Thermus* in the gene trees (false positive rate 0.07%), indicating our procedure has a very high specificity. To estimate the false negative rate, we used *Reclinomonas americana*, the hitherto least derived mitochondrial genome as the positive control. Considering that some of the *R. americana* mitochondrial genes have diverged too far to reliably identify their homologs, we used a subset of 50 *R. americana* mitochondrial genes that were present in at least 2 other mitochondrial genomes. 46 out of 50 genes were recovered by our procedure (false negative rate 8%), indicating that our procedure is also very sensitive.

Reconstructing the metabolism of proto-mitochondria

The 394 mitochondria-derived nuclear genes are presumably all present in proto-mitochondria. To reconstruct the metabolism of proto-mitochondria, we assigned 394 families to Clusters of Orthologous Groups (COGs) [18] and mapped them onto the Kyoto Encyclopedia of Genes and Genomes (KEGG) pathways [19] (Figure 2, Table S1). Pathways including pyruvate metabolism, TCA cycle, electron transport and ribosomal biogenesis, as well as fatty acids biosynthesis, beta-oxidation, branched-chain amino acids degradation (Leucine, Valine, Isoleucine) and the biosynthesis of ubiquinone, biotin and one carbon unit pool, were almost completely recovered, all of which are functional in modern mitochondria. Conversely, functional categories such as DNA replication and transcription were largely absent in our reconstructed metabolism, and the heterotrophic carbohydrate metabolisms such as glycolysis and pentose phosphate pathway were entirely missing. Therefore, our reconstruction depicts streamlined proto-mitochondria highly similar to modern mitochondria, while Gabaldon et al. study suggests an ancestor with more diverse functions (Figure 3) [16].

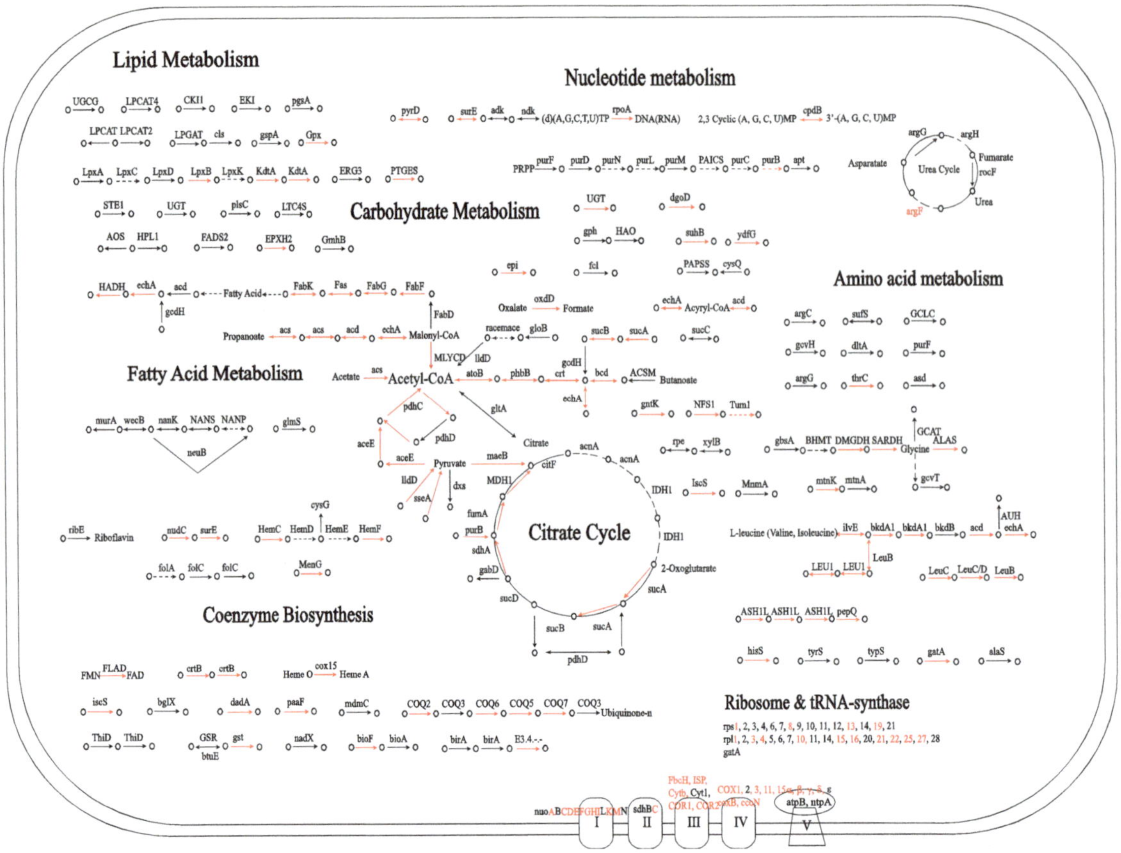

Figure 2. Reconstructed metabolism of proto-mitochondria. Black solid lines represent genes identified only in our reconstruction. Dotted lines represent missing genes in an otherwise complete pathway in our reconstruction. Red solid lines represent genes also present in Gabaldon et al. 2007.

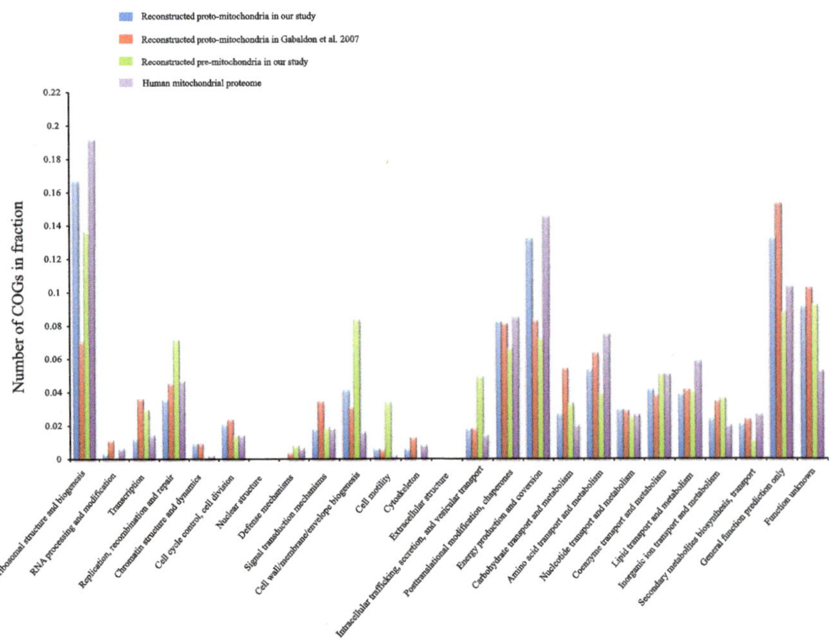

Figure 3. Function breakdown of reconstructed metabolisms of mitochondria ancestors. Within each class, from left to right are 1) the reconstructed proto-mitochondria in our study, 2) the reconstructed proto-mitochondria by Gabaldon et al. 2007, 3) the reconstructed pre-mitochondria in our study, 4) human mitochondrial proteome.

Compared to Gabaldon et al. study, our analysis is more sensitive. This is evident by the fact that despite reconstructing much leaner proto-mitochondria (our 394 vs. Gabaldon's 842 gene families), we were able to fill many of the gaps in well-characterized mitochondrial pathways that were left open in Gabaldon's reconstruction. For example, in the pyruvate metabolism and TCA cycle, we identified a pyruvate dehydrogenase E3 subunit (*lpdA/pdhD*, COG1249), a succinyl-CoA synthetase (*sucD*, COG0074) and a succinate dehydrogenase (*sdh2*, COG0479), which are essential for the pathways but were all missing in Gabaldon's reconstruction. Similarly for ATP synthesis, we added one F0F1-type ATP synthase subunit (*atpC*, COG0355), 3 NADH dehydrogenases (*nuoB*, COG0377; *nuoM*, COG1008; *nuoL*, COG1009) and four cytochrome c components (*fbcC*, COG2857; *cyb561*, COG3038; COG3474; *cycM*, COG5274), completing a functional electron transport chain. For the assembly of iron-sulfur cluster, we added *iscU* (COG0822) and *iscA* (COG0820), two critical scaffold proteins upon which the cluster is assembled and transferred. In terms of the translation machinery, we identified 19 additional ribosomal proteins, along with 8 translation factors (*IF-2*, COG0532; *EF-P*, COG0231 and 6 GTPase (COG0012, COG0206, COG0050, COG2262, COG0218, COG1159)) and 3 aminoacyl-tRNA synthetases (COG0124, COG0162, COG0180). Also, we added a *COQ3* (COG2227) enzyme involved in the ubiquinone biosynthesis, essentially recovering a fully functional *de novo* ubiquinone biosynthesis pathway. In the biotin metabolism, we added a biotin-protein ligase (*birA*, COG4285), the key enzyme that connects the biotin metabolism with fatty acid biosynthesis.

Our reconstruction also has higher specificity. Of the 156 gene families we identified that are present in human nuclear genome, 104 (66.7%) families have been identified in the human mitochondrial proteome, compared with 121 out of 355 (34.1%) families in Gabaldon et al. 2007. Similar results were also observed in yeast and Arabidopsis (Table 1). In addition, 56.1% (221 out of 394) of the families identified in our study contain at least one gene with a N-terminal mitochondria targeting signal, compared to 30.7% (258 out of 842) in Gabaldon et al. 2007. Since mitochondria-derived nuclear genes are often recruited back to mitochondria, the higher percentage of mitochondria-localized nuclear gene families in our reconstruction indicates higher specificity of our results.

The reasons for the increased sensitivity and specificity of our results could be at least twofold. First, our phylogenomic analysis used a much larger genome dataset representing a substantially broader range of taxon sampling (1,613 genomes in our study compared to 144 in Gabaldon's 2007 study, Table 1). In particular, many more genomes were included for the *Rickettsiales* order that has shown close relationship to mitochondria (67 in our study compared to 2 in Gabaldon's 2007 study, Table 1). The phylogenetic diversity of eukaryotic genomes was also greatly increased in our sampling, including 6 novel phyla that had not been sampled in previous studies (Figure S1) [15,16]. Having a broader taxon sampling improved the power of phylogenetic analysis, which enabled us to more reliably trace the evolutionary history of gene families. Second, in Gabaldon study, nuclear genes that clustered with beta- and gammaproteobacteria were also identified as mitochondria-derived nuclear genes. The rationale is to increase the recovery rate of the *R. americana* mitochondrial genes as most of the ribosomal proteins were found to cluster with beta- and gammaproteobacteria in their phylogenetic analyses [16]. However, with increased alphaproteobacterial sampling we only identified two *R. americana* ribosomal proteins with such spurious patterns. Therefore, we think the criterion used in Gabaldon study is unnecessarily relaxed and could increase the false positive rate, leading to an overestimate of the number of mitochondria-derived nuclear genes. For example, genes involved in the pentose phosphate pathway all clustered with gammaproteobacteria or a mixture of gamma- and alphaproteobacteria and were identified as mitochondria-derived in Gabaldon study [16].

Novel insights into mitochondria-derived eukaryotic metabolisms

Our reconstruction provides several novel insights into mitochondria-derived eukaryotic metabolisms. Of particular interest are a number of genes involved in the eukaryotic lipid metabolism (Table 2). For example, we identified 3 enzymes involved in steroid biosynthesis, including a squalene/phytoene synthase (COG1562), a sterol-C5-desaturase (COG3000) and a 1-deoxy-D-xylulose-5-phosphate synthase (COG1154) as mitochondria-derived, suggesting that mitochondrial ancestor has contributed to eukaryotic steroid biosynthesis. Mitochondrial origin of these enzymes is supported by functional studies. Mitochondria are known to play an essential role in the biosynthesis of steroids by providing sites for the onset of the process [20]. In return, steroids are also critical in maintaining the mitochondrial morphology [21]. Indeed, studies in *C. neoformans* and *T. brucei* indicated that mutants of squalene synthase and sterol desaturase were defective in mitochondrial membrane integrity [22,23].

In addition, we identified a ceramide glycosyltransferase (COG1215) involved in the glycosphingolipids biosynthesis, carrying out ceramides glycosylation reactions. Interestingly, this enzyme is located at the "mitochondria-associated membrane", a specific ER subdomain that bridges between ER and mitochondria [24]. Both glycosphingolipids and ceramides are ubiquitously present as essential membrane components in almost all eukary-

Table 1. Comparison between proto-mitochondrial reconstructions of this study and Gabaldon et al. 2007.

		This study	Gabaldon et al. 2007
Number of genomes	Total	1613	144
	Alphaproteobacteria/*Rickettsiales*	171/67	11/2
	Eukaryotes	30	16
Nuclear gene families in mitochondria proteome	Human	66.7%	34.1%
	Yeast	69.5%	46.8%
	Arabidopsis	63.3%	42.6%
Gene families with mitochondrial-targeted signal		56.1%	30.7%

Table 2. List of mitochondria-derived nuclear genes involved in eukaryotic lipid metabolism.

Gene family	COG	Description	Lipid metabolized	Cellular localization	Identified in Gabaldon et al. 2007
Group_236	COG2867	cyclase/dehydrase	Oligoketide	Unknown	N
Group_267	COG1215	ceramide glucosyltransferase	Sphingolipid	ER (MAM)	Y
Group_268	COG1562	squalene/phytoene synthase	Cholesterol	ER	Y
Group_946	COG1154	1-deoxy-D-xylulose-5-phosphate synthase	Terpernoid	ER	Y
Group_1713	COG3660	hypothetical protein	Likely lipid A	Mitochondria	Y
Group_1971	COG5597	sqdD glycosyl transferase	Sulfolipid	Unknown	N
Group_2416	COG1044	lpxD UDP-3-O- 3-hydroxymyristoyl glucosamine N-acyltransferase	Lipid A	Mitochondria	N
Group_2710	COG4689	acetoacetate decarboxylase	Ketone body	Unknown	N
Group_3620	COG1043	lpxA UDP-N-acetylglucosamine acyltransferase	Lipid A	Mitochondria	N
Group_3864	COG3000	Sterol C5 desaturase	Sterol	ER	N
Group_4604	COG1519	kdtA 3-deoxy-D-manno-octulosonic-acid transferase	Lipid A	Mitochondria	Y
Group_4794	COG0763	lpxB lipid-A-disaccharide synthase	Lipid A	Mitochondria	N

otic cells and mitochondria, but are rarely identified in bacteria. Accordingly, the substrates and glycolipid products of the bacterial and eukaryotic glycosyltransferases were thought to be very different [25]. Therefore, the bacterial origin of this gene indicates acquisition of novel function by eukaryotes for synthesizing its own endomembranes and for the crosstalk and lipid trafficking between mitochondria and ER.

Our results also shed light on the mitochondria's contribution to other eukaryotic metabolisms. For instance, we identified several genes (*purD*, COG0151; *purM*, COG0150; *mutT*, COG1051; *pyrD*, COG0167) involved in the *de novo* nucleotide biosynthesis as mitochondria-derived. Both *purD* and *purM* belong to the family of glycinamide ribonucleotide transformylase and catalyze different steps in the *de novo* purine biosynthesis. Mitochondria also contribute to the cytosolic purine biosynthesis by providing formate as the one-carbon unit. Consistently, the entire formate biosynthesis pathway was identified as mitochondria-derived in our results. On the other hand, *pyrD* is a mitochondria-localized protein critical for the pyrimidine biosynthesis [26]. *purD* and *purM* have been previously identified as of mitochondrial origin [13] but were missing in the results of Gabaldon et al. 2007. We also identified a number of genes (*glmS*, COG0449; *wecB*, COG1940; *neuB*, COG2089; *murA*, COG0766) involved in UDP-sugar biosynthesis. UDP-sugar provides essential modifications to various target proteins such as nuclear pore proteins and cytoskeleton components [27]. Thus it is tempting to speculate that these mitochondria-derived genes might participate in controlling the activities of these eukaryote-specific complexes.

Reconstructing the metabolism of pre-mitochondria

To gain insights into the circumstances that surrounded the initial endosymbiosis event, it is imperative to reconstruct the metabolism of pre-mitochondria. To increase the power of ancestral reconstruction, we sequenced 5 novel linages of *Rickettsiales* (endosymbiont of acanthamoeba UWC8, *Candidatus* Caedibacter acanthamoebae, *Candidatus* Paracaedibacter acanthamoebae, *Candidatus* Paracaedibacter symbiosus, NHP bacterium) [28,29]. These newly sequenced *Rickettsiales* species, together with 44 other alphaproteobacteria and 6 eukaryote representatives were used to reconstruct a Bayesian tree based on the concatenated protein sequence alignment of 55 mitochondrial and nuclear marker genes (Figure 4, hereafter referred as species tree). Mitochondria were placed within the *Rickettsiales* order, as a sister clade to the *Rickettsiaceae*, *Anaplasmataceae* and *Candidatus* Midichloriaceae families, all subtended by the *Holosporaceae* family. We note that the 5 newly sequenced *Rickettsiales* are all closely related to mitochondria. In particular, the 4 *Holosporaceae* species, together with previously sequenced *Candidatus* Odyssella thessalonicensis form an outgroup to the LCA of mitochondria and the other *Rickettsiales*, and therefore are extremely informative for reconstructing the ancestral state of pre-mitochondria.

Using tree in Figure 4 and a Bayesian character mapping method, we predicted the gene complement of pre-mitochondria to contain 887 COGs (Table S1). Based on the approximate linear relationships between the number of gene families, the number of genes and the genome size (Figure S2), we estimated the pre-mitochondrial genome to be 1.5–1.6 Mb, with 1100–1300 genes. This is typical of an obligate intracellular bacterium and suggests that genome reduction was well under way before mitochondria split from alphaproteobacteria, i.e., the origin of mitochondria. Figure 5 shows the reconstructed metabolism of pre-mitochondria. Compared to the highly specialized proto-mitochondria, pre-mitochondria were capable of much more diversified metabolisms. In addition to the major pathways involved in translation (13.6%), cell wall, LPS and membrane biogenesis (8.3%), energy production (7.2%), and DNA replication, recombination and repair (7.1%) (Figure 3), they were predicted to possess multiple key metabolic pathways including glycolysis, TCA cycle, pentose phosphate pathway, and fatty acid biosynthesis pathway. Also, pre-mitochondria possessed a large number of genes involved in synthesizing various cofactors, such as riboflavin, folate, biotin and ubiquinone. On the other hand, similar to most *Rickettsiales*, pre-mitochondria possessed a limited number of genes for synthesizing certain amino acids (Glutamine, Leucine, Valine and Isoleucine) from metabolic intermediates, but were deficient in *de novo* amino acid biosynthesis. Therefore pre-mitochondria had to obtain essential amino acids from the host. Accordingly, at least 5 amino acid transporters were predicted in pre-mitochondria.

Pre-mitochondria were predicted to lack most of the genes involved in the *de novo* nucleotide biosynthesis pathway, except for a few genes such as *purD* and *pyrD* that were also present in proto-mitochondria. Among *Rickettsiales*, the *de novo* nucleotide

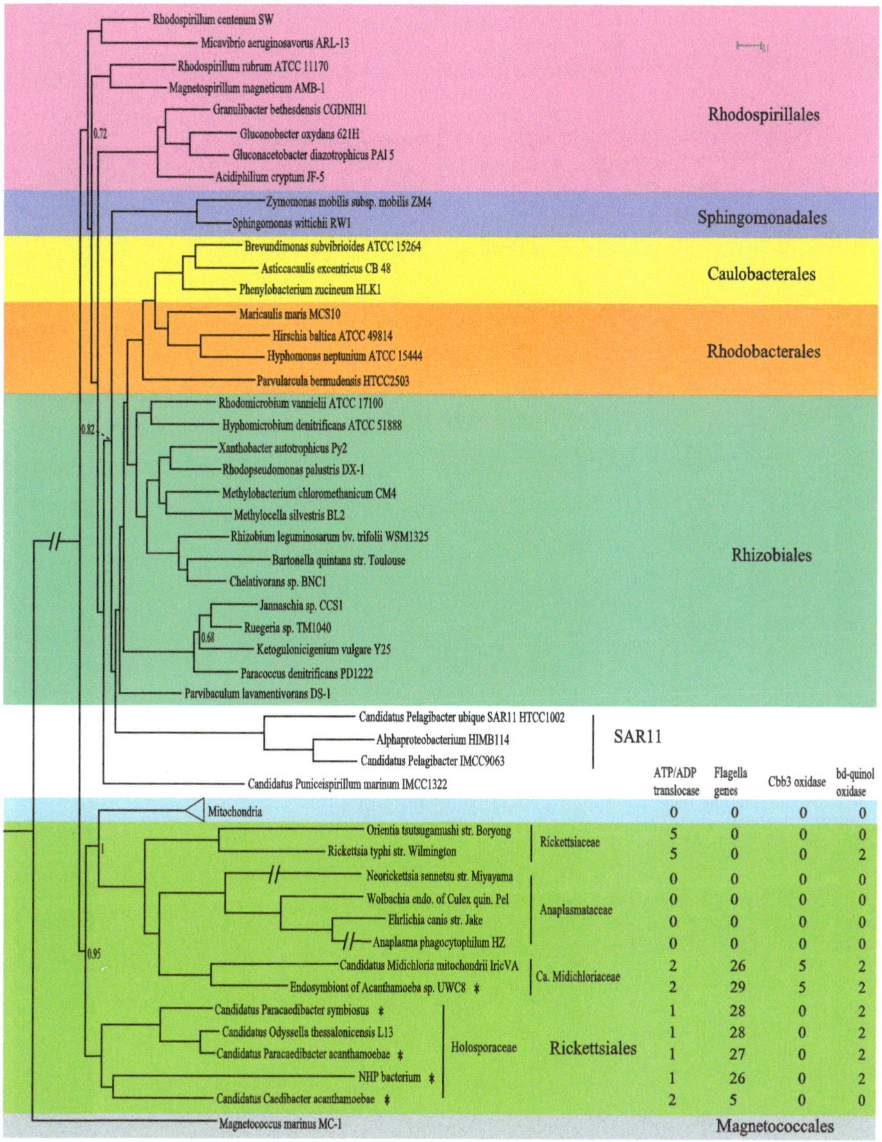

Figure 4. A species tree of alphaproteobacteria and mitochondria. The Bayesian consensus tree was based on concatenated protein sequences of 55 mitochondrial and nuclear genes. The tree was rooted using 8 beta- and gammaproteobacteria as the outgroup, as described in the methods. Asterisks indicate 5 newly sequenced *Rickettsiales* species. The posterior probability support values of the internal nodes are greater than 0.9 unless as indicated in the tree. Branches of several lineages are shortened for display purpose. The distribution of ATP/ADP translocase, flagella genes, cbb3 cytochrome oxidase and bd-quinol cytochrome oxidase in *Rickettsiales* and mitochondria is displayed besides the lineages.

biosynthesis pathway is present in the family *Anaplasmataceae* but absent in all other lineages. Hence one interpretation is that the *de novo* nucleotide biosynthesis pathway was gained in *Anaplasmataceae*, as suggested by the Bayesian reconstruction method. Because of their isolated lifestyles and the lack of selection pressures for diversifying due to relatively constant intracellular environment, obligate intracellular bacteria such as *Anaplasmataceae* are thought to have limited opportunities for gene acquisition [30]. Therefore, we think acquiring the entire *de novo* nucleotide biosynthesis pathway, including 12 purine and 6 pyrimidine biosynthesis genes, is extremely unlikely in *Anaplasmataceae*. Instead, we think it is more likely that the nucleotide biosynthesis pathway was present in the ancestor of *Rickettsiales* and mitochondria and was subsequently lost multiple times in both *Rickettsiales* (except for *Anaplasmataceae*) and mitochondria.

Pre-mitochondrion possessed flagella

A recent study has suggested that the mitochondrial ancestor possessed a flagellum [9]. However, this prediction was based on the presence of 26 flagellar genes in only one *Rickettsiales* species, *Candidatus* Midichloria mitochondrii. Interestingly, most of these 26 flagellar genes were found in 4 out of the 5 *Rickettsiales* endosymbionts we sequenced and also in the recently sequenced endosymbiont *Candidatus* Odyssella thessalonicensis (Figure 4). The phylogenetic tree of the concatenated flagellar genes is congruent with the *Rickettsiales* species tree, indicating that flagellar genes have been inherited vertically in *Rickettsiales* species. Consistently, they form syntenic gene clusters (Figure 6). It is therefore not surprising that pre-mitochondria were predicted to possess 25 COGs encoding the core flagellar components including the basal body, motor, hook, rod, filament and export

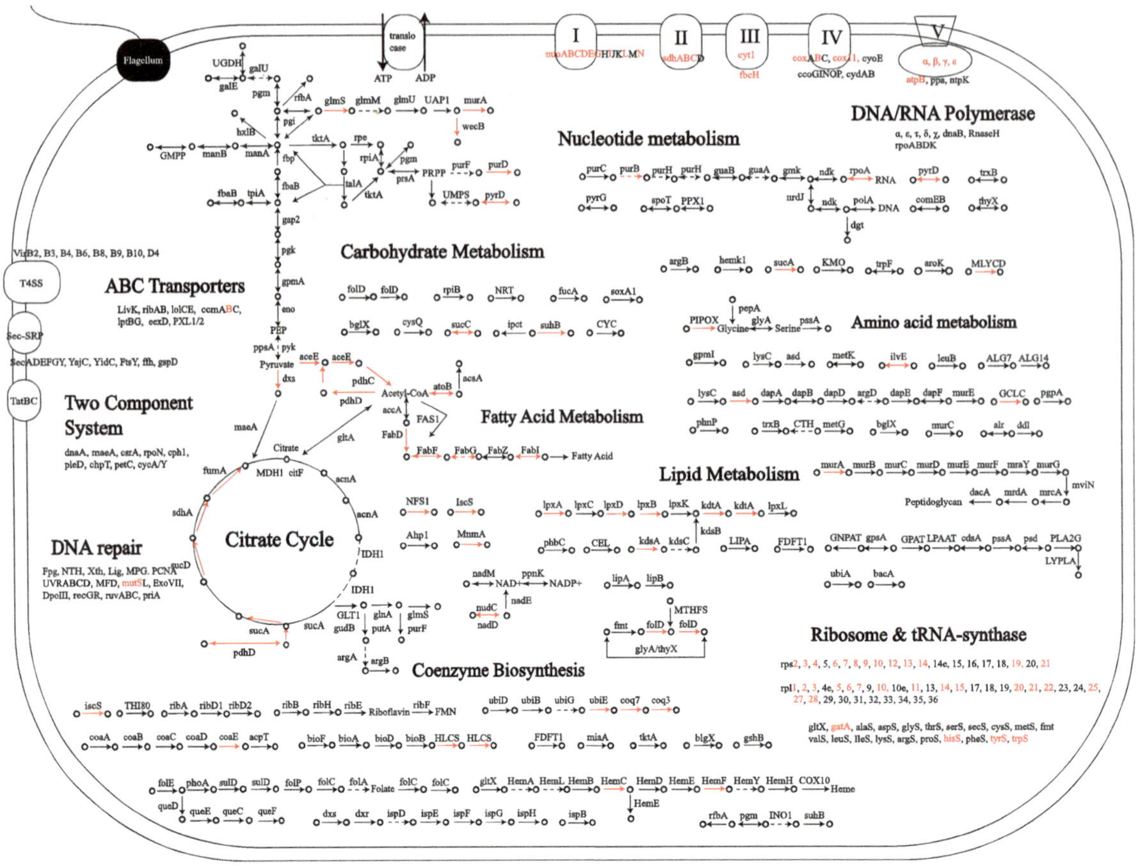

Figure 5. Reconstructed metabolism of pre-mitochondria. Black solid lines represent genes identified in our reconstruction while dotted lines represent missing genes in an otherwise complete pathway. Red lines represent genes present in proto-mitochondria in Figure 2.

apparatus [31]. Electron microscopy of the NHP bacterium has shown flagella at the basal end of its cell [32]. One recent study also observed flagella in two endosymbionts of Paramecium belonging to the *Lyticum* genus of the *Midichloriaceae* family [33]. Take together, these results strongly support the presence of flagella in pre-mitochondria.

Pre-mitochondrion was capable of respiration at low oxygen condition

A key yet debated point among different hypotheses is whether the initial endosymbiotic condition was aerobic or anaerobic [4,5]. Interestingly, it has been suggested that pre-mitochondria could respire at low oxygen condition because it possessed a cbb3-type cytochrome oxidase that mainly function under the microoxic condition [9]. Like the flagellar genes, this prediction was largely based on the presence of cbb3-type cytochrome oxidase in *Candidatus* Midichloria mitochondrii alone. Of the 5 *Rickettsiales* endosymbionts we sequenced, cbb3 oxidases were found only in the endosymbiont of acanthamoeba UWC8, the sister clade of *Candidatus* Midichloria mitochondrii. Although our reconstruction predicted the presence of cbb3 oxidase in the LCA of mitochondria and *Rickettsiales* (COG2010, COG2993, COG3278, COG0348, COG2217), the posterior probabilities were only 0.56. Therefore, we cannot exclude the alternative interpretation that cbb3 oxidases were absent in pre-mitochondria and were only gained in the *Candidatus Midichloriaceae* lineage. Interestingly, however, we also predicted the presence of a cytochrome bd-type quinol oxidase (COG1271, COG1294) in pre-mitochondria. The bd-type oxidases are widely distributed in the *Holosporaceae* and *Ca.* Midichloriaceae families (Figure 4), strongly suggesting that they were present in pre-mitochondria (posterior probability 0.89) and have been lost in mitochondria and other *Rickettsiales* lineages. Like cbb3-type oxidases, the bd-type oxidases are functional under limited oxygen level. Our study therefore provides strong and additional support to the hypothesis that pre-mitochondria were capable of oxidative phosphorylation under low oxygen condition.

Pre-mitochondrion was an energy parasite

Intriguingly, pre-mitochondria were predicted to encode the plastid/parasite type of ATP/ADP translocase (COG3202; posterior probability 0.93), a hallmark protein of many obligate intracellular bacteria that is used to import ATP from the host, although remote homologs of this gene belonging to the major facilitator superfamily (MFS) transporters are also present in free-living bacteria [e.g., 34]. The ATP/ADP translocase commonly functions as an ATP/ADP antiporter that exchanges bacterial ADP for the host cell ATP [35]. In addition, it has been shown that some intracellular bacteria, including Chlamydia and Rickettsia, encode additional isoforms of this protein for the uptake of nucleotides to compensate for their inability to synthesize nucleotides *de novo* [36,37]. Consistently, this gene family is absent in the *Anaplasmataceae* family of *Rickettsiales*, members of which all possess complete *de novo* nucleotide biosynthesis pathway [13,38,39]. Previous studies have suggested that there were ancient lateral gene transfers of this gene between

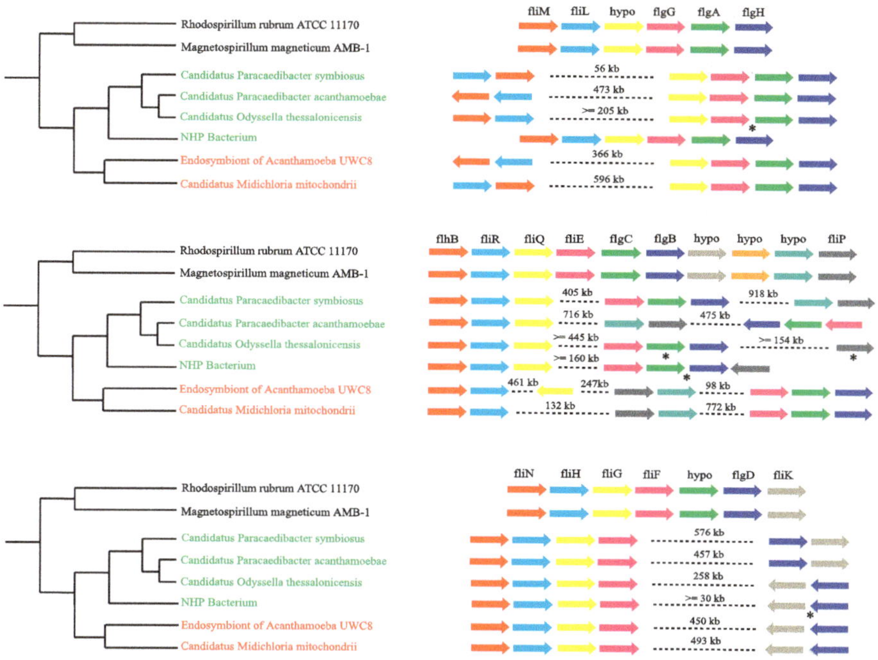

Figure 6. The synteny of flagellar genes. *Holosporaceae, Candidatus* Midichloriaceae, and free-living alphaproteobacteria representatives are highlighted in green, red and black respectively. Each arrow represents a gene in the cluster. Genome rearrangements are shown as dotted lines between two genes, with the distance between them shown above the lines. Because of the incomplete nature of some genome assemblies, the exact distance between two genes could not be determined. In this case, a minimum distance was estimated as the sum of distances of each gene to the end of the contig it was located in. For the same reason, the orientation of some genes could not be determined (indicated by asterisks below the genes).

the ancestors of *Chlamydiales*, *Rickettsiales* and plastids [35,40] followed by largely vertical inheritance [41]. In consistence, our phylogenetic analysis shows that the gene tree is largely congruent with the species tree of the *Rickettsiales* order (Figure S3), suggesting that this gene has been vertically inherited in *Rickettsiales* and thus was most likely present in their last common ancestor and by inference, pre-mitochondria. If the nucleotide biosynthesis pathway was also present in pre-mitochondria as we have predicted, then this gene most likely functioned as an ATP/ADP exchanger instead of a nucleotide transporter in pre-mitochondria.

Remarkably, the plastid/parasite ATP/ADP translocase is evolutionarily unrelated to and functionally distinct from the ATP/ADP translocase in modern mitochondria, which exhibits an opposite polarity by exporting ATP into the host cytosol [42,43,44]. Therefore, our reconstruction posits pre-mitochondrion as an "energy scavenger" and suggests an energy parasitism between the endosymbiont and its host at the origin of mitochondria, as first proposed by Andersson et al. [5,45]. This is in sharp contrast with the current role of mitochondria as the cell's energy producer and contradicts the traditional endosymbiotic theory that the symbiosis was driven by the symbiont supplying the host ATP [46,47]. The replacement of plastid/parasite ATP/ADP translocase by mitochondrial ATP/ADP translocase occurred subsequently, resulting in a reverse flow of ATP between mitochondrion and its host. This remarkable transformation in energy metabolism might mark the transition of mitochondria from a parasitic endosymbiont to a mutualistic organelle [48].

A recent systematic survey of bacterial symbiosis has shown that mutualisms can originate either directly from environmental free-living bacteria or from intracellular parasites [49]. A key difference between these two evolutionary routes is that to initiate symbiosis, free-living bacteria need to offer immediate benefits to the host while intracellular parasitic bacteria do not [50]. Our results suggest that mitochondria most likely originated from an obligate intracellular parasite and not from a free-living bacterium. This has important implications for our understanding of the origin of mitochondria. It implies that at the beginning of the endosymbiosis, the bacterial symbiont provided no benefits whatsoever to the host. Therefore we argue that the benefits proposed by various hypotheses (e.g., oxygen scavenger hypothesis and hydrogen hypothesis) are irrelevant in explaining the establishment of the initial symbiosis. Instead, it might be more appropriate to apply them to explain the transition of mitochondria from a parasite to a mutualistic organelle at a later stage.

Materials and Methods

Identification of mitochondria-derived nuclear genes

The phylogenetic distribution of all sequenced eukaryotic genomes was retrieved from GenomeOnline database (GOLD, http://genomesonline.org/). 30 eukaryotic genomes representing a broad range of eukaryotic diversity (10 phyla) were selected for identifying the mitochondria-derived nuclear genes (*Allomyce macrogynus, Arabidopsis thaliana, Batrachochytrium dedrobatidis, Caenorhabditis elegans, Chlamydomonas reinhardtii, Cryptococcus neoformans, Dictyostelium discoideum, Drosophila melanogaster, Encephalitozoon intestinalis, Entamoeba histolytica, Enterocytozoon bieneusi, Giardia lamblia, Homo sapiens, Leishmania major, Micromonas pusilla, Monosiga brevicollis, Naegleria gruberi, Nectria haematococca, Nematostella vectensis, Nosema ceranae, Phytophthora infestans, Plasmodium falciparum, Saccharomyce cerevisiae, Schistosoma mansoni, Spizellomyces punctatus, Strongy-*

locentrotus purpuratus, Tetrahymena thermophila, Thalassiosira pseudonana, Trichoplax adhaerens, Trypanosoma brucei).

For every single gene of 30 eukaryotic nuclear genomes, an initial BLASTP search was performed against 2,742 complete bacterial, archaeal, and mitochondrial genomes. A eukaryotic gene was retained for further phylogenetic analysis if its top five hits contained an alphaproteobacterial or mitochondrial sequence (e-value cutoff 1e-4). All eukaryotic genes passing the initial BLAST search were clustered into families using the Markov Cluster Algorithm [51]. Families that were present in at least two eukaryotic lineages were selected for phylogenetic analysis. For each retained protein family, its homologs from all complete bacterial genomes were retrieved by BLASTP search (e-value cutoff 1e-15). Protein sequences were aligned using MAFFT [52] and trimmed using ZORRO [53]. Phylogenetic trees were constructed using FastTree 2 [54]. When possible, each individual tree was rooted using three different rooting methods, rooting with Archaea or *Deinococcus* as the outgroup or midpoint rooting. Each of the rooted trees was scanned for a bipartition where eukaryotic genes clustered with alphaproteobacterial or mitochondrial homologs. A partition was retained as one gene family if it contained at least two eukaryotes and two alphaproteobacterial species. Paralogs, if existed in a family, were separated and each was treated as a new family of mitochondria-derived nuclear genes.

Functional annotation of mitochondria-derived nuclear genes

Mitochondria-derived nuclear genes were classified into Clusters of Orthologous Groups (COGs) by hidden Markov model search using HMMer3 [55]. To reconstruct metabolic pathways, genes were mapped onto Kyoto Encyclopedia of Genes and Genomes (KEGG) database using KEGG Automatic Annotation Server (KAAS) [56] with "bi-directional best hit (BBH)" as the assignment method.

Genome sequencing, assembly and annotation

4 endosymbionts of acanthamoebae (endosymbiont of acanthamoeba UWC8, *Candidatus* Caedibacter acanthamoebae, *Candidatus* Paracaedibacter acanthamoebae, *Candidatus* Paracaedibacter symbiosus) and one pathogen of shrimp NHP bacterium were sequenced by a combination of 454, Illumina and Pacific Biosciences SMRT sequencing. The genomes were assembled using Newbler 2.5.3, CLCGenomicWorkbench 6.0.1 or Celera assembler 8.2 and closed manually by resolving repeats using paired-end information and comparing assemblies using Mauve [57]. PCR and Sanger sequencing were used to close gaps between contigs when necessary. The genomes have been deposited at DDBJ/EMBL/GenBank as follows: endosymbiont of acanthamoeba UWC8 (CP004403), *Candidatus* Caedibacter acanthamoebae (CP008936-CP008940), *Candidatus* Paracaedibacter acanthamoebae (CP008941-CP008942), *Candidatus* Paracaedibacter symbiosus (JQAK00000000) and NHP bacterium (JQAJ00000000). Protein-coding genes were predicted using the GLIMMER software package [58].

Alphaproteobacteria and mitochondria species tree

We selected a set of 49 alphaproteobacterial representatives using a tree-based greedy algorithm to maximize their phylogenetic diversity [59]. A set of 6 eukaryotic lineages (*Cryptococcus neoformans, Arabidopsis thaliana, Nematostella vectensis, Spizellomyces punctatus, Monosiga brevicollis, Phytophthora infestans*) were selected as eukaryotic representatives. Phyla *Alveolata, Amoebozoa, Euglenozoa* and *Diplomonadida* were not included because they had extremely long branches and therefore were prone to long branch attraction (LBA).

We selected a set of 26 mitochondria-encoded genes (*cob, cox2, cox3, nad1, nad2, nad3, nad4, nad4L, nad5, nad6, nad9, rpl2, rpl5, rpl6, rpl16, rps1, rps2, rps3, rps4, rps7, rps8, rps11, rps12, rps13, rps14, rps19*) and 29 mitochondria-derived nuclear genes (*cox11, sdhB, sucD, petA, erpA, hesB, ybjS, nuoC, nuoD, nuoF, nuoG, nuoI, rpl3, grpE, groEL, dnaK, clpB, clpP, hslV, engA, gidA, trmE, AFG1, apaG, bioC, hemN, ksgA, mraW, hypothetical*) as the marker genes for reconstructing the species tree. The list of mitochondria-encoded genes was taken from [13]. Three genes (*atp6, atp9, atpA*) were excluded from the original list because of their potential involvement in lateral gene transfer [60]. Three other genes (*cox1, yejR, yejU*) were excluded because of the potential paralogy problem. The nuclear genes were selected because they were single-copy genes and were distributed in at least two eukaryotic phyla. Protein sequences of 55 mitochondrial and nuclear marker genes from selected genomes were identified, aligned, trimmed and concatenated using AMPHORA2 [61]. From the concatenated protein sequence alignment, a Bayesian tree was reconstructed using PhyloBayes [62] with the -CAT -GTR options as recommended in the manual. Two independent MCMC chains were run and the chains were considered converged when the maxdiff dropped below 0.3, as suggested in the manual. The trees were sampled every 10 cycles and the beginning one fifth of the trees from each chain were discarded as burn-in. The tree was rooted using 8 beta- and gamma-proteobacteria as outgroup (*Nitrosomonas sp. Is79A3, Ralstonia solanacearum GMI1000, Dechloromonas aromatica RCB, Chromobacterium violaceum ATCC 12472, Francisella tularensis subsp. tularensis FSC198, Legionella pneumophila str. Lens, Escherichia coli str. K-12 substr. MG1655* and *Pseudomonas aeruginosa PA7*). The outgroup species were taken from a previous study [14]. We excluded *Buchnera* because it had a long branch and might cause the LBA artifact.

Pre-mitochondria ancestral state reconstruction

Using the species tree, pre-mitochondria were reconstructed with a Bayesian character mapping inference algorithm implemented in BayesTraits V2 [63]. The Bayesian method has been shown to be superior to both the parsimony method and maximum likelihood method in accounting for uncertainties in both model parameters and phylogeny [63,64]. The mitochondrial genes were compiled by combining the mitochondria-derived nuclear genes with the mitochondria-encoded genes. Mitochondrial and 148,485 genes of 49 alphaproteobacterial genomes were first assigned to 4,873 COGs by hidden Markov model search using HMMer3 [55]. Genes that could not be assigned to a COG were then clustered into families using the Markov Cluster Algorithm [51], creating 3,210 "expanded COGs". The presence/absence of each COG in each species was treated as a binary trait and was used for the ancestral state reconstruction. Gamma distribution was adopted as the prior distribution with its parameter estimated from an initial maximum likelihood analysis. The "hyperprior" option was used to reduce the uncertainty in choosing priors in the MCMC. A total number of 1,050,000 iterations were performed, with the first 50,000 cycles discarded as burn-in. The average value of each binary state in the remaining 1,000,000 cycles was then taken as the probability of the presence of each COG in the reconstructed ancestral state.

Supporting Information

Figure S1 Overview of mitochondria-derived nuclear genes. The eukaryotic species tree was reconstructed based on the concatenated 29 ribosomal proteins conserved among all three domains. Within each species, from innermost to outermost are the numbers of 1) nuclear genes, 2) mitochondrial genes, 3) mitochondria-derived nuclear genes, 4) mitochondria-derived nuclear gene families. The heights of the bars were scaled for display purposes. Each color in the tree represents a different eukaryotic phylum. 6 novel phyla that had not been sampled by previous studies were indicated by asterisks. Lineages highlighted in red represent amitochondriate eukaryotes. Branches with dots represent the ones with bootstrap support $>=80$ (100 replicates).

Figure S2 Correlation a) between the number of COGs and the number of genes, and b) between the number of COGs and the genome size of 49 alphaproteobacteria representatives. In both graphs, the red dots represent pre-mitochondria.

Figure S3 A maximum-likelihood tree inferred from amino acid sequences of the ATP/ADP translocase in *Chlamydiales* (blue), *Rickettsiales* (orange), *Bacteroidetes* (yellow) and plastids (green). The tree was rooted by homologs in Microsporidia (*Encephalitozoon intestinalis, Encephalitozoon cuniculi, Enterocytozoon bieneusi* and *Nosema ceranae*). Branches of several lineages are shortened for display purpose. Bootstrap values (out of 100 replicates) are above 80 unless as indicated in the tree.

Author Contributions

Conceived and designed the experiments: MW. Performed the experiments: ZW. Analyzed the data: ZW MW. Contributed to the writing of the manuscript: ZW MW.

References

1. Lang BF, Gray MW, Burger G (1999) Mitochondrial genome evolution and the origin of eukaryotes. Annu Rev Genet 33: 351–397.
2. Andersson SG, Zomorodipour A, Andersson JO, Sicheritz-Ponten T, Alsmark UC, et al. (1998) The genome sequence of *Rickettsia prowazekii* and the origin of mitochondria. Nature 396: 133–140.
3. Gray MW, Burger G, Lang BF (1999) Mitochondrial evolution. Science 283: 1476–1481.
4. Martin W, Muller M (1998) The hydrogen hypothesis for the first eukaryote. Nature 392: 37–41.
5. Andersson SG, Karlberg O, Canback B, Kurland CG (2003) On the origin of mitochondria: a genomics perspective. Philos Trans R Soc Lond B Biol Sci 358: 165–177; discussion 177–169.
6. Embley TM, Martin W (2006) Eukaryotic evolution, changes and challenges. Nature 440: 623–630.
7. Koonin EV (2010) The origin and early evolution of eukaryotes in the light of phylogenomics. Genome Biol 11: 209.
8. Andersson SG, Karlberg O, Canback B, Kurland CG (2003) On the origin of mitochondria: a genomics perspective. Philosophical Transactions of the Royal Society of London Series B, Biological Sciences 358: 165–177; discussion 177–169.
9. Sassera D, Lo N, Epis S, D'Auria G, Montagna M, et al. (2011) Phylogenomic evidence for the presence of a flagellum and cbb(3) oxidase in the free-living mitochondrial ancestor. Mol Biol Evol 28: 3285–3296.
10. Lang BF, Burger G, O'Kelly CJ, Cedergren R, Golding GB, et al. (1997) An ancestral mitochondrial DNA resembling a eubacterial genome in miniature. Nature 387: 493–497.
11. Fitzpatrick DA, Creevey CJ, McInerney JO (2006) Genome phylogenies indicate a meaningful alpha-proteobacterial phylogeny and support a grouping of the mitochondria with the Rickettsiales. Mol Biol Evol 23: 74–85.
12. Williams KP, Sobral BW, Dickerman AW (2007) A robust species tree for the alphaproteobacteria. J Bacteriol 189: 4578–4586.
13. Wu M, Sun LV, Vamathevan J, Riegler M, Deboy R, et al. (2004) Phylogenomics of the Reproductive Parasite Wolbachia pipientis wMel: A Streamlined Genome Overrun by Mobile Genetic Elements. PLoS Biol 2: E69.
14. Rodriguez-Ezpeleta N, Embley TM (2012) The SAR11 group of alpha-proteobacteria is not related to the origin of mitochondria. PLoS ONE 7: e30520.
15. Gabaldon T, Huynen MA (2003) Reconstruction of the proto-mitochondrial metabolism. Science 301: 609.
16. Gabaldon T, Huynen MA (2007) From endosymbiont to host-controlled organelle: the hijacking of mitochondrial protein synthesis and metabolism. PLoS Comput Biol 3: e219.
17. Driscoll T, Gillespie JJ, Nordberg EK, Azad AF, Sobral BW (2013) Bacterial DNA sifted from the *Trichoplax adhaerens* (Animalia: Placozoa) genome project reveals a putative rickettsial endosymbiont. Genome Biol Evol 5: 621–645.
18. Tatusov RL, Galperin MY, Natale DA, Koonin EV (2000) The COG database: a tool for genome-scale analysis of protein functions and evolution. Nucleic Acids Res 28: 33–36.
19. Kanehisa M, Goto S (2000) KEGG: kyoto encyclopedia of genes and genomes. Nucleic Acids Res 28: 27–30.
20. Duarte A, Poderoso C, Cooke M, Soria G, Maciel FC, et al. (2012) Mitochondrial Fusion Is Essential for Steroid Biosynthesis. Plos One 7.
21. Prince FP, Buttle KF (2004) Mitochondrial structure in steroid-producing cells: three-dimensional reconstruction of human Leydig cell mitochondria by electron microscopic tomography. Anat Rec A Discov Mol Cell Evol Biol 278: 454–461.
22. Ingavale SS, Chang YC, Lee H, McClelland CM, Leong ML, et al. (2008) Importance of mitochondria in survival of *Cryptococcus neoformans* under low oxygen conditions and tolerance to cobalt chloride. Plos Pathogens 4.
23. Perez-Moreno G, Sealey-Cardona M, Rodrigues-Poveda C, Gelb MH, Ruiz-Perez LM, et al. (2012) Endogenous sterol biosynthesis is important for mitochondrial function and cell morphology in procyclic forms of *Trypanosoma brucei*. International Journal for Parasitology 42: 975–989.
24. Ardail D, Popa I, Bodennec J, Louisot P, Schmitt D, et al. (2003) The mitochondria-associated endoplasmic-reticulum subcompartment (MAM fraction) of rat liver contains highly active sphingolipid-specific glycosyltransferases. Biochem J 371: 1013–1019.
25. Holzl G, Leipelt M, Ott C, Zahringer U, Lindner B, et al. (2005) Processive lipid galactosyl/glucosyltransferases from *Agrobacterium tumefaciens* and *Mesorhizobium loti* display multiple specificities. Glycobiology 15: 874–886.
26. Desler C, Lykke A, Rasmussen LJ (2010) The effect of mitochondrial dysfunction on cytosolic nucleotide metabolism. J Nucleic Acids 2010.
27. Hanover JA (2001) Glycan-dependent signaling: O-linked N-acetylglucosamine. FASEB J 15: 1865–1876.
28. Horn M, Fritsche TR, Gautom RK, Schleifer KH, Wagner M (1999) Novel bacterial endosymbionts of Acanthamoeba spp. related to the *Paramecium caudatum* symbiont *Caedibacter caryophilus*. Environ Microbiol 1: 357–367.
29. Fritsche TR, Horn M, Seyedirashti S, Gautom RK, Schleifer KH, et al. (1999) In situ detection of novel bacterial endosymbionts of Acanthamoeba spp. phylogenetically related to members of the order Rickettsiales. Applied and Environmental Microbiology 65: 206–212.
30. Moreno E (1998) Genome evolution within the alpha Proteobacteria: why do some bacteria not possess plasmids and others exhibit more than one different chromosome? FEMS Microbiology Reviews 22: 255–275.
31. Liu R, Ochman H (2007) Stepwise formation of the bacterial flagellar system. Proc Natl Acad Sci U S A 104: 7116–7121.
32. Bradley-Dunlop DJ, Pantoja C, Lightner DV (2004) Development of monoclonal antibodies for detection of necrotizing hepatopancreatitis in penaeid shrimp. Dis Aquat Organ 60: 233–240.
33. Boscaro V, Schrallhammer M, Benken KA, Krenek S, Szokoli F, et al. (2013) Rediscovering the genus Lyticum, multiflagellated symbionts of the order Rickettsiales. Sci Rep 3: 3305.
34. Vahling CM, Duan Y, Lin H (2010) Characterization of an ATP translocase identified in the destructive plant pathogen "*Candidatus* Liberibacter asiaticus". Journal of Bacteriology 192: 834–840.
35. Schmitz-Esser S, Linka N, Collingro A, Beier CL, Neuhaus HE, et al. (2004) ATP/ADP translocases: a common feature of obligate intracellular amoebal symbionts related to Chlamydiae and Rickettsiae. J Bacteriol 186: 683–691.
36. Tjaden J, Winkler HH, Schwoppe C, Van Der Laan M, Mohlmann T, et al. (1999) Two nucleotide transport proteins in *Chlamydia trachomatis*, one for net nucleoside triphosphate uptake and the other for transport of energy. J Bacteriol 181: 1196–1202.
37. Audia JP, Winkler HH (2006) Study of the five *Rickettsia prowazekii* proteins annotated as ATP/ADP translocases (Tlc): Only Tlc1 transports ATP/ADP, while Tlc4 and Tlc5 transport other ribonucleotides. J Bacteriol 188: 6261–6268.

38. Brayton KA, Kappmeyer LS, Herndon DR, Dark MJ, Tibbals DL, et al. (2005) Complete genome sequencing of *Anaplasma marginale* reveals that the surface is skewed to two superfamilies of outer membrane proteins. Proceedings of the National Academy of Sciences of the United States of America 102: 844–849.
39. Mavromatis K, Doyle CK, Lykidis A, Ivanova N, Francino MP, et al. (2006) The genome of the obligately intracellular bacterium *Ehrlichia canis* reveals themes of complex membrane structure and immune evasion strategies. J Bacteriol 188: 4015–4023.
40. Greub G, Raoult D (2003) History of the ADP/ATP-translocase-encoding gene, a parasitism gene transferred from a Chlamydiales ancestor to plants 1 billion years ago. Appl Environ Microbiol 69: 5530–5535.
41. Amiri H, Karlberg O, Andersson SG (2003) Deep origin of plastid/parasite ATP/ADP translocases. Journal of Molecular Evolution 56: 137–150.
42. Andersson SG, Kurland CG (1998) Reductive evolution of resident genomes. Trends in Microbiology 6: 263–268.
43. Wolf YI, Aravind L, Koonin EV (1999) Rickettsiae and Chlamydiae: evidence of horizontal gene transfer and gene exchange. Trends Genet 15: 173–175.
44. Winkler HH, Neuhaus HE (1999) Non-mitochondrial ATP transport. Trends in Biochemical Sciences 24: 64–68.
45. Amiri H, Karlberg O, Andersson SGE (2003) Deep origin of plastid/parasite ATP/ADP translocases. Journal of Molecular Evolution 56: 137–150.
46. John P, Whatley FR (1975) *Paracoccus denitrificans* and the evolutionary origin of the mitochondrion. Nature 254: 495–498.
47. Whatley JM, John P, Whatley FR (1979) From extracellular to intracellular: the establishment of mitochondria and chloroplasts. Proc R Soc Lond B Biol Sci 204: 165–187.
48. Kurland CG, Andersson SGE (2000) Origin and evolution of the mitochondrial proteome. Microbiology and Molecular Biology Reviews 64: 786-+.
49. Sachs JL, Skophammer RG, Regus JU (2011) Evolutionary transitions in bacterial symbiosis. Proc Natl Acad Sci U S A 108 Suppl 2: 10800–10807.
50. Ewald PW (1987) Transmission modes and evolution of the parasitism-mutualism continuum. Annals of the New York Academy of Sciences 503: 295–306.
51. Enright AJ, Van Dongen S, Ouzounis CA (2002) An efficient algorithm for large-scale detection of protein families. Nucleic Acids Res 30: 1575–1584.
52. Katoh K, Misawa K, Kuma K, Miyata T (2002) MAFFT: a novel method for rapid multiple sequence alignment based on fast Fourier transform. Nucleic Acids Res 30: 3059–3066.
53. Wu M, Chatterji S, Eisen JA (2012) Accounting for alignment uncertainty in phylogenomics. PLoS One 7: e30288.
54. Price MN, Dehal PS, Arkin AP (2010) FastTree 2–approximately maximum-likelihood trees for large alignments. PloS one 5: e9490.
55. Eddy SR (1998) Profile hidden Markov models. Bioinformatics 14: 755–763.
56. Moriya Y, Itoh M, Okuda S, Yoshizawa AC, Kanehisa M (2007) KAAS: an automatic genome annotation and pathway reconstruction server. Nucleic Acids Res 35: W182–185.
57. Darling AE, Mau B, Perna NT (2010) progressiveMauve: multiple genome alignment with gene gain, loss and rearrangement. PLoS One 5: e11147.
58. Delcher AL, Bratke KA, Powers EC, Salzberg SL (2007) Identifying bacterial genes and endosymbiont DNA with Glimmer. Bioinformatics 23: 673–679.
59. Steel M (2005) Phylogenetic diversity and the greedy algorithm. Syst Biol 54: 527–529.
60. Hilario E, Gogarten JP (1993) Horizontal transfer of ATPase genes–the tree of life becomes a net of life. Biosystems 31: 111–119.
61. Wu M, Scott AJ (2012) Phylogenomic analysis of bacterial and archaeal sequences with AMPHORA2. Bioinformatics 28: 1033–1034.
62. Lartillot N, Philippe H (2004) A Bayesian mixture model for across-site heterogeneities in the amino-acid replacement process. Mol Biol Evol 21: 1095–1109.
63. Pagel M, Meade A, Barker D (2004) Bayesian estimation of ancestral character states on phylogenies. Syst Biol 53: 673–684.
64. Vanderpoorten A, Goffinet B (2006) Mapping uncertainty and phylogenetic uncertainty in ancestral character state reconstruction: an example in the moss genus Brachytheciastrum. Syst Biol 55: 957–971.

Protein Domain Analysis of Genomic Sequence Data Reveals Regulation of LRR Related Domains in Plant Transpiration in *Ficus*

Tiange Lang[1], Kangquan Yin[2,3], Jinyu Liu[1], Kunfang Cao[1,4], Charles H. Cannon[1,5], Fang K. Du[2]*

1 Key Laboratory of Tropical Forest Ecology, Xishuangbanna Tropical Botanical Garden, Chinese Academy of Sciences, Menglun, Mengla, Yunnan Province, China, **2** College of Forestry, Beijing Forestry University, Beijing, China, **3** School of Life Science, Tsinghua University, Beijing, China, **4** State Key Laboratory for Conservation and Utilization of Subtropical Agro-Bioresources, and College of Forestry, Guangxi University, Nanning, Guangxi, China, **5** Department of Biological Sciences, Texas Tech University, Lubbock, Texas, United States of America

Abstract

Predicting protein domains is essential for understanding a protein's function at the molecular level. However, up till now, there has been no direct and straightforward method for predicting protein domains in species without a reference genome sequence. In this study, we developed a functionality with a set of programs that can predict protein domains directly from genomic sequence data without a reference genome. Using whole genome sequence data, the programming functionality mainly comprised DNA assembly in combination with next-generation sequencing (NGS) assembly methods and traditional methods, peptide prediction and protein domain prediction. The proposed new functionality avoids problems associated with *de novo* assembly due to micro reads and small single repeats. Furthermore, we applied our functionality for the prediction of leucine rich repeat (LRR) domains in four species of *Ficus* with no reference genome, based on NGS genomic data. We found that the LRRNT_2 and LRR_8 domains are related to plant transpiration efficiency, as indicated by the stomata index, in the four species of *Ficus*. The programming functionality established in this study provides new insights for protein domain prediction, which is particularly timely in the current age of NGS data expansion.

Editor: Fengfeng Zhou, Shenzhen Institutes of Advanced Technology, China

Funding: This work was funded by National Natural Science Foundation of China (grant number 61271447) to TL; National Natural Science Foundation of China (grant number 41201051), 111 Project (grant number B13007) and Program for Changjiang Scholars and Innovative Research Team in University (grant number IRT13047) to FKD. The funders had no role in study design, data collection and analysis, decision to publish, or preparation of the manuscript.

Competing Interests: The authors have declared that no competing interests exist.

* Email: dufang325@gmail.com

Introduction

With the advent of next-generation sequencing (NGS) technology, a massive amount of DNA data is currently being produced in both model and non-model species. However, there are many problems associated with *de novo* assembly, i.e., when there is no reference genome on which to map reads, especially when the genome structure is complex with large parts of repetitive elements, as it is often the case in plant species [1]. In such cases, the DNA reads can only be assembled to scaffold or contig level [2]. Thus, methods based on an analysis of the fragments are needed.

A protein domain is a conserved part of a protein sequence which has a specific structure and function. The typical length of a protein domain is from about 25 to 500 amino acids. For some protein domain analysis, the whole protein sequence is not required [3]. Hence, some of the problems associated with full-length assembly without a reference genome can be avoided by protein domain analysis.

In the present study, fig trees belonging to the *Ficus* genus of the Moraceae family were examined to verify the above hypothesis. The *Ficus* genus has been found to have great diversity in tropical and subtropical areas, which is linked to geographical evolution within the genus [4,5]. *Ficus altissima* Blume, *Ficus tinctoria* G. Forst, *Ficus langkokensis* Drake and *Ficus fistulosa* Reinw. ex Blume usually have overlapping distributions. However, their ecological niches are different due to their physiology. *F. altissima* and *F. tinctoria* are semi-epiphytic and their leaves are coriaceous. As a result, they can tolerate environments with drought episodes [6]. In contrast, *F. langkokensis* and *F. fistulosa* grow in relatively humid habitats, such as waterside rocks, and their leaves are thin coriaceous [7]. The ecological differences in the growing areas of these different *Ficus* species might thus exert different types of drought stress pressures, leading to different responses in stomatal development and morphology [8]. Hence, it would be valuable to develop a model that predicts the peptide domains of proteins for genes potentially involved in responses to drought stress, using genomic data.

One of the strategies used by plants to respond to drought stress events is plant transpiration efficiency. In the model plant *Arabidopsis*, plant transpiration efficiency is a quantitative trait, which has been shown to be controlled by several genes based on quantitative trait loci (QTLs) mapping studies [9]. To date, only a few contributing genes have been identified, one of which is the *ERECTA* gene, which explains 21–46% of the total phenotypic variation in Δ(leaf carbon isotopic discrimination) [9]. In *Arabidopsis*, ERECTA is one of the best studied receptor like kinases (RLKs) with leucine rich repeat (LRR) domains, which not only participates in plant transpiration efficiency but also regulates aerial architecture, stomatal patterning and confers resistance to the pathogenic bacteria *Ralstonia solanacearum*, the necrotrophic fungi *Plectosphaerella cucumerina* and *Pythium irregulare* [10,11]. Structurally, the protein encoded by the *ERECTA* gene in *Arabidopsis* has one LRRNT_2 protein domain at the N-terminal, two LRR_8 protein domains in the middle part, and one Pkinase domain at the C-terminal (Fig. 1A). The LRR_8 domains form the hydrophobic core of the proteins, and they are frequently involved in the formation of protein-protein interactions [11,12]. The LRRNT_2 domain of the protein encoded by *ERECTA* in *Arabidopsis* has LRRs flanked by cysteine rich sequences (Fig. 1B).

In contrast to model species, the molecular mechanism of plant transpiration efficiency still remains unclear in many plant and tree species, especially those without reference genomes. Improving functional annotation of assembled data obtained from NGS technology may provide new insights into genes potentially involved in this important trait. In this study, our first objective was to develop a method for obtaining high quality contigs from low coverage NGS data. Secondly, we attempted to predict protein domains from contigs obtained via the above method. Finally, we utilized our programming functionality to predict LRR domains homologous to those from the *Arabidopsis ERECTA* gene in four *Ficus* species that respond differently to drought environments and examined the relationship between LRR domain numbers and plant transpiration efficiency.

Materials and Methods

DNA extraction and genome sequence

Leaf material of four species, *F. altissima*, *F. tinctoria*, *F. langkokensis* and *F. fistulosa*, was collected from the Xishuangbanna Tropical Botanical Garden, Yunnan Province, P. R. China (101°25′E, 21°41′N) in April 2013 and stored in a paper bag with silica gel until DNA extraction. The four species had been

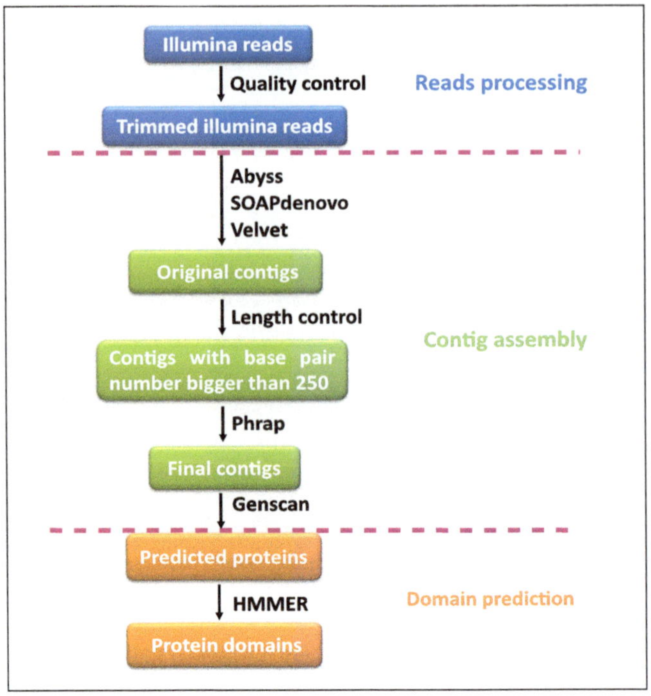

Figure 2. The proposed programming functionality for predicting protein domains directly from genomic sequence data without a reference genome. The Illumina reads were first trimmed with quality control methods. Then, assembly software ABySS, SOAPdenovo and Velvet were used separately to obtain original contigs. Next, length control methods were used to select contigs larger than 250 base pairs. Afterwards, the assembly software Phrap was used to obtain final contigs and Genscan was used to predict peptides from these contigs. Finally, Hmmsearch was used to predict protein domains.

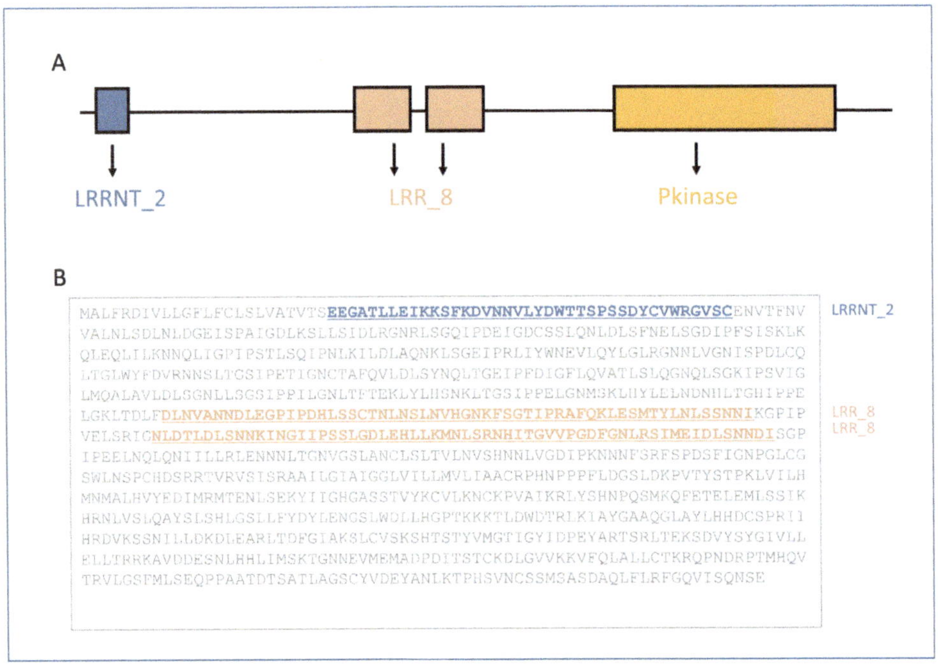

Figure 1. Protein domain structure of the protein encoded by the *ERECTA* gene in *Arabidopsis thaliana*. A. From the N- to C-terminal, the protein is composed of one LRRNT_2 domain, two LRR_8 domains and one Pkinase domain. B. Amino acids of the protein. The LRRNT_2 domain and two LRR_8 domains are underlined. Leucine repeats can be found in the latter domains.

Table 1. Results from the assembly software.

Species	#fastq reads	Coverage	Software	#contig_250	max_len(bp)	#pep	max_len(aa)	#LRRNT_2	#LRR_8
FA	2,185,253,886	4.86	Abyss	26,816	1,968	10,846	606	5	72
			SOAP	26,898	1,906	10,735	578	11	71
			Velvet	123,763	6,407	23,086	702	22	120
			Phrap	114,596	6,914	21,901	880	19	132
FT	2,197,543,362	4.88	Abyss	54,144	2,739	8,595	436	3	40
			SOAP	59,831	2,524	8,419	306	4	33
			Velvet	170,753	9,251	15,319	418	6	59
			Phrap	154,710	10,755	14,807	467	9	62
FL	1,993,136,266	4.43	Abyss	7,679	2,002	2,426	506	1	24
			SOAP	8,321	3,430	2,611	506	3	23
			Velvet	86,717	6,718	6,479	534	3	32
			Phrap	84,287	6,665	6,822	550	3	45
FF	869,615,244	1.93	Abyss	7,087	5,558	2,669	772	2	14
			SOAP	7,049	7,064	2,609	772	0	12
			Velvet	12,129	7,511	3,092	772	0	17
			Phrap	13,972	9,203	3,827	1,536	2	19

FA, FT, FL and FF stands for *Ficus altissima, Ficus tinctoria, Ficus langkokensis* and *Ficus fistulosa*, respectively.
#fastq reads: number of fastq reads from Illumina Hiseq2000.
#contig_250: number of predicted contigs longer than 250 base pairs.
max_len (bp): number of base pairs (bp) of the contigs predicted with maximum length.
#pep: number of peptides predicted.
max_len (aa): number of amino acids (aa) of the peptides predicted with maximum length.
#LRRNT_2: number of LRRNT_2 domains predicted.
#LRR_8: number of LRR_8 domains predicted.

transplanted in 1990 from the natural Xishuangbanna Tropical Forest, Yunna Province, P. R. China (101°57′E, 21°48′N). Genomic DNA of each individual was extracted from dried leaves using the DNeasy Plant Kit (Qiagen). DNA quality was checked on 2% agarose gels stained with ethidium bromide using a UV-Vis spectrometer (Bio-Rad Molecular Imager ChemiDoc XRS+ Imaging System) coupled with a Qubit fluorometer (ds DNA BR, Invitrogen). 40 ug RNA-free genomic DNA were used for the library construction. Library preparation (400-bp and 150-bp paired-end reads) and sequencing on an Illumina HiSeq2000 were performed by the Beijing Genomics Institute.

Quality control methods

The raw data from the Illumina Hiseq2000 were trimmed with two programs for performing quality control written in the Practical Extraction and Report Language (PERL). The first program was used to remove nucleotides with a Phred score lower than 20 (Script S1). The second program was used to delete fastq reads with length less than 20 base pairs as well as "orphanage" reads (single reads not in a pair) created by the first program (Script S2).

Sequence assembly

To generate a better genome assembly, we used a combination of four popular assembly software packages: ABySS, SOAPdenovo, Velvet and Phrap. ABySS, SOAPdenovo and Velvet were used to align the trimmed Illumina fastq reads to obtain contigs. These contigs were then aligned again with Phrap to improve the alignment.

First of all, ABySS (http://www.bcgsc.ca/platform/bioinfo/software/abyss) which allows *de novo*, parallel, paired-end sequence assembly for short sequence reads was used to construct alignments [13–15] on our *Ficus* genome data. We employed 25 as the k-mer length and 10 as the minimum number of pairs needed for two joined contigs.

Secondly, SOAPdenovo (http://soap.genomics.org.cn/soapdenovo.html) [16], which is particularly designed to assemble Illumina GA short reads, was used for building the contigs. The detailed parameter set was as follows: k-mer length 25; average insert size 250; cutoff value of pair number for a reliable connection between two contigs of pre-scaffolds 3; and minimum alignment length between a read and contig required for a reliable read location 32.

Thirdly, Velvet (http://www.ebi.ac.uk/~zerbino/velvet/) [17], which is a sequence assembler for very short sequence reads, was also applied for the sequence alignment. We set the k-mer length as 25 and the average insert size as 250.

Finally, Phrap (http://www.phrap.org/) [18], which is a program for assembling shotgun DNA sequence data was further applied on the sequence to increase the maximum length and remove redundancy. We analyzed the results of ABySS, SOAPdenovo and Velvet by Phrap (for parameters see Table S1 and some connection Script S3).

Gene structure identification

GENSCAN (http://genes.mit.edu/GENSCANinfo.html) was used to identify complete gene structures in genomic DNA. It is a GHMM-based program that can be used to predict the location of genes and their exon-intron boundaries in genomic sequences are from a variety of organisms. The "Arabidopsis.smat" file was downloaded and used as parameter file for the *Ficus* genome data [19].

Protein domain prediction

HMMER 3.0 (http://hmmer.janelia.org/) was used for searching sequence databases for homologs of protein sequences and making protein sequence alignments [20]. It employs methods using probabilistic models called profile hidden Markov models (profile HMMs). We used hmmscan to predict protein domains in the gene *ERECTA*, which were then predicted in *Ficus* and *Populus* by hmmsearch.

Experimental analysis

Stomata index evaluation. The study was conducted in the Xishuangbanna Tropical Botanical Garden in Yunnan Province, P. R. China (101°25′E, 21°41′N) in August 2013. Four to six trees of each species showing good growth performance were sampled. We collected three mature and well-exposed leaves from each tree. To obtain a better view of the stomata, we removed the main vein of leaves and then boiled them in hot alkaline buffer to remove the mesophyll. Treated leaves were examined under a light microscope (DM2500, Leica, Germany). The numbers of stomata and epidermal cells were counted using ImageJ (National Institutes of

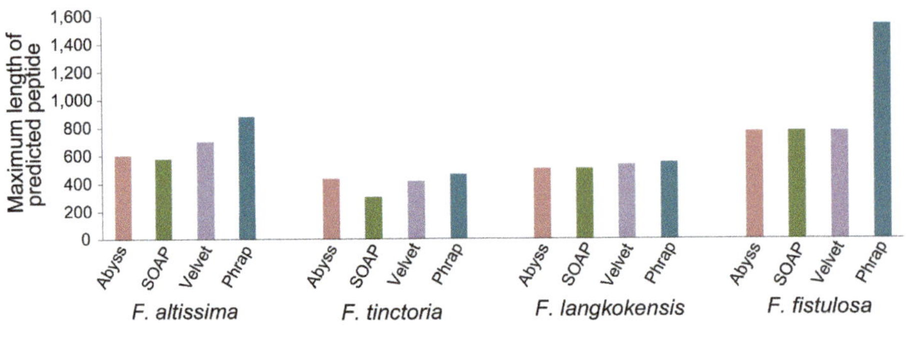

Figure 3. Maximum length (number of amino acids) of peptides predicted by the programming functionality. The Illumina reads for *F. altissima (FA), F. tinctoria (FT), F. langkokensis (FL)* and *F. fistulosa (FF)* were assembled by ABySS, SOAPdenovo and Velvet. Phrap was used to assemble the contigs from ABySS, SOAPdenovo and Velvet, and then Genscan was used to predict peptides from these contigs. The maximum length of the peptides could be increased by Phrap in *FA, FT, FL* and *FF*.

Table 2. Redundancy removed by Phrap.

Species	#contig_250 from Abyss, SOAP and Velvet	#base pairs	#contig_250 not used by Phrap	#base pairs	#contig_250 used by Phrap	#base pairs	#contig_250 after Phrap	#base pairs	Percent of redundancy removed by Phrap
FA	177477	79652086	86943	37241030	90534	42411056	27653	13333742	36.51
FT	284698	124099037	95706	40299597	188992	83799440	59004	26821922	45.91
FL	102717	40680387	75107	29210626	27610	11469761	9180	3886983	18.64
FF	26265	9447858	7416	2344792	18849	7103066	6556	2670964	46.91

FA, FT, FL and FF stands for *Ficus altissima*, *Ficus tinctoria*, *Ficus langkokensis* and *Ficus fistulosa*, respectively.
#contig_250: number of contigs longer than 250 base pairs.
#base pairs: number of base pairs.

Health, Bethesda, MD, USA, http://rsb.info.nih.gov/ij/index.html). We used the formula (100×stomatal density)/(stomatal density + epidermal cell density) to calculate the stomata index as in Hara *et al.*, (2009) [21]. We repeated the experiments six times.

Results and Discussion

Genomic contigs assembled from Illumina genomic sequence data

The Illumina Hiseq2000 genomic sequence data in fastq format for each of *F. altissima, F. tinctoria, F. langkokensis* and *F. fistulosa* was about 15 Gigabytes, and thus the coverage was about 15x [22]. The quality control methods improved the quality of the Illumina genomic sequence data for all four species, where the number of reads was about 135 million for each species. For the gene and protein domain prediction, it was most appropriate to use the contigs that had a base pair number >250 as the input for Phrap as the typical number of nucleotides in a DNA fragment encoding a protein domain of the LRR family is about 250 (Fig. 2). The maximum lengths of contigs predicted by Phrap in *F. altissima, F. tinctoria, F. langkokensis* and *F. fistulosa* were 6914 base pairs, 10755 base pairs, 6665 base pairs and 9203 base pairs, respectively. These numbers showed that the genomic contigs finally assembled could be used for the gene prediction (Table 1).

Peptides and protein domains predicted from genomic contig data

Genscan was used to predict peptide sequences from the genomic contigs assembled by Phrap for *F. altissima, F. tinctoria, F. langkokensis and F. fistulosa* (Fig. 2). The maximum lengths of peptides predicted by Genscan for the four species (880, 467, 550 and 1576 amino acids, respectively) were employed since longer peptides enable better protein domain prediction. These numbers indicated that the predicted peptides could be used to predict protein domains (Table 1).

HMMER was used to predict protein domains from the peptides predicted from Genscan for the four species (Fig. 2). The numbers of LRRNT_2 domains predicted from the contigs assembled by Phrap were 19, 9, 3, and 2 in *F. altissima, F. tinctoria, F. langkokensis and F. fistulosa*, respectively, whereas the numbers of LRR_8 domains were 132, 62, 45, and 19, respectively (Table 1).

Phrap can improve assembly of contigs and remove identical segments of genomic sequences

Phrap cannot directly work on Illumina fastq reads. However, it can increase the maximum length of the contigs assembled through ABySS, SOAPdenovo and Velvet, which can directly work on Illumina fastq reads (Table 1). Thus, the maximum length of the peptides predicted by Genscan may also be increased (Fig. 3).

Assembly through Phrap can remove redundancy of the contigs, including the identical genomic sequence segments predicted by ABySS, SOAPdenovo and Velvet. The percentage of redundancy removed from *F. altissima, F. tinctoria, F. langkokensis* and *F. fistulosa* was 36.51, 45.91, 18.64 and 46.91 respectively (Table 2). The Phrap software can be used to combine identical DNA fragments into one sequence, thus avoid the effect of gene expression difference produced by NGS methods. Programs like CAP3 and TIGR Assembler may also offer similar functions. However, it is important to choose correct parameters for different species. When applying on the contigs which are assembled by NGS assembly methods, we found that Phrap has more suitable

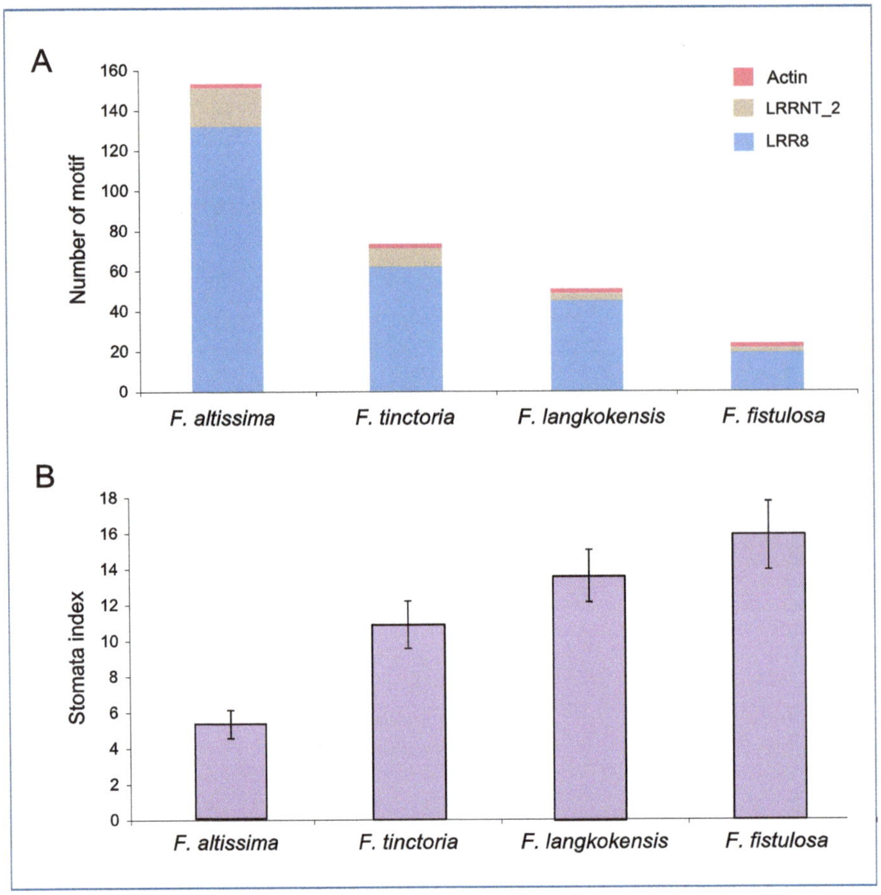

Figure 4. Number of LRRNT_2, LRR_8 and actin domains predicted in *F. altissima (FA), F. tinctoria (FT), F. langkokensis (FL)* and *F. fistulosa (FF)* (A); and stomata index in *FA, FT, FL* and *FF* (B). As the number of LRRNT_2 and LRR_8 domains decreased for *FA, FT, FL* and *FF*, the stomata index increased.

Table 3. Physiological, anatomical and stomata response data in *Ficus*.

Species		#stomata	#epidermal cells	Stomatal density	Epidermal cell density	Stomatal index
FA	M	12.91667	231.1944	326.5458	5844.819	5.301273
	SD	2.061553	20.15769	52.11805	509.606	0.79305
	SE	0.343592	3.359615	8.686342	84.93433	0.132175
FT	M	20.84848	169.0303	527.0699	4273.25	10.90198
	SD	4.016538	13.41754	101.542	339.2083	1.331365
	SE	0.699189	2.335693	17.67619	59.04859	0.231761
FL	M	15.66667	99.47619	396.0685	2514.854	13.61349
	SD	1.932184	7.35268	48.84747	185.8829	1.467321
	SE	0.421637	1.604486	10.65939	40.56297	0.320196
FF	M	19.125	99.2	483.4985	2507.872	15.90947
	SD	3.879433	10.0584	98.07582	254.2861	1.932021
	SE	0.969858	2.597068	24.51895	65.65639	0.498846

FA, FT, FL and FF stands for *Ficus altissima, Ficus tinctoria, Ficus langkokensis* and *Ficus fistulosa*, respectively.
M, SD, and SE: mean, standard deviation and standard error, respectively.
#stomata: number of stomata.
#epideman cells: number of epidermal cells.

parameters to be adjusted comparing to other programs by testing different combinations of parameters values. The raw data sequences used here were submitted to NCBI under accession number SRP041276.

The numbers of LRRNT_2 and LRR_8 domains in *Ficus* correlate with stomata index values

To test whether the LRRNT_2 domains and LRR_8 domains are related to transpiration efficiency, we used our programming functionality to predict their numbers in the four species of *Ficus*. The mean values of the stomata index for *F. altissima, F. tinctoria, F. langkokensis and F. fistulosa* were 5.3, 10.9, 13.6 and 15.9, respectively (Table 3). As the stomata index values increased in these species the numbers of LRRNT_2 and LRR_8 domains decreased accordingly (Fig. 4). To eliminate the contingency in protein domain selection, we used the actin domain from actin1 protein in *Arabidopsis thaliana* (NCBI accession number NP_850284.1) for control analysis. Actin is a house-keeping protein expressed in every plant cell as a component of the cytoskeleton [23], and thus provides a good control. Among all the peptides predicted for the four *Ficus* species, one actin domain was found to be longer than 100 amino acids and another was shorter than 50 amino acids (Fig. 4). These results suggest that the transpiration efficiency could be related to the LRRNT_2 and LRR_8 domains in *Ficus*.

The *ERECTA* gene has not only a positive regulatory role on respiration in drought conditions but also benefits plants in the absence of water shortage [9]. Therefore, the protein domains in the *ERECTA* gene might show a cumulative positive evolution. The LRR_8 domain has more LRRs than the LRRNT_2 domain, and thus may have a more important role in protein-protein interactions (Fig. 1). Hence, this could explain why the number of LRR_8 domains was more than that of LRRNT_2 domains (Fig. 4).

Conclusion

The programming functionality in this study was proved to be a useful tool in biological studies by showing that the LRRNT_2 and LRR_8 domains were potentially related to plant transpiration efficiency, as we can see from the stomata index in *F. altissima*, *F. tinctoria*, *F. langkokensis*, and *F. fistulosa*. The main benefit of the functionality is that it overcomes many of the complex problems associated with *de novo* assembly. However, with the increasing read lengths produced by NGS and improvements in third-generation sequencing, such problems may also be solved with the rapid developments of *de novo* assembly methods. The main limitation of the functionality is GENSCAN prediction step, which requires a suitable model. In addition, it is hard for some species to choose a perfect model to predict the gene structure. Confronting with this situation, researchers normally prefer to pick a widely used model which turns out to have more or less shortage. Nevertheless, methods of whole genome protein domain analysis will still help researchers to better understand some mechanisms of biological function from the perspective of genetic sequence, if combined with a large amount of NGS data.

Supporting Information

Table S1 Table for Phrap parameters.

Script S1 Perl program used to remove the nucleotides which have Phred score lower than a specific value.

Script S2 Perl program used to delete the fastq reads which have length less than a specific value as well as to erase the "orphange" reads (single reads without pair).

Script S3 Perl programs used for dealing with the assembly files which were created by Phrap as well as for making statistic analysis.

Acknowledgments

We would like to thank the following colleagues from the Xishuangbanna Tropical Botanical Garden (XTBG), Chinese Academy of Sciences (CAS): Bo Pan for collecting samples and Jun Yang for providing experimental equipment.

Author Contributions

Conceived and designed the experiments: TGL KQY KFC CHC FKD. Performed the experiments: TGL JYL. Analyzed the data: TGL KQY FKD. Contributed reagents/materials/analysis tools: TGL KQY JYL. Wrote the paper: TGL KQY FKD.

References

1. Butler J, MacCallum I, Kleber M, Shlyakhter IA, Belmonte MK, et al. (2008) ALLPATHS: de novo assembly of whole-genome shotgun microreads. Genome Res 18: 810–820.
2. Grabherr MG, Haas BJ, Yassour M, Levin JZ, Thompson DA, et al. (2011) Full-length transcriptome assembly from RNA-Seq data without a reference genome. Nat Biotechnol 29: 644–652.
3. Corpet F, Servant F, Gouzy J, Kahn D (2000) ProDom and ProDom-CG: tools for protein domain analysis and whole genome comparisons. Nucleic Acids Res 28: 267–269.
4. Berg CC (1989) Classification and distribution of Ficus. Experientia 45: 605–611.
5. Harrison RD (2005) Figs and the diversity of tropical rainforests. Bioscience 55: 1053–1064.
6. Hao GY, Sack L, Wang AY, Cao KF, Goldstein G (2010) Differentiation of leaf water flux and drought tolerance traits in hemiepiphytic and non-hemiepiphytic Ficus tree species. Funct Ecol 24: 731–740.
7. Wu ZY, Zhou ZK, Gilbert MG (2004) Flora of China. Beijing: Science Press. Vol. 5: 21 p.
8. Hamanishi ET, Thomas BR, Campbell MM (2012) Drought induces alterations in the stomatal development program in Populus. J Exp Bot 63: 4959–4971.
9. Masle J, Gilmore SR, Farquhar GD (2005) The ERECTA gene regulates plant transpiration efficiency in Arabidopsis. Nature 436: 866–870.
10. Torii KU, Mitsukawa N, Oosumi T, Matsuura Y, Yokoyama R, et al. (1996) The arabidopsis ERECTA gene encodes a putative receptor protein kinase with extracellular leucine-rich repeats. Plant Cell 8: 735–746.
11. Kobe B, Kajava AV (2001) The leucine-rich repeat as a protein recognition motif. Curr Opin Struct Biol 11: 725–732.
12. Wei T, Gong J, Jamitzky F, Heckl WM, Stark RW, et al. (2008) LRRML: a conformational database and an XML description of leucine-rich repeats (LRRs). BMC Struct Biol 8: 47.
13. Simpson JT, Wong K, Jackman SD, Schein JE, Jones SJ, et al. (2009) ABySS: a parallel assembler for short read sequence data. Genome Res 19: 1117–1123.
14. Robertson G, Schein J, Chiu R, Corbett R, Field M, et al. (2010) De novo assembly and analysis of RNA-seq data. Nat Methods 7: 909–912.
15. Birol I, Jackman SD, Nielsen CB, Qian JQ, Varhol R, et al. (2009) De novo transcriptome assembly with ABySS. Bioinformatics 25: 2872–2877.
16. Li R, Yu C, Li Y, Lam TW, Yiu SM, et al. (2009) SOAP2: an improved ultrafast tool for short read alignment. Bioinformatics 25: 1966–1967.
17. Zerbino DR, Birney E (2008) Velvet: algorithms for de novo short read assembly using de Bruijn graphs. Genome Res 18: 821–829.
18. Machado M, Magalhães WC, Sene A, Araújo B, Faria-Campos AC, et al. (2011) Phred-Phrap package to analyses tools: a pipeline to facilitate population genetics re-sequencing studies. Investig Genet 2: 3.
19. Burge C, Karlin S (1997) Prediction of complete gene structures in human genomic DNA. J Mol Biol 268: 78–94.

20. Finn RD, Clements J, Eddy SR (2011) HMMER web server: interactive sequence similarity searching. Nucleic Acids Res 39: W29–37.
21. Hara K, Yokoo T, Kajita R, Onishi T, Yahata S, et al. (2009) Epidermal cell density is autoregulated via a secretory peptide, EPIDERMAL PATTERNING FACTOR 2 in Arabidopsis leaves. Plant Cell Physiol 50: 1019–1031.
22. Deepak O, Genome Size Variation and Plant Systematics (1998) Annals of Botany 82 (Supplement A): 75–83.
23. Shah DM, Hightower RC, Meagher RB (1983) Genes encoding actin in higher plants: intron positions are highly conserved but the coding sequences are not. J Mol Appl Genet 2: 111–126.

16

Complete Chloroplast Genome Sequence of Poisonous and Medicinal Plant *Datura stramonium*: Organizations and Implications for Genetic Engineering

Yang Yang[1]ᵠ, Dang Yuanye[1]ᵠ, Li Qing[2], Lu Jinjian[1], Li Xiwen[1,3]*, Wang Yitao[1]*

1 State Key Laboratory of Quality Research in Chinese Medicine, Institute of Chinese Medical Sciences, University of Macau, Macau, China, **2** Department of Pharmacy, Shanghai Changzheng Hospital, Second Military Medical University, Shanghai, China, **3** Institute of Chinese Materia Medica, China Academy of Chinese Medical Sciences, Beijing, China

Abstract

Datura stramonium is a widely used poisonous plant with great medicinal and economic value. Its chloroplast (cp) genome is 155,871 bp in length with a typical quadripartite structure of the large (LSC, 86,302 bp) and small (SSC, 18,367 bp) single-copy regions, separated by a pair of inverted repeats (IRs, 25,601 bp). The genome contains 113 unique genes, including 80 protein-coding genes, 29 tRNAs and four rRNAs. A total of 11 forward, 9 palindromic and 13 tandem repeats were detected in the *D. stramonium* cp genome. Most simple sequence repeats (SSR) are AT-rich and are less abundant in coding regions than in non-coding regions. Both SSRs and GC content were unevenly distributed in the entire cp genome. All preferred synonymous codons were found to use A/T ending codons. The difference in GC contents of entire genomes and of the three-codon positions suggests that the *D. stramonium* cp genome might possess different genomic organization, in part due to different mutational pressures. The five most divergent coding regions and four non-coding regions (*trnH-psbA*, *rps4-trnS*, *ndhD-ccsA*, and *ndhI-ndhG*) were identified using whole plastome alignment, which can be used to develop molecular markers for phylogenetics and barcoding studies within the Solanaceae. Phylogenetic analysis based on 68 protein-coding genes supported *Datura* as a sister to *Solanum*. This study provides valuable information for phylogenetic and cp genetic engineering studies of this poisonous and medicinal plant.

Editor: Zhong-Hua Chen, University of Western Sydney, Australia

Funding: This work was supported by Macau Science and Technology Development Fund (http://www.fdct.gov.mo/) (077/2011/A3, 074/2012/A3), Research committee of University of Macau (http://www.umac.mo/research/research_committee.html) (MYRG208A (Y3-L4)-ICMS11-WYT, MRG012/WYT/2013/ICMS, MRG013/WYT/2013/ICMS) and the National Natural Science Foundation of China (81202860, 81303160). The funders had no role in study design, data collection and analysis, decision to publish, or preparation of the manuscript.

Competing Interests: The authors have declared that no competing interests exist.

* Email: xiweijia2004@aliyun.com (LXW); ytwang@umac.mo (WYT)

ᵠ These authors contributed equally to this work.

Introduction

Scopolamine is an important tropane alkaloid from Solanaceae plants widely used as anticholinergic agent that acts on the parasympathetic nervous system [1]. It is widely used as sedative in clinical practice including preanesthetic medication for general anesthesia, and also for manic psychosis, motion sickness, parkinsonism and organophosphorus pesticide poisoning [2,3]. Due to its activity in exciting the respiratory center and sedative effect on the cerebral cortex, scopolamine is used to rescue respiratory failure caused by extremely heavy epidemic encephalitis, accompanied by severe frequent tics in such condition [4,5]. Recently scopolamine also exhibited great potential as a drug for use in withdrawal for heroin addicts [6]. Scopolamine occurres in all plant organs and was traditionally extracted from flowers of *Datura* species. It was recently reported that the maximum concentrations were found in the stems and leaves of juvenile plants [3,7]. But the concentration is still quite low and its supply cannot meet the market demand. Therefore significant attention has been paid to its commercial production using biotechnologies.

Over the last decades, engineering techniques have been intensively investigated as a possible tool for the production of scopolamine in different plant species that produce tropane alkaloids,including overexpression of genes involved in the biosynthesis of scopolamine [1,8,9] as well as biotransforming hyoscyamine into scopolamine in hairy root cultures [9–11]. However production was too low for commercialization. Because of the complicated metabolic pathway of biosynthesis, it has become clear that unorganized plant tissue cultures are frequently unable to produce scopolamine at the same levels as the intact plant [8].

Plastids of higher plant are cellular organelles with circular genomes of 120–160 kb in size present in 1,000–10,000 copies per cell [12], and are maternally inherited in most angiosperm plant species [13]. Chloroplast transformation offers a higher level expression of foreign genes in intact plant compared with hair root cultures. In the past two decades, more than forty transgenes have been stably integrated and expressed in the tobacco cp genome to confer important agronomic traits or produce commercial products including biomaterials and recombinant proteins [8]. Chloroplast engineering, either alone or in combination with traditional cultivation techniques, may provide the means to develop novel sources of plants to solve tropane alkaloid biosynthesis, the century old problem. Great progress has been made in the study of discovering rate-limiting enzymes in the key steps of catalysis for tropane alkaloids synthesis [1,10].

However the lack of plastid genome data available in public databases limits further studies of cp transformation. *Datura stramonium* has been one of the major plant sources for extracting scopolamine. It is a good model plant to study at the biochemical and molecular level. We here analyzed and characterized the cp genome of *D. stramonium*, providing the basic genetic information for cp engineering. Comparison of the genome structures with other plant species was also determined. These data should also contribute to a better understanding in future studies of evolution within the asteridae clade and species identification of this poisonous and medicinal plant.

Materials and Methods

Genome Sequencing Preparation

Chloroplast DNA (cp DNA) was extracted from approximately 100 g fresh young leaves of *Datura stramonium* using a sucrose gradient centrifugation method that was improved by Li *et al.* [14]. The concentration of the DNA for each cp genome was estimated by measuring A260 with an ND-2000 spectrometer (Nanodrop technologies, Wilmington, DE, USA), and visual approximation was performed using gel electrophoresis. Pure cpDNA was sequenced using a 454/Roche FLX high-throughput sequencing platform.

Genome Assembly and Annotation

The Sff-file obtained was pre-processed, including the trimming of low-quality sequences. *De novo* assembly was performed using version 2.5 of the GS FLX system software. The position and direction of the contigs were identified using the cp genome sequence of *Nicotiana sylvestris* (NC_007500) as the reference sequence. The boundaries of IR-LSC and IR-SSC were confirmed using PCR amplification. We used the online program DOGMA (Dual Organellar GenoMe Annotator) [15] to annotate the cp genome. The position of each gene was determined using a blast method with the complete cp genome sequence of *N. sylvestris* as a reference sequence. Minor revisions were performed according to the start and stop codons. The tRNA genes were identified using DOGMA and tRNAscan-SE [16]. The nomenclature of cp genes followed the ChloroplastDB [17]. The circular cp genome map was drawn by the OGDRAW program [18]. To analyze the characteristics of variations in synonymous codon usage by neglecting the influence of amino acid composition, the relative synonymous codon usage values (RSCU) were determined using MEGA5.2 [19]. The final cp genome of *Datura stramonium* has been deposited to GenBank (accession number NC_018117).

Genome Comparison and Sequence Analysis

The pairwise alignments of cp genomes were performed using MUMmer [20]. The mVISTA program in Shuffle-LAGAN mode [21] was used to compare the cp genome of *Datura stramonium* with three other cp genomes using the genome sequence of *Datura stramonium* as reference. We used DnaSP v5 [22] to calculate the substitution rates. Simple sequence repeats (SSRs) were detected using MISA (http://pgrc.ipkgatersleben.de/misa/), with thresholds of eight repeat units for mononucleotide SSRs, four repeat units for di- and trinucleotide SSRs and three repeat units for tetra-, penta- and hexanucleotide SSRs. All of the repeats found were manually verified, and the redundant results were removed. We investigated the distribution of SSRs located in LSC, SSC and IR regions. The proportions of different nucleotides (A, T, C, G) were calculated and different chloroplast SSR types (CSTs) found among SSRs were discovered. To determine the repeat structure, REPuter [23] was used to visualize both forward and palindrome repeats. The settings for the minimal repeat size was 30 bp and the identity of repeats was no less than 90% (hamming distance = 3). Low complexity and nested repeats were ignored. Tandem repeats were analyzed with the aid of Tandem Repeats Finder (TRF) v4.04 [24] and the parameters were set according to Nie *et al* [25].

Phylogenetic Analysis

In order to identify the phylogenetic position of *Datura* within the asterid lineages, 42 complete cp genome sequences are downloaded from the Genbank of NCBI database (Table S1). Protein-coding gene sequences (Table S2) were aligned using the ClustalW2 algorithm [26]. Pairwise sequence divergences were calculated using Kimura two-parameter (K2P) model [27]. And 68 protein-coding genes (Table S2) shared by all studied plastid genomes were extracted for phylogenetic analysis. Each gene was aligned using the ClustalW and the alignment was edited manually. Maximum likelihood (ML) analysis was performed using RAxML v7.0 [28] using the GTR+I+G nucleotide substitution model under the best fit parameters determined by Modeltest ver. 3.7 [29]. Maximum Parsimony (MP) analysis was performed using PAUP ver. 4.0b10 [30] taking the cp genome sequence of *Cycas taitungensis* (NC_009618) as the outgroup. MP searches included 1,000 replicates of random taxon addition and a heuristic search using tree bisection and reconnection (TBR) branch swapping (Multrees option in effect). Both of these analyses, we using 1000 bootstrap replicates.

Result

Genome Features

The complete cpDNA genome of *Datura stramonium* is 155,871 bp in length (GeneBank: NC_018117) with a typical quadripartite structure of land plant cp genomes. The cp genome are divided into a LSC (86,302 bp) and a SSC (18,367 bp) regions separated by a pair of inverted repeat regions (IRa and IRb) of 25,602 bp (Table 1, Figure 1). The overall GC content of the whole cp genome sequence is 37.9% which is similar to those of the other reported asteridae cp genomes [31–35]. However the GC content is unevenly distributed in the entire cp genome. It is highest in the IR regions (43.1%), median in the LSC region (36.0%) and lowest in the SSC regions (32.3%).

The positions of all the genes identified in the *D. stramonium* cp genome and functional categorization of these genes are presented in Figure 1. This cp genome encodes 132 predicted functional genes, of which 113 genes are unique, including 80 protein-coding genes, 29 transfer RNA (tRNA) genes and four rRNAs (Table S3). In addition, seven tRNA, all rRNA and eight protein-coding genes

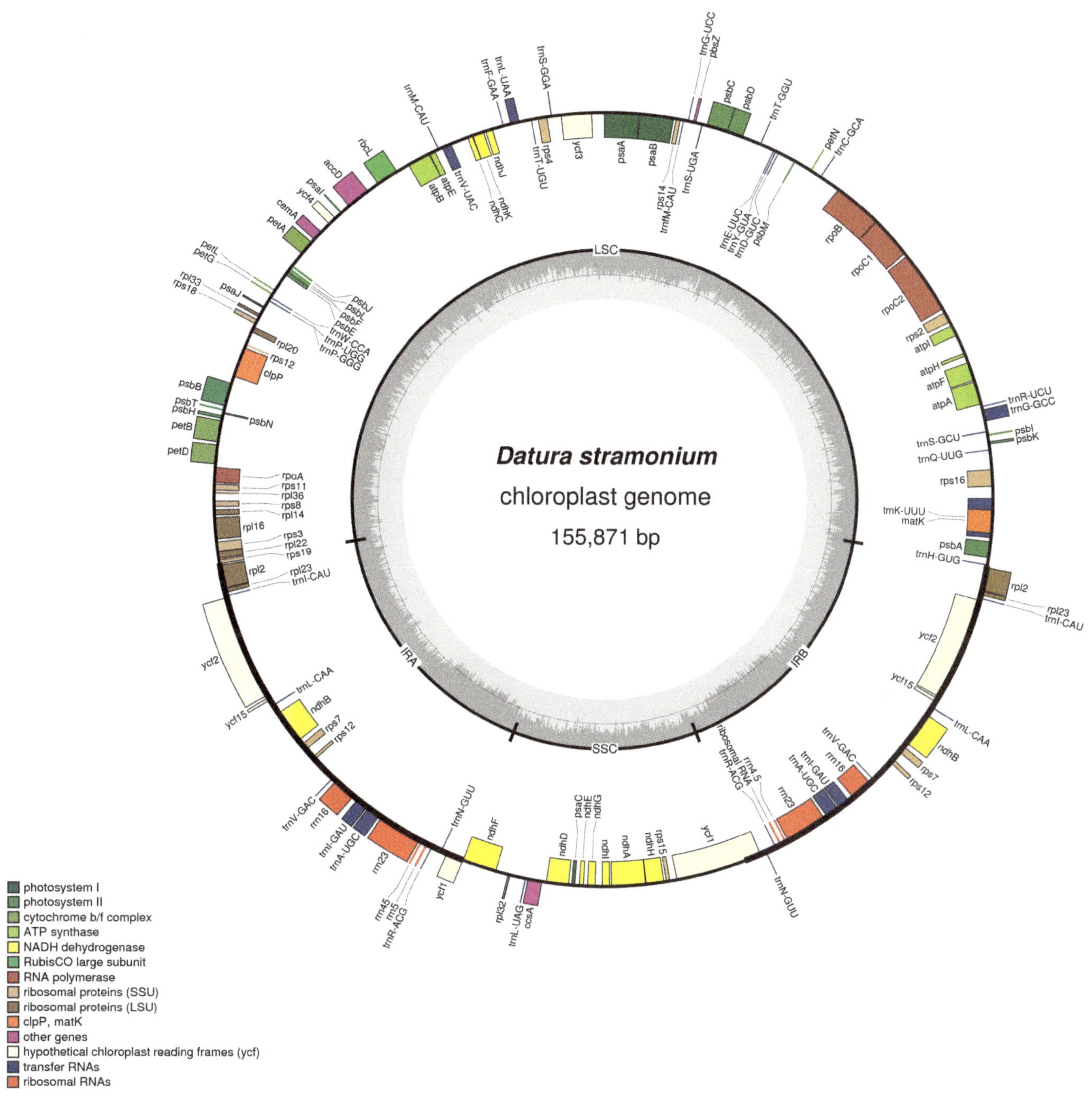

Figure 1. Gene map of the *Datura stramonium* chloroplast genome. Genes drawn inside the circle are transcribed clockwise, and those outside are counterclockwise. Genes belonging to different functional groups are color-coded. The darker gray in the inner circle corresponds to GC content, while the lighter gray corresponds to AT content.

are duplicated in the IR regions. The LSC region contains 62 protein-coding and 25 tRNA genes but one tRNA and 11 protein-coding genes in the SSC region. There are altogether 14 intron-containing genes, 11 (nine protein-coding and two tRNA genes) of which contain one intron and three (*rps12*, *clpP* and *ycf3*) of which contain two introns (Table S4). The *rps12* gene is trans-spliced and the 5′ end located in the LSC region and the two duplicated 3′ end are in the IR regions. The *ndhA* gene has the longest intron (1,155 bp).

Protein-coding regions accounted for 59.7% of the whole genome sequence, while rRNA and tRNA regions accounted for 4.5% and 5.8%, respectively. The remaining regions are non-coding sequences, including introns, intergenic spacers and pseudogenes. Moreover, the total length of all the 88 protein-coding genes is 80,316 bp and these genes comprise 26,772 codons. Frequency of codon usage was calculated in the *D. stramonium* cp genome, and summarized in Table 2. A total of 10.6% of all codons (2,848) encodes leucine, and 1.1% of which (305) encodes cysteine, which are the most and least prevalent amino acids, respectively. Within protein-coding sequences (CDS), the percentage of AT content of the first, second and third codon positions are 54.3%, 61.8% and 69.4%, respectively (Table 1).

Table 1. Base composition in the *Datura stramonium* chloroplast genome.

	T (U) (%)	C (%)	A (%)	G (%)	Length (bp)
LSC	32.7	18.4	31.3	17.6	86,302
SSC	34.0	16.8	33.7	15.5	18,367
IRa	28.3	20.7	28.6	22.4	25,601
IRb	28.6	22.4	28.3	20.7	25,601
Total	31.5	19.2	30.6	18.6	155,871
CDS	31.2	17.9	30.6	20.3	80,316
1st position	23.7	18.9	30.6	26.8	26,772
2nd position	32.5	20.3	29.3	17.9	26,772
3rd position	37.6	14.3	31.8	16.3	26,772

Such bias towards a higher AT representation at the third codon position was also observed in other land plant cp genomes [25,31,36,37]. There were 96.7% (29/30) of all the types of preferred synonymous codons (RSCU>1) ending with A or U and 90.6% (29/32) of non-preferred synonymous codons (RSCU <1) ending with G or C. In addition, A- and/or U-ending codons account for 69.3% of all codons within CDS. The usage of start codon (AUG) and UGG coding trp has no bias (RSCU = 1).

SSR Analysis

The simple sequence repeats (SSR), also called microsatellites, are a group of tandem repeated sequences which consist of 1–6 nucleotide repeat units [38]. A total of 160 SSR loci were detected in *D. stramonium* cp genome including 109 mononucleotide, 40 dinucleotide, 3 trinucleotide and 8 tetranucleotide repeat units. However, only 53 loci were identified in 19 CDS. Among them, 5 genes were found to harbor at least two SSRs, including *atpA*, *ycf3*, *accD*, *rbcL* and *clpP*. We also detected perfect SSRs longer than 8 bp in *D. stramonium* together with 41 other cp genomes to determine whether there was any homology between the isolated SSR fragments and previously reported sequences (Figure 2). *Arabidopsis thaliana* had the maximum amount of SSRs (335) while the smallest number (127) occurred in *Oryza nivara*. Mononucleotide and dinucleotide repeat units are the prevalent types in all species, ranging from 91 (*Magnolia grandiflora*) to 234 (*Arabidopsis thaliana*) and 20 (*Oryza nivara*) to 85 (*Quercus rubra*) in quantity, respectively. The number of trinucleotides is slightly lower than that of tetranucleotides, and only rarely are pentanucleotides or hexanucleotides observed in these 41 cp genomes. Most of SSRs detected in these cp genomes were A (28.2%) and T (35.2%) mononucleotide SSRs while C or G repeats were rarely found. We also detected the distribution of SSRs in the CDS of studied cp genomes (Table S5). The CDS accounts for approximately 51% of the total length, whereas the SSR proportion ranges from 19% to 41%. Average total number of SSRs identified in CDS is 56 accounting for 30% of all SSRs in these whole cp genomes. In addition, the majority of SSRs are located in LSC region (63.2–66.9%) in 10 Solanales cp genomes.

Repeat Analysis

For the repeat structure analysis, there are 33 large repeats of 30 bp or larger in *Datura stramonium* cp genome (Table 3). Eleven forward, nine palindromic and thirteen tandem repeats were identified. There are three repeat motifs detected in the CDS of *ycf2* gene and the *IGS* (*rps12.trnV-GAC*)). In 11 repeats there were two repeat motifs while in other 20 repeats only one motif was found (Table 3). Most repeats are located in the intergenic or intronic regions, while some of them are in protein-coding regions. Most of the repeats exhibit lengths between 30 and 60 bp, while the two longest repeats respectively occurred in rrn4.5-rrn5 (66 bp) and IGS (*rps12.trnV-GAC*). Eight forward, six palindromic and eight tandem repeats were distributed in the LSC region.

Comparison with Other cp Genomes in the Solanales Order

There are currently ten complete cp genome sequences in the Solanales order available in genbank. The gene order and organization of *Datura* are almost identical to those of *N. tabacum* (NC_001879) and other species. The average size of the Solanales cp genomes is 156,422 bp in length. *Ipomoea purpurea* has the largest genome size that is approximately 6.2 kb larger than that of *D. stramonium*, which is mainly attributed to the difference in the length of the IR regions. The genome size of *S. tuberosum* is smallest and is approximately 575 bp smaller than that of *D. stramonium*. This variation in sequence length is mainly caused by the divergence in the length of the LSC region (Table S6). We compared four cp genomes from four different genera in Solanales and observed approximately identical gene order and organization among them (Figure 3). The overall sequence identity of the four cp genomes was plotted using *D. stramonium* as reference. The average sequence divergence of coding regions in *Ipomoea purpurea* is 1.47%, while 1.06% and 1.09% in *Nicotiana undulate* and *Solanum tuberosum*, respectively (Table S7). This study found that the ten most divergent coding regions were *ycf1, clpP, cemA, accD, rpl32, rpl22, matK, ccsA, ndhF* and *rpl36* based on the p-distance measurements. These genes are mainly located in single copy regions. In addition, sequences in non-coding regions exhibit a higher divergence than those in coding regions and the most divergent regions localize in the intergenic spacers among the four cp genomes. In our alignment, these highly divergent regions included *trnH-psbA*, *rps4-trnS*, *ndhD-ccsA* and *ndhI-ndhG*.

The non-synonymous (Ka) to synonymous (Ks) rate ratio (denoted by Ka/Ks) among *Datura*, *Ipomoea*, *Nicotiana* and *Solanum* was calculated and is shown in Figure 4. In IRs region, the Ka/Ks ratio of different genes was all lower than that in the SSC and LSC regions. In four species, most of the ratios of genes were below 1.0, except the value of *atpA*, *rpoC2*. The Ka/Ks values of *atpA*, *rpoC2* and *psbC* (except in *Nicotiana undulate*) among four species are all over 1.0, which means the positive selection was exerting an influence on these genes in the evolution of Solanoideae. In contrast, ratios in gene of *Datura stramonium* were variable from 0 to 0.99 (exclude *ndhD*, 1.54), indicating these

Table 2. The codon–anticodon recognition pattern and relative synonymous codon usage (RSCU) for the *Datura stramonium* chloroplast genome.

Amino acid	Codon	Count	RSCU	tRNA
Phe	UUU	955	1.27	
Phe	UUC	551	0.73	trnF-GAA
Leu	UUA	875	1.84	trnL-UAA
Leu	UUG	579	1.22	trnL-CAA
Leu	CUU	612	1.29	
Leu	CUC	207	0.44	
Leu	CUA	378	0.80	trnL-UAG
Leu	CUG	197	0.42	
Ile	AUU	1105	1.47	
Ile	AUC	461	0.61	trnI-GAU
Ile	AUA	684	0.91	trnI-CAU
Met	AUG	622	1.00	trn(f)M-CAU
Val	GUU	526	1.46	
Val	GUC	182	0.50	trnV-GAC
Val	GUA	539	1.46	trnV-UAC
Val	GUG	197	0.55	
Ser	UCU	591	1.70	
Ser	UCC	341	0.98	trnS-GGA
Ser	UCA	407	1.17	trnS-UGA
Ser	UCG	210	0.60	
Pro	CCU	428	1.52	
Pro	CCC	206	0.73	trnP-GGG
Pro	CCA	334	1.18	trnP-UGG
Pro	CCG	162	0.57	
Thr	ACU	540	1.58	
Thr	ACC	267	0.78	trnT-GGU
Thr	ACA	410	1.20	trnT-UGU
Thr	ACG	148	0.43	
Ala	GCU	616	1.75	
Ala	GCC	245	0.70	
Ala	GCA	400	1.14	trnA-UGC
Ala	GCG	143	0.41	
Tyr	UAU	783	1.60	
Tyr	UAC	198	0.40	trnY-GUA
Stop	UAA	44	1.52	
Stop	UAG	25	0.86	
His	CAU	482	1.53	
His	CAC	147	0.47	trnH-GUG
Gln	CAA	705	1.49	trnQ-UUG
Gln	CAG	244	0.51	
Asn	AAU	1003	1.52	
Asn	AAC	317	0.48	trnN-GUU
Lys	AAA	1052	1.48	trnK-UUU
Lys	AAG	371	0.52	
Asp	GAU	860	1.60	
Asp	GAC	217	0.40	trnD-GUC
Glu	GAA	1036	1.47	trnE-UUC
Glu	GAG	370	0.53	
Cys	UGU	223	1.46	

Table 2. Cont.

Amino acid	Codon	Count	RSCU	tRNA
Cys	UGC	82	0.54	trnC-GCA
Stop	UGA	18	0.62	
Trp	UGG	475	1.00	trnW-CCA
Arg	CGU	341	1.26	trnR-ACG
Arg	CGC	100	0.37	
Arg	CGA	394	1.46	
Arg	CGG	126	0.47	
Arg	AGA	487	1.80	trnR-UCU
Arg	AGG	174	0.64	
Ser	AGU	420	1.21	
Ser	AGC	119	0.34	trnS-GCU
Gly	GGU	578	1.26	
Gly	GGC	199	0.43	trnG-GCC
Gly	GGA	740	1.61	trnG-UCC
Gly	GGG	324	0.70	

gene may have already been under purify selection prior to the evolution between *D. stramonium* and *N. tabacum*. *Nicotiana undulate* was relatively closest to the reference species *Nicotiana tabacum* among all four species. The Ka/Ks ratio was 0 in 45 of 62 compared gene, which displayed that the rate of synonymous and non- synonymous was not existed among *N. undulate* and *N. tabacum*. The evolution in species level of Solanoideae is highly conservative.

Variation between the coding sequences of *D. stramonium* and *Ipomoea purpurea*, *Nicotiana undulate* or *Solanum tuberosum* was also analyzed by comparing each individual gene as well as the overall sequences (Table S7). The four rRNA genes are the most conserved, while the most divergent coding regions are *accD*, *cemA*, *psbT*, *clpP*, and *ycf1*.

IR Contraction and Expansion

The size variation of angiosperm cp genomes is primarily due to expansion and contraction of the IR region and the single copy (SC) boundary regions. Detailed comparison at the junction of the IR/SC boundaries among *Atropa belladonna*, *Nicotiana tomentosiformis*, *Nicotiana tabacum*, *Solanum bulbocastanum*, *Solanum lycopersicum*, *Datura stramonium* was presented in Figure 5. Despite the similar length of the IR regions in the six species, from 25,342 bp to 25,906 bp, some IR expansions and contractions were observed. *Rps19* and *ycf1* pseudogenes of various lengths were located at the IRb/LSC and IRb/SSC boundaries, respectively. The border of IRb-LSC junction was located within the *rps19* gene in these cp genomes except in *N. tabacum*, resulting in the formation of the *rps19* pseudogenes. In *D.*

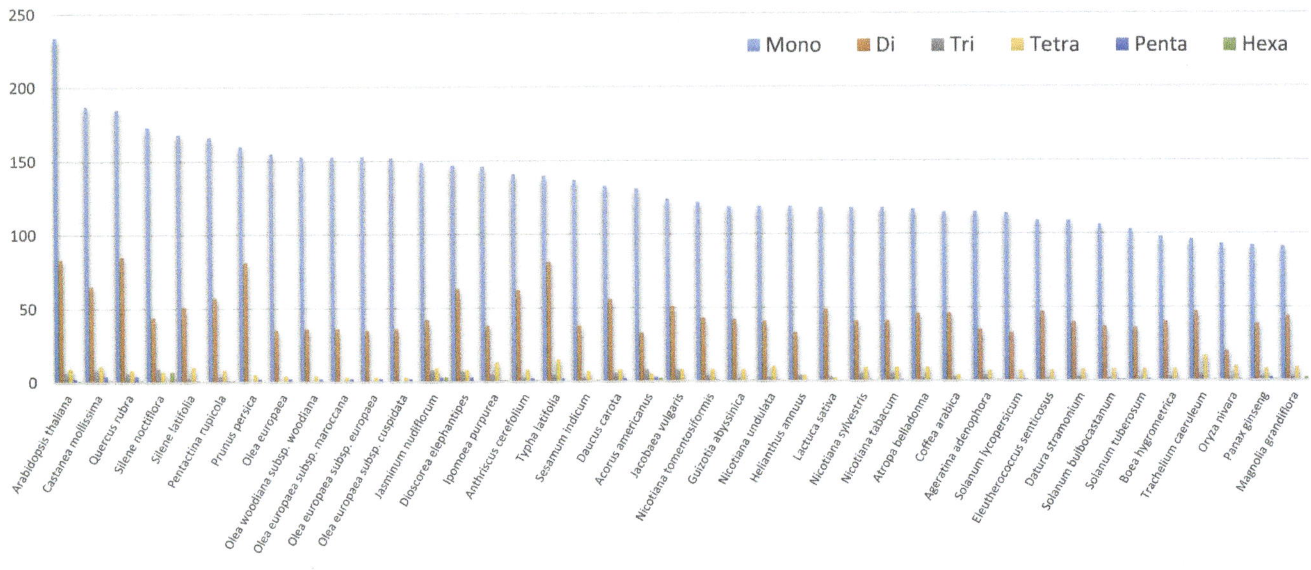

Figure 2. Distribution of SSRs present in the 41 asteridae chloroplast genomes.

Table 3. Repeated sequences in the *Datura stramonium* chloroplast genome.

Repeat number	Size (bp)	Type	Location	Repeat Unit	Region
1	50	F	*psaB* (CDS), *psaA* (CDS)	GAGAAAAATAAATGCAATAGCTAAATGGTGATGGGCAATATCAGTCAGCC	LSC
2	39	F	*ycf3* (intron), IGS (*rps12, trnV-GAC*)	CCAGAACCGTACGTGAGATTTTCACCTCATACGGCTCCT	LSC, IRa
3	39	F	*ycf3* (intron), *ndhA* (intron)	ACAGAACTGTACGTGAGATTTTCACCTCATACGGCTCCT	LSC, SSC
4	39	F	IGS (*rps12, trnV-GAC*), *ndhA* (intron)	TCAGAACCGTACATGAGATTTTCACCTCATACGGCTCCT	IRa, SSC
5	40	F	*trnF-GAA*	TCAGAGGACTGAAAATCCTCGTGTTACCACTCCAAATCTG	LSC
6	34	F	IGS (*rrn4.5, rrn5*)	TCATTGTTCAAATCTTTGACAACACGAAAAAACC	SSC
7	35	F	*ycf2* (CDS)	AATATTGATGATAGTGACGATATTGATGATAGTGA	IRa
8	30	F	*ycf3* (intron), IGS (*rps12, trnV-GAC*)	GTGAGATTTTCACCTCATACGGCTCCTCCC	LSC, IRa
9	31	F	*trnS-GCU, trnS-UGA*	CAACGGAAAGAGAGGGATTCGAACCCTCGGT	LSC
10	31	F	*trnG-GCC, trnG-UCC*	CGATGCGGGTTCGATTCCCGCTACCCGCTCT	LSC
11	31	F	IGS (*psbC, trnS-UGA*), *clpP* (intron)	CTTTTTTCTTTTTTGTTTTCAACTCATTTTA	LSC
12	56	P	*petD* (intron)	GTATAAGTGAACTAGATAAAACGGAATCTTGATTCCGTTTTATCTAGTTCACTTAT	LSC
13	48	P	IGS (*psbT, psbN*)	CAGTTGAAGTACTGAGCCTCCCGATATCGGGAGGCTCAGTACTTCAAC	LSC
14	39	P	*ycf3* (intron), IGS (*trnV-GAC, rps12*)	CCAGAACCGTACGTGAGATTTTCACCTCATACGGCTCCT	LSC, IRb
15	39	P	*ndhA* (intron), IGS (*trnV-GAC, rps12*)	ACAGAACTGTACGTGAGATTTTCACCTCATACGGCTCCT	SSC, IRb
16	30	P	*trnS-GCU, trnS-GGA*	AACGGAAAGAGAGGGATTCGAACCCTCGGT	LSC
17	34	P	IGS (*rrn4.5, rrn5*), IGS (*rrn5, rrn4.5*)	TCATTGTTCAAATCTTTGACAACACGAAAAAACC	IRa, IRb
18	35	P	*ycf2* (CDS)	AATATTGATGATAGTGACGATATTGATGATAGTGA	IRa, IRb
19	32	P	IGS (*trnE-UUC, trnT-GGU*)	CTTTTTTTATTTAGAAATTTGTAAATAAAAAA	LSC
20	30	P	*trnS-UGA, trnS-GGA*	AAAGGAGAGAGAGGGATTCGAACCCTCGAT	LSC
21	44	T	IGS (*trnK-UUU, rps16*)	CTACTTAATTTAAAATTTAAAA (*2)	LSC
22	40	T	IGS (*atpH, atpI*)	TTATTCATTTTTATTATTTAT (*2)	LSC
23	33	T	IGS (*rps2, rpoC2*)	CATTATTCTTTTCTATT (*2)	LSC
24	45	T	IGS (*trnT-GGU, psbD*)	ATTAATTTCATCTATATTATATA (*2)	LSC
25	35	T	IGS (*trnT-UGU, trnL-UAA*)	TTCTATATTGGATTCTA (*2)	LSC
26	50	T	IGS (*trnP-GGG, psaJ*)	ATTATATAGAAAATACTTATATACA (*2)	LSC
27	41	T	*rps18* (CDS)	TAAATCCAAGCGACCTTTTCT (*2)	LSC
28	47	T	*clpP* (intron)	GATAAAGCAAAGAGAAAAGAA (*2)	LSC
29	58	T	*ycf2* (CDS)	ATATTGATGATAGTGACG (*3)	IRa
30	62	T	IGS (*rps12, trnV-GAC*)	TATTATATTAGTATTTTCTATT (*3)	IRa
31	66	T	IGS (*rrn4.5, rrn5*)	CATTGTTCAAATCTTTGACAACACGAAAAAAC (*2)	IRa
32	39	T	*ndhF* (CDS)	AATAAAAACCTAAAATTCCT (*2)	SSC
33	55	T	*ycf1* (CDS)	TTCCTTTCTTTGATTCTCCTCTTTTTT (*2)	SSC

*copy number.
'F' is forward, 'P' is palindromic, and 'T' is Tandem; IGS: Intergenic spacer; CDS: protein-coding regions.

stramonium, a short *rps19* pseudogene of 60 bp was created at the IRa-LSC border. The same pseudogene was 60 bp in *A. belladonna* and *N. tomentosiformis*, 39 bp in *S. bulbocastanum* and 88 bp in *S. lycopersicum*, respectively. The IRb-SSC border extended into the *ycf1* genes to create long *ycf1* pseudogenes in all of the cp genomes except in *Solanum bulbocastanum* where the IRb-SSC border expanded to duplicate a part of *ndhF* gene. The length of *ycf1* pseudogene was 1,469 bp in *A. belladonna*, 1,016 bp in *N. tomentosiformis*, 1,052 bp in *N. tabacum*, 1,117 bp in *S. bulbocastanum*, 1,139 bp in *Solanum lycopersicum* and 1,118 bp in *D. stramonium*. In *N. tabacum*, *S. lycopersicum* and *D. stramonium* cp genomes, the *ycf1* pseudogene and the *ndhF* gene overlapped by 11 bp, 3 bp and 20 bp, respectively. The IRa-SSC border was located within the CDS of *ycf1* gene and the length of the part of this gene in IR region was significantly different among the six cp genomes. The *trnH* gene was all located in the LSC regions and has 3–6 bp apart from the IRa-LSC border.

Phylogenetic Analysis

Phylogenetic analysis was performed on a 42-taxon 68-gene data matrix using MP and ML methods. The sequence alignment data comprised 41,127 characters after the gaps were excluded to avoid alignment ambiguities due to length variation. The MP analysis resulted in a single most-parsimonious tree (Figure S1) with a consistency index (CI) of 0.53 (excluding uninformative characters) and a retention index (RI) of 0.68. Bootstrap analysis showed that 36 of the 40 nodes were supported by values ≥70%, and all of the nodes had a bootstrap value >50%. The ML analyses, using a single model for all of the genes (GTR +G+I), produced a single tree (Figure 6) with -lnL (unconstrained) = 356757.56. The ML bootstrap values were also high, with values of ≥90% for 31 of the 39 nodes and only one support value < 70%. ML and MP trees exhibited the same topology within asteridae lineage and phylogenetic position of *Datura* was found between *Solanum* and *Atropa* in this study.

Discussion

Comparative Analysis of the cp Genome Organization

Datura stramonium, also known as jimson weed, devil weed or thorn apple, has been used for mystic and religious purposes as a mystical sacrament which brings about powerful visions and opens the user to communication with spirit world [39]. Especially *D. stramonium* had a long history of medicinal use in Asian countries since two thousand years ago. During the Three Kingdom Period (220–265 A.D.),its use as the first anesthetic for surgery was recorded in literatures [40]. Modern studies showed that *D. stramonium* had varieties of pharmacological effects including antiasthmatic [41], antiepileptic [42], antioxidant [43], antimicrobial [44,45], antifungal [46] and anti-inflammatory [47] activities. Approximately 400 complete cp genome sequences have been sequenced in GenBank. However most of these sequences are focused on economically important plants, such as *Solanum lycopersicum*, *Oryza sativa* and *Nicotiana tabacum*. In contrast, only few cp genome sequences have been reported for medicinal purposes such as *Salvia miltiorrhiza* and *Panax ginseng*, and still no cp genome sequences have been reported for *Datura*. The availability of the complete cp DNA sequences from *D. stramonium* provides us an improved evolutionary understanding of the chloroplast genome itself and it can also serve as a medicinal improvement tool. The *D. stramonium* cp genome has a typical angiosperm organization with a pair of IRs separating the LSC and SSC regions and exhibits identical gene order and content to the sequenced Solanales cp genomes [48], emphasizing the highly conserved nature of these land plant cp genomes [49]. The cp genome of *D. stramonium* has no significant difference compared with other Solanales genomes except *Ipomoea purpurea* (162,046 bp, Table S6). The GC content could be one of the most important factors in the evolution of genomic structures [50].

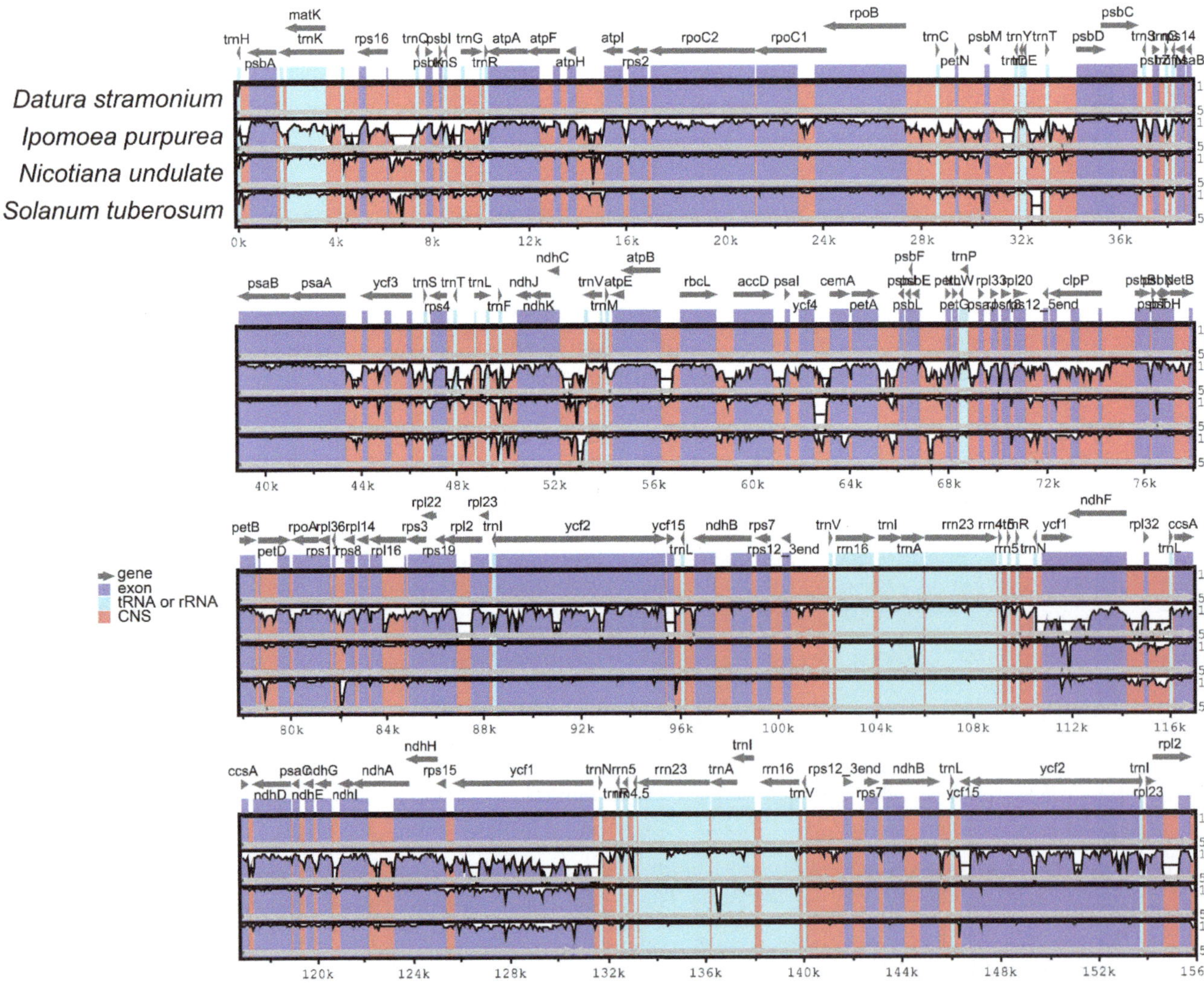

Figure 3. Comparison of four chloroplast genomes using mVISTA program. Grey arrows and thick black lines above the alignment indicate genes with their orientation and the position of the IRs, respectively. A cut-off of 70% identity was used for the plots, and the Y-scale represents the percent identity between 50–100%. Genome regions are color-coded as protein-coding (exon), rRNA, tRNA and conserved noncoding sequences (CNS).

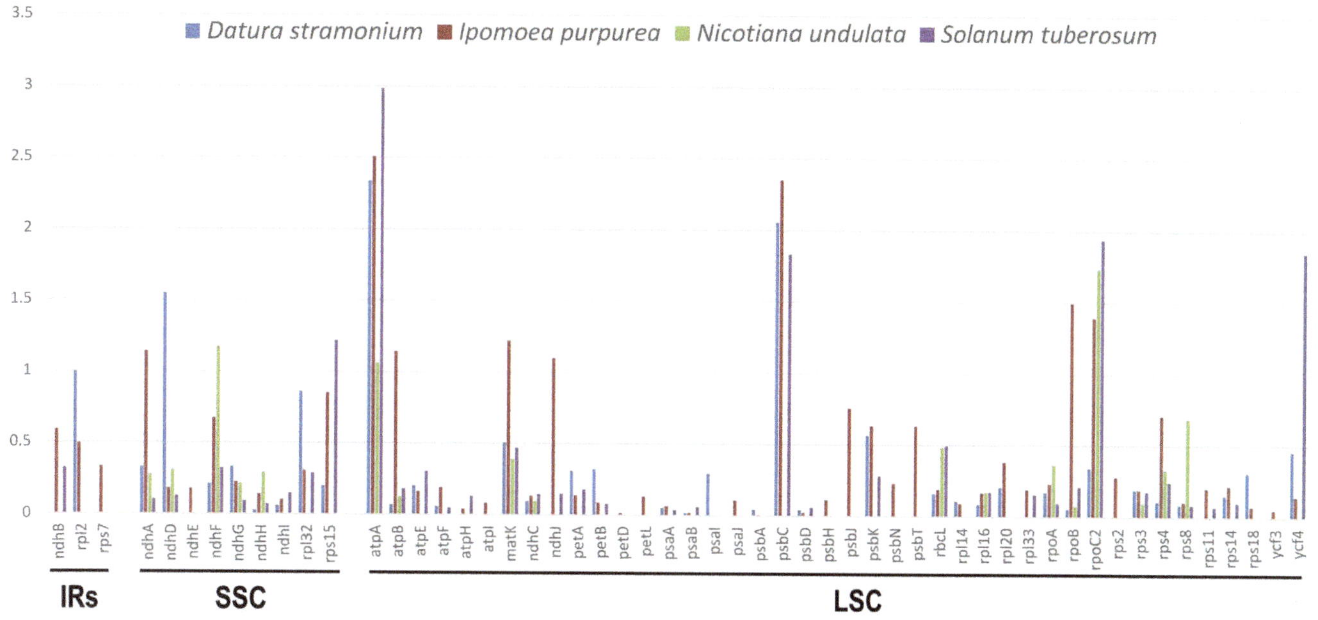

Figure 4. The Ka/Ks ratio of Datura stramonium, Ipomoea purpurea, Nicotiana undulate and Solanum tuberosum for comparison with Nicotiana tabacum.

We found that the GC content was unevenly distributed in the entire cp genome of *D. stramonium* and the divergence of conserved nature between IR and SC regions might be partly due to the different GC content. In addition, the *ycf15* gene was completely annotated in *D. stramonium* cp genome while most recent studies supported the conclusion that *ycf15* is not a functional gene in protein-coding process [51–53].

In the universal genetic code, codons mainly differ at the third position. Though many synonymous codons needed to regulate the translation process, but only particular codons are preferred. Results in this study showed that the synonymous codons usage was not at the same frequencies and the patterns of synonymous codon usage also varied significantly among genes, which were consistent with previous investigations [54]. Codon bias of cp genes has been reported to be towards codons ended with A or T due to the compositional bias towards AT rich content [55,56]. Since all cp genomes have high AT content, AT biased mutational pressure is believed to be the factor responsible for codon usage bias. Previous studies demonstrated that there existed a significant relationship between codon usage bias and gene expression level [57,58], which suggested stronger natural selection constraints on highly expressed genes to optimize translation efficiency using major codons [59]. Information about the rare and preferred codons can be effectively used for enhancing gene expression by optimizing synonymous codons, which may provide us a further understanding of synthesis and metabolism of secondary metabolites in *D. stramonium*.

The intron plays an important role in the regulation of gene expression. Some recent studies have found that many introns improve exogenous gene expression at specific positions and times, resulting in the expected agronomic characters. Therefore, introns can be a useful tool to improve transformation efficiency [60]. A total of 14 intron-containing genes were detected and 11 of which contain one intron but 3 of which have two introns, which are similar to the cp genome of *Nicotiana tabacum*. These results are helpful for further transformation studies in *D. stramonium*.

Cp SSRs have frequently been used in species identification and genetic analysis at individual or group levels because of their high reproductivity, codominant inheritance, relatively high polymorphism, and relative abundance in genomes. There are altogether 160 cp SSRs discovered in *D. stramonium* cp genome. These markers will allow us to improve our understanding of the population structure and genetic diversity of this species that are essential for molecular breeding and cp genetic engineering. In this study, we also investigated the distribution of SSR in 41 cp genomes of Asteridae. The average number of SSRs in the CDS regions accounted for 30% of all discovered SSRs and the average SSR proportion located in LSC regions was 64.95% in these studied cp genomes. This result indicates that SSRs are less abundant in CDS than in non-coding regions and that they are unevenly distributed within Lamiales cp genomes, which provides more information for choosing effective molecular markers and detecting both intra- and interspecific polymorphisms within this order [61,62]. In addition, mononucleotide, dinucleotide, and trinucleotide repeats were composed of A or T at a higher level. This may contribute to a bias in base composition, which was consistent with the overall A-T richness (62.3%) of the Asteridae cp genome. The bias may have a close relationship with the easier changes to A-T rather than G-C in the genome [63]. An interesting finding was that the first seven SSR loci with largest number of mononucleotide repeat were distributed in Fagales, Rosales, Caryophyllales and Brassicales, and the four groups were closely related within asteridae and formed into a clade in maximum parsimony tree in this study.

In the analysis of repeat structure, 11 forward, 9 palindromic and 13 tandem repeats were revealed. Among these repeats, 72.7% of all forward repeats were distributed in the LSC region, 66.7% and 61.5% in palindromic and tandem repeats, respectively. In addition, most of all repeats are discovered located in the intergenic spacers or introns (Table 3). Short dispersed repeats are considered to be one of the major factors promoting cp genome rearrangements [64]. It was demonstrated that there existed a correlation between the abundance of short dispersed repeats and

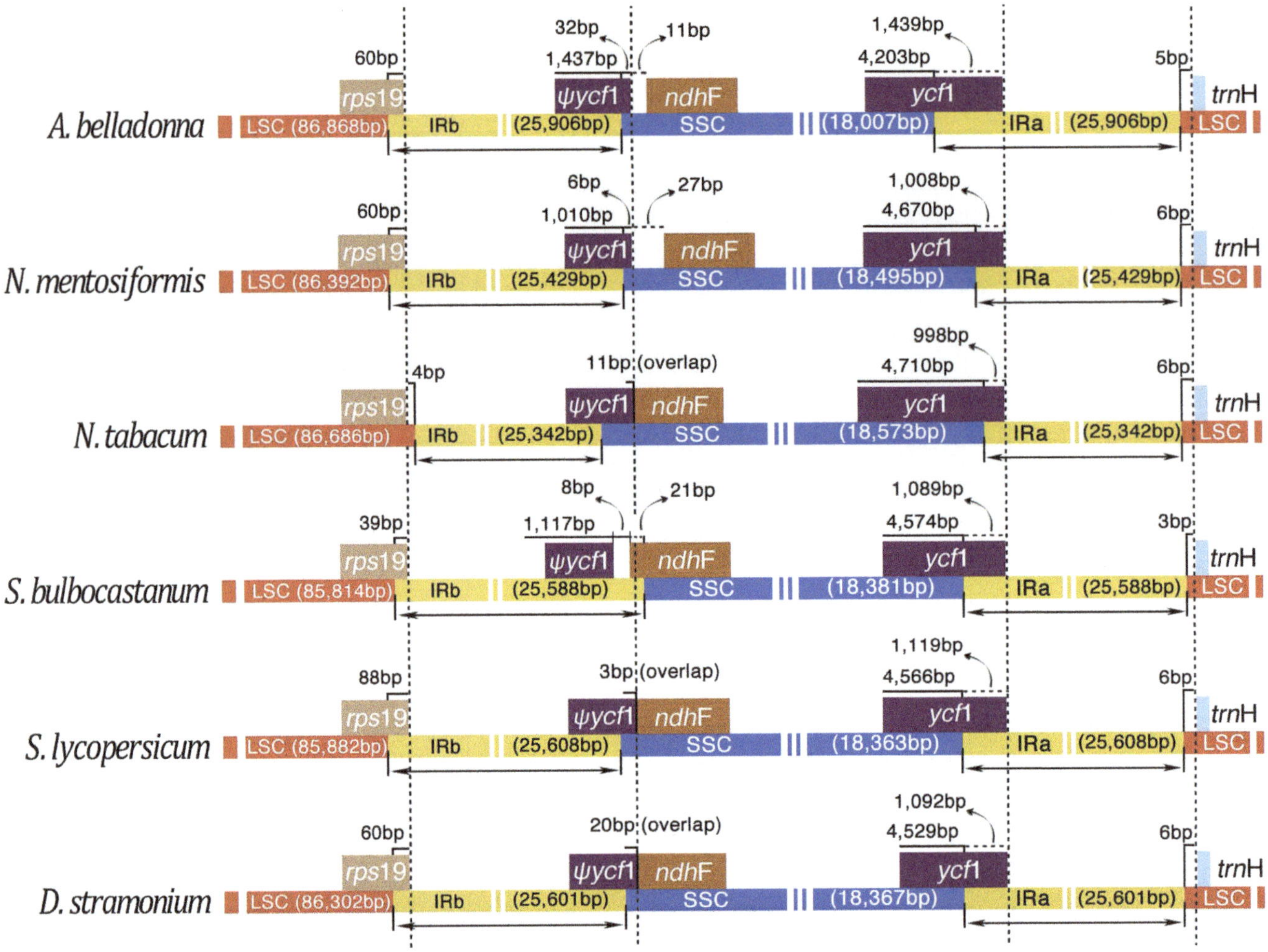

Figure 5. Comparison of the borders of LSC, SSC and IR regions among six chloroplast genomes. The IRb/SSC border extended into the *ycf1* genes to create various lengths of *ycf1* pseudogenes among six chloroplast genomes. The *ycf1* pseudogene and the *ndhF* gene overlapped in the *N. tabacum*, *S. lycopersicum* and *D. stramonium* cp genomes from 3 bp to 20 bp, respectively. Various lengths of *rps19* pseudogenes were created at the IRa/LSC borders except *N. tabacum*. This figure is not to scale.

the extent of gene rearrangements [65]. Most of these repeats always occur near the rearrangements hotspots and may mediate these regions [66,67]. In addition, short repeat motifs may facilitate inter-molecular recombination and create diversity of chloroplast genomes in a population [68]. Therefore repeats found in this study provide valuable information for phylogeny of *Datura* and population studies of *D. stramonium*.

Differences in cp genome size are mainly caused by the contraction and expansion of the IR regions [63,69]. However comparison of the IR boundary among six Solanaceae species showed that the size of the IR regions has no significant relationship with the length of the complete cp genome sequence (Figure 5). Correlation analysis indicated that the length of the IR regions had a positive correlation with that of ycf1 gene located in IR region ($R2 = 0.9$, $P<0.05$). All *trnH* genes in the six Solanaceae species were found located in LSC region whereas this gene was completely located in the IR region in monocot cp genomes [70]. We also compared four Solanales cp genomes using mVISTA and observed an approximately identical gene order and organization among them (Figure 3). The comparison demonstrates that the two IR regions are less divergent than the LSC and SSC regions. The five most divergent coding regions are *accD*, *cemA*, *psbT*, *clpP* and *ycf1*. The *ycf1* gene is considered as the most variable locus with unknown function in recent study, and is confirmed that it was more variable than the *matK* gene in the Orchidaceae [71]. The *ψycf1* (pseudogenes) located in the IRb region is conservative while the *ycf1* located in the SSC with highly variable. Dong *et al* used the two regions of *ycf1* (*ycf1*-a and *ycf1*-b) as a new tool to solve the phylogenetic problems at species level and for DNA barcoding of some closely related flowering plant species because of their high variability [72]. In addition, non-coding regions exhibit a higher divergence than coding regions and the most divergent regions localize in the intergenic spacers. These highly divergent regions including *trnH-psbA*, *rps4-trnS*, *ndhD-ccsA* and *ndhI-ndhG* can be used to develop markers or specific barcodes [73] that would maximize the ability to differentiate species within the Solanales. Data analysis concerning sequence divergence (Table S7) and Ka/Ks ratio (Figure 4) also supported that the IR regions are more conserved compared with SSC and LSC regions.

Phylogenetic Implications

Chloroplast genomes have shown a substantial power in studies of phylogenetics, evolution and molecular systematics. During the

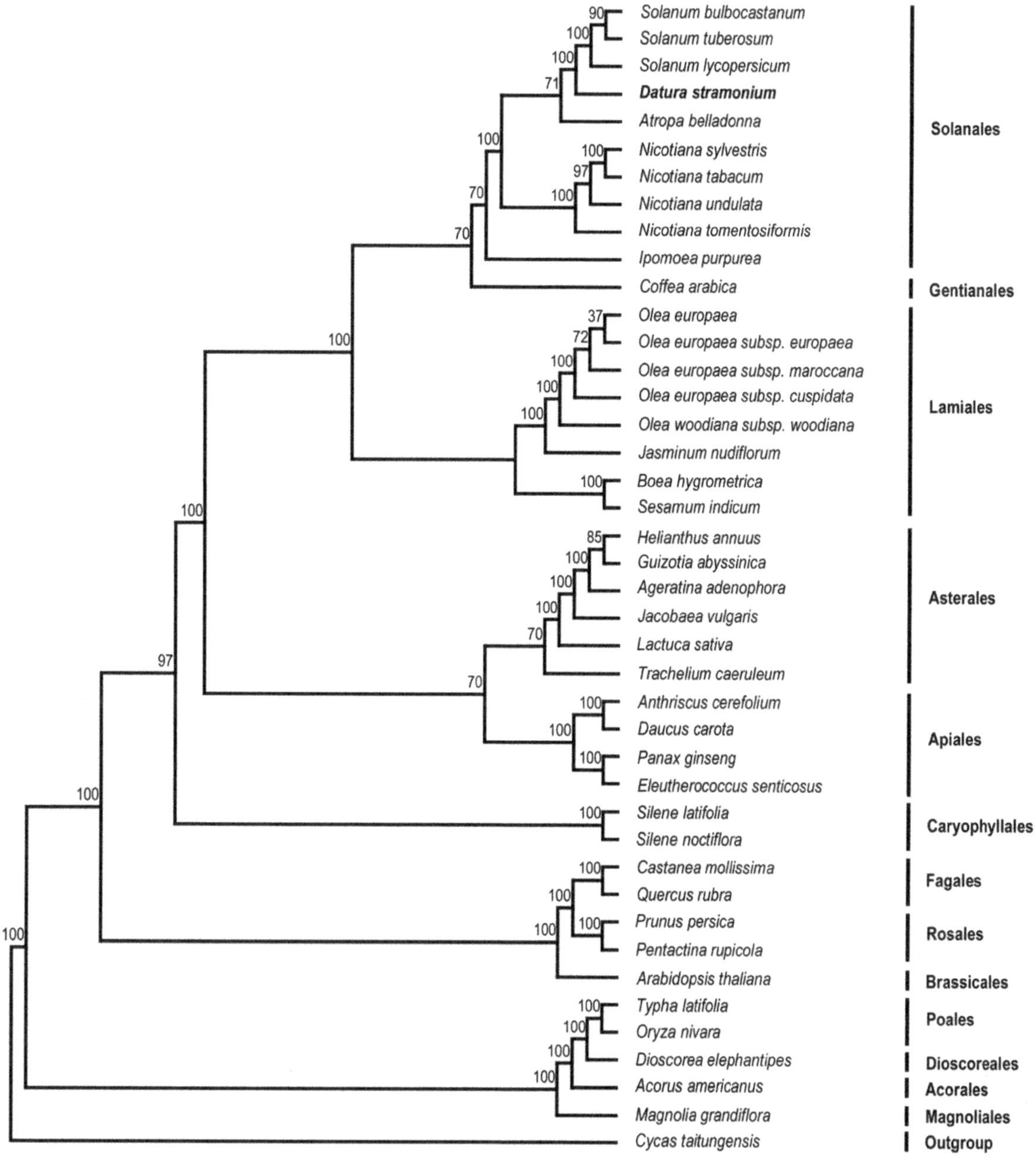

Figure 6. The ML phylogenetic tree of the asteridae clade based on 68 protein-coding genes. The ML tree was obtained with the - lnL of 356757.56 using the GTR+I+G nucleotide substitution model. Number above each node are bootstrap support values. *Cycas taitungensis* was set as outgroup.

last decade, there have been many analyses to address phylogenetic questions at deep nodes based on comparison of multiple protein-coding genes [74–76] and complete sequences in chloroplast genomes [77,78], enhancing our understanding of enigmatic evolutionary relationships among angiosperms. However, further development of Asteridae phylogeny is typically limited due to sporadic taxon sampling. Phylogenetic analysis using maximum likelihood and maximum parsimony were performed based on 68 shared genes (Table S2) in 42 sequenced genomes, including the cp genome of *D. stramonium* sequenced in this study, to examine the position of *D. stramonium* and relationships within the Asteridae. Both trees have provided strong support for the position of *D. stramonium* as a sister to *Solanum lycopersicum*, followed by *Atropa belladonna* (Figure 6 and Figure S1). The difference between the MP and ML trees involves the position of *Silene*, which is likely to be caused by long-branch attraction [79]. The asteridae, the largest and diverse subclass in angiosperm, includes more than 60,000 species and is widely distributed throughout the world. More taxon samplings should be required to clarify accurate relationships among asteridae.

Implication for Chloroplast Genetic Engineering

Chloroplasts are distributed throughout the differentiated cells of plant organs and tissues. Over evolutionary time, cp genomes have given up most of their genes and cellular functions to become the energy transduction and metabolic center of plant cell [80].

The high copy number of chloroplast genomes makes it possible to provide an engineering of multiple foreign genes for the production of a metabolic pathway with a high transformation rate in contrast with nuclear transformation. Great progress in chloroplast engineering has been achieved since the first chloroplast genetic transformation succeeded two decades ago [81].

Although a number of plant species are transformable, plastid transformation is now routinely carried out only in tobacco [82]. In addition, while gene regulation is generally conserved, expression of a foreign gene may vary between different plant species [83]. The expression of a transgene can be affected by various factors such as the promoter, the 5′ untranslated region (UTR), the downstream box, the N-terminal amino acid sequence, the codon usage, the 3′ UTR and genes located upstream and downstream [83]. The efficiency of transformation in most plants remains too low. This study showed that *Datura stramonium* had an identical plastid genome structure and similar sequence relative to *Nicotiana tabacum*. The two plant species are very closely related in evolutionary relationship. In addition, many transformable species are from Solanaceous including tomato [84], petunia [85], eggplant [86] and potato [87]. *D. stramonium* may have great potential to become a model medicinal plant to carry out plant transformation. The availability of the complete cp genome sequence of *D. stramonium* is helpful to recognize the optimal regions for transgene integration and to develop site-specific cp transformation vectors.

Datura stramonium naturally grows in warm and temperate regions and does have a low tolerate for cold environments. They are not especially susceptible to pests, but will suffer from mealy bugs and aphids. In addition, *Datura* are propagated by seed. Young seedlings are very tolerant of poor soil and even drought but cannot tolerate herbicide. The plastid genome is an attractive location for the engineering of pest-resistance and herbicide-tolerance traits. Expression of insecticidal proteins and herbicide-tolerant enzymes from the chloroplast genome has proven to be a very efficient strategy for successful resistance management and weed control [88–92]. Plastid engineering should be particularly useful to develop resistant to abiotic and biotic stresses in molecular breeding of *D. stramonium*.

Conclusions

This is the first study of complete cp genome of *Datura* species which can extract scopolamine. The gene order and genome organization of *D. stramonium* are similar to those of cp genomes in the Solanales. There are no significant structural rearrangements of Solanales cp genomes during the evolutionary process. Further, the repeated sequences, SSR and protein-coding gene sequence were determined. Phylogenetic relationships among 42 angiosperms provide a strong support for the position of *D. stramonium*. In addition, the data presented in this paper will facilitate the further biological study in the field of phylogenomics and plant biotechnology of this important poisonous and medicinal plant.

Supporting Information

Figure S1 The MP phylogenetic tree of the asteridae clade based on 68 protein-coding genes. The MP tree has a length of 59,852, with a consistency index of 0.53 and a retention index of 0.68. Number above each node are bootstrap support values. *Cycas taitungensis* was set as outgroup.

Table S1 The list of accession numbers of the chloroplast genome sequences used in this study.

Table S2 Average pairwise sequence distance of protein-coding genes among 42 chloroplast genomes.

Table S3 Genes present in the Datura stramonium chroloplast genome.

Table S4 The genes with introns in the Datura stramonium chloroplast genome and the length of the exons and introns.

Table S5 Distribution of SSRs present in the CDS among 41 asteridae chloroplast genomes.

Table S6 Size comparison of 10 cp genomes in the order of Solanales.

Table S7 Comparison of homologues between the *Datura stramonium* and *Ipomoea purpurea* (*Ip*), *Nicotiana undulate* (*Nu*) or *Solanum tuberosum* (*St*) chloroplast genomes using the percent identity of protein-coding sequences.

Acknowledgments

The authors would like to thank the reviewers for their valuable comments and suggestions. We are also grateful to Robert Henry for his critical reading of the manuscript.

Author Contributions

Conceived and designed the experiments: LQ LXW YY. Performed the experiments: LQ LXW YY. Analyzed the data: YY LXW. Contributed reagents/materials/analysis tools: LXW WYT. Wrote the paper: YY LXW LJJ DYY.

References

1. Zhang L, Ding R, Chai Y, Bonfill M, Moyano E, et al. (2004) Engineering tropane biosynthetic pathway in Hyoscyamus niger hairy root cultures. Proc Natl Acad Sci U S A 101: 6786–6791.
2. Weissman B, Raveh L (2011) Multifunctional drugs as novel antidotes for organophosphates' poisoning. Toxicology 290: 149–155.
3. Jakabova S, Vincze L, Farkas A, Kilar F, Boros B, et al. (2012) Determination of tropane alkaloids atropine and scopolamine by liquid chromatography-mass spectrometry in plant organs of Datura species. Journal of Chromatography A 1232: 295–301.
4. Lacy BE, Wang F, Bhowal S, Schaefer E (2013) On-demand hyoscine butylbromide for the treatment of self-reported functional cramping abdominal pain. Scandinavian Journal of Gastroenterology 48: 926–935.
5. Klinkenberg I, Blokland A (2010) The validity of scopolamine as a pharmacological model for cognitive impairment: A review of animal behavioral studies. Neuroscience and Biobehavioral Reviews 34: 1307–1350.
6. Liu S, Li LH, Shen WW, Shen XY, Yang GD, et al. (2013) Scopolamine Detoxification Technique for Heroin Dependence: A Randomized Trial. Cns Drugs 27: 1093–1102.
7. Rasila Devi M, Bawari M, Paul S, Sharma G (2011) Neurotoxic and medicinal properties of Datura stramonium L.–review. Assam University Journal of Science and Technology 7: 139–144.
8. Palazon J, Navarro-Ocana A, Hernandez-Vazquez L, Mirjalili MH (2008) Application of metabolic engineering to the production of scopolamine. Molecules 13: 1722–1742.

9. Moyano E, Jouhikainen K, Tammela P, Palazon J, Cusido RM, et al. (2003) Effect of pmt gene overexpression on tropane alkaloid production in transformed root cultures of Datura metel and Hyoscyamus muticus. J Exp Bot 54: 203–211.
10. Pramod KK, Singh S, Jayabaskaran C (2010) Biochemical and structural characterization of recombinant hyoscyamine 6 beta-hydroxylase from Datura metel L. Plant Physiology and Biochemistry 48: 966–970.
11. Moyano E, Palazon J, Bonfill M, Osuna L, Cusido RM, et al. (2007) Biotransformation of hyoscyamine into scopolamine in transgenic tobacco cell cultures. Journal of Plant Physiology 164: 521–524.
12. Bendich AJ (1987) Why Do Chloroplasts and Mitochondria Contain So Many Copies of Their Genome. Bioessays 6: 279–282.
13. Hagemann R (2004) The sexual inheritance of plant organelles. Molecular biology and biotechnology of plant organelles: Springer. pp.93–113.
14. Li XW, Hu ZG, Lin XH, Li Q, Gao HH, et al. (2012) [High-throughput pyrosequencing of the complete chloroplast genome of Magnolia officinalis and its application in species identification]. Acta pharmaceutica Sinica 47: 124–130.
15. Wyman SK, Jansen RK, Boore JL (2004) Automatic annotation of organellar genomes with DOGMA. Bioinformatics 20: 3252–3255.
16. Schattner P, Brooks AN, Lowe TM (2005) The tRNAscan-SE, snoscan and snoGPS web servers for the detection of tRNAs and snoRNAs. Nucleic Acids Research 33: W686–W689.
17. Cui LY, Veeraraghavan N, Richter A, Wall K, Jansen RK, et al. (2006) ChloroplastDB: the chloroplast genome database. Nucleic Acids Research 34: D692–D696.
18. Lohse M, Drechsel O, Bock R (2007) OrganellarGenomeDRAW (OGDRAW): a tool for the easy generation of high-quality custom graphical maps of plastid and mitochondrial genomes. Current Genetics 52: 267–274.
19. Tamura K, Peterson D, Peterson N, Stecher G, Nei M, et al. (2011) MEGA5: Molecular Evolutionary Genetics Analysis Using Maximum Likelihood, Evolutionary Distance, and Maximum Parsimony Methods. Molecular Biology and Evolution 28: 2731–2739.
20. Kurtz S, Phillippy A, Delcher AL, Smoot M, Shumway M, et al. (2004) Versatile and open software for comparing large genomes. Genome Biology 5.
21. Frazer KA, Pachter L, Poliakov A, Rubin EM, Dubchak I (2004) VISTA: computational tools for comparative genomics. Nucleic Acids Research 32: W273–W279.
22. Librado P, Rozas J (2009) DnaSP v5: a software for comprehensive analysis of DNA polymorphism data. Bioinformatics 25: 1451–1452.
23. Kurtz S, Choudhuri JV, Ohlebusch E, Schleiermacher C, Stoye J, et al. (2001) REPuter: the manifold applications of repeat analysis on a genomic scale. Nucleic Acids Research 29: 4633–4642.
24. Benson G (1999) Tandem repeats finder: a program to analyze DNA sequences. Nucleic Acids Research 27: 573–580.
25. Nie XJ, Lv SZ, Zhang YX, Du XH, Wang L, et al. (2012) Complete Chloroplast Genome Sequence of a Major Invasive Species, Crofton Weed (Ageratina adenophora). Plos One 7.
26. Thompson JD, Higgins DG, Gibson TJ (1994) Clustal-W - Improving the Sensitivity of Progressive Multiple Sequence Alignment through Sequence Weighting, Position-Specific Gap Penalties and Weight Matrix Choice. Nucleic Acids Research 22: 4673–4680.
27. Kimura M (1980) A Simple Method for Estimating Evolutionary Rates of Base Substitutions through Comparative Studies of Nucleotide-Sequences. Journal of Molecular Evolution 16: 111–120.
28. Stamatakis A (2006) RAxML-VI-HPC: maximum likelihood-based phylogenetic analyses with thousands of taxa and mixed models. Bioinformatics 22: 2688–2690.
29. Posada D, Crandall KA (1998) MODELTEST: testing the model of DNA substitution. Bioinformatics 14: 817–818.
30. Swofford DL, Sullivan J (2003) Phylogeny inference based on parsimony and other methods using PAUP*. The Phylogenetic Handbook: A Practical Approach to DNA and Protein Phylogeny, cáp 7: 160–206.
31. Yi DK, Kim KJ (2012) Complete Chloroplast Genome Sequences of Important Oilseed Crop Sesamum indicum L. Plos One 7.
32. Mariotti R, Cultrera NGM, Diez CM, Baldoni L, Rubini A (2010) Identification of new polymorphic regions and differentiation of cultivated olives (Olea europaea L.) through plastome sequence comparison. Bmc Plant Biology 10.
33. Zhang TW, Fang YJ, Wang XM, Deng X, Zhang XW, et al. (2012) The Complete Chloroplast and Mitochondrial Genome Sequences of Boea hygrometrica: Insights into the Evolution of Plant Organellar Genomes. Plos One 7.
34. Kim KJ, Lee HL (2004) Complete chloroplast genome sequences from Korean ginseng (Panax schinseng Nees) and comparative analysis of sequence evolution among 17 vascular plants. DNA Research 11: 247–261.
35. Shinozaki K, Ohme M, Tanaka M, Wakasugi T, Hayashida N, et al. (1986) The Complete Nucleotide-Sequence of the Tobacco Chloroplast Genome - Its Gene Organization and Expression. Embo Journal 5: 2043–2049.
36. Tangphatsornruang S, Sangsrakru D, Chanprasert J, Uthaipaisanwong P, Yoocha T, et al. (2010) The chloroplast genome sequence of mungbean (Vigna radiata) determined by high-throughput pyrosequencing: structural organization and phylogenetic relationships. DNA Reserach 17: 11–22.
37. Clegg MT, Gaut BS, Learn GH Jr, Morton BR (1994) Rates and patterns of chloroplast DNA evolution. Proc Natl Acad Sci U S A 91: 6795–6801.
38. Chen C, Zhou P, Choi YA, Huang S, Gmitter Jr FG (2006) Mining and characterizing microsatellites from citrus ESTs. Theoretical and Applied Genetics 112: 1248–1257.
39. Gaire BP, Subedi L (2013) A review on the pharmacological and toxicological aspects of Datura stramonium L. Journal of Chinese Integrative Medicine 11: 73–79.
40. Fan Y (2007) Hou Han Shu: Hua tuo biographies. Beijing: Zhonghua book company press. 82 p.
41. Pretorius E, Marx J (2006) Datura stramonium in asthma treatment and possible effects on prenatal development. Environmental Toxicology and Pharmacology 21: 331–337.
42. Peredery O, Persinger MA (2004) Herbal treatment following post-seizure induction in rat by lithium pilocarpine: Scutellaria lateriflora (Skullcap), Gelsemium sempervirens (Gelsemium) and Datura stramonium (Jimson Weed) may prevent development of spontaneous seizures. Phytotherapy Research 18: 700–705.
43. Nimal Christhudas IVS, Praveen Kumar P, Agastian P (2013) In vitro α-glucosidase inhibition and antioxidative potential of an endophyte species (streptomyces sp. Loyola UGC) isolated from datura stramonium L. Current Microbiology 67: 69–76.
44. Eftekhar F, Yousefzadi M, Tafakori V (2005) Antimicrobial activity of Datura innoxia and Datura stramonium. Fitoterapia 76: 118–120.
45. Sharma A, Patel VK, Chaturvedi AN (2009) Vibriocidal activity of certain medicinal plants used in Indian folklore medicine by tribals of Mahakoshal region of central India. Indian Journal of Pharmacology 41: 129–133.
46. Mdee LK, Masoko P, Eloff JN (2009) The activity of extracts of seven common invasive plant species on fungal phytopathogens. South African Journal of Botany 75: 375–379.
47. Sonika G MR, Deepak J (2010) Comparative studies on anti-inflammatory activity of Coriandrum sativum, Datura stramonium and Azadirachta indica. Asian J Exp Biol Sci 1(1): 151–154.
48. Raubeson LA, Peery R, Chumley TW, Dziubek C, Fourcade HM, et al. (2007) Comparative chloroplast genomics: analyses including new sequences from the angiosperms Nuphar advena and Ranunculus macranthus. Bmc Genomics 8.
49. Wicke S, Schneeweiss GM, dePamphilis CW, Muller KF, Quandt D (2011) The evolution of the plastid chromosome in land plants: gene content, gene order, gene function. Plant Molecular Biology 76: 273–297.
50. Bellgard M, Schibeci D, Trifonov E, Gojobori T (2001) Early detection of G + C differences in bacterial species inferred from the comparative analysis of the two completely sequenced Helicobacter pylori strains. J Mol Evol 53: 465–468.
51. Goremykin VV, Hirsch-Ernst KI, Wölfl S, Hellwig FH (2003) Analysis of the Amborella trichopoda chloroplast genome sequence suggests that Amborella is not a basal angiosperm. Molecular biology and evolution 20: 1499–1505.
52. Schmitz-Linneweber C, Maier RM, Alcaraz JP, Cottet A, Herrmann RG, et al. (2001) The plastid chromosome of spinach (Spinacia oleracea): complete nucleotide sequence and gene organization. Plant Molecular Biology 45: 307–315.
53. Steane DA (2005) Complete nucleotide sequence of the chloroplast genome from the Tasmanian blue gum, Eucalyptus globulus (Myrtaceae). DNA Research 12: 215–220.
54. Nair RR, Nandhini MB, Monalisha E, Murugan K, Sethuraman T, et al. (2012) Synonymous codon usage in chloroplast genome of Coffea arabica. Bioinformation 8: 1096.
55. Morton BR (2003) The role of context-dependent mutations in generating compositional and codon usage bias in grass chloroplast DNA. J Mol Evol 56: 616–629.
56. Wolfe KH, Morden CW, Palmer JD (1992) Function and evolution of a minimal plastid genome from a nonphotosynthetic parasitic plant. Proc Natl Acad Sci U S A 89: 10648–10652.
57. Iannacone R, Grieco PD, Cellini F (1997) Specific sequence modifications of a cry3B endotoxin gene result in high levels of expression and insect resistance. Plant Molecular Biology 34: 485–496.
58. Rouwendal GJA, Mendes O, Wolbert EJH, deBoer AD (1997) Enhanced expression in tobacco of the gene encoding green fluorescent protein by modification of its codon usage. Plant Molecular Biology 33: 989–999.
59. Bulmer M (1988) Are Codon Usage Patterns in Unicellular Organisms Determined by Selection-Mutation Balance. Journal of Evolutionary Biology 1: 15–26.
60. Xu J, Feng D, Song G, Wei X, Chen L, et al. (2003) The first intron of rice EPSP synthase enhances expression of foreign gene. Sci China C Life Sci 46: 561–569.
61. Powell W, Morgante M, Andre C, Mcnicol JW, Machray GC, et al. (1995a) Hypervariable Microsatellites Provide a General Source of Polymorphic DNA Markers for the Chloroplast Genome. Current Biology 5: 1023–1029.
62. Powell W, Morgante M, McDevitt R, Vendramin GG, Rafalski JA (1995b) Polymorphic simple sequence repeat regions in chloroplast genomes: applications to the population genetics of pines. Proc Natl Acad Sci U S A 92: 7759–7763.
63. Li X, Gao H, Wang Y, Song J, Henry R, et al. (2013) Complete chloroplast genome sequence of Magnolia grandiflora and comparative analysis with related species. Sci China Life Sci 56: 189–198.
64. Qian J, Song J, Gao H, Zhu Y, Xu J, et al. (2013) The complete chloroplast genome sequence of the medicinal plant Salvia miltiorrhiza. PLoS One 8: e57607.

65. Pombert JF, Otis C, Lemieux C, Turmel M (2005) Chloroplast genome sequence of the green alga Pseudendoclonium akinetum (Ulvophyceae) reveals unusual structural features and new insights into the branching order of chlorophyte lineages. Molecular Biology and Evolution 22: 1903–1918.
66. Chumley TW, Palmer JD, Mower JP, Fourcade HM, Calie PJ, et al. (2006) The complete chloroplast genome sequence of Pelargonium x hortorum: Organization and evolution of the largest and most highly rearranged chloroplast genome of land plants. Molecular Biology and Evolution 23: 2175–2190.
67. Haberle RC, Fourcade HM, Boore JL, Jansen RK (2008) Extensive rearrangements in the chloroplast genome of Trachelium caeruleum are associated with repeats and tRNA genes. Journal of Molecular Evolution 66: 350–361.
68. Kawata M, Harada T, Shimamoto Y, Oono K, Takaiwa F (1997) Short inverted repeats function as hotspots of intermolecular recombination giving rise to oligomers of deleted plastid DNAs (ptDNAs). Current Genetics 31: 179–184.
69. Ravi V, Khurana JP, Tyagi AK, Khurana P (2008) An update on chloroplast genomes. Plant Systematics and Evolution 271: 101–122.
70. Huotari T, Korpelainen H (2012) Complete chloroplast genome sequence of Elodea canadensis and comparative analyses with other monocot plastid genomes. Gene 508: 96–105.
71. Neubig KM, Whitten WM, Carlsward BS, Blanco MA, Endara L, et al. (2009) Phylogenetic utility of ycf1 in orchids: a plastid gene more variable than matK. Plant Systematics and Evolution 277: 75–84.
72. Dong W, Liu J, Yu J, Wang L, Zhou S (2012) Highly Variable Chloroplast Markers for Evaluating Plant Phylogeny at Low Taxonomic Levels and for DNA Barcoding. PLoS ONE 7: e35071.
73. Li X, Yang Y, Henry RJ, Rossetto M, Wang Y, et al. (2014) Plant DNA barcoding: from gene to genome. Biological Reviews. doi:10.1111/brv.12104
74. De las Rivas J, Lozano JJ, Ortiz AR (2002) Comparative analysis of chloroplast genomes: Functional annotation, genome-based phylogeny, and deduced evolutionary patterns. Genome Research 12: 567–583.
75. Lemieux C, Otis C, Turmel M (2000) Ancestral chloroplast genome in Mesostigma viride reveals an early branch of green plant evolution. Nature 403: 649–652.
76. Moore MJ, Soltis PS, Bell CD, Burleigh JG, Soltis DE (2010) Phylogenetic analysis of 83 plastid genes further resolves the early diversification of eudicots. Proceedings of the National Academy of Sciences of the United States of America 107: 4623–4628.
77. Moore MJ, Bell CD, Soltis PS, Soltis DE (2007) Using plastid genome-scale data to resolve enigmatic relationships among basal angiosperms. Proceedings of the National Academy of Sciences of the United States of America 104: 19363–19368.
78. Goremykin VV, Hirsch-Ernst KI, Wolfl S, Hellwig FH (2004) The chloroplast genome of Nymphaea alba: Whole-genome analyses and the problem of identifying the most basal angiosperm. Molecular Biology and Evolution 21: 1445–1454.
79. Bergsten J (2005) A review of long-branch attraction. Cladistics 21: 163–193.
80. Heifetz PB, Tuttle AM (2001) Protein expression in plastids. Current Opinion in Plant Biology 4: 157–161.
81. Boynton JE, Gillham NW, Harris EH, Hosler JP, Johnson AM, et al. (1988) Chloroplast Transformation in Chlamydomonas with High-Velocity Microprojectiles. Science 240: 1534–1538.
82. Wang HH, Yin WB, Hu ZM (2009) Advances in chloroplast engineering. Journal of Genetics and Genomics 36: 387–398.
83. Hanson MR, Gray BN, Ahner BA (2013) Chloroplast transformation for engineering of photosynthesis. Journal of Experimental Botany 64: 731–742.
84. Ruf S, Hermann M, Berger IJ, Carrer H, Bock R (2001) Stable genetic transformation of tomato plastids and expression of a foreign protein in fruit. Nature Biotechnology 19: 870–875.
85. Zubko MK, Zubko EI, van Zuilen K, Meyer P, Day A (2004) Stable transformation of petunia plastids. Transgenic Research 13: 523–530.
86. Singh AK, Verma SS, Bansal KC (2010) Plastid transformation in eggplant (Solanum melongena L.). Transgenic Research 19: 113–119.
87. Sidorov VA, Kasten D, Pang SZ, Hajdukiewicz PTJ, Staub JM, et al. (1999) Stable chloroplast transformation in potato: use of green fluorescent protein as a plastid marker. Plant Journal 19: 209–216.
88. Kota M, Daniell H, Varma S, Garczynski SF, Gould F, et al. (1999) Overexpression of the Bacillus thuringiensis (Bt) Cry2Aa2 protein in chloroplasts confers resistance to plants against susceptible and Bt-resistant insects. Proceedings of the National Academy of Sciences of the United States of America 96: 1840–1845.
89. De Cosa B, Moar W, Lee SB, Miller M, Daniell H (2001) Overexpression of the Bt cry2Aa2 operon in chloroplasts leads to formation of insecticidal crystals. Nature Biotechnology 19: 71–74.
90. Lutz KA, Knapp JE, Maliga P (2001) Expression of bar in the plastid genome confers herbicide resistance. Plant Physiology 125: 1585–1590.
91. Ye GN, Colburn SM, Xu CW, Hajdukiewicz PT, Staub JM (2003) Persistence of unselected transgenic DNA during a plastid transformation and segregation approach to herbicide resistance. Plant Physiol 133: 402–410.
92. Dufourmantel N, Dubald M, Matringe M, Canard H, Garcon F, et al. (2007) Generation and characterization of soybean and marker-free tobacco plastid transformants over-expressing a bacterial 4-hydroxyphenylpyruvate dioxygenase which provides strong herbicide tolerance. Plant biotechnology journal 5: 118–133.

17

Genome–Scale Reconstruction of Metabolic Networks of *Lactobacillus casei* ATCC 334 and 12A

Elena Vinay-Lara[1], Joshua J. Hamilton[2], Buffy Stahl[3], Jeff R. Broadbent[4], Jennifer L. Reed[2], James L. Steele[1]*

1 Department of Food Science, University of Wisconsin-Madison, Madison, Wisconsin, United States of America, **2** Department of Chemical and Biological Engineering, University of Wisconsin-Madison, Madison, Wisconsin, United States of America, **3** DuPont Nutrition and Health, Madison, Wisconsin, United States of America, **4** Utah State University Department of Nutrition, Dietetics, and Food Sciences, Logan, Utah, United States of America

Abstract

Lactobacillus casei strains are widely used in industry and the utility of this organism in these industrial applications is strain dependent. Hence, tools capable of predicting strain specific phenotypes would have utility in the selection of strains for specific industrial processes. Genome-scale metabolic models can be utilized to better understand genotype-phenotype relationships and to compare different organisms. To assist in the selection and development of strains with enhanced industrial utility, genome-scale models for *L. casei* ATCC 334, a well characterized strain, and strain 12A, a corn silage isolate, were constructed. Draft models were generated from RAST genome annotations using the Model SEED database and refined by evaluating ATP generating cycles, mass-and-charge-balances of reactions, and growth phenotypes. After the validation process was finished, we compared the metabolic networks of these two strains to identify metabolic, genetic and ortholog differences that may lead to different phenotypic behaviors. We conclude that the metabolic capabilities of the two networks are highly similar. The *L. casei* ATCC 334 model accounts for 1,040 reactions, 959 metabolites and 548 genes, while the *L. casei* 12A model accounts for 1,076 reactions, 979 metabolites and 640 genes. The developed *L. casei* ATCC 334 and 12A metabolic models will enable better understanding of the physiology of these organisms and be valuable tools in the development and selection of strains with enhanced utility in a variety of industrial applications.

Editor: Mikael Rørdam Andersen, Technical University of Denmark, Denmark

Funding: This work was funded by the DOE Great Lakes Bioenergy Research Center (DOE BER Office of Science DE-FC02-07ER64494) and by the National Research Initiative Competitive Grant no. 2011-67009-30043 from the USDA National Institute of Food and Agriculture. Elena Vinay-Lara gratefully acknowledges the scholarship from CONACyT to pursue her postgraduate studies. Josh J. Hamilton received support through a National Science Foundation Graduate Research Fellowship (DGE-0718123). The funders had no role in study design, data collection and analysis, decision to publish, or preparation of the manuscript.

Competing Interests: BS and JLS received salary and/or funding from DuPont, a supplier of bacterial cultures to the food industry. Peggy Steele, a member of JLS's family, is also employed by DuPont.

* Email: jlsteele@wisc.edu

Introduction

Lactobacillus casei are Gram positive, non-sporulating, nutritionally fastidious and strictly fermentative bacteria that produce lactic acid as their principal metabolic end product [1]. *L. casei* strains are used in the manufacture of fermented foods, as probiotics, and has significant potential for the production of biofuels and co-products with a broad range of applications such as food ingredients [2], [3], [4]. The utility of *L. casei* in all of these applications is strain dependent, due to significant variation in gene content that is present in this species. Comparative genome analysis of seventeen *L. casei* strains has revealed this species has an average of 2,800 orthologous genes (±151), of which 1,715 are shared by all the strains, and 4,220 non-conserved orthologous genes with an average of 119 unique genes found in each strain [5]. This large variation in gene content allows for the selection of strains with unique and industrially important characteristics. Therefore, detailed comparisons of strain specific differences at the genomic and phenotypic level are needed.

Genome-scale metabolic models can serve as tools for evaluating the metabolic potential of strains based on genomic data. These models are becoming a common tool for linking genomic data to the biochemical reaction networks that govern cellular processes [6]. They can be utilized to understand genotype-phenotype relationships and to compare behaviors between different organisms [7]. Additionally, genome-scale models can provide a framework for the integration of transcriptomic, proteomic, and metabolomic data, thus offering a unique global view of cellular physiology and microbial responses to environmental and genetic changes. The integration of such datasets can help reduce the multiple possible flux distributions by reducing the solution space. [8]. Genome-scale models are typically analyzed using constraint-based methods, which impose constraints that consider network stoichiometry, thermodynamics, flux capacity and sometimes transcriptional regulation. These constraints limit metabolic flux distributions and cellular phenotypes [9]. Constraint-based models are rapidly becoming essential tools for understanding microbial metabolism and for developing microorganisms for use in a variety of applications. Furthermore, it is now possible to rapidly construct these models using automated reconstruction technologies [8], [10], thereby enabling the adoption of constraint-based models by a larger scientific community.

To date, *in silico* models are available for four species of lactic acid bacteria (LAB): *Lactococcus lactis* [11], [12], *Lactobacillus plantarum* [13], *Streptococcus thermophilus* [14], and *Lactobacillus reuteri* [15]. The first LAB model was used to simulate the shift from homo- to heterolactic fermentation, identify an *in silico* minimal medium and to improve diacetyl production [11]. Additionally, a genome-scale metabolic model for *Lactococcus lactis* MG1363 was developed; this model couples the carbon and nitrogen metabolism of the strain with catabolic pathways that lead to flavor formation in cheese [12]. The genome-scale metabolic reconstruction of *L. plantarum* was successfully used to evaluate nutrient requirements of the WCFS1 strain [13]. The genome-scale metabolic reconstruction of *S. thermophilus* [14] revealed metabolic differences between *S. thermophilus* and other LAB. The most recent LAB genome-scale metabolic models have been developed for three strains of *Lactobacillus reuteri*, the model strain JCM1112, and two human-derived strains, 6475 and 55730; these models were utilized to identify probiotic features of these strains [15]. Genome-scale metabolic models of LAB have allowed for species and strain comparison, provided guidance for metabolic engineering, and stimulated the development of new hypotheses concerning mechanisms by which probiotics survive in hosts. The availability of genome-scale metabolic models for *L. casei* strains would enhance current understanding of how strain specific characteristics influence the utility of these species in various industrial applications.

In this study, genome-scale metabolic models were constructed for two fully sequenced strains of *L. casei*: *L. casei* ATCC 334, a cheese isolate, and 12A a corn silage isolate. A draft genome sequence for 12A was available (GenBank accession number AFYJ00000000) and this genome was closed for the purposes of this study. These models were then utilized to conduct a detailed comparison of these two strains.

Methods

Genome Sequencing

Genomic DNA of *L. casei* 12A was prepared using the DNA Isolation Bacterial CTAB protocol as requested by the Department of Energy (DOE) Joint Genome Institute (JGI). This protocol can be found at http://my.jgi.doe.gov/general/protocols.html. Genomic DNA quality was confirmed by agarose gel electrophoresis and was submitted to the Department of Energy Joint Genome Institute for genome sequencing, where Ilumina HiSeq 2000 technology was used. JGI delivered a draft genome containing 397 DNA Scaffolds. This draft genome was assembled in collaboration with DuPont Inc., utilizing a previously available 12A draft genome as a scaffold [5]. The assembly contained 19 contigs with 531 fold coverage. The contigs were ordered using the complete genome of *L. casei* ATCC 334 as a scaffold [16]. The remaining gaps were sequenced at the University of Wisconsin Biotechnology Center.

Genome Annotation

The complete genome sequence and annotation of *L. casei* ATCC 334 (GenBank accession number NC_008526 and NC_008502) [16] and 12A are freely available online at National Center for Biotechnology Information (NCBI) (http://www.ncbi.nlm.nih.gov) and at the RAST Server (Rapid Annotations using Subsystems Technology) (http://rast.nmpdr.org/). The genome sequence for *L. casei* 12A was imported into the RAST server (http://rast.nmpdr.org/) for gene calling and annotation with subsequent manual inspection and curation. This information was used in the metabolic reconstruction and validation processes, as described below.

Reconstruction of *L. casei* metabolic networks

Figure 1 summarizes the overall process of reconstructing the two *L. casei* metabolic networks. Genome-scale metabolic draft models for both *L. casei* strains were generated in 2011 using the RAST server [17] and the Model SEED database [18]. We then identified and removed thermodynamically infeasible cycles (e.g. cycles resulting in free ATP production) [19] and mass- and charge-balanced all reactions. The resulting models were then tested by comparing model predictions to amino acid requirement and carbohydrate utilization data [5]. The models were modified as needed using the SMILEY algorithm [20], which suggests reaction additions from a metabolic reaction database (in this case the Model SEED database) to match growth phenotypes. Reactions from the SEED database were downloaded in 2011, and unbalanced reactions were removed from the database prior to employing SMILEY. A copy of the database used for gapfilling is provided in Table S5 in Excel and in Table S9 in SBML format. Additional reactions not suggested by this algorithm were manually added as needed. Reactions suggested by SMILEY, or manually identified, were only added when sufficient evidence and information was available from experimental data, NCBI, KEGG, JGI, and SEED databases. To compare the metabolic networks we identified orthologous genes between the two genomes, which we defined as bidirectional best-BLAST hits with >90% DNA identity (alignments were run in SEED). A total of 511 pairs of orthologs between the two strains were identified that were included in the models.

Flux Balance Analysis

Flux balance analysis (FBA) identifies a steady-state metabolic flux distribution that maximizes a stated objective, such as cellular growth rate or product formation [21]. FBA was used to interrogate the properties and capabilities of both *L. casei* metabolic models. FBA simulations were performed by maximizing biomass, with an *in-silico* growth medium containing nutrients in the chemically defined medium (CDM) [22] with additional trace minerals (Table S2). To predict amino acid requirements, each model was allowed uptake of all 20 amino acids. Then, exchange fluxes for each amino acid were individually set to zero and the maximum growth rate was calculated. A maximum growth rate of zero indicated the amino acid was essential. For carbohydrate utilization, only essential amino acids were allowed to be taken up by the models. The set of essential amino acids included arginine, isoleucine, leucine, phenylalanine, tryptophan, tyrosine and valine for both strains, with the addition of glutamate for *L. casei* 12A simulations. We first maximized growth using just the essential amino acids (without a carbohydrate present). We then introduced each carbohydrate to the *in silico* media. An increase in maximum growth rate in the presence of a particular carbohydrate indicated the ability of the model to utilize the carbohydrate as a carbon source.

Experiments for amino acid requirements

Working cultures were prepared from frozen stocks by two sequential transfers in de Man, Rogosa and Sharpe (MRS) medium and CDM broth prepared as described by Christensen and Steele [22]. Cultures were incubated statically at 37°C for 24 h and 18 h, respectively. To determine experimental amino

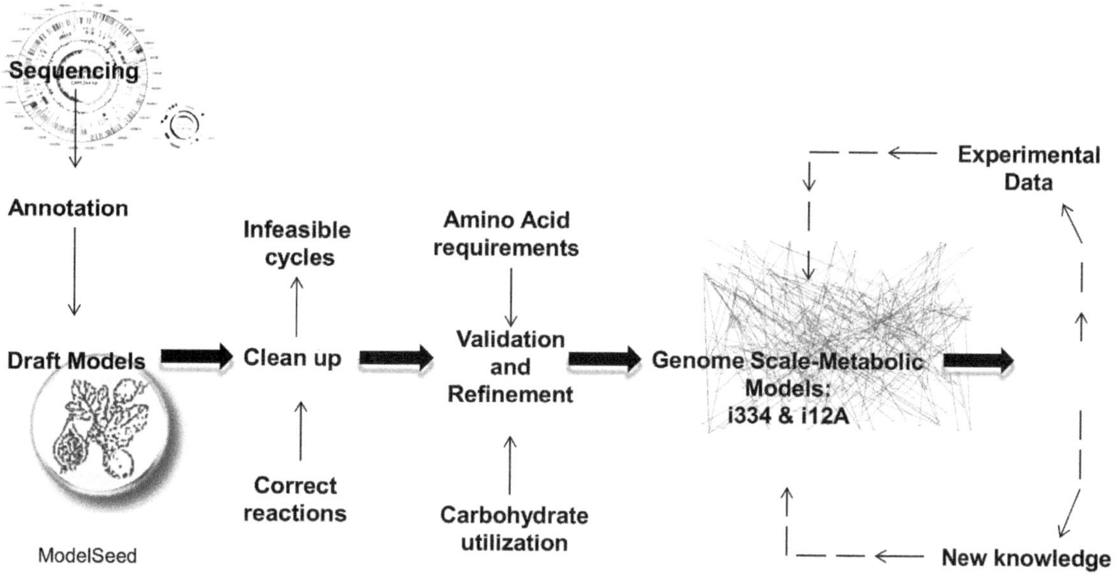

Figure 1. Schematic representation of the approach utilized in the reconstruction of iLca334_548 and iLca12A_640 genome-scale metabolic models. Dashed lines represent iterative processes.

acid requirements, CDM broth was prepared with the omission of individual amino acids, inoculated with 1% of the working cultures, and incubated at 37°C under static conditions. Optical density at 600 nm was determined at intervals during the 48 h incubation.

Identification of Functional Network Differences

We used CONGA to identify functional differences between the two metabolic networks [23]. This constraint-based method identifies gene deletions that affect maximum growth rates differently in two models. Simulations were performed under four different growth conditions, varying the carbon source (glucose or galactose) and the amino acids present (all amino acids or essential amino acids). To control for differences in amino acid requirements, glutamate was included as essential amino acid in both models. For each condition, CONGA maximized differences between each model's maximal growth rate when selected gene deletions were imposed (i.e. maximize μ_1- μ_2, subject to common gene deletions).

Results

Genome sequencing

The closed *L. casei* 12A genome consists of a circular chromosome of 2,907,945 bp. The general features of the 12A genome are presented in Table 1. Compared to *L. casei* ATCC 334, the 12A genome is 12,681 bp larger. The number of pseudogenes in 12A is 152, a relatively high number compared to the 82 in *L. casei* ATCC 334, which could indicate that *L. casei* 12A has undergone more extensive gene decay. Comparative genome analysis between the strains indicates that they share 2,222 orthologous genes; *L. casei* 12A has 507 unique genes compared to *L. casei* ATCC334, whereas *L. casei* ATCC334 has 423 unique genes compared to *L. casei 12A*. The complete genome sequence of 12A was deposited at GenBank with the accession number CP006690.

Genome-scale metabolic networks

The genome-scale metabolic models of *L. casei* ATCC 334 and 12A each contain an extracellular and cytoplasm compartment, and exchange and transport reactions allow metabolites to be imported into or secreted from these compartments. The model for *L. casei* ATCC 334 (*i*Lca334_548) consists of 548 genes, 905 metabolic reactions and 959 metabolites. The model for *L. casei* 12A (*i*Lca12A_640) consists of 640 genes, 932 metabolic reactions and 979 metabolites. The properties of the models are summarized in Table 2. These models are made available in Table S6 and Table S7 in Excel format and SBML format (Table S10 and Table S11).

Amino acid requirements

Single amino acid omission experiments in both *L. casei* strains showed the strains have similar amino acids requirements, where the branched-chain amino acids (leucine, isoleucine and valine), the aromatic amino acids (tyrosine, phenylalanine and tryptophan), and arginine are essential for both strains. However, glutamate is essential for *L. casei* 12A growth, while in ATCC 334 the absence of glutamate only caused a growth defect.

Initial draft models from Model SEED were used to predict the amino acid requirements for both strains by FBA. In the case of the ATCC 334 model, 17 out of 20 simulations agreed with the *in vivo* results; the draft model incorrectly predicted tryptophan to be non-essential (false positive) while cysteine and methionine were incorrectly predicted to be essential (false negative). The draft 12A model predictions agreed with the *in vivo* data for 16 of the 20 simulations. The inconsistencies for tryptophan, methionine and cysteine observed with the ATCC 334 model were also observed with the 12A model. In addition, the 12A model predicted that glutamate was non-essential, while the *in vivo* data indicated it was essential. SMILEY [20] was used to fix false negative predictions in the draft models. To fix both model's requirement for cysteine, the algorithm suggested adding reactions carried out by sulfite dehydrogenase (E.C. 1.8.2.1), thiosulfate dehydrogenase (E.C. 1.8.2.2) and cystathionine β-synthase (E.C. 4.2.1.22). Homoserine O-acetyltransferase (E.C. 2.3.1.31) and O-acetylhomoserine sulf-

Table 1. Important features of the *L. casei* 12A genome.

Chromosome	
Length, bp	2,907,945
Number of CDs	2,900
G + C content, %	46.44
Genes total number	2,833
Protein Coding Genes	2,746
with function prediction	2,222
without function prediction	524
Number of pseudogenes	152
Number of tRNA genes	57

hydrylase (E.C. 2.5.1.49) reactions were added to the *L. casei* ATCC 334 model to fix the model's requirement for methionine. While correcting the methionine requirement in the *L. casei* 12A model involved adding methionine synthase (E.C. 2.1.1.13) and O-succinylhomoserine sulfhydrylase (E.C. 4.2.99.9). To correct the discrepancy between the model predictions and the *in vivo* data requirements for tryptophan, some reactions that were initially gap filled by Model SEED (and identified using Cyto-SEED [24] were removed. The anthranilate synthase (E.C. 4.1.3.21) and tryptophan synthase (E.C. 4.2.1.20) reactions were removed from the 334 and 12A networks, respectively, since there was no evidence to support their inclusion based on genome annotations and experimental data. In the case of the *L. casei* 12A model, it was not possible to fix the discrepancy between the model and *in vivo* data requirement for glutamate since the 12A genome contains a gene (GI:410525452) annotated as glucosamine-fructose-6-phosphate aminotransferase (E.C. 2.6.1.16), which converts glutamine into glutamate. After making these corrections the *L. casei* ATCC 334 model was able to correctly predict the amino acid requirements and the *L. casei* 12A model was able to correctly predict 19 out of the 20 amino acid requirements (Table 3).

Carbohydrate utilization

Carbohydrate utilization plays a key role in the ecology and industrial utility of *L. casei*; therefore it was important to ensure the models were in agreement with the *in vivo* carbohydrate utilization data. The draft metabolic models of ATCC 334 and 12A were tested against data for utilization of 56 carbohydrates, where the utilization of these carbon sources by ATCC 334 and 12A strains has been reported previously [5]. Both strains were unable to utilize 25 carbohydrates, including β-cyclodextrin, γ-cyclodextrin, amylopectin, amylase, arabinogalactan, carboxymethyl cellulose, D-ribitol, D-arabinose, D-arabitol, dextrin, D-maltitol, D-xylitol, D-xylose, fucose, galacturonic acid, glucuronic acid, heparin, lignin, meso-erythritol, mucin, phytic acid, rhamnose, sialic acid, stachyose, xylan and α-cyclodextrin. For all 25 of these un-usable carbohydrates, the drafts models accurately predicted no growth. The 31 carbohydrates that at least one of the strains could utilize *in vivo* are presented in Table 4. The draft models did not perform as well in predicting use of carbohydrates where growth was observed *in vivo*. The poorer performance of the drafts models was mostly due to missing transporters, likely caused by limitations in carbohydrate transporter annotation, causing these transporters to be missed from draft reconstructions. A total of 7 enzymes and 14 transporters were added to the *L.*

Table 2. Overview of metabolic models of *L. casei* 12A and ATCC 334

Categories	**Strains**	
	ATCC 334	12A
Genes	548 (37)	640 (129)
Metabolites	959 (9)	979 (29)
Intracellular	827(9)	838(20)
Extracellular	132	141(9)
Reactions	1040 (23)	1076 (59)
Exchange	133	142(9)
Metabolic	905(22)	932(49)
With $^{+}$GPRs	818	847
Without GPRs	87	85
*Non-metabolic	2(1)	2(1)

. Numbers in parentheses indicate unique genes, metabolites and reactions present only in one *L. casei* model.
*Non-metabolic reactions include biomass and sink reactions.
$^{+}$Gene – Protein- Reactions (GPRs).

Table 3. The amino acid requirements of *L. casei* ATCC 334 and 12A as determined by *in vivo* experiments and flux balance analysis of the models prior to and after model refinement.

Amino acids	*L. casei* **ATCC 334**			*L. casei* **12A**		
	In vivo	*Draft Model*	*Refined Model*	*In vivo*	*Draft Model*	*Refined Model*
Alanine	NR	NR	NR	NR	NR	NR
Arginine	R	R	R	R	R	R
Asparagine	NR	NR	NR	NR	NR	NR
Aspartate	W	NR	NR	W	NR	NR
Cysteine	NR	**R**	NR	NR	**R**	NR
Glutamate	W	NR	NR	R	**NR**	**NR**
Glutamine	NR	NR	NR	NR	NR	NR
Glycine	NR	NR	NR	NR	NR	NR
Histidine	NR	NR	NR	NR	NR	NR
Isoleucine	R	R	R	R	R	R
Leucine	R	R	R	R	R	R
Lysine	NR	NR	NR	NR	NR	NR
Methionine	NR	**R**	NR	NR	**R**	NR
Phenylalanine	R	R	R	R	R	R
Proline	NR	NR	NR	NR	NR	NR
Serine	NR	NR	NR	NR	NR	NR
Threonine	NR	NR	NR	NR	NR	NR
Tryptophan	R	**NR**	R	R	**NR**	R
Tyrosine	R	R	R	R	R	R
Valine	R	R	R	R	R	R
Accuracy		*17/20*	*20/20*		*16/20*	*19/20*

Discrepancies between experimental data and simulations are in bold.
R = Amino Acid required for growth (maximal optical density observed ≤0.05).
NR = Amino Acid not required for growth (maximal optical density observed ≥0.15).
W = Weak growth in the absence of the amino acid (maximal optical density observed was 0.09).

casei ATCC 334 model and 7 enzymes and 17 transporters were added to the *L. casei* 12A model to fix false negative predictions (Table S8). For the carbohydrate amygdalin-6P, we were unable to identify a route with genomic evidence to degrade its by-product, mandelonitrile; we introduced a sink reaction to consume this metabolite. A metabolic map of carbohydrate metabolism is presented in Figure 2; this metabolic map indicates how some relevant carbon sources are transported into the cytoplasm and how they are integrated into glycolysis.

Identification of Functional Network Differences

CONGA identifies gene deletion sets, sets of one or more genes whose deletion are predicted to be lethal by only one of two models. Gene deletion sets are classified based on the reason for differences in growth predictions: genetic differences, in which gene-protein-reaction (GPR) relationships differ in the models; orthology differences, in which genes encoding enzymes with identical functions were not assigned as orthologs; and metabolic differences, where one organism has additional reactions that grant it unique capabilities [23].

The CONGA results were insensitive to carbon source (indicating there are no differences between glucose or galactose catabolism in the two strains). Therefore, the results are summarized based on amino acid availability. For the first condition (with only essential amino acids), CONGA found five gene deletion sets that were lethal only in *i*Lca334_548 (e.g., no biomass formation was predicted by *i*Lca334_548) and five gene deletion sets that were lethal only in *i*Lca12A_640. For the second condition (all amino acids), two gene deletion sets were lethal only to *i*Lca334_548; and four gene deletion sets were lethal only to *i*Lca12A_640. These deletion sets were a subset of those found in the first condition (where only essential amino acids are present). CONGA results are summarized in Table 5 and the full results are presented in Table S3 and Table S4.

Four of the ten unique gene deletions sets identified by CONGA were attributed to genetic differences, which arise due to differences in the number of isozymes between the two models for essential reactions. Three of these four gene deletion sets were lethal in *i*Lca334_548. This highlights the presence of more isozymes in *i*Lca12A_640, suggesting it is more robust to gene deletions than *i*Lca334_548.

Using CONGA we also identified one metabolic difference, in which deletion of the enzyme 5, 10-methylenetetrahydrofolate (5,10-CH_2-THF) dehydrogenase (E.C. 1.5.1.5) is lethal only in the *i*Lca334_548 model. This means that the *i*Lca12A_640 model has a unique mechanism (shown in Figure 3) for recovering from this gene deletion. Briefly, 5,10-CH_2-THF is a precursor to 10-formyltetrahydrofolate (10-CHO-THF) a cofactor involved in purine biosynthesis, an essential activity for cellular growth. The deletion of 5,10-CH_2-THF dehydrogenase prevents the biosynthesis of THF in *L. casei* ATCC 334. Our model for *L. casei* 12A predicts that this deletion can be rescued by the actions of 5,10-

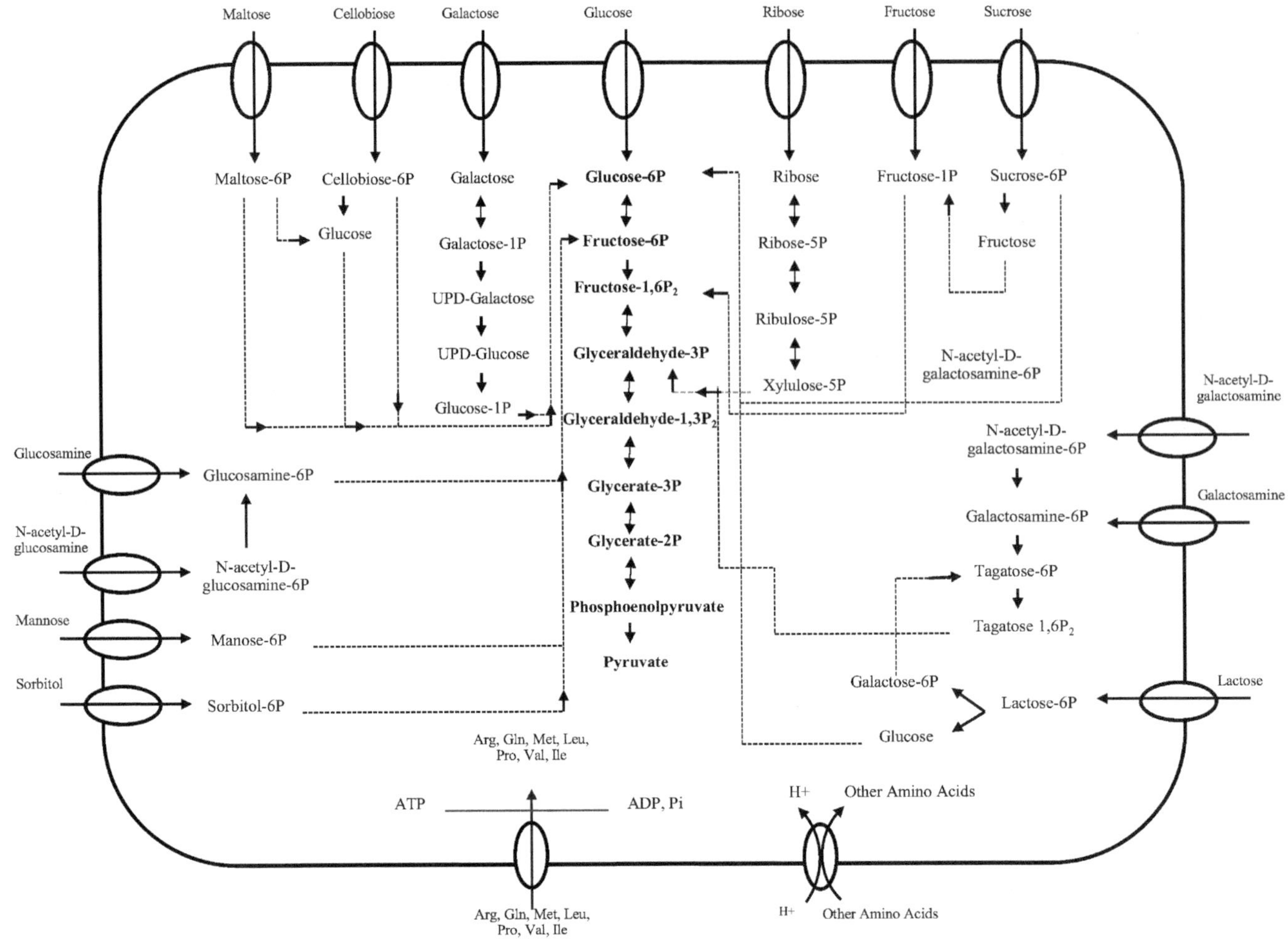

Figure 2. Metabolic map of carbohydrate metabolism of *L. casei* 12A and ATCC 334. Glycolytic metabolites are listed in bold.

CH_2-THF:3-methyl-2-oxobutanoate (E.C. 2.1.2.11) and formate: THF ligase (E.C. 6.3.4.3), in which 5,10-CH_2-THF is converted directly to THF, producing 2-dehydropantoate as a by-product. The reactions pantoate 2-oxidoreductase (E.C. 1.1.1.169) and pantothenate amidohydrolase (E.C. 3.5.1.22) convert 2-dehydropantoate to pantothenate, which can be secreted by a transporter unique to the *i*Lca12A_640 model, or some pantothenate can also be used for CoA biosynthesis, important for biomass formation.

Discussion

Genome-scale metabolic networks

In this study, we assembled a draft genome-scale metabolic model for two strains of *L. casei*: ATCC 334 and 12A, using the Model Seed pipeline; followed by model curation to ensure consistent predictions for amino acid requirements and carbohydrate utilization. The *in silico* reconstructions, *i*Lca334_548 and *i*Lca12A_640 appear to have the largest metabolic networks in comparison to other LAB metabolic networks published to date, with 1,040 and 1,076 reactions, respectively.

Our experimental results show one difference in amino acid requirements between the two *L. casei* strains, glutamate is essential for *L. casei* 12A but not in *L. casei* ATCC334. However, differences between the strains are more apparent based on their carbohydrate utilization.

Amino acid requirements

Compared to other LAB with reported metabolic models, our strains have fewer amino acid requirements than some *Lactobacillus* species (e.g. *L. reuteri* JM112 and *L. plantarum* WCFS1). On the other hand, our strains have considerably more requirements than *Lc. lactis* and *S. thermophilus* (Table S1). The rich environments from which our *Lactobacillus* strains were isolated reduce the need for synthesizing amino acids.

The branched-chain amino acid (isoleucine, leucine and valine) biosynthesis pathways in *L. casei* ATCC 334 and 12A appear to be absent based on genome annotation, which was confirmed by the growth experiments indicating their essentiality.

Aromatic amino acids (tryptophan, phenylalanine and tyrosine) had incomplete biosynthesis pathways in the two *L. casei* strains. In the biosynthesis pathway of tryptophan, the enzyme anthranilate synthase (E.C. 4.1.3.27) was not found in the genome annotation of *L. casei* ATCC 334 strain; presumably this reaction was added to the draft metabolic models as a mean to complete the pathway. Experimental data indicating tryptophan must be supplied support the idea that this pathway is incomplete. In *L. casei* 12A the reaction catalyzed by tryptophan synthase (E.C. 4.2.1.20) was found to confer the model, the ability to synthesize tryptophan, therefore it was removed base on experimental data. Since tryptophan was required for growth by both strains, a reaction was deleted from both metabolic models. The require-

Table 4. Carbohydrate utilization of *L. casei* ATCC 334 and 12A as determined by *in vivo* experiments and flux balance analysis of the models prior to and after model refinement.

Carbohydrate	*L. casei* **ATCC 334**			*L. casei* **12A**		
	+In vivo	**Draft Model*	**Refined Model*	*+In vivo*	**Draft Model*	**Refined Model*
Amygdalin	G	**NG**	G	G	**NG**	G
D-cellobiose	G	G	G	G	G	G
D-fructose	G	**NG**	G	G	**NG**	G
D-galactose	G	**NG**	G	G	**NG**	G
D-glucosamine	G	**NG**	G	G	**NG**	G
D-glucose	G	G	G	G	G	G
D-lactitol	G	**NG**	G	G	**NG**	G
D-maltose	G	G	G	G	G	G
D-mannose	G	G	G	G	G	G
D-melezitose	G	**NG**	G	G	**NG**	G
D-raffinose	NG	NG	NG	G	**NG**	G
D-ribose	G	**NG**	G	G	G	G
D-sorbitol/glucitol	G	**NG**	G	G	G	G
D-turanose	G	**NG**	G	G	**NG**	G
Galactosamine	G	**NG**	G	G	**NG**	G
Gluconic Acid	G	G	G	G	G	G
Inulin	G	**NG**	G	G	**NG**	G
Isomaltose	G	**NG**	G	G	**NG**	G
Lactose	G	**NG**	G	G	**NG**	G
Lactulose	G	**NG**	G	G	**NG**	G
Maltotriose	G	**NG**	G	G	**NG**	G
Myo-inositol	NG	NG	NG	NG	**G**	**G**
N-acetyl-D-galactosamine	G	**NG**	G	G	**NG**	G
N-acetyl-D-glucosamine	G	**NG**	G	G	**NG**	G
Panose	NG	NG	NG	G	**NG**	G
Polydextrose	G	**NG**	G	G	**NG**	G
Pullulan	NG	NG	NG	G	**NG**	G
Sucrose	G	**NG**	G	G	**NG**	G
Accuracy		*35/54*	*54/54*		*33/54*	*53/54*

Discrepancies between experimental data and simulations are in bold.
G = Growth in the presence of the carbohydrate.
NG = No growth in the presence of the carbohydrate.
*For simulations, G represents increased biomass production in the presence of the carbohydrate; NG represents no change in biomass production in the presence of the carbohydrate.
+In vivo data are based on the study cited by Broadbent et al. (2012) [5].

ment for phenylalanine and tyrosine for growth was expected, as an enzyme needed for their biosynthesis, prephenate dehydratase, was not found in the genome annotation of either strain.

L. casei strains are reported to require arginine for growth [25]; therefore a requirement for the amino acid was expected in both cases. The enzyme arginine dihydrolase (E.C. 3.5.3.6), involved in

Table 5. Number of gene deletion sets found by CONGA under four different conditions between *i*Lca334_548 and *i*Lca12A_640.

Conditions	*Lethal in iLca334_548*	*Lethal in iLca12A_640*
	Genetic = 3	Genetic = 1
Glucose or Galactose + Essential A.A.	Metabolic = 1	Orthology = 4
	Orthology = 1	
Glucose or Galactose + All A.A.	Genetic = 2	Orthology = 4

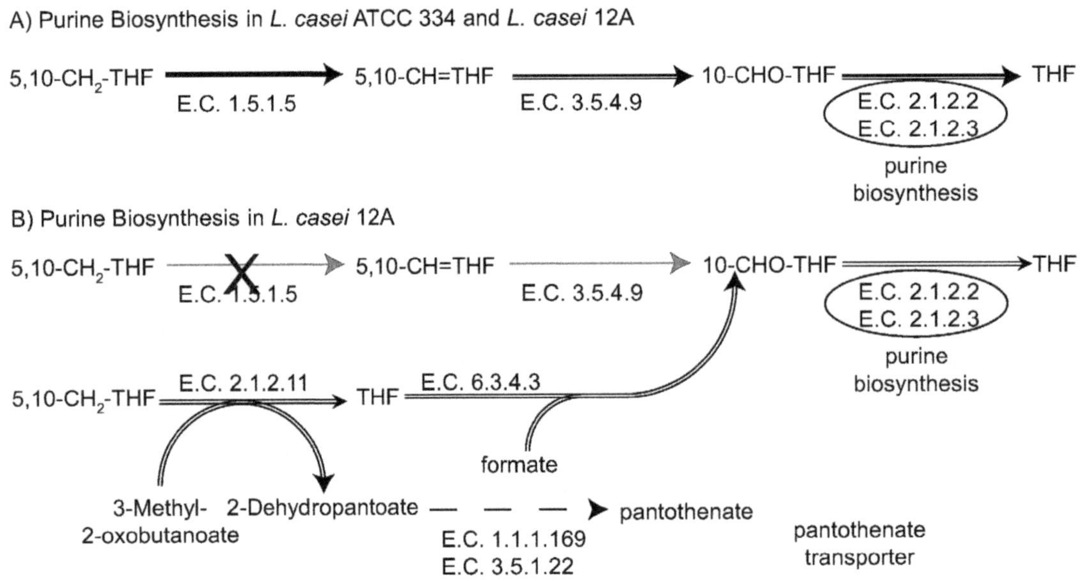

Figure 3. Metabolic differences in the two *L. casei* strains. (A): Pathway for the synthesis of tetrahydrofolate (THF) from 5, 10-methylenetetrahydrofolate (5,10-CH2-THF) and its role in purine biosynthesis. This pathway is common to both strains. (B): Additional pathway for the conversion of 5,10-CH_2-THF to THF active in the *i*Lca12A_640 model. With the exception of the panthtothenate transporter, the reactions are found in both models. (A and B): Thick arrows indicate flux in both models. Double arrows represent flux in the *i*Lca12A_640 model. The black 'X' indicates a gene deletion identified by CONGA lethal in *i*Lca334_548 but not *i*Lca12A_640, and gray arrows indicate inactive reactions arising from the deletion. The dashed arrow represents two separate steps. Reactions and metabolites corresponding to the given E.C. numbers and metabolite identifiers are given in the Supporting Material.

arginine biosynthesis, was not found in the genome annotation of either strain.

The omission of glutamate in the medium caused a severe defect in growth for *L. casei* ATCC 334 and no growth for *L. casei* 12A. This has been well documented in other LAB, due to an incomplete citric acid cycle and limited supply of α-ketoglutarate [26], but the metabolic models in both cases predicted growth in the absence of this amino acid. This can be explained by the presence of glutamine. In both metabolic models glutamine can be converted into glutamate using the reaction catalyzed by glucosamine-fructose-6-phosphate aminotransferase (E.C. 2.6.1.16). When this enzyme is deleted from *i*Lca12A_640, glutamate is predicted to be an essential amino acid, however, this deletion would also make glutamine a predicted essential amino acid, but glutamine is not required experimentally, therefore the reaction was not removed. Experimental studies suggest that the interconversion of glutamine to glutamate results in low yields of glutamate [27], therefore glutamate is likely needed even in the presence of glutamine.

Carbohydrate utilization

Comparison of the two *L. casei* models highlights the presence of additional enzymes that gives the *L. casei* 12A model the ability to utilize D-rafinnose, panose and pullulan. For instance, the 12A model possesses an ABC transporter predicted to function for raffinose uptake. This model also includes a pullulanase and a neopullulanase, important in pullulan and panose degradation into maltotriose and maltose and glucose, respectively. As discussed by Broadbent *et al.* (2012), the ability of strains to utilize different carbohydrates predicts the adaptation of the strains to diverse ecological niches; D-rafinnose, panose and pullulan are components commonly found in plant material, the environment from which *L. casei* 12A was isolated. On the other hand, as observed in growth experiments and confirmed by the metabolic models, *L. casei* ATCC 334, can utilize fewer of the carbohydrates tested (see Table 4). This could be related to evolution of cheese isolates by gene decay [5], suggesting that dairy environments are richer in less complex carbon sources than those found in plant environments.

One of the most abundant stereoisomers of inositol is *myo*-inositol (MI), a component of phytic acid, which can serve as a phosphate storage molecule in plant seeds [28]. Most LAB species cannot utilize this carbon source, although Yebra *et al.* (2007) reported the identification of a gene cluster which enables *L. casei* BL23 to utilize myo-inositol under aerobic conditions. The catabolic pathway of MI is absent in *L. casei* ATCC 334, and experimental studies and prediction by the metabolic models confirmed no growth on MI. Surprisingly, the genome-scale metabolic model for *L. casei* 12A predicted growth using MI as a carbon source, implying the catabolic pathway for MI is complete in this strain. Further analysis in the genome annotation of *L. casei* 12A reveals the presence of all the enzymes (E.C. 1.1.1.18, 4.2.1.44, 3.7.1, 5.3.1, 2.7.1.92, 4.1.2.29 and 5.3.1.1) needed for converting MI to glyceraldehyde-3P. The environment from which the 12A strain was isolated could have played a role in the retention of the pathway. The reason why *L. casei* 12A was unable to utilize MI is speculated to be due to regulatory effects that genome-scale metabolic models do not capture.

Identification of Functional Network Differences

A unique transporter (pantothenate sodium symporter) in *i*Lca12A_640 for pantothenate was found to confer the model with the ability to overcome the deletion of 5,10-CH_2-THF dehydrogenase. This transporter prevents the accumulation of pantothenate inside the cell, therefore allowing the reactions, catalyzed by the enzymes E.C. 2.1.2.11 and 6.3.4.3 to be active, yielding to the synthesis of 10-CHO-THF.

The pantothenate transporter in *i*Lca12A_640 (added by Model SEED) lacks a gene association, however further analysis of *L. casei* 12A genome annotation reveals the presence of a pantothenate energy-coupling factor (ECF) transporter (GI 410523916). An ortholog of this gene was found in *L. casei* ATCC 334, which is annotated as a predicted membrane protein which may be the reason it was missed during the automated reconstruction of the model. This provides an example of how CONGA can be used to aide in further model development.

CONGA simulations performed under the glucose and galactose conditions, both six carbon carbohydrates, identified the same differences between the models. Simulations with ribose where also conducted (data not shown), but no new gene deletions sets were found. Even though we were able to identify genetic, orthology and metabolic differences between the models, it appears that pathways in catabolism of these three sugars and central metabolism are shared by the strains.

The genome-scale metabolic models of two *L. casei* strains provide a comprehensive overview of the metabolism of the strains. These models are valuable tools for the development of strains with enhanced utility in a variety of industrial applications (e.g. food ingredients), as well as in the design of new metabolic engineering approaches in the production of commodity chemicals such as biofuels.

Supporting Information

Table S1 Comparison of essential amino acids (listed as E) determined from experiments among lactic acid bacteria, for which metabolic models have been built.

Table S2 Compounds allowed to be taken up during flux balance analysis in both *L. casei* models. Flux represents uptake rate, units are mmol/gDW/h.

Table S3 Type of gene deletion sets, with glucose or galactose with essential amino acids condition, found in *i*Lca12A_640 and *i*Lca334_548.

Table S4 Type of gene deletion sets, with glucose or galactose and all amino acids, found in *i*Lca12A_640 and *i*Lca334_548.

Table S5 Complete list of reactions and compounds present in GapfillMatrix.

Table S6 Complete list of reactions and compounds present in iLca12A_640.

Table S7 Complete list of reactions and compounds present in iLca334_548.

Table S8 Enzymes and transporters added to *L. casei* ATCC 334 and 12A models to fix false negative predictions. Addition of an enzyme/transporter is indicated with filled circles; open circles indicate no addition of enzyme/ transporter.

Table S9 GapfillMatrix in SBML format.

Table S10 *i*Lca12A_640 in SBML format.

Table S11 *i*Lca334_548 in SBML format.

Acknowledgments

We thank Dr. Eric Neeno-Eckwall for his contribution updating the *L. casei* 12A genome sequence at GenBank and ASAP, and Cameron Cotten and Brent Denton, for their contributions to the balancing of the metabolic networks.

Author Contributions

Conceived and designed the experiments: JLS JLR JRB. Performed the experiments: EVL JJH BS. Analyzed the data: EVL JJH JLR BS. Wrote the paper: EVL JJH BS JRB JLR JLS.

References

1. Axelsson L (2004) Lactic acid bacteria: clasification and physiology. In: Salminen S, von Wright A OA, editor. Lactic acid bacteria macrobiological and funtional Aspects. New York, NY: Marcel Dekker Inc. pp.1–66.
2. Stiles ME, Holzapfel WH (1997) Lactic acid bacteria of foods and their current taxonomy. Int J Food Microbiol 36: 1–29. Available: http://www.ncbi.nlm.nih.gov/pubmed/9168311.
3. Goldin BR, Gorbach S (1992) Probiotics for humans. In: Fuller R . London, editor. Probiotics: the scientific basis. Chapman and Hall. pp.355–372.
4. Mayra-Makinen A BM (1998) Industrial use and production of lactic acid bacteria. In: Salminen S WA, editor. Lactic acid bacteria—microbiology and functional aspects. New York, NY: Marcel Dekker Inc. pp.73–102.
5. Broadbent JR, Neeno-Eckwall EC, Stahl B, Tandee K, Cai H, et al. (2012) Analysis of the Lactobacillus casei supragenome and its influence in species evolution and lifestyle adaptation. BMC Genomics 13: 533. Available: http://www.ncbi.nlm.nih.gov/pubmed/23035691. Accessed 2012 Nov 8.
6. Bordbar A, Monk JM, King Z, Palsson BO (2014) Constraint-based models predict metabolic and associated cellular functions. Nat Rev Genet 15: 107–120. Available: http://www.ncbi.nlm.nih.gov/pubmed/24430943. Accessed 2014 Jan 21.
7. Lewis NE, Nagarajan H, Palsson BO (2013) Constraining the metabolic genotype-phenotype relationship using a phylogeny of in silico methods. Nat Rev Microbiol 10: 291–305. doi:10.1038/nrmicro2737.Constraining.
8. Reed JL (2012) Shrinking the metabolic solution space using experimental datasets. PLoS Comput Biol 8: e1002662. Available: http://www.pubmedcentral.nih.gov/articlerender.fcgi?artid=3431291&tool=pmcentrez&rendertype=abstract. Accessed 2014 Mar 10.
9. Zomorrodi AR, Suthers PF, Ranganathan S, Maranas CD (2012) Mathematical optimization applications in metabolic networks. Metab Eng 14: 672–686. Available: http://www.ncbi.nlm.nih.gov/pubmed/23026121. Accessed 2014 Jan 27.
10. Agren R, Liu L, Shoaie S, Vongsangnak W, Nookaew I, et al. (2013) The RAVEN toolbox and its use for generating a genome-scale metabolic model for Penicillium chrysogenum. Plos Comput Biol 9: e1002980. Available: http://www.pubmedcentral.nih.gov/articlerender.fcgi?artid=3605104&tool=pmcentrez&rendertype=abstract. Accessed 2014 Jan 29.
11. Oliveira AP, Nielsen J, Förster J (2005) Modeling Lactococcus lactis using a genome-scale flux model. BMC Microbiol 5: 39. Available: http://www.pubmedcentral.nih.gov/articlerender.fcgi?artid=1185544&tool=pmcentrez&rendertype=abstract. Accessed 2012 Nov 28.
12. Flahaut N a L, Wiersma A, van de Bunt B, Martens DE, Schaap PJ, et al. (2013) Genome-scale metabolic model for Lactococcus lactis MG1363 and its application to the analysis of flavor formation. Appl Microbiol Biotechnol 97: 8729–8739. Available: http://www.ncbi.nlm.nih.gov/pubmed/23974365. Accessed 2014 Jul 17.
13. Teusink B, Wiersma A, Molenaar D, Francke C, de Vos WM, et al. (2006) Analysis of growth of Lactobacillus plantarum WCFS1 on a complex medium using a genome-scale metabolic model. J Biol Chem 281: 40041–40048. Available: http://www.ncbi.nlm.nih.gov/pubmed/17062565. Accessed 2012 Nov 28.
14. Pastink MI, Teusink B, Hols P, Visser S, de Vos WM, et al. (2009) Genome-scale model of Streptococcus thermophilus LMG18311 for metabolic comparison of lactic acid bacteria. Appl Environ Microbiol 75: 3627–3633. Available: http://www.pubmedcentral.nih.gov/articlerender.fcgi?artid=2687286&tool=pmcentrez&rendertype=abstract. Accessed 2012 Oct 30.
15. Saulnier DM, Santos F, Roos S, Mistretta T-A, Spinler JK, et al. (2011) Exploring metabolic pathway reconstruction and genome-wide expression profiling in Lactobacillus reuteri to define functional probiotic features. PLoS

One 6: e18783. Available: http://www.pubmedcentral.nih.gov/articlerender.fcgi?artid=3084715&tool=pmcentrez&rendertype=abstract. Accessed 2012 Nov 7.

16. Makarova K, Slesarev A, Wolf Y, Sorokin A, Mirkin B, et al. (2006) Comparative genomics of the lactic acid bacteria. Proc Natl Acad Sci U S A 103: 15611–15616. Available: http://www.pubmedcentral.nih.gov/articlerender.fcgi?artid=2660350&tool=pmcentrez&rendertype=abstract.
17. Aziz RK, Bartels D, Best A a, DeJongh M, Disz T, et al. (2008) The RAST Server: rapid annotations using subsystems technology. BMC Genomics 9: 75. Available: http://www.pubmedcentral.nih.gov/articlerender.fcgi?artid=2265698&tool=pmcentrez&rendertype=abstract. Accessed 2012 Jul 13.
18. Henry CS, DeJongh M, Best A a, Frybarger PM, Linsay B, et al. (2010) High-throughput generation, optimization and analysis of genome-scale metabolic models. Nat Biotechnol 28: 977–982. Available: http://www.ncbi.nlm.nih.gov/pubmed/20802497. Accessed 2012 Jul 16.
19. Pinchuk GE, Hill E a, Geydebrekht O V, De Ingeniis J, Zhang X, et al. (2010) Constraint-based model of Shewanella oneidensis MR-1 metabolism: a tool for data analysis and hypothesis generation. Plos Comput Biol 6: e1000822. Available: http://www.pubmedcentral.nih.gov/articlerender.fcgi?artid=2891590&tool=pmcentrez&rendertype=abstract. Accessed 2013 Mar 6.
20. Reed JL, Patel TR, Chen KH, Joyce AR, Applebee MK, et al. (2006) Systems approach to refining genome annotation. Proc Natl Acad Sci U S A 103: 17480–17484. Available: http://www.pubmedcentral.nih.gov/articlerender.fcgi?artid=1859954&tool=pmcentrez&rendertype=abstract.
21. Orth JD, Palsson BØ (2011) What is flux balance analysis? Nat Biotechnol 28: 245–248. doi:10.1038/nbt.1614.
22. Christensen JE, Steele JL (2003) Impaired growth rates in milk of Lactobacillus helveticus peptidase mutants can be overcome by use of amino acid supplements. J Bacteriol 185: 3297–3306. doi:10.1128/JB.185.11.3297.
23. Hamilton JJ, Reed JL (2012) Identification of functional differences in metabolic networks using comparative genomics and constraint-based models. Plos Ones 7: e34670. Available: http://www.pubmedcentral.nih.gov/articlerender.fcgi?artid=3359066&tool=pmcentrez&rendertype=abstract. Accessed 2012 Oct 29.
24. DeJongh M, Bockstege B, Frybarger P, Hazekamp N, Kammeraad J, et al. (2012) CytoSEED: a Cytoscape plugin for viewing, manipulating and analyzing metabolic models created by the Model SEED. Bioinformatics 28: 891–892. Available: http://www.pubmedcentral.nih.gov/articlerender.fcgi?artid=3307106&tool=pmcentrez&rendertype=abstract.
25. Bringel F, Hubert J (2003) Extent of genetic lesions of the arginine and pyrimidine biosynthetic pathways in Lactobacillus plantarum, L. paraplantarum, L. pentosus, and L. casei: prevalence of CO2 -dependent auxotrophs and characterization of deficient arg genes in L. plantarum. Appl Environ Microbiol 69: 2674–2683. doi:10.1128/AEM.69.5.2674.
26. Teusink B, Enckevort FHJ Van, Francke C, Wiersma A, Wegkamp A, et al. (2005) In silico reconstruction of the metabolic pathways of Lactobacillus plantarum: comparing predictions of nutrient requirements with those from growth experiments. Appl Environ Microbiol 71: 7253–7262. doi:10.1128/AEM.71.11.7253.
27. Vermeulen N, Gänzle MG, Vogel RF (2007) Glutamine deamidation by cereal-associated lactic acid bacteria. J Appl Microbiol 103: 1197–1205. Available: http://www.ncbi.nlm.nih.gov/pubmed/17897224. Accessed 2013 Apr 20.
28. Yebra MJ, Zúñiga M, Beaufils S, Pérez-Martínez G, Deutscher J, et al. (2007) Identification of a gene cluster enabling Lactobacillus casei BL23 to utilize myo-inositol. Appl Environ Microbiol 73: 3850–3858. Available: http://www.pubmedcentral.nih.gov/articlerender.fcgi?artid=1932728&tool=pmcentrez&rendertype=abstract. Accessed 2013 Mar 30.

18

Reconstruction of Ancestral Gene Orders Using Probabilistic and Gene Encoding Approaches

Ning Yang[3], Fei Hu[1,2], Lingxi Zhou[1,2], Jijun Tang[1,2]*

1 Tianjin Key Laboratory of Cognitive Computing and Application, Tianjin University, Tianjin, China, **2** Department of Computer Science and Engineering, University of South Carolina, Columbia, South Carolina, United States of America, **3** School of Automation, Northwestern Polytechnical University, Xi'an, shaanxi, China

Abstract

Current tools used in the reconstruction of ancestral gene orders often fall into event-based and adjacency-based methods according to the principles they follow. Event-based methods such as GRAPPA are very accurate but with extremely high complexity, while more recent methods based on gene adjacencies such as InferCARsPro is relatively faster, but often produces an excessive number of chromosomes. This issue is mitigated by newer methods such as GapAdj, however it sacrifices a considerable portion of accuracy. We recently developed an adjacency-based method in the probabilistic framework called PMAG to infer ancestral gene orders. PMAG relies on calculating the conditional probabilities of gene adjacencies that are found in the leaf genomes using the Bayes' theorem. It uses a novel transition model which accounts for adjacency changes along the tree branches as well as a re-rooting procedure to prevent any information loss. In this paper, we improved PMAG with a new method to assemble gene adjacencies into valid gene orders, using an exact solver for traveling salesman problem (TSP) to maximize the overall conditional probabilities. We conducted a series of simulation experiments using a wide range of configurations. The first set of experiments was to verify the effectiveness of our strategy of using the better transition model and re-rooting the tree under the targeted ancestral genome. PMAG was then thoroughly compared in terms of three measurements with its four major competitors including InferCARsPro, GapAdj, GASTS and SCJ in order to assess their performances. According to the results, PMAG demonstrates superior performance in terms of adjacency, distance and assembly accuracies, and yet achieves comparable running time, even all TSP instances were solved exactly. PMAG is available for free at http://phylo.cse.sc.edu.

Editor: Gabriel Moreno-Hagelsieb, Wilfrid Laurier University, Canada

Funding: The authors were funded by NSF IIS 1161586 and an internal grant from Tianjin University, China. The funders had no role in study design, data collection and analysis, decision to publish, or preparation of the manuscript.

Competing Interests: The authors have declared that no competing interests exist.

* Email: jtang@cse.sc.edu

Introduction

Overview

Evolutionary biologists have had a long tradition in reconstructing genomes of extinct ancestral species. Mutations in a genomic sequence are made up not only at the level of base-pair changes but also by rearrangement operations on chromosomal structures such as inversions, transpositions, fissions and fusions [2]. Over the past few years, ancestral gene-order inference has brought profound predictions of protein functional shift and positive selection [3].

Methods for ancestral genome reconstruction either assume a given phylogeny that represents the evolutionary history among given species or search the most appropriate tree along with a set of ancestral genomes to fit the observed data. Depending on how the gene-order data is interpreted and handled, methods for solving the latter can be partitioned into two groups: event-based methods and adjacency-based methods. Event-based methods typically search for the set of ancestral gene orders that minimizes the sum of rearrangement distances over all edges of the given phylogeny. However, methods seeking exact solutions following such paradigm (such as GRAPPA [4], MGR [5,6]) have already encountered huge difficulties in handling modern genomes due to their NP-hard complexity. In consequence, methods such as GASTS [7] were developed to provide heuristic solutions. In addition to event-based methods, a number of adjacency-based methods have been proposed such as SCJ [8], InferCARsPro [9] and GapAdj [10]. Instead of explicitly considering a predefined set of rearrangement events, these methods take gene adjacencies into account and treat them as binary characters with present and absence states. In this way, by viewing the gene order as a set of gene adjacencies, the goal is to determine which adjacencies are contained in the genome.

We recently developed an adjacency-based method in the probabilistic framework called PMAG [1] to reconstruct ancestral genomic orders given a phylogeny. In this paper, we improved PMAG to introduce a better algorithm that can assemble gene orders with fewer contigs, hence provided better accuracy. This new algorithm is also faster, enabling us to handle larger datasets. Through simulation experiments, we verify the usefulness of our biased transition model and re-rooting procedure we incorporate in the program. Then the performance of PMAG is evaluated against other existing methods including InferCARsPro, GapAdj, SCJ and GASTS under a wide range of settings.

Genome rearrangement

Given a set of n genes $\{g_1,g_2,\cdots,g_n\}$, a genome can be represented by an *ordering* of these genes. To indicate the strandedness of genes, each gene is assigned with an orientation that is either positive, written g_i, or negative, written $-g_i$. Two genes i and j are said to be *adjacent* in genome G if i is immediately followed by j, or, equivalently, $-j$ is immediately followed by $-i$. A *breakpoint* of two genomes is defined as an adjacency appears in one but not in the other.

Let G be the multi-chromosomal genome with signed ordering $\{a_1,a_2,\cdots,a_n\}$, $\{b_1,b_2,\cdots,b_m\},\cdots$ ($\{\cdots\}$indicates a chromosome). An *inversion* (also called *reversal*) between indices i and j ($i \leq j$) of chromosome a, produces a chromosome a' with linear ordering

$$a_1,a_2,\cdots,a_{i-1},-a_j,-a_{j-1},\cdots,-a_i,a_{j+1},\cdots,a_n.$$

For example, the identity genome $\{1,2,3,4,5,6\}$ is transformed into $\{1,-4,-3,-2,5,6\}$ when the gene block $\{2,3,4\}$ is inverted.

A *transposition* on a chromosome a acts on three indices i,j,k, with $i \leq j$ and $k \notin [i,j]$, picking up the interval $a_i,\cdots,a_j$ and inserting it immediately after a_k. Thus the chromosome a of the genome is replaced by (assume $k>j$):

$$a_1,\cdots,a_{i-1},a_{j+1},\cdots,a_k,a_i,a_{i+1},\cdots,a_j,a_{k+1},\cdots,a_n.$$

For example, genome $\{1,2,3,4,5,6\}$ is transformed into $\{1,5,2,3,4,6\}$ when the gene block $\{2,3,4\}$ is moved in front of gene 6.

A *translocation* on a genome G acts on two chromosomes $a=\{a_1,a_2,\cdots,a_n\}$ and $b=\{b_1,b_2,\cdots,b_m\}$. Given two indices i,j, it picks up the interval $a_i,\cdots,a_n$ and $b_j,\cdots,b_m$ and then changes their places. Thus the two chromosomes a,b of genome G become

$$\{a_1,a_2,\cdots,a_{i-1},b_j,\cdots,b_m\},\{b_1,b_2,\cdots,b_{j-1},a_i,\cdots,a_n\}$$

Yancopoulos [11] proposed a universal double-cut-and-join operation that accounts for inversions, transposition and translocations which resulted in a new genomic distance that can be computed in linear time. In particular, a DCJ operation consists of cutting two connections (breakpoints) in the genome, and rejoining the resulting four unconnected ends in two new pairs. Although there is no direct biological evidence for DCJ operations, these operations are very attractive because they provide a simpler and unifying model for genome rearrangement.

Later the Single-Cut-or-Join (SCJ) [12] operation was proposed as a basis for a new rearrangement distance between multichromosomal genomes, leading to very fast algorithms. The SCJ operation is modeled on the two most fundamental rearrangement operations—the cutting and joining of adjacencies. A cutting operation breaks an adjacency into two telomeres, and a joining operation is performed in the opposite way by pairing two telomeres into an adjacency. Any cutting or joining applied to the genome will be called a Single-Cut-or-Join (SCJ) operation. Since the genome is represented as a set of adjacencies, a cutting can also be viewed as the removal of an adjacency from the set while the joining is the addition of an adjacency.

Event-based methods typically iterate over each internal node to solve for the median genomes until the sum of rearrangement events over all edge distances (tree score) is minimized. The median problem can be formalized as follows: given a set of m genomes with permutations $\{x_i\}_{1\leq i\leq m}$ and a distance measurement d, find another permutation x_t such that the median score defined as $\sum_{i=1}^{m} d(x_i,x_t)$ is minimized. However solving even the simplest case when m equals to three is NP-hard for most distance measurements [13,14]. Among all existing median solvers, the best is the DCJ median solver proposed by Xu and Sankoff (ASMedian [15]) based on the concept of adequate sub-graph and decomposes a multiple breakpoint graph [16] into smaller and easier graphs. Although ASMedian could remarkably scale down the computational expenses of median searching, it yet runs very slow when the genomes are distant. Based on ASMedian, Xu developed its heuristic algorithm GASTS that can quickly score a fixed phylogenetic tree and enables us to attack previously unapproachable problems by GRAPPA and MGR, as demonstrated on a set of vertebrate genome with over 2,000 genes.

Ancestral gene-order reconstruction based on gene adjacencies

Over recent years, a collection of models based on the study of gene adjacencies have been proposed for solving various rearrangement problems. In these models, each gene adjacency is considered as a binary character and the problem of reconstructing ancestral genomes can then be reduced as deciding, for every adjacency, whether an ancestral genome contains the adjacency.

The breakpoint-like model, single-cut-or-join (SCJ), utilized the Fitch's algorithm to reconstruct ancestors based on binary characters in terms of gene adjacencies; however, the characters are not independent, since conflicting adjacencies cannot belong simultaneously to the same genome. The SCJ's strategy of solving the conflict is simply to initialize the root with absence thus any ambiguity state will be resolved at the root as absence. This method is regarded as the only known distance for which the ancestral genome reconstruction problem has a polynomial time solution [8].

The other type of model that handles gene adjacencies relies on two separate steps. First, the weight or probability that a gene adjacency is present in a genome is computed independently. Then those gene adjacencies are assembled into a valid ancestral genome. InferCAR [17] and its probabilistic version InferCARsPro are the pioneering methods based on this model. In this model, all combinations of gene adjacencies are considered, and their probabilities are computed by a variant of the Fitch's parsimony algorithm. Finally, a greedy heuristic is used for to assemble the genes into a valid genome.

Later by relaxing the constraint of gene adjacency to gapped adjacency, GapAdj is proposed with the computation of a rigorous score for each potential ancestral adjacency (a,b), reflecting the maximum number of times a and b can be adjacent for any setting of ancestral genomes, as well as an algorithm to generate more reliable amount of chromosomes. Simulation experiments conducted by GapAdj show that GapAdj often ended up with a completely assembled genome, but resulted in a higher error rate that InferCAR.

Algorithmic details

Our recent method *Probabilistic Method of Ancestral Genomics* (PMAG) [1] is also based on adjacencies and uses a probabilistic framework. It requires a given topology of the input genomes (assumed to be the phylogeny) and places the known genomes at the leaves. PMAG first encodes the gene orders into binary sequences and estimates the parameters in the transition model for adjacency changes. It then checks each ancestral (internal) node in turn, and each will be computed independently by going through the following three steps: it first re-roots the input tree to have the target ancestor as the root of a new tree; it then uses a probabilistic inference tool to compute the conditional probabilities of all adjacencies; at last it uses a greedy algorithm similar to that presented in [17] to assemble a valid gene order. The last step is very critical, but the greedy algorithm tends to produce excessive number of contigs, indicating that it is very easy to get trapped in local optima. In this paper, we introduce a new assembly algorithm, which not only improves the accuracy of the assembled gene orders, but also reduce the number of contigs to be very close to the true result. The following are the details of our algorithm.

Re-rooting the phylogeny

Before we can infer the ancestral genome of an internal node, we must first re-root the given phylogeny tree to that node, making it the root of the new tree, which is a standard procedure and has already been used in [1,9]. The underlying rationale is that the calculation of probabilities follows a bottom-up manner such that only the species in the sub-tree of the target node are considered, it will result in loss of information if the node is not the root. As we are dealing with binary trees, the re-rooting procedure will need some extra work to preserve the tree structure as demonstrated in Figure 1. In this figure, as the ancestral node we have interest in is genome $A1$, to re-root the tree on this genome, we have to add an auxiliary node $A1'$, but set the branch length between $A1$ and $A1'$ (dashed edge) as always 0.

Obtaining probabilities of adjacencies

A gene order can be expressed as a sequence of adjacency information that specifies the presence or absence of all the adjacencies [18,19]. Denote the head of a gene i by i^h and its tail by i^t. We refer $+i$ as an indication of the direction from head to tail ($i^h \to i^t$) and otherwise $-i$ as ($i^t \to i^h$). We further write 1 (0) to indicate the presence (absence) of the adjacency and we consider only those adjacencies and telomeres that appear at least once in the input genomes.

Since we are handling binary sequences with two characters, we use a general time-reversible framework to simulate the transitions from presence (1) to absence (0) and vice versa. Since each genome contains $n+O(1)$ adjacencies and telomeres where n is the gene number and $O(1)$ equals to the number of linear chromosomes in the genome, thus the probability that an adjacency changes from presence (1) to absence (0) in the sequence is $\frac{2}{n+O(1)}$ under one operation. Since there are up to $\binom{2n+2}{2}$ possible adjacencies and telomeres, the probability for an adjacency changing from absence (0) to presence (1) is $\frac{2}{2n^2+O(n)}$. Therefore, we come to the conclusion that the transition from 1 to 0 is roughly $2n$ times more likely than that from 0 to 1.

To show how the transition model and the re-rooting procedure can respectively influence the performance of PMAG, we compare PMAG to its three variants through simulations (see details later):

- Naive: The naive version of PMAG with a neutral model of adjacency changes and fixed tree topology for all ancestral nodes.
- Naive+Model: Naive method cooperating with the biased transition model.
- Naive+Re-rooting: Naive method cooperating with the re-rooting procedure.

Figure 2 summarizes the comparison result using the smaller datasets among the four methods with tree diameters from $1n$ (easy case) to $5n$ (very difficult). In general, higher tree diameter effectively increased the difficulties and hence reduced the portion of correct adjacencies all methods can recover. Unsurprisingly the Naive method is the least accurate in all cases, and both Naive+Model and Naive+Re-rooting can independently enhance the accuracy of Naive method. By incorporating both mechanisms, PMAG not only inherited both improvements, but also obtained additional improvements as well, suggesting the transition model and the re-rooting procedure be useful and indispensable for our method.

Computing probabilities of gene adjacencies and assemble gene orders

Once we have the tree topology and binary sequences encoding the input gene orders, we use the extended probabilistic approach for sequence data described by Yang [20] to infer the ancestral gene orders at the root node, as described in detail in [1].

In the binary sequences, each site k represents an adjacency with character either 0 (absence) or 1 (presence); for each site at the root node, we seek to calculate the conditional probability of observing that adjacency. Suppose x is the root of a given tree, then the conditional probability that node x has the character s_x at site k, given D_x representing the observed data at site k in all

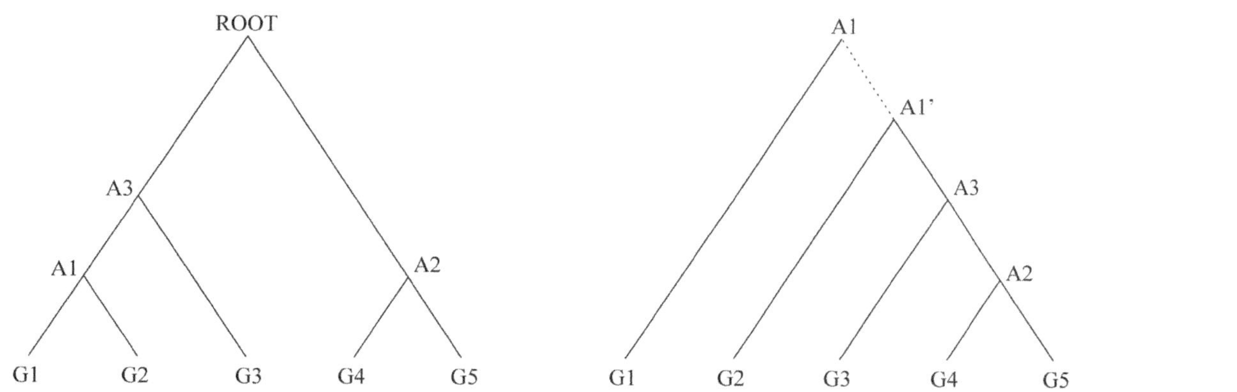

Figure 1. Re-rooting the phylogeny tree from the original root to the ancestral node under inference which is $A1$ in this case. Auxiliary node A' is added to preserve its binary structure.

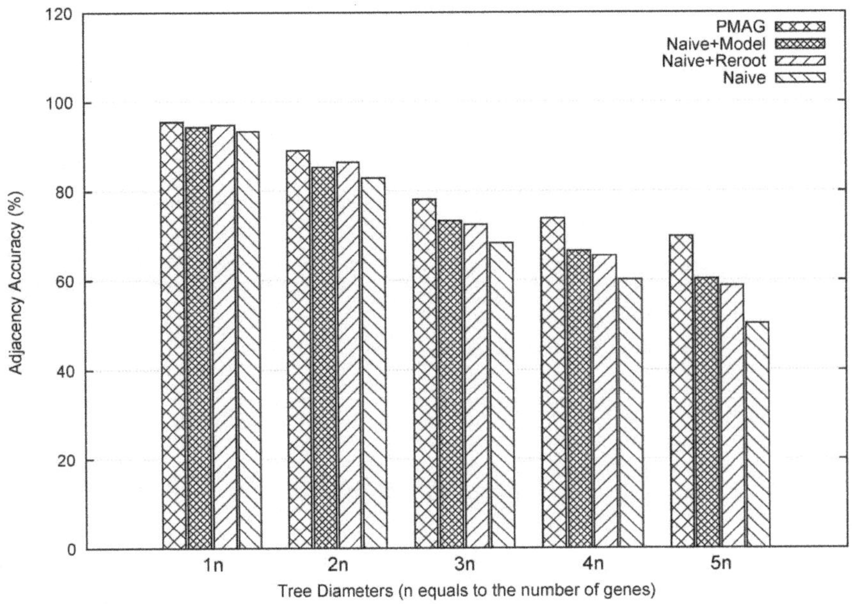

Figure 2. Comparison of adjacency accuracy between PMAG and its three premature versions. Datasets were simulated to have 10 genomes and 500 genes. X-axis represents the tree diameters from 1 to 5 times the number of genes.

leaves of the sub-tree rooted at x, is

$$P(s_x|D_x) = \frac{P(s_x)P(D_x|s_x)}{P(D_x)} = \frac{\pi_{s_x} L_x(s_x)}{\sum_{s_x} \pi_{s_x} L_x(s_x)}$$

where π_{s_x} is the character frequency for s_x.

For a site k, its conditional probability in the form of $L_x(s_x)$ is defined as the probability of observing the leaves that belong to the sub-tree rooted at x, given that the character of site k at node x is s_x. It can be calculated recursively in a post-order traversal fashion suggested by Felsenstein [1,21] as:

$$L_x(s_x) = \begin{cases} 1 & \text{if } x \text{ is a leaf with character } = s_x \text{ at the site} \\ 0 & \text{if } x \text{ is a leaf with character } \neq s_x \text{ at the site} \\ \left[\sum_{s_f} p_{s_x s_f}(t_f) L_f(s_f)\right] \times \left[\sum_{s_g} p_{s_x s_g}(t_g) L_g(s_g)\right] & \text{otherwise} \end{cases}$$

where f and g are the two direct descendants of x. $p_{ij}(t)$ defines the transition probability that character i changes to j after an evolutionary distance t. As the true branch lengths are not available, we take advantage of the widely-used maximum-likelihood estimation from the binary sequences at the leaves to estimate the branch length.

Following the deduction of transition probability in [21], our transition-probability matrix can be written as

$$p_{ij}(t) = \pi_j + e^{-t}(\delta_{ij} - \pi_j)$$

Here the δ_{ij} is 1 if $i=j$, otherwise δ_{ij} is 0.

We use RAxML [22] as it has a method to handle binary sequences efficiently and follows the steps suggested by Yang [20]. We modified RAxML so that it takes into account the biased transition model.

Our improvement over [1] is a better algorithm to assemble gene adjacencies and telomere into a valid gene order, with the requirement that each gene appears exactly once in the ancestral genome. In general, a higher probability of the presence state implies an adjacency or telomere should be more likely to be included in the ancestor; however, the decision on choosing an adjacency or telomere cannot be solely made upon its own probability as each gene can only be selected once. In the original PMAG, ancestral adjacencies are assembled by the greedy heuristic based on the adjacency graph proposed by *Ma et al.* There are two issues with the greedy approach: 1) it can only achieve a good approximation for closely related genomes; 2) it tends to create new contigs instead of connecting genes, resulted in an excessive number of contigs. In this paper, we develop the following algorithm based on the observation from [23], i.e. it can transform the problem of obtaining gene orders from (conflict) adjacencies into an instance of Traveling Salesman Problem (TSP). Although the TSP is NP-hard, it is a widely studied problem with very good solvers exist.

Specifically, we will transform genes into cities and adjacency probabilities into edge weights, and our goal is to find a tour that traverses all genes with the largest combined probabilities along the tour. As most TSP solver aims at finding a tour with minimum cost, to use probabilities as edge weights, we convert them by taking their logarithmic values. Suppose for an ancestral node K and a set of m adjacencies $A=\{a_1,a_2,...,a_m\}$ and n telomeres $T=\{t_1,t_2,...,t_n\}$ from leaf species, each with probabilities $P=\{p_{a_1},...,p_{a_m},p_{t_1},...,p_{t_n}\}$, we can create the TSP graph G by first splitting each gene g to two cities, denoted as g^h and g^t respectively, and representing each telomere t by a unique vertex e_i, where $1 \leq i \leq n$. To ensure a valid tour, we must connect g^h and g^t in a tour; thus we set the cost between g^h and g^t as $-\infty$. For any adjacency $(f,g) \in A$, we add an edge between f^t and g^h; similar edges are added for other combinations of orientations $(-f,g)$, $(f,-g)$ and $(-f,-g)$, as well as genes connecting to telomeres. For the rest of edges, as we could not find a valid probability, it means these edges should have a very low chance to

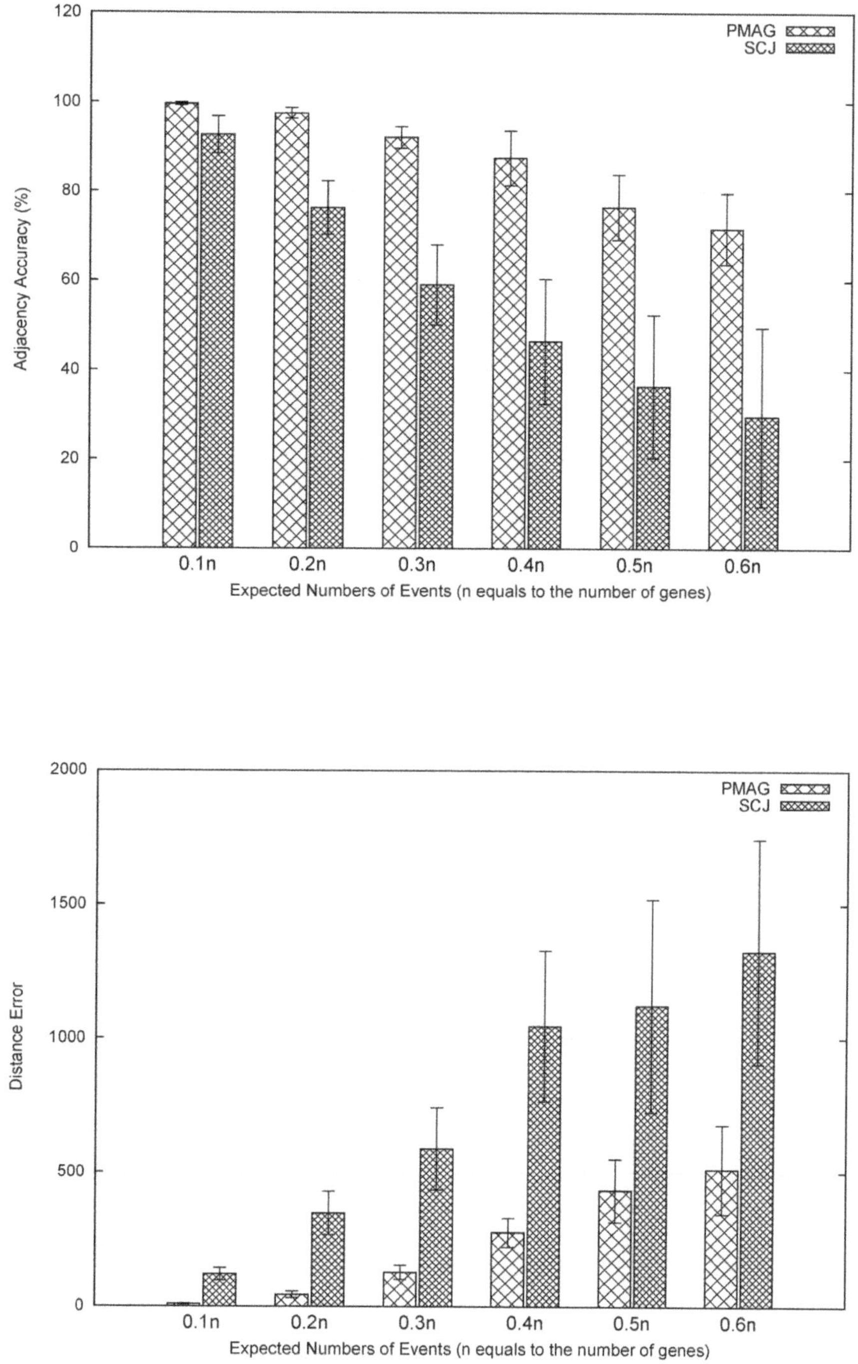

Figure 3. Comparison of adjacency accuracy (top) and distance accuracy (bottom) between PMAG and SCJ. Datasets were produced by the simulator provided in SCJ program that contain 32 genomes, each with five chromosomes and a total of 2,000 genes. X-axis represents the expected number of events from 0.1 to 0.6 times the number of genes.

be present in the ancestral genome; thus we set the edge weights to $+\infty$ to exclude them from the solution. In the solution path, multiple contiguous caps are shrunk into a single one, and a gene segment between two caps is taken as a chromosome.

One of the best and most used TSP solver is Concorde [24], which we integrated it into MAG. For the solution path, multiple contiguous extremities are shrunk to a single one and a gene segment between two extremities is taken as a contig. Our construction of TSP topology is in a spirit similar to GapAdj, however GapAdj requires additional procedures and parameters to adjust the contig number. Instead, our inference of the ancestral genome is uniform and directly from the solution of TSP, minimizing the risk of introducing artifacts.

Experimental Results

Experimental design

Since actual ancestors are rarely known for sure, it is difficult to evaluate ancestral reconstruction methods with real datasets. In

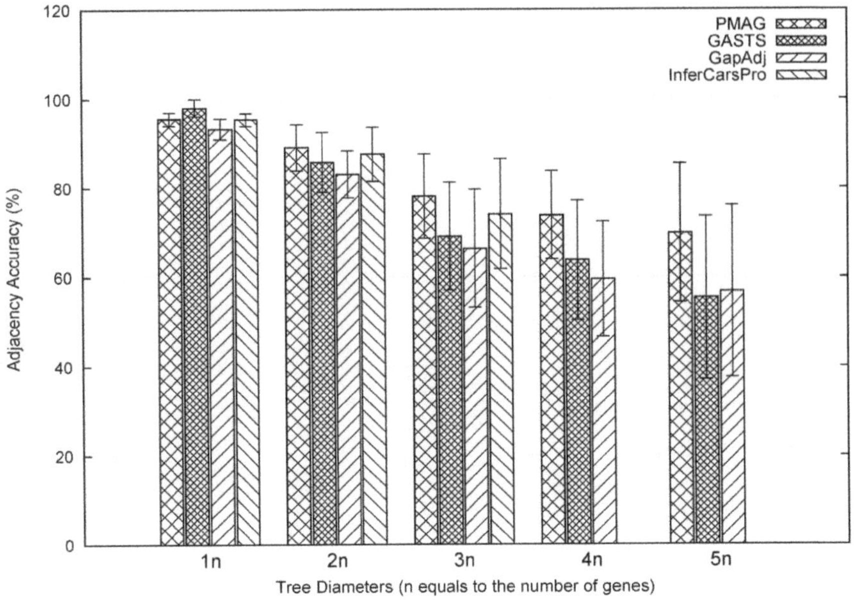

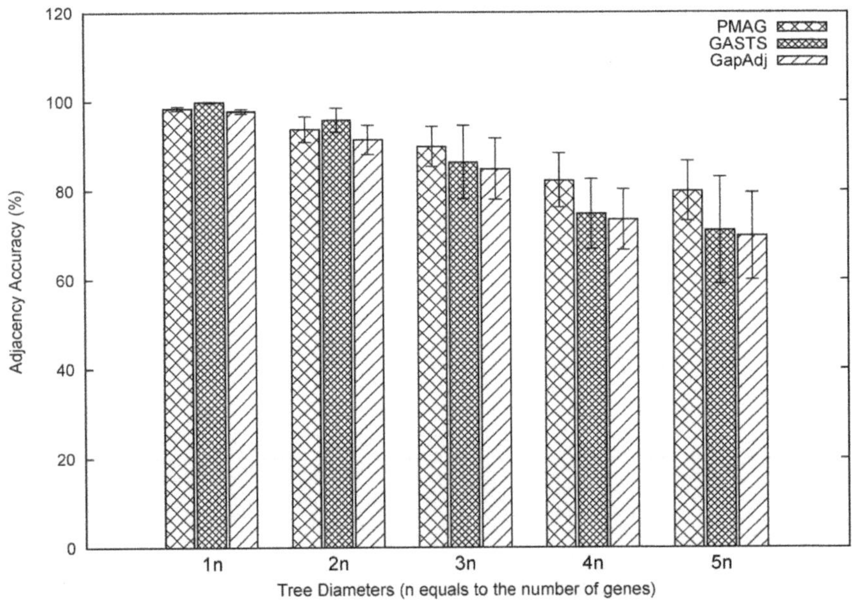

Figure 4. Comparison of adjacency accuracy between PMAG, InferCARsPro, GASTS and GapAdj: (top) datasets contain 10 genomes and 500 genes; (bottom) datasets contain 20 genomes and 2000 genes. Standard deviations are given at the top of bars. X-axis represents the tree diameters from 1 to 5 times the number of genes.

order to carry out a complete evaluation over a group of methods under a wide range of configurations, we conducted a collection of simulation experiments following the standard steps of such tests that have been extensively adopted in genome rearrangement studies [19,25].

In particular, a group of tree topologies were first generated with respect to the expected tree diameters. An initial gene order was assigned at the root so it can evolve down to the leaves following the tree topology mimicking the natural process of evolution, by carrying out a number of predefined evolutionary events. In this way, we obtained the complete evolutionary history of the model tree and the whole set of genomes it has.

Normally we utilized the simulator proposed by Lin *et al.* [26] to produce birth-death tree topologies. Since SCJ has its own simulator, we used that simulator for a fair comparison in the tests involving SCJ. With a model tree, we can produce genomes of any size by simply adjusting four main parameters: the number of genomes m, the number of chromosome c, the number of genes n,

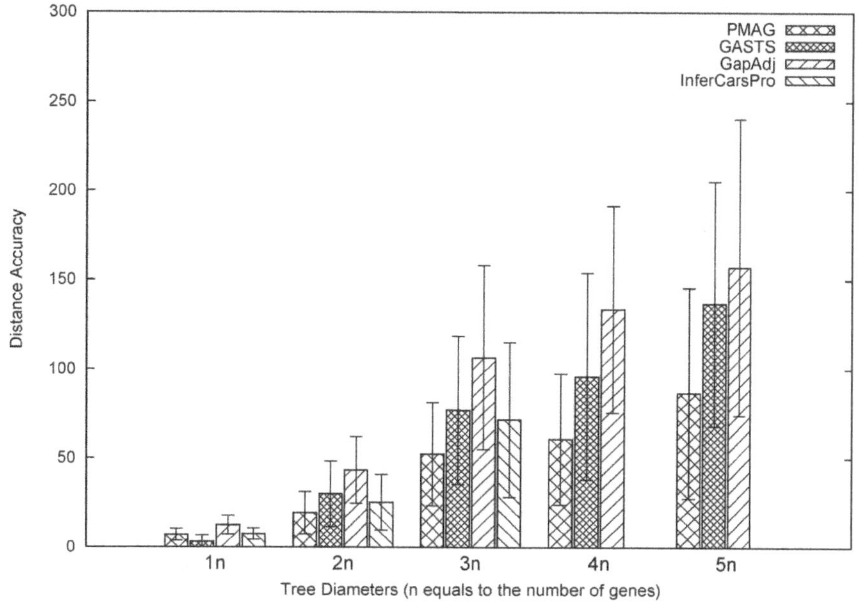

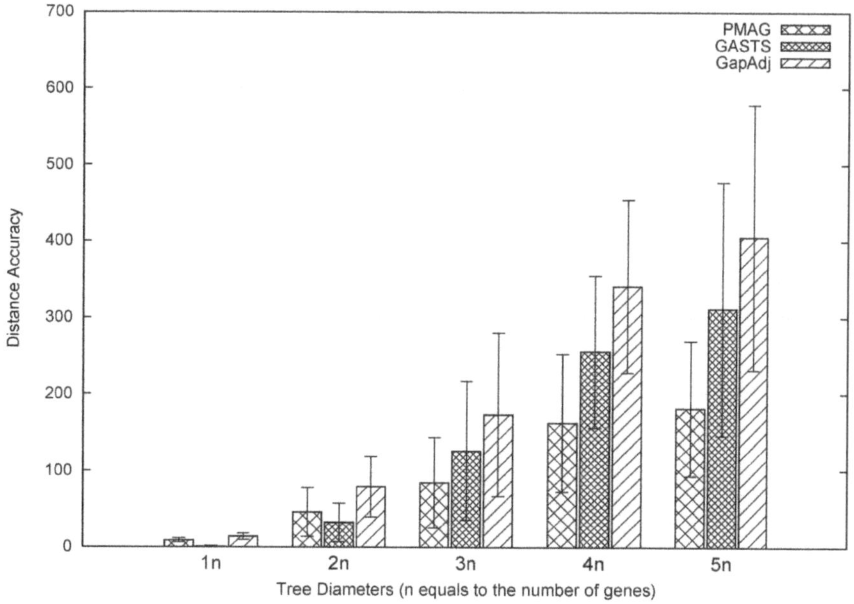

Figure 5. Comparison of distance accuracy between PMAG, InferCARsPro, GASTS and GapAdj: (top) datasets contain 10 genomes and 500 genes; (bottom) datasets contain 20 genomes and 2000 genes. Standard deviations are given at the top of bars. X-axis represents the tree diameters from 1 to 5 times the number of genes.

and the tree diameter d (equivalent to the branch length l in SCJ's simulator). For the following simulation experiments, we generated datasets of two different sizes. First we generated a smaller datasets with 10 genomes, each with 500 genes and five chromosomes to closely mimic the rearrangement scenarios in bacterial genomes with multi-chromosomes. We also produced datasets of larger size contains 20 genomes, each with 2,000 genes and five chromosomes. Along each branch, we performed 80% random inversions and 20% random translocations to account for intra- and inter-chromosomal rearrangements.

The following three measurements were used to assess the predicted ancestral genomes. We first calculated the adjacency accuracy C as the total number of correctly inferred adjacencies (i.e. those also appear in the true ancestral genomes) divided by the total number of adjacencies in both true genome and predicted genome. Second, we calculated distance accuracy D defined as the DCJ distance between a predicted ancestor and its corresponding

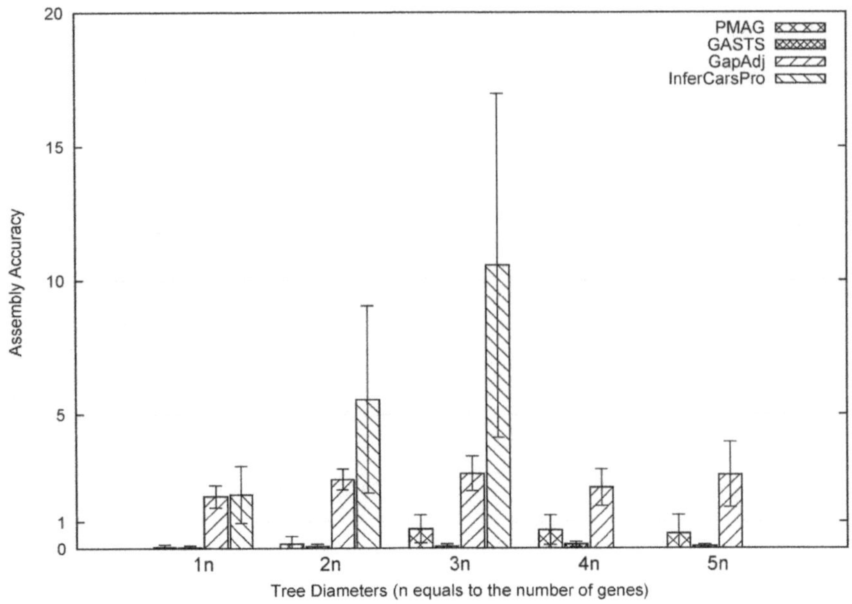

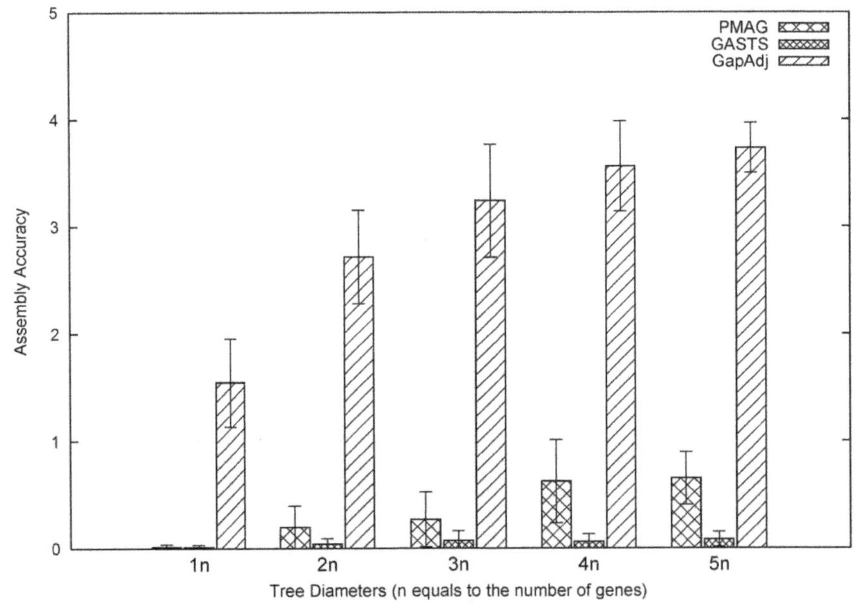

Figure 6. Comparison of assembly accuracy between PMAG, InferCARsPro, GASTS and GapAdj: (top) datasets contain 10 genomes and 500 genes; (bottom) datasets contain 20 genomes and 2000 genes. X-axis represents the tree diameters from 1 to 5 times the number of genes.

Table 1. Comparison of assembly accuracy between PMAG and SCJ under different expected numbers of evolutionary events along a tree branch. (n equals to the number of genes).

Tree Diameter	0.1n	0.2n	0.3n	0.4n	0.5n	0.6n
PMAG	0	0.13	0.30	0.45	0.48	0.63
SCJ	157	476	785	1031	1280	1413

Table 2. Time consumptions of four methods (including the previous greedy version of PMAG in analyzing large datasets under different tree diameters. (n equals to the number of genes).

Tree Diameter	PMAG	GapAdj	GASTS	PMAG-Greedy
1n	13 min	10 min	1 min	1 min
2n	13 min	12 min	10 min	3 min
3n	15 min	12 min	45 min	3 min
4n	18 min	14 min	120 min	4 min
5n	20 min	16 min	159 min	5 min

true genome. Apparently for the genome rearrangement study, distance accuracy is more appropriate as it not only considers the adjacency changes, but also takes differences in genome structures into account. Finally, to assess the assembly capabilities, we computed assembly accuracy A as the absolute differences of the number of chromosomes between a predicted ancestor and its corresponding truth. For each dataset, the average of each measurement across all ancestors was computed and for each tree diameter, we produced 10 datasets and reported their average, as well as their standard deviation.

Evaluation of PMAG against SCJ

Simulator embedded in the SCJ was used, and the measurement of difficulty became the branch length l, denoting the expected number of evolutionary events along an edge of the tree which is sampled from an uniform distribution on the set $\{1,2,3...d\}$, where d equals to $l \times n$ and n is the number of genes. As before, those events consisted of 80% of inversions and 20% translocations. Since SCJ and PMAG are both fast enough, we therefore generated a set of larger dataset containing 32 genomes, each with five chromosomes and a total of 2,000 genes.

Figure 3 demonstrates the adjacency accuracy and the distance accuracy of PMAG and SCJ respectively. This figure clearly suggests that PMAG can significantly outperform SCJ in all tested cases.

Evaluation of PMAG against other methods

In this section, we picked three main competitors from both event-based and adjacency-based methods, and compared them with PMAG. In particular we supplied InferCARsPro with multichromosomal genomic distances as its branch lengths computed by GRIMM [27]. Moreover, in GapAdj, the cutoff value and maximal iterations were set to 0.6 and 25 as suggested by the authors. The event-based method GASTS was simply run by providing the true tree and the input genomes. Results of InferCARsPro under large tree diameters were missing as it failed to finish the tests in three days.

Figure 4 shows the results measured by the adjacency accuracies. When the tree diameters were $1n$, all methods were able to produce highly accurate ancestral genomes (>90%) and the differences among methods were not significant. In particular, GASTS was the most accurate method, while the performances of PMAG and InferCARsPro were similar, and both were better than GapAdj. As the tree diameters went larger, GASTS quickly became unreliable which is consistent with the experimental findings reported in the study of GASTS [7]. In all tests, PMAG showed great robustness against disturbance and achieved the highest adjacency accuracy when the tree diameter grows greater than $2n$. Figure 5 shows the results measured by the distance accuracies. In general, the relative performances of various methods in distance measurement are very similar to the adjacency accuracies.

Comparison of performances on assembly

The final step of adjacency-based methods often involves assembly of adjacencies into contiguous segments which can be viewed as chromosomes or more precisely contigs. Previous methods InferCARsPro employing a greedy algorithm for assembly often ends up with an excessive number of contigs. Later the assembly accuracy was improved by GapAdj using the concept of gapped adjacencies with a sacrifice of accuracy.

Our measurement of accuracy only counts adjacencies correctly recovered. However, for two assembled gene orders with similar adjacency accuracy, the one with the number of contigs close to the number of chromosomes should be viewed as having better accuracy. Thus, we summarized the number of contigs produced by various methods and computed the averages of assembly accuracy for all cases in Figure 6. From the figure, the event-based method GASTS without the need for assembly of gene adjacencies produced the most relevant number of contigs. Among the adjacency-based methods, PMAG showed much better assembly performance, and its performance was very close to GASTS. As expected, the greedy assembly used in InferCARsPro produced the least relevant number of contigs. By examining Figure 4, we found that although PMAG returned more contigs than GASTS, its distance and adjacency accuracies were better, indicating that GASTS had a tendency to introduce bad adjacencies in order to keep the number of contigs small.

Table 3. Comparison of the adjacency accuracy between PMAG and its greedy version. The number of genomes is 20 and the number of genes is 2,000.

Tree Diameter	1n	2n	3n	4n	5n
PMAG	98.7	93.5	89.7	82.2	79.5
PMAG-Greedy	98.5	93.2	88.6	80.2	77.8

Table 4. Comparison of the assembly accuracy between PMAG and its greedy version. The number of genomes is 20 and the number of genes is 2,000.

Diameter	1n	2n	3n	4n	5n
PMAG	0.03	0.2	0.3	0.6	0.6
PMAG-Greedy	2.1	5.5	8.5	15.9	17.6

Table 1 shows the assembly accuracies of PMAG and SCJ. From the table, PMAG yielded very accurate amount of contigs; however, since SCJ is overly conservative, it missed a large portion of true adjacencies and produced a massive amount of contigs. In other words, SCJ has difficulty in assembling gene orders due to its overly simplified cost to weigh adjacencies.

Time efficiency

All tests were conducted on a workstation with 2.4 GHz CPUs and 4 GB RAM. In general, SCJ is undoubtedly the fastest and can return results in just a few seconds. In the experiments with small datasets, InferCARsPro required the most amount of time and the other three methods can always finish within a minute.

We summarized the time consumptions of PMAG, GapAdj, GASTS and the previous greedy version of PMAG in handling large datasets in table 2. From the table, the running time of PMAG and GapAdj were very close and stable, and tree diameters did not remarkably slow down these programs. On the other hand, GASTS severely suffered from large tree diameters, suggesting its potential limitation in handling genomes that are distant to each other.

Comparing PMAG with Its Previous Version

We compared PMAG with its greedy version to evaluate the new TSP approach. Table 3 and Table 4 showed the adjacency and assembly accuracy, respectively. These tables suggested that although the TSP solver was about 10 times slower than the greedy solver (Table 2), the new PMAG method had achieved improved adjacency accuracy with much better performance in term of the number of recovered contigs.

Conclusions and Future Work

In this paper, we introduced the adjacency-based method PMAG in the probabilistic framework for ancestral gene-order inference. PMAG determines the state for each adjacency in the binary encoding to be either present or absent in an ancestral genome according to its conditional probability. Ancestral gene orders are finally assemblies by connecting individual adjacencies into continuous regions using a TSP approach Experimental results reveal that PMAG can not only accurately infer ancestral genomes, and also did a good job in assembling adjacencies into valid gene orders. Finally, PMAG is fast and also stable across a wide range of configurations.

However, much work remains to be done. As PMAG relies on gene adjacencies, how to recover adjacencies lost in evolution (thus not shown in leave genomes) is an interesting problem. In the current implementation, these adjacencies have no definite edge weight and they are all set as $+\infty$. As a result, the TSP tour is prevented from passing through them, although they may be better choices. Our experiments showed that these adjacencies account for about 25% of errors in PMAG; thus, we need to devise a method that can assign better edge weights to missing adjacencies. Since each internal node can be computed independently, the speed of PMAG can be further improved by utilizing the presence of multiple computing cores in modern CPUs by placing each node's computation on a core.

Author Contributions

Conceived and designed the experiments: NY FH JT. Performed the experiments: NY FH LZ. Analyzed the data: NY FH JT LZ. Contributed reagents/materials/analysis tools: NY FH. Wrote the paper: NY FH JT.

References

1. Hu F, Zhou L, Tang J (2011) Reconstructing Ancestral Genomic Orders Using Binary Encoding and Probabilistic Models. Proceedings of the 9th International Symposium on Bioinformatics Research and Applications (ISBRA), 17–27.
2. Kent W, Baertsch R, Hinrichs A, Miller W, Haussler D (2003) Evolutions cauldron: duplication, deletion, and rearrangement in the mouse and human genomes. Proceedings of the National Academy of Sciences, 100(20):11484–11489.
3. Muller K, Borsch T, Legendre L, Porembski S, Theisen I, et al. (2008) Evolution of carnivory in Lentibulariaceae and the Lamiales. Plant Biology, 6(4):477–490.
4. Moret BME, Wyman S, Bader DA, Warnow T, Yan M (2001) A New Implmentation and Detailed Study of Breakpoint Analysis. Proceedings of the 6th Pacific Symposium on Biocomputing (PSB), 583–594.
5. Bourque G, Pevzner P (2002) Genome-scale evolution: reconstructing gene orders in the ancestral species. Genome Research, 12(1): 26–36.
6. Alexeyev M, Pevzner P (2009) Breakpoint graphs and ancestral genome reconstructions. Genome Research, 19 (5): 943–957.
7. Xu W, Moret BME (2011) GASTS: Parsimony scoring under rearrangements. Proceedings of the 11th Workshops on Algorithms in Bioinformatics (WABI), 351–363.
8. Biller P, Feijao P, Meidanis J (2012) Rearrangement-based phylogeny using the Single-Cut-or-Join operation. IEEE/ACM Transactions on Computational Biology and Bioinformatics, 10 (1), 122–134.
9. Ma J (2010) A probabilistic framework for inferring ancestral genomic orders. Proceedings of the IEEE International Conference on Bioinformatics and Biomedicine (BIBM), 179–184.
10. Gagnon Y, Blanchette M, El-Mabrouk N (2012) A flexible ancestral genome reconstruction method based on gapped adjacencies. BMC bioinformatics 13 (Suppl 19): S4.
11. Yancopoulos S, Attie O, Friedberg R (2005) Efficient sorting of genomic permutations by translocation, inversion and block interchange. Bioinformatics, 21(16): 3340–3346.
12. Feijao P, Meidanis J (2009) SCJ: a variant of breakpoint distance for which sorting, genome median and genome halving problems are easy. Proceedings of the 9th Workshop on Algorithms in Bioinformatics (WABI), 85–96.
13. Caprara A (1999) Formulations and hardness of multiple sorting by reversals. Proceedings of the 3rd Annual International Conference on Computational Molecular Biology (RECOMB), 84–93.
14. Tannier E, Zheng C, Sankoff D (2009) Multichromosomal median and halving problems under different genomic distances. BMC Bioinformatics, 10(1): 120.
15. Xu W, Sankoff D (2008) Decompositions of multiple breakpoint graphs and rapid exact solutions to the median problem. Proceedings of the 8th Workshop on Algorithms in Bioinformatics (WABI), 25–37.
16. Hannenhalli S, Pevzner P (1995) Transforming cabbage into turnip: polynomial algorithm for sorting signed permutations by reversals. Proceedings of the 27th Annual ACM Symposium on Theory of Computing, 178–189.
17. Ma J, Zhang L, Suh B, Raney B, Burhans R, et al. (2006) Reconstructing contiguous regions of an ancestral genome. Genome Research, 16(12):1557–1565.
18. Hu F, Gao N, Tang J (2011) Maximum likelihood phylogenetic reconstruction using gene order encodings. IEEE Symposium on Computational Intelligence in Bioinformatics and Computational Biology (CIBCB), 117–122.

19. Lin Y, Hu F, Tang J, Moret BME (2013) Maximum likelihood phylogenetic reconstruction from high-resolution whole-genome data and a tree of 68 eukaryotes. Proceedings of the 18th Pacific Symp. on Biocomputing (PSB), 285–296.
20. Yang Z, Kumar S, Nei M (1995) A new method of inference of ancestral nucleotide and amino acid sequences. Genetics, 141(4):1641–1650.
21. Felsenstein J (1981) Evolutionary trees from DNA sequences: a maximum likelihood approach. Journal of Molecular Evolution, 17(6):368–376.
22. Stamatakis A (2006) RAxML-VI-HPC: maximum likelihood-based phylogenetic analyses with thousands of taxa and mixed models. Bioinformatics, 22(21):2688–2690.
23. Tang J, Wang L (2005) Improving genome rearrangement phylogeny using sequence-style parsimony. Proceedings of 5th IEEE Symposium on Bioinformatics and Bioengineering, 137–144.
24. Applegate D, Bixby R, Chvatal V, Cook W (2003) Concorde TSP solver. Available: http://www.tsp.gatech.edu/concorde. Accessed September 4th 2014.
25. Jahn K, Zheng C, Kovac J, Sankoff D (2012) A consolidation algorithm for genomes fractionated after higher order polyploidization. BMC Bioinformatics, 13 (Suppl 19): S8.
26. Lin Y, Rajan V, Moret BME (2011) Fast and accurate phylogenetic reconstruction from high-resolution whole-genome data and a novel robustness estimator. Journal of Computational Biology, 18(9): 1131–1139.
27. Tesler G (2002) Efficient algorithms for multichromosomal genome rearrangements. Journal of Computer and System Sciences, 65(3): 587–609.

19

Carrot yellow leaf virus Is Associated with Carrot Internal Necrosis

Ian P. Adams[1], Anna Skelton[1], Roy Macarthur[1], Tobias Hodges[1,3], Howard Hinds[2], Laura Flint[1], Palash Deb Nath[4], Neil Boonham[1], Adrian Fox[1]*

1 Centre for Crop Protection, Food and Environment Research Agency, Sand Hutton, York, United Kingdom, **2** RootCrop Ltd., Hoveringham, Nottinghamshire, United Kingdom, **3** University of York, York, United Kingdom, **4** Department of Plant Pathology, Assam Agricultural University, Jorhat, India

Abstract

Internal necrosis of carrot has been observed in UK carrots for at least 10 years, and has been anecdotally linked to virus infection. In the 2009 growing season some growers had up to 10% of yield with these symptoms. Traditional diagnostic methods are targeted towards specific pathogens. By using a metagenomic approach with high throughput sequencing technology, other, as yet unidentified causes of root necrosis were investigated. Additionally a statistical analysis has shown which viruses are most closely associated with disease symptoms. Carrot samples were collected from a crop exhibiting root necrosis (102 Affected: 99 Unaffected) and tested for the presence of the established carrot viruses: *Carrot red leaf virus (CtRLV), Carrot mottle virus (CMoV)*, Carrot red leaf associated viral RNA (CtRLVaRNA) and *Parsnip yellow fleck virus (PYFV)*. The presence of these viruses was not associated with symptomatic carrot roots either as single viruses or in combinations. A sub-sample of carrots of mixed symptom status was subjected to MiSeq sequencing. The results from these tests suggested *Carrot yellow leaf virus* (CYLV) was associated with symptomatic roots. Additionally a novel Torradovirus, a novel Closterovirus and two novel Betaflexiviradae related plant viruses were detected. A specific diagnostic test was designed for CYLV. Of the 102 affected carrots, 98% were positive for CYLV compared to 22% of the unaffected carrots. From these data we conclude that although we have yet to practically demonstrate a causal link, CYLV appears to be strongly associated with the presence of necrosis of carrots.

Editor: Zhengguang Zhang, Nanjing Agricultural University, China

Funding: This work was funded by the Horticultural Development Company, a levy board of the Agricultural and Horticultural Development Board, as project FV382a. The funders had no role in study design, data collection and analysis, or preparation of the manuscript. Permission was required to allow submission of the manuscript.

Competing Interests: Howard Hinds is the sole owner and employee of RootCrop Ltd. Through the author HH, RootCrop Ltd. had a role in the study design, data collection and analysis, decision to publish, and preparation of the manuscript. The specific role of author, Howard Hinds, is articulated in the "author contributions" section. There are no patents, products in development, or marketed products to declare.

* Email: adrian.fox@fera.gsi.gov.uk

Introduction

For at least 10 years UK growers have reported carrot roots exhibiting internal necrosis around the root core extending from crown to tip, and these have been anecdotally associated with the presence of viruses. The 2009 growing season saw some growers with up to 10% of yield affected by these symptoms, although symptom development appeared to be locally significant, with many growers reporting no evidence of root symptoms in crops. It is difficult to grade out affected carrots because the symptoms tend to be internal. Results of a limited survey in 2010 [1] suggested a possible association between the presence of root necrosis symptoms and virus infection. However, a large proportion of the carrots tested in this earlier study were negative when tested for PYFV or the Carrot Motley Dwarf complex (CMD) of viruses. This finding raised the question of other viruses being a cause of the development of carrot root necrosis.

Globally more than 30 viruses are known to affect carrot [2]. The principal viruses known to affect commercial carrot crops in the UK are *Parsnip yellow fleck virus* (PYFV) and the CMD Complex consisting of *Carrot red leaf virus* (CtRLV), *Carrot mottle virus* (CMoV) and Carrot red leaf associated RNA (CtRLVaRNA). The importance of PYFV and CtRLV as viruses causing economic damage have been recognised for over 20 years due to the foliar symptoms (CtRLV) and viral die-back of seedlings (PYFV) [3]. In the UK these viruses affect carrot crops only sporadically but when they do occur they can be devastating. Other carrot viruses are known to occur in the UK, however, their effects are not clear.

Carrot yellow leaf virus (CYLV) (Genus *Closterovirus*, Family *Closteroviridae*) was first isolated from carrot samples showing yellowing foliage from Japan [4] and described on the basis of particle morphology; measurement by Electron Microscopy (1,600×12 nm, 3.7 nm Helical pitch); being limited to phloem and having characteristics of closterovirus infection. Bem and Murant [5] later described a series of viruses found in the UK from hogweed (*Heracleum sphondylium*), among which were the filamentous viruses Hogweed 2 virus with a particle size of 700–750 nm, and Hogweed 6 virus (HV6) with a particle size of 1400 nm, and being transmissible by aphids these were tentatively assigned to the genus Closterovirus [5]. Hogweed 2 virus, with shorter particles, was characterised as *Heracleum latent virus*

(HLV) [6]. Hogweed 6 virus was subsequently shown to be transmitted by aphids including *Cavariella* spp. and to act as the helper virus for the transmission of HLV [7]. Murant again reported the length of HV6 as 1400 nm and that mechanical inoculation of HV6 was unsuccessful [8]. During a subsequent survey of umbellifers in the Netherlands [3] the virus previously reported by Murant [6,7,8] as HV6 was considered to be CYLV on the basis of host range similarity and the ability to facilitate co-transmission of HLV.

A virus isolated from carrot in the Netherlands was reported as 'resembling CYLV' due to host symptoms; the presence of closterovirus-like particles; and a host range that was different to *Beet yellows virus* [9]. Murant reported that sap inoculation of CYLV had been unsuccessful [8], van Dijk and Bos [9] reported poor sap transmissibility into *Nicotiana benthamiana*, but subsequent attempts to mechanically transmit this virus back into carrot were unsuccessful.

An isolate of an unknown closterovirus from a German carrot sample exhibiting foliar yellowing symptoms was shown to be CYLV through molecular characterisation [10]. There is no literature linking any carrot viruses with necrotic root symptoms.

Detection of carrot viruses is currently carried out using conventional PCR methods, which give efficient, specific detection of single targets. With the use of degenerate primer sets they can be used to detect a number of pathogens of the same genus [10]. However, such targeted testing will not reveal the presence of unexpected or unknown viruses. Even multi-target approaches such as micro-array based methods [11] are unlikely to reveal the presence of complete unknowns, unless cross-hybridisation to known close relatives occurs. A more efficient approach would be to use a 'non-targeted' method such as next generation (high throughput) sequencing for diagnosis of viral pathogens. These techniques have been successfully deployed in plant pathology for the detection of novel viruses [12,13,14] or for the diagnosis of unusual strains of plant viruses [15]. Such approaches are rapidly becoming more cost effective as the high throughput platforms develop. Previous reports utilising this technology have tended to identify the presence of a novel or unusual virus in single or pooled samples and then use the sequence generated to design targeted diagnostics to validate the finding from the original sample. Putting these findings into a broader context of field pathology is more challenging.

To definitively link a pathogenic cause to an observed symptom it is necessary to demonstrate Koch's postulates, the isolation from a diseased individual of a pure culture of a pathogen which is then used to induce symptoms in a previously healthy host. These requirements, first described in 1890 [16], were intended to set a standard methodology for proof of a causal relationship. As viruses are obligate pathogens, it is not possible to obtain a 'pure culture', in addition some viruses can be difficult to transmit and the specific transmission mechanism of a new virus may not be known. For diseases induced by a single virus species Koch's postulates may be satisfied in their broadest interpretation i.e. a pathogen is isolated from a symptomatic plant into an experimental host and then back-inoculated into the original host species to try and replicate the original symptom. Attempts have also been made to look at causation in light of developments in molecular detection [16]. However, where a complex of viruses may be affecting a host or where there may be environmental or agronomic influences on symptom development (temperature, moisture, time from exposure, time in ground/crop growth stage, etc) trying to link detection of pathogen/s with a symptom using a conventional cause-and-effect relation is often not possible. Therefore statistical approaches have been employed to demonstrate the possible influence of single or multiple pathogens on the expression of symptoms within a sampled population [17].

This paper describes a study of the potential causes of carrot internal necrosis using RT_PCR of common carrot viruses and next generation sequencing in carrots with and without symptoms of necrosis and a statistical approach to associate particular viruses with the incidence of necrotic symptoms.

Materials and Methods

Ethics statement

The carrots were obtained with the permission of Rodger Hobson, Hobson's Farming and no further permissions were required. The samples were taken post-harvest so there was no damage to endangered or protected species.

Source of carrot samples

The crop of carrots sampled were grown at Gothic Back Field, York, UK (OS Grid Ref: SE 65486 445297; Latitude 53.899711, Longitude -1.0048628). On the grading line carrots were cut and examined for symptoms, of these 3% of the 3300 individual carrots examined contained necrotic symptoms. Some of these carrots had surface necrosis which could have been graded out, others had only internal necrosis, and there were also carrots sampled with a combination of internal and external symptoms. From these samples 102 carrot roots were selected which were affected by disease (i.e. necrotic/symptomatic) and 99 carrot roots were selected which were un-affected (non-necrotic/asymptomatic). As the carrots were sampled on the grading line no assessment of foliar symptoms could be made.

Sap inoculations

Five plants of each of the standard indicators *Nicotiana benthamiana, N. debneyi, N. hesperis, N. tobaccum* (cv White Burley), *N. occidentalils* (P1), *Chenopodium quinoa*, and *C. amaranticolor*, Tomato and the umbelliferous plants Carrot, Chervil and Coriander were inoculated from carrot samples using the methods described in Hill [18]. Control and non-inoculated plants were maintained for all species. Plants were maintained in a green house with a mean temperature of 22°C with an 18 hr. photoperiod and assessed for symptoms weekly.

Reverse Transcription PCR

RNA was extracted from carrot roots by magnetic bead extraction using Invimag Virus DNA/RNA mini-kit (Invitek GMBH). Conventional RT-PCR was carried out for the presence of the four carrot viruses known to be common in the UK namely *Parsnip yellow fleck virus* [19] and the viruses of the Carrot Motley Dwarf complex, *Carrot red leaf virus, Carrot Mottle virus* [20] and Carrot red leaf associated viral RNA [21]. All RT-PCR reactions were carried out using Verso 1-Step RT-PCR ReddyMix Kit (Thermo Scientific) on a GeneAmp 9700 (Applied Biosystems) (Annealing Temperature 50°C).

High Throughput Sequencing

RNA was extracted from 12 affected and 12 unaffected carrots using an RNeasy kit (Qiagen, UK). TruSeq RNA Indexed sequencing libraries (Illumina) were then prepared following the manufacturers recommended protocols, before being sequenced on 2 500 cycle v2 flow cells using MiSeq Sequencer (Illumina). The resulting sequences were trimmed to remove low-quality nucleotides from the 3′ end, using a Phred score threshold of 30 and the *bwa* trimming approach implemented in SolexaQA [22], and assembled using Trinity [23]. Contigs produced were then compared to the GenBank protein database using BLASTx + [24] and viral reads extracted using MEGAN [25,26]. Open reading frames were identified using Vecor NTi v11 (Invitrogen, UK) and alignments and phylogenetic trees produced using Mega5 with 500 bootstrapped replicates [27]. To determine the number of viral reads in the affected and unaffected samples reads were mapped back to the genomes of identified viruses using bwa *aln sampe* [28] and the numbers of matched read pairs extracted using Samtools [29]. To normalize mapped read counts against the length of genome and total number of reads, the values are reported as mapped reads per kilobase of viral genome per million reads (RPKM), an approach originally introduced for the comparison of mRNA abundance in differential expression analyses [30]. The fastq data produced during the project were submitted to the short read achive acc: SRP042501.

Real-time PCR

Real-time (TaqMan) primers and probes were designed using Primer Express 2 with sequences from GenBank and derived from the sequencing in this study when available (Applied Biosystems). Real-time RT-PCR was performed on previously extracted RNA in 96 well plates on an ABI 7900 instrument (Applied Biosystems). Reactions consisted of 1 × buffer A (Applied Biosystems), 0.2 mM of each dNTP, 5.5 mM $MgCl_2$, 0.025 U/µl AmpliTaq Gold (Applied Biosystems), 0.4 U/µl Revertaid (Fermentas), 300 nM of each primer, 100 nM of probe and 1 µl of extracted RNA (concentration as extracted) to give a final reaction volume of 25 µl. The cycling conditions used were: 30 min at 48°C, 10 min at 95°C, then 40 times, 15 sec at 95°C and 1 min at 60°C. Negative controls consisted of water replacing the template. Results were scored as positive or negative for CYLV based on presence or absence of amplification after 40 cycles.

Statistical analysis

The extent to which necrosis may be caused by viruses was assessed by counting the proportion of carrots that are necrotic and testing equal numbers of necrotic and non-necrotic carrots for the presence of viruses. Hence we observed three proportions:

$P(N)$: the proportion of carrots that were necrotic.

$P(V|N)$: the proportion of necrotic carrots that contained a virus,

$P(V|\sim N)$: the proportion of non-necrotic carrots that contained a virus.

Additionally estimates of three proportions were derived:

$P(V)$: the proportion of carrots that contained virus,

$P(N|V)$: the proportion of carrots with a virus that are necrotic

$P(N|\sim V)$: the proportion of carrots without a virus that are necrotic.

Estimates were derived from the law of total probability

$$P(V)=P(V|N)P(N)+P(V|\sim N)(1-P(N)) \quad \text{(Equation 1)}$$

and Bayes' Theorem

$$P(N|V)=\frac{P(V|N)P(N)}{P(V)}$$
$$P(N|\sim V)=\frac{P(\sim V|N)P(N)}{1-P(V)} \quad \text{(Equation 2)}$$

The size of the uncertainty associated with observed proportions was estimated using a Modified Jeffreys interval [1], where given x 'positives' out on n observations the probability p underlying the observed proportion is with confidence 1-α

$$B(\alpha/2,x+0.5,n-x+0.5)\leq p\leq B(1-\alpha/2,x+0.5,n-x+0.5)$$

if 0<x<n,

where $B(\alpha,b,c)$ is the α quantile of the $Beta(b,c)$ distribution

$$p\leq 1-\alpha^{1/n}$$

If x=0, and

$$p\leq \alpha^{1/n}$$

If x=n. (Equation 3)

The size of the predicted effect of removing a virus, on the prevalence of necrosis expressed as the proportional reduction in prevalence was estimated using

$$E=\frac{P(N)-P(N|\sim V)}{P(N)} \quad \text{(Equation 4)}$$

The uncertainty associated with derived estimates was estimated by generating independent random (uniform (0,1)) quantiles for each of the observed proportions (Equation 3) and calculating derived values using Equations 1 and 2. 95% confidence intervals were taken from the 2.5th and 97.5th percentiles of 10000 derived values.

Results

Sampling

A sample of 3300 carrots were examined for necrotic root symptoms. The prevalence of symptoms was estimated from this sample to be approximately 3%. The 102 necrotic/symptomatic carrot samples found within the sample were taken for analysis. These roots exhibited a range of symptoms which included cases of internal and external necrosis. (See Figure 1). 99 asymptomatic/non-necrotic carrots were also taken for analysis.

Reverse-Transcriptase PCR Testing

The sample of affected (symptomatic) and unaffected carrots were initially tested using conventional RT-PCR assays for the presence of the carrot viruses which are known to be common in the UK (PYFV, CMoV, CtRLV, CtRLVaRNA). Results from this testing are presented in Figure 2 which shows the percent of affected or unaffected carrots which contained the viruses. Approximately equal proportions of affected and unaffected carrots were positive for PYFV and CMD viruses (37% affected, 38% unaffected). The two groups also contained similar incidences of CtRLV (33% affected, 27% unaffected) and CMoV (9% affected, 14% unaffected). No CtRLVaRNA was detected from carrots of either symptom status. As RT-PCR results from the two groupings were broadly comparable no influence of the viruses upon the incidence of necrosis symptoms was detected. A subsample of carrots was subsequently tested using high throughput sequencing to investigate the presence of non-target pathogens.

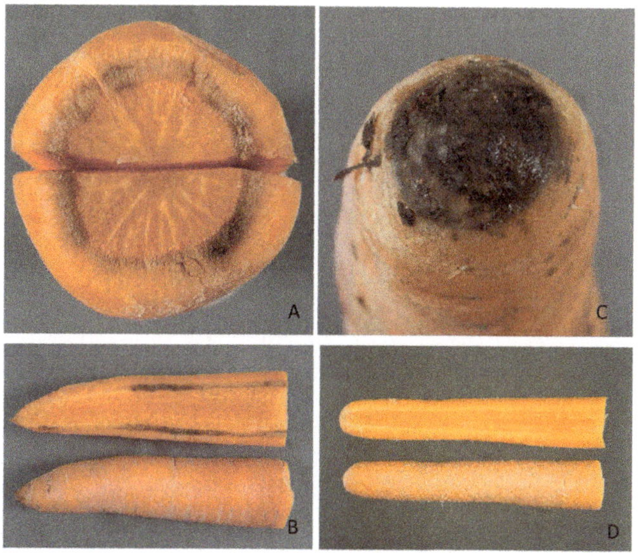

Figure 1. Examples of symptoms of 'affected' carrot samples (a) Cross section of carrot root showing internal necrosis around the root core. (b) Internal necrosis of carrot root along the root core. *(c)* External necrosis of the root tip. (d) Unaffected carrot.

High Throughput Sequencing

Indexed pair end reads (16225604 in total) with a sequence length of 2×250 bp were obtained from 12 affected and 12 unaffected carrot RNA extracts. A phylogram (figure 3), produced in MEGAN [25,26] details the viruses found in the affected and unaffected samples. *Carrot yellow leaf virus* (CYLV) was by far the most prevalent virus in the affected samples (RPKM = 221.3) but it was not common in the unaffected samples (RPKM = 2.2). Comparison of the MEGAN pylograms for affected and unaffected carrots did not reveal any fungi or bacteria unique to the affected carrots.

Full genomes for CYLV were assembled from reads obtained from all but one of the affected samples, while only small (< 1000 bp) fragments were assembled from reads obtained from the unaffected samples. The genome sequences (KF533698–KF533708) were conserved between samples (>98% nucleotide identity) and closely related (95% nucleotide identity) to the reference genome for CYLV [31] (Acc NC013007.1). CYLV was not detected in one of the affected samples, but this sample was found to be infected with a novel virus related to CYLV. This virus, tentatively named Carrot closterovirus 1 (CtCV-1) for the purposes of this study, has a genome of 19923 nucleotides (KF533697). Open reading frame analysis identified 9 open reading frames analogous to 9 of the open reading frames found in CYLV. CtCV-1 lacks ORF 4 found in CYLV but a related virus *Mint virus 1* also lacks this open reading frame. Examination of the amino acid sequences of the putative coat, polymerase and HSP70h proteins suggest that CtCV-1 is a distinct member of the genus *Closterovirus*. When compared to its closest sequenced relative CYLV, the polymerase of CtCV-1 is 84% homologous, the HSP70h 64% homologous and the two coat proteins 69% and 46% homologous respectively. The species demarcation for closteroviruses specifies less than 75% homology in these values [32]. The coat proteins of CtCV-1 (24.7 kDa, 23.3 kDa) are also slightly larger than those of CYLV (24.5 kDa, 22.7 kDa). Figure 4 shows a phylogenetic tree constructed using the HSP70h sequences of CtCV-1, CYLV and related closteroviruses, again confirming CtCV-1 as a distinct member of the genus *Closterovirus*.

Almost complete genomes of CtRLV were recovered from 3 of the unaffected samples (KF533716–KF533718) but not from the

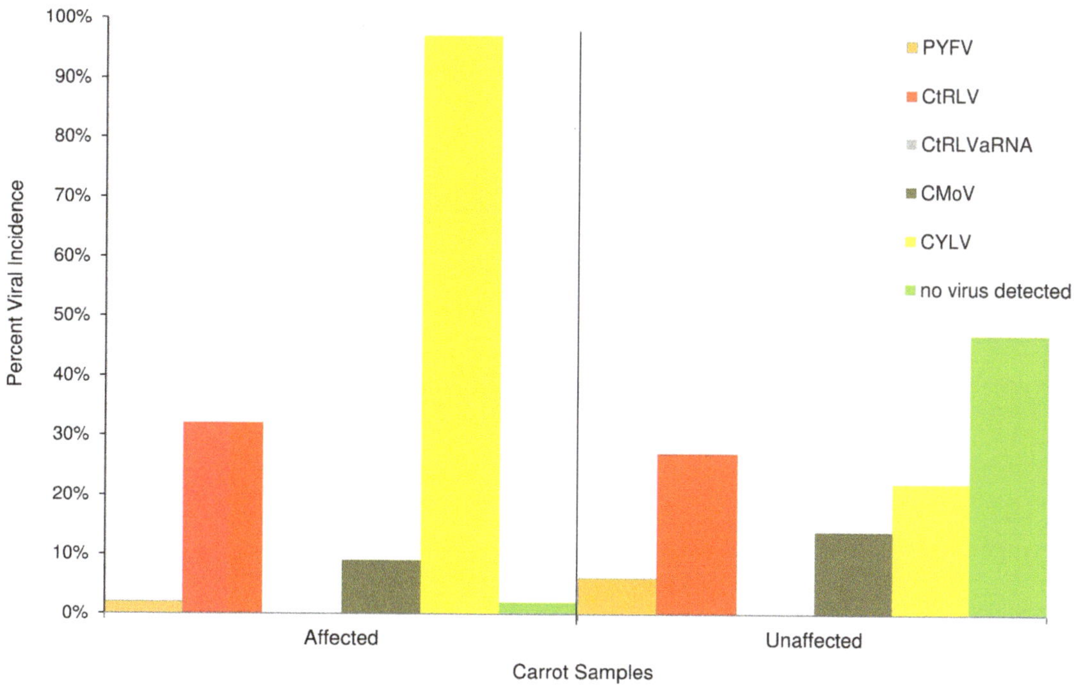

Figure 2. Percent virus incidence in carrot roots from the total field sample of 102 affected and 99 unaffected carrots for the presence of viruses including CYLV, presented as a percentage of carrots with necrosis symptoms (affected) and without necrosis symptoms (unaffected) where virus was detected by PCR or TaqMan.

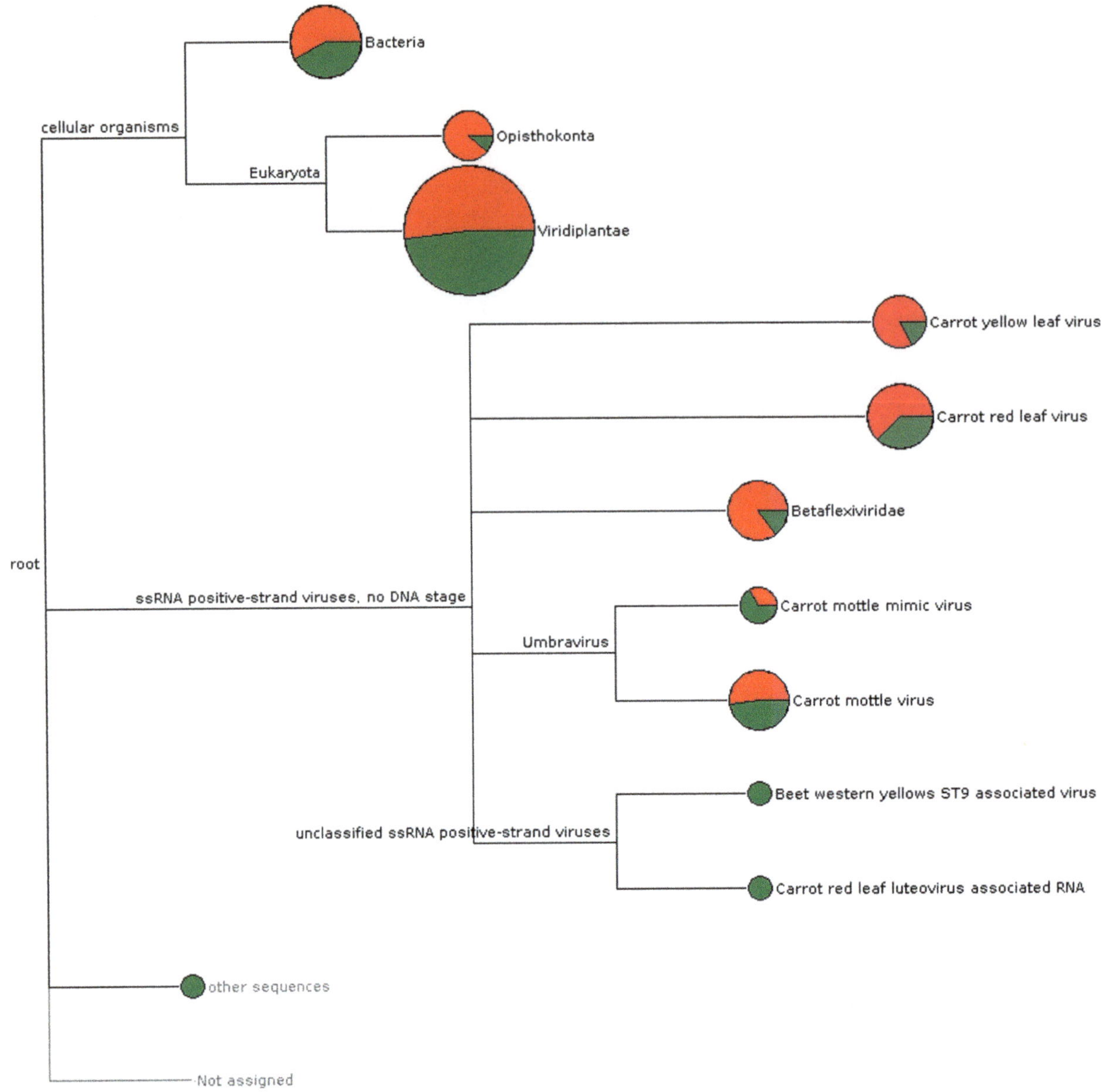

Figure 3. MEGAN derived Phylogram showing the putative identification of contigs in the Sequencing data. Circle size is derived from the number of contigs assigned to each taxa Red: affected, Green: unaffected.

affected samples. Small fragments of CtRLVaRNA were recovered from one of the unaffected samples (KF533715) which was also infected with CtRLV. A large fragment (2 kb, KF533709) was recovered from one unaffected sample, which was also infected with CtRLV. The analysis of this fragment showed 94% sequence identity to Beet western yellows virus associated RNA (BWY-VaRNA) and correspond to 75% of the complete genome.

Complete genomes of CMoV were recovered from 1 affected and 2 unaffected samples (KF533712–KF533714). These sequences have between 91–96% identity to the complete genome of CMoV (Acc: FJ88473).

Over 10,000 reads of a novel unclassified ssRNA positive strand virus were found in the unaffected and affected samples. Large fragments (6.9 k and 4.7 k nucleotides) of a bipartite viral genome were found in one affected and one unaffected sample and smaller fragments in another unaffected sample. The fragments from the two different samples were>99% identical suggesting that they were infected with the same virus. Analysis suggests that the 6.9 k nucleotide fragment is the RNA1 genome of a novel *Torradovirus* tentatively named Carrot torradovirus 1 (CTV-1) (KF533719). It contains an open reading frame coding for a 2214 amino acid (249 kDa) polypeptide. BLAST analysis of this putative protein showed it to contain RNA helicase and RdRp domains and have 40% homology to the equivalent protein sequences from *Tomato marchitez virus* [33] and Tomato chocolate spot virus [34]. The 4.7 k nucleotide fragment appears to be the RNA2 genome of a novel torradovirus (KF533720). This contains two open reading frames ORF1 encoding a putative 202 amino acid (22 kDa) protein with 43% homology to the RNA2 ORF1 from Lettuce necrotic leaf curl virus, a recently reported torradovirus (KC855266). This rate of homology within ORF1 leads to a clear demarcation with the genus *Torradovirus*. The second ORF

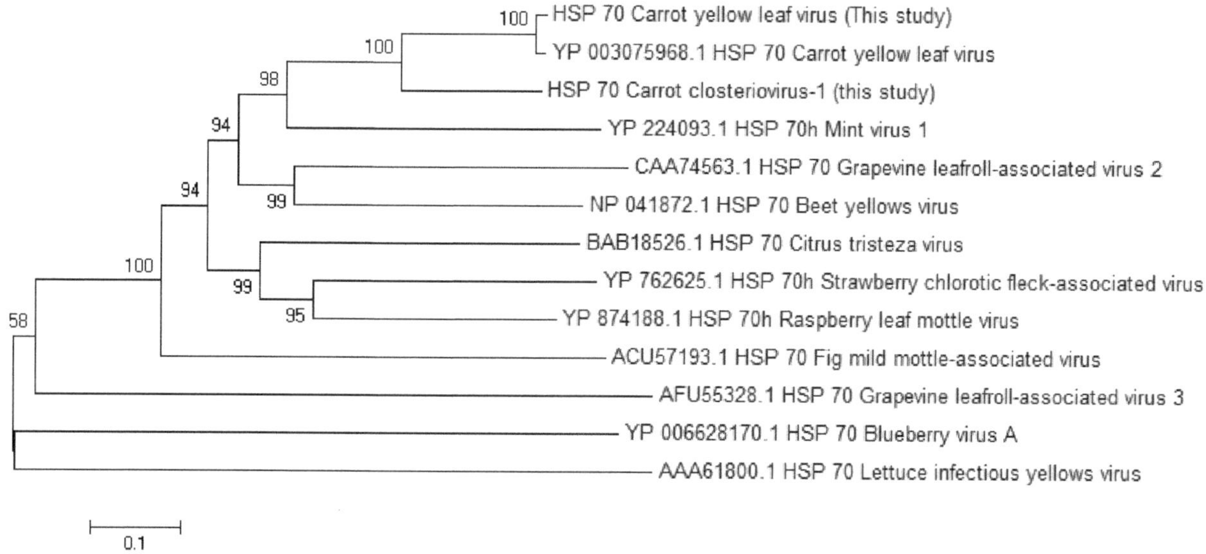

Figure 4. Bootstrapped neighbour joining tree of HSP70 proteins from viruses within the family *Closteroviridae* constructed with MEGA5 using 500 replicates.

encodes a putative 1167 amino acid (130 kDa) polyprotein. This polyprotein appears to contain movement and coat protein domains and have 35% homology to the RNA2 ORF2 from *Tomato torrado virus*. A phylogenetic tree (figure 5) constructed using the RNA2 polyprotein sequences from torradoviruses and other viruses from the *Secoviridiae* shows the closest related torradovirus to be Lettuce necrotic leaf curl virus.

Over 1000 reads with identity to the family *Betaflexiviridae* were found mainly in the affected samples. Table 1 details the relative abundance (expressed as RPKM) of the identified viruses in the affected and unaffected samples. Further examination revealed that 5 affected samples contained large fragments (>5 k nucleotides) and 3 other affected and 1 unaffected samples had fragments between 1000–5000 nucleotides. Analysis of the 5 larger fragments suggests that they are derived from 2 distinct viruses 1 from 4 samples tentatively named Carrot chordovirus-1 (CtChV-1) (KF533711), 1 from the fifth sample, tentatively named Carrot chordovirus-2 (CtChV-2) (KF533710). Both viruses have genomes of approximately 8.5 k nucleotides and encode 3 putative proteins expected in the *Betaflexiviridae*. The putative coat protein and polymerase nucleotide sequences have less than 45% identity to any previously sequenced virus suggesting they may constitute a new genus [32], within the *Betaflexiviridae*, tentatively named Chordovirus. Comparison of the putative coat protein and polymerase nucleotide and amino acid sequences suggests that the two viruses are in the same genus (>45% nucleotide identity) but distinct viruses 40% amino acid homology within the coat protein. Figure 6 shows a phylogenetic tree produced using the coat protein sequences of related members of the family *Betaflexiviridae* and further providing evidence that these viruses may constitute a novel genus.

Sequencing follow-up

Following the outcomes of sequencing, real-time RT-PCR (TaqMan) assays were designed to CYLV and CtCV-1 as follows: CYLV Forward: 5′-AAGATTCTCTTGTAACGAAGGTTTCC, reverse: 5′-GCCGCCTCCACGATCAC, Probe: 5′ Fam-AGACCTCACTATGCTAAACCCGAGCCGG-Tamra. CtCV-1 Forward: 5′-GCCTCCCGCTTGTTGGA reverse: 5′-AGCCGCCAACGTCTATGAAG Probe 5′ Fam-AATAGGACCGTCGCGAGTTTCTGCTCTG-Tamra.

These assays were then used to test the nucleic acid extracts of the 24 carrot sub-sample which had been analysed by sequencing. Of these only 1 of the affected carrots contained CtCV-1, with all 12 affected carrots testing positive for CYLV. Three of the unaffected carrot roots tested positive for CYLV, though in two of these cases the virus was detected at weak levels (>39 Ct).

On the basis of this finding the nucleic acid extracts from the field samples (affected and unaffected carrots) which had been previously tested using RT-PCR were tested for the presence of CYLV using the real-time RT-PCR assay. The results (Figure 2) show that of the carrots affected by necrotic symptoms 98% (99 of 102 carrots tested) were found to be positive, whilst in the unaffected sub-sample 22% (22 of 99 carrots tested) contained the virus. It was also found that the 3 affected samples which tested negative for CYLV tested positive for CtCV-1.

Attempts were made to sap inoculate CYLV from infected carrots into healthy carrots and other species of indicator plants. No infection was detected.

Statistical analysis

The crop of carrots sampled showing 3% of necrotic symptoms i.e. 3300 carrots were sliced to obtain 100 symptomatic samples. Some of these carrots had surface necrosis which could have been graded out, others had only internal necrosis, and there were also carrots sampled with a combination of internal and external symptoms with approximately even numbers of each within the sample set.

The statistical analysis is presented in Table 2. The results for carrot roots found to be positive for the established carrot viruses (PYFV, CtRLV, CtRLaVRNA and CMoV) show a similar proportion of carrots with and without necrosis indicating that necrosis is probably independent of infection with these viruses. With one exception (*Carrot yellow leaf virus*, CYLV) there is no association between virus status and the prevalence of necrosis. CYLV-positive carrots have an estimated prevalence of necrosis of 12.0% (8.4–17.1%) while the prevalence of necrosis in CYLV-negative carrots is estimated to be 0.1% (0.0–0.3%). The estimated

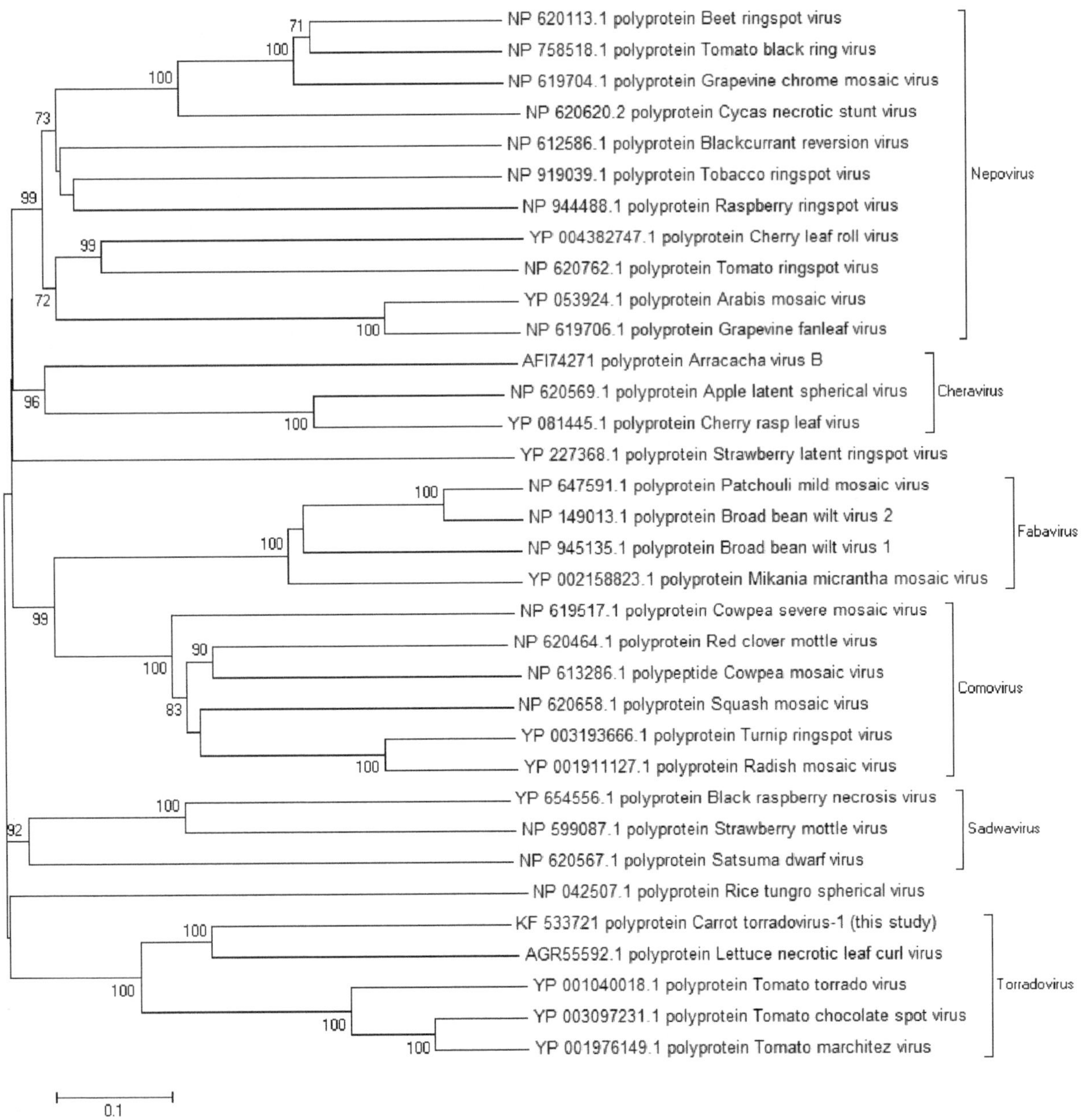

Figure 5. Bootstrapped neighbour joining tree of RNA2 polyproteins from viruses within the family *Secoviridae* constructed with MEGA5 using 500 replicates.

reduction in necrosis prevalence associated with the removal of a virus can be seen in table 2 (Column E%). Removing CYLV from the population is estimated to have a large potential effect, with an estimated reduction of 96% (89.6–98.8%). of necrosis. Because necrosis without CYLV, even in the presence of other viruses, is estimated to be rare (0.0–0.3%) the removal of CYLV alone may be sufficient to greatly reduce the prevalence of necrosis if CYLV is indeed causative.

Discussion

Carrot roots exhibiting internal necrosis have been a growing problem in the UK. Due to a lack of methods allowing the rapid screening of both symptomatic and asymptomatic roots for the presence of a broad range of pathogens, progress on identifying the causal pathogen has been limited.

Of the thirty or so viruses known to have carrot as a host, at least twelve are known to be present in the UK. However, these viruses are not amongst those regularly tested for by diagnostic labs either due to unknown prevalence, poor symptomatic recognition, or more commonly, poor availability of targeted diagnostics. As a result this study has applied high throughput sequencing to screen carrots to help identify a putative causative agent for internal necrosis and a range of previously un-described viruses.

Testing for the four most common viruses (PYFV, CMoV, CtRLV, CtRLVaRNA) in affected and unaffected carrot samples

Table 1. Relative abundance (Reads per kilobase of viral genome per million sequenced reads-RPKM) of select viruses in sequenced affected and unaffected samples.

virus	unaffected	affected
CYLV	2.2	221.3
CtCV-1	0.0	22.0
CTV-1 RNA1	36.4	3.6
CTV-1 RNA2	15.9	5.1
CtRLV	16.2	1.4
CtChV-1	0.0	2.3
CtChV-2	0.0	2.1
CMoV	4.0	4.2

did not provide any evidence for a link between any single or group of viruses and necrosis.

The data obtained following high-throughput sequencing showed that *Carrot yellow leaf virus* was present in eleven of the affected samples and was by far the most common virus recovered warranting further study. The prevalence of this virus was much lower in the unaffected samples. The statistical analysis clearly indicates a link between this unexpected virus finding and the presence of necrosis in carrot roots at this site. Indeed if CYLV is the causal pathogen of carrot internal necrosis, removing CYLV from the sampled carrot population would give an estimated effect of reducing the incidence of necrosis by 96%. Demonstrating a mathematical statistical relationship does not show that there is biological causative relationship but it does point towards where further investigations should be carried out.

Attempts were made to sap inoculate CYLV into carrot or other indicator plants to carry out Koch's postulates but this proved unsuccessful. Previously Murant [8] reported that sap inoculation had been unsuccessful whereas van Dijk and Bos [9] reported poor sap transmissibility into *Nicotiana benthamiana*. Discussion with the authors of the 2009 paper on CLYV [31] also confirmed that they had been unable to sap inoculate CYLV. CYLV is aphid transmitted. In order to transmit CYLV using captive aphids live carrots with attached leaves would be required. In the current study the necrotic symptoms were determined by cutting the carrot root on the grading line when the carrot had already been harvested and the leaves removed. To date it has not been possible to obtain a live symptomatic carrot plant, therefore it has not been possible to further characterize the effects of CYLV on carrot root necrosis.

Due to the low incidence of expression (3%) and localised nature of these symptoms, it was decided that the strategy most likely to yield informative results was to focus in depth on an affected crop from a single site, sampled at the point where symptoms were evident (i.e. on the processing line). It is appreciated that caution should be applied in extrapolating from a single sampled site to other affected sites and future work will include a multi-site survey to confirm the applicability of these findings in a broader context. The statistical approach used in this study, based upon testing approximately equal numbers of affected and unaffected individuals for the presence of pathogens can be applied where infection and disease are often anecdotally linked but lack an empirically observed basis. This could be of particular use when applied to diseases thought to be caused by obligate pathogens such as viruses, viroids, phytoplasmas, or fungal obligates such as rusts or powdery and downy mildews.

Sequencing also identified sequences from 3 of the known carrot infecting viruses CtRLV, CtRLVaRNA and CMoV. CMoV was

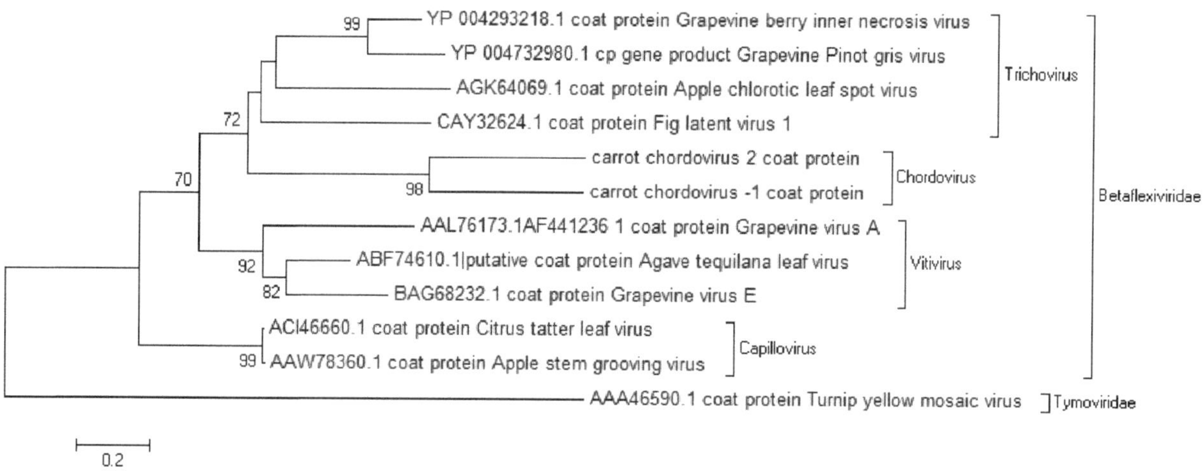

Figure 6. Bootstrapped neighbour joining tree of coat proteins from viruses within the family *Betaflexiviridae* constructed with MEGA5 using 500 replicates.

Table 2. Estimates of prevalence of virus and necrosis and the effect of virus removal on reducing necrosis (values in brackets are 95% confidence intervals).

Virus	P(V|N) (%)	P(V|~N) (%)	P(V) (%)	P(N|V) (%)	P(N|~V) (%)	E (%)
PYFV	2.0 (0.4–6.1)	6.1 (2.6–12.1)	5.9 (2.5–12.0)	1.0 (0.1–4.2)	3.1 (2.5–3.8)	−4.2 (−10.8–1.3)
CtRLV	32.4 (23.9–41.8)	27.3 (19.2–36.6)	27.4 (19.7–36.3)	3.5 (2.2–5.5)	2.8 (2.1–3.6)	6.8 (−11.3–21.8)
CtRLVaRNA	0.0 (0.0–2.9)	0.0 (0.0–3.0)	0.0 (0.0–3.5)	NE	NE	NE
CMoV	8.8 (4.5–15.5)	14.1(8.3–22.0)	14 (8.3–21.9)	1.9 (0.8–4.2)	3.2 (2.5–3.9)	−6.0 (−17–3.7)
CYLV	97.1 (92.4–99.2)	22.2 (14.9–31.1)	24.5 (17.1–33.1)	12.0 (8.4–17.1)	0.1 (0.0–0.3)	96.1 (89.6–98.8)
Any positive	98 (93.9–99.6)	52.5 (42.7–62.2)	53.9 (44.3–63.5)	5.5 (4.2–7.1)	0.1 (0.0–0.4)	95.7 (86.5–99.1)
CtRLV+CMoV	0.0 (0.0–2.9)	4.0 (1.4–9.3)	3.9 (1.4–9.1)	0.0 (0–17)	3.1 (2.5–3.7)	−4.1 (−9.2–0.4)
PYFV+CMoV	0.0 (0.0–2.9)	1.0 (0.1–4.6)	1.0 (0.1–4.6)	0.0 (0–17)	3.0 (2.4–3.6)	−1.0 (−4.2–2.5)
PYFV+CtRLV	0.0 (0.0–2.9)	1.0 (0.1–4.6)	1.0 (0.1–4.6)	0.0 (0–17)	3.0 (2.4–3.6)	−1.0 (−4.1–2.6)
PYFV+CYLV	1.0 (0.1–4.5)	0.0 (0.0–3.0)	0.0 (0–3.7)	100 (0.4–100)	3.0 (2.4–3.6)	1.0 (−2.5–3.7)
CtRLV+CYLV	26.5 (18.7–35.6)	7.1 (3.2–13.4)	7.7 (3.9–13.7)	10.4 (5.2–21.7)	2.4 (1.9–3.0)	20.4 (10.2–30.4)
CMoV+CYLV	2.9 (0.8–7.6)	0.0 (0.0–3.0)	0.1 (0.1–3.5)	100 (2.2–100)	2.9 (2.4–3.5)	2.9 (−1.3–6.5)
PYFV, CtRLV +CMoV	0.0 (0.0–2.9)	1.0 (0.1–4.6)	1.0 (0.1–4.5)	0.0 (0.0–24.3)	3.0 (2.4–3.6)	−1.0 (−4.2–2.3)
CtRLV, CMoV +CYLV	4.9 (1.9–10.4)	1.0 (0.1–4.6)	1.1 (0.3–4.5)	13.1 (2.4-61)	2.9 (2.3–3.5)	3.8 (−0.8–9.1)
PYFV, CMoV +CYLV	1.0 (0.1–4.5)	0.0 (0.0–3.0)	0.0 (0.0–3.6)	100 (0.4–100)	3.0 (2.4–3.6)	1.0 (−2.8–3.8)

Virus is presented both singly and in combinations. **P(V|N)**: proportion of necrotic carrots with the virus, **P(V|~N)**: proportion of non-necrotic carrots with the virus, **P(V)** prevalence of virus across all carrots, **P(N|V)**: proportion of carrots with the virus that are necrotic, **P(N|~V)**: proportion of carrots without the virus that are necrotic. **E**: estimated effect of removing the virus on the prevalence of necrosis expressed as a proportional reduction in the prevalence of necrosis, **NE**: not estimated.

detected in both affected and unaffected samples by both methods. CtRLV was found by PCR and sequencing to be present in both affected and unaffected samples. CtRLVaRNA was not detected by the conventional PCR [21] but small fragments were detected by sequencing. This might be due to specificity issues with the PCR assay. Therefore the new sequence may prove useful for improving the currently used primer sets. Conversely PYFV was detected by conventional PCR in 2% of affected and 6% of unaffected samples but none was found by sequencing, perhaps indicating the PCR approach is more sensitive than sequencing.

BWYVaRNA sequences were recovered from a sample also containing CtRLV. As far as we are aware, this is the first example of an association between these two viruses and deserves further examination as to whether BWYVaRNA is encapsidated by the CtRLV coat protein.

Four new viruses were identified in the sequencing data. A new closterovirus tentatively named CtCV-1, closely related to CYLV was found in necrotic carrots in the absence of CYLV. This suggests that CtCV-1 may also have a role in carrot necrosis and is worthy of further study. A novel torradovirus CTV-1 was found in affected and at a higher abundance in unaffected carrots whilst two novel betaflexiviruses CtChV-1 and CtChV-2 were found at low abundance predominantly in affected carrots. These viruses do not appear to be correlated to the necrotic root symptoms central to this study, but may be worthy of further investigation as they may be associated with other carrot diseases or be causing a reduction of crop yield. The novel viruses found within this small study demonstrate the limited knowledge of viral populations in carrots and it can be speculated in other crops also. The impact these viruses are having on crops are certainly not clear, however, these studies provide some of the knowledge (sequence data) and tools (specific tests) to enable us to investigate these effects in the future.

The overall aim of the project was to investigate viral causes of carrot root necrosis. The outcomes of this work show that, although potentially damaging to carrots in terms of lowering yield and causing growth defects, infection by the four established carrot viruses did not correlate with internal necrosis. There is a clear statistical association between the presence of internal necrosis and infection with CYLV. A closely related yet distinct virus tentatively named Carrot closterovirus-1 (CtCV-1) may yet be associated with necrosis, all be it with a lower incidence. This is the first report of a root necrosis symptom in carrot being associated with *Carrot yellow leaf virus*. On the basis of these findings, work is ongoing to demonstrate a biological causal relationship (Koch's Postulates) between CYLV and root necrosis.

Acknowledgments

This work was funded by the Horticultural Development Company, a levy board of the Agricultural and Horticultural Development Board, as project FV382a.

Author Contributions

Conceived and designed the experiments: IPA NB AF AS RM LF. Performed the experiments: IPA AS TH HH LF PDN. Analyzed the data: IPA AS RM TH NB AF LF PDN. Contributed reagents/materials/analysis tools: RM TH. Wrote the paper: IPA AS RM TH NB HH AF LF PDN.

References

1. Fox A (2011) Symptomatic survey of virus complexes of carrot. Horticultural Development Company. Available: http://www.hdc.org.uk/project/carrot-symptomatic-survey-virus-complexes-4. Accessed 2014 June 25.

2. Brunt AA, Crabtree K, Dallwitz MJ, Gibbs AJ, Watson L, et al. (1996) Plant Viruses Online: Descriptions and Lists from the VIDE Database. Available: http://pvo.bio-mirror.cn/refs.htm. Accessed 2014 June 25.

3. Dijk P, Bos L (1989) Survey and biological differentiation of viruses of wild and cultivated umbelliferae in the Netherlands. Netherlands Journal of Plant Pathology 95: 1–34.
4. Yamashita S, Ohki S, Doi Y, Yora K (1976) Identification of two viruses associated with the carrot yellow leaf syndrome. Annual Phytopathology Society of Japan 42: 382–383.
5. Bem F, Murant AF (1979) Transmission and differentiation of six viruses infecting hogweed (*Heracleum sphondylium*) in Scotland. Annals of Applied Biology 92: 237–242.
6. Bem F, Murant AF (1979) Host range, purification and serological properties of *heracleum latent virus*. Annals of Applied Biology 92: 243–256.
7. Murant AF (1982) Helper-dependent transmission of *Heracleum latent virus* (HLV) by aphids. Report of the scottish Crop Research Institute.191.
8. Murant AF (1983) *Heracleum latent virus* (HLV) and *Heracleum virus 6* (HV6). Report of the scottish Crop Research Institute189.
9. Van Dijk P, Bos L (1985) Viral dieback of carrot and other umbelliferae caused by the Anthriscus strain of parsnip yellow fleck virus, and its distinction from carrot motley dwarf. Netherlands Journal of Plant Pathology 91: 169–187.
10. Botermans M, van de Vossenberg BTLH, Verhoeven JTJ, Roenhorst JW, Hooftman M, et al. (2013) Development and validation of a real-time RT-PCR assay for generic detection of pospiviroids. Journal of Virological Methods 187: 43–50.
11. Boonham N, Walsh K, Smith P, Madagan K, Graham I, et al. (2003) Detection of potato viruses using microarray technology: towards a generic method for plant viral disease diagnosis. Journal of Virological Methods 108: 181–187.
12. Adams IP, Glover RH, Monger WA, Mumford R, Jackeviciene E, et al. (2009) Next-generation sequencing and metagenomic analysis: a universal diagnostic tool in plant virology. Molecular Plant Pathology 10: 537–545.
13. Harju V, Skelton A, Forde S, Bennett S, Glover RH, et al. (2011) New virus detected on *Nasturtium officinale*, Watercress New Disease Reports.
14. Kreuze JF, Perez A, Untiveros M, Quispe D, Fuentes S, et al. (2009) Complete viral genome sequence and discovery of novel viruses by deep sequencing of small RNAs: a generic method for diagnosis, discovery and sequencing of viruses. Virology 388: 1–7.
15. Adams IP, Miano DW, Kinyua ZM, Wangai A, Kimani E, et al. (2013) Use of next-generation sequencing for the identification and characterization of *Maize chlorotic mottle virus* and *Sugarcane mosaic virus* causing maize lethal necrosis in Kenya. Plant Pathology 62: 741–749.
16. Fredericks DN, Relman DA (1996) Sequence-based identification of microbial pathogens: a reconsideration of Koch's postulates. Clinical Microbiology Reviews 9: 18–33.
17. Carnegie SF, McCreath M (2010) Mosaic Virus Symptoms in Potato Crops and the Occurrence of Growth Cracking in Tubers. Potato Research 53: 17–24.
18. Hill SA (1984) Methods in Plant Virology. Oxford, UK: Blackwell Scienctific.
19. Morgan D (2004) Parsnip yellow fleck virus: Development of a disease management strategy. Horticultural Development Company. Available: http://www.hdc.org.uk/project/parsnip-yellow-fleck-virus-development-disease-management-strategy-4. Accessed 2014 June 25.
20. Vercruysse P, Gibbs M, Tirry L, Höfte M (2000) RT-PCR using redundant primers to detect the three viruses associated with carrot motley dwarf disease. Journal of Virological Methods 88: 153–161.
21. Morton A, Spence NJ, Boonham N, Barbara DJ (2003) Carrot red leaf associated RNA in carrots in the United Kingdom. Plant Pathology 52: 795–795.
22. Cox MP, Peterson DA, Biggs PJ (2010) SolexaQA: At-a-glance quality assessment of Illumina second-generation sequencing data. BMC Bioinformatics 11: 485.
23. Grabherr MG, Haas BJ, Yassour M, Levin JZ, Thompson DA, et al. (2011) Full-length transcriptome assembly from RNA-Seq data without a reference genome. Nature Biotechnology 29: 644–652.
24. Camacho C, Coulouris G, Avagyan V, Ma N, Papadopoulos J, et al. (2009) BLAST+: architecture and applications. BMC Bioinformatics 10: 421.
25. Huson DH, Auch AF, Qi J, Schuster SC (2007) MEGAN analysis of metagenomic data. Genome Research 17: 377–386.
26. Huson DH, Mitra S, Ruscheweyh HJ, Weber N, Schuster SC (2011) Integrative analysis of environmental sequences using MEGAN4. Genome Research 21: 1552–1560.
27. Tamura K, Peterson D, Peterson N, Stecher G, Nei M, et al. (2011) MEGA5: Molecular Evolutionary Genetics Analysis Using Maximum Likelihood, Evolutionary Distance, and Maximum Parsimony Methods. Molecular Biology and Evolution 28: 2731–2739.
28. Li H, Durbin R (2009) Fast and accurate short read alignment with Burrows-Wheeler transform. Bioinformatics 25: 1754–1760.
29. Li H, Handsaker B, Wysoker A, Fennell T, Ruan J, et al. (2009) The Sequence Alignment/Map format and SAMtools. Bioinformatics 25: 2078–2079.
30. Mortazavi A, Williams BA, McCue K, Schaeffer L, Wold B (2008) Mapping and quantifying mammalian transcriptomes by RNA-Seq. Nature Methods 5: 621–628.
31. Menzel W, Goetz R, Lesemann DE, Vetten HJ (2009) Molecular characterization of a closterovirus from carrot and its identification as a German isolate of *Carrot yellow leaf virus*. Archives Virology 154: 1343–1347.
32. King E, Kowitz E, Adams M, Carstens EB (2011) Virus Taxonomy: Ninth Report of the International Committee on Taxonomy of Viruses. London: Elsevier.
33. Verbeek M, Dullemans AM, van den Heuvel JF, Maris PC, van der Vlugt RA (2008) *Tomato marchitez virus*, a new plant picorna-like virus from tomato related to tomato torrado virus. Archives of Virology 153: 127–134.
34. Verbeek M, Dullemans A, van den Heuvel H, Maris P, van der Vlugt R (2010) *Tomato chocolate virus*: a new plant virus infecting tomato and a proposed member of the genus Torradovirus. Archives of Virology 155: 751–755.

20

Molecular Evolution and Expansion Analysis of the NAC Transcription Factor in *Zea mays*

Kai Fan[1], Ming Wang[1], Ying Miao[2], Mi Ni[1], Noreen Bibi[1], Shuna Yuan[1], Feng Li[1], Xuede Wang[1]*

1 Institute of Crop Science, College of Agriculture and Biotechnology, Zhejiang University, Hangzhou, P.R. China, **2** Center for Molecular Cell and Systems Biology, College of Life Science, Fujian Agriculture and Forestry University, Fuzhou, P.R. China

Abstract

NAC (*NAM*, *ATAF1*, *2* and *CUC2*) family is a plant-specific transcription factor and it controls various plant developmental processes. In the current study, 124 NAC members were identified in *Zea mays* and were phylogenetically clustered into 13 distinct subfamilies. The whole genome duplication (WGD), especially an additional WGD event, may lead to expanding ZmNAC members. Different subfamily has different expansion rate, and NAC subfamily preference was found during the expansion in maize. Moreover, the duplication events might occur after the divergence of the lineages of *Z. mays* and *S. italica*, and segmental duplication seemed to be the dominant pattern for the gene duplication in maize. Furthermore, the expansion of ZmNAC members may be also related to gain and loss of introns. Besides, the restriction of functional divergence was discovered after most of the gene duplication events. These results could provide novel insights into molecular evolution and expansion analysis of NAC family in maize, and advance the NAC researches in other plants, especially polyploid plants.

Editor: Zhong-Hua Chen, University of Western Sydney, Australia

Funding: This work was supported by the Ministry of Agriculture (2009ZX08009-061B) and the National Natural Science Foundation of China (no. 31171616). The funders had no role in study design, data collection and analysis, decision to publish, or preparation of the manuscript.

Competing Interests: The authors have declared that no competing interests exist.

* Email: xdwang@zju.edu.cn

Introduction

Transcription factors (TFs) are a group of regulatory proteins which can regulate the expression of target genes through binding to specific *cis*-acting elements in the promoters of target genes [1,2]. Although the genes encoding the transcription factors just accounts for a little portion in the whole genome, transcription factors are important in the regulated networks [3]. Recently it has been reported that numerous transcription factors can control many critical biological processes during plant development and growth, such as TCP, WRKY, MYB, AT-hook and E2F [4,5,6].

The NAC (*NAM*, *ATAF1*, *2* and *CUC2*) family is one of the largest families of plant-specific transcription factors and exists widely in various kinds of plants [7,8]. The NAC gene was firstly reported to be related to forming shoot apical meristem and primordium in *Petunia hybia* [9], and then more and more NAC members have been investigated in a wide range of plants. Nowadays more than 100 NAC TFs have been isolated in *Arabidopsis thaliana* and *Oryza sativa* [10,11]. Besides, the NAC family has also been found in other plant species such as *Glycine max* [12], *Solanum tuberosum* [13], *Musa acuminata* [14], *Citrus* [15], *Setaria italica* [16], *Malus domestica* [17] and *Vitis vinifera* [18]. The NAC proteins are characterized by a highly conserved N-terminal region (NAC domain) with a relatively highly divergent C-terminus region (TAR region: Transcriptional Activation Region) [11,19]. The NAC domain (nearly 160 amino acid residues) can be further divided into five subdomains labeled from A to E and it is involved in DNA binding [20], while the TAR region usually is related to the regulation diversity and it may determine some specific functions [21]. The NAC family has been found to be involved in regulating plant growth and development, such as flower development [22], lateral root formation [23], embryogenesis [24], leaf senescence [25], secondary wall thickening [26], cell metabolism [27], seed development [28] and hormone signaling [29]. Moreover, NAC family also plays pivotal roles in response to many biotic and abiotic stresses including fungal infection [30], pathogen disease [31], salt [32], temperature [33], drought [34] and osmotic [35]. Additionally, NAC transcription factors are related to the crop yield and quality [36,37].

As a major cereal crop in the world, *Zea mays* is not only a primary food resource, but also an important crude material. With maize genome sequencing completed, an excellent opportunity is coming to conduct whole-genome annotation, evolution and comparative genomic study in maize [38]. Moreover, previous lines of evidence have already demonstrated that NAC family may be an excellent candidate to regulate the plant development and growth, without the exception for maize. ZmNAC41 and ZmNAC100 were reported to be related to the maize defense network [39], and ZmSNAC1 can enhance tolerance to dehydration in transgenic Arabidopsis [40]. Meanwhile, ZmSWNs (NAC members) can regulate the secondary wall thickening in maize [41]. However, the related researches about the NAC family are limited in maize. Besides, duplication event has been extensively existed in many plants, but the related reports, especially in a specific family, are lacked in maize. Thus, a systematic molecular evolution and expansion analysis of the ZmNAC members is urgently required to comprehensively understand the genetics, evolution, basic function and expansion history of the NAC family in maize. In the present study, we identified NAC family in maize; and we conducted a relatively detailed research about the phylogenetics, genomic-dynamic, chromosomal localization, expansion history and expression analysis to evaluate molecular evolution and expansion history of ZmNAC family.

Materials and Methods

Isolation and structural analysis of the ZmNAC proteins

All NAC proteins of Arabidopsis were collected from TAIR (http://www.arabidopsis.org/index.jsp). The genome sequences of *Z. mays* were downloaded from the Phytozome database (http://www.phytozome.net/). Through searching and alignment, only the sequences containing at least four out of five conserved NAC subdomains (from A to E) were considered and used for phylogenetic analysis [15,42]. Meanwhile, the SMART and Pfam tool confirmed that all ZmNACs contained the conserved NAM domain (PF02365) (SMART: http://smart.embl-heidelberg.de/; Pfam: http://pfam.xfam.org/). Then the genomic schematic diagram of ZmNACs was visualized using GSDS tool (http://gsds.cbi.pku.edu.cn/). Protein primary and secondary structure were calculated by ProtParam (http://web.expasy.org/protparam/) and SOPMA (http://npsa-pbil.ibcp.fr/cgi-bin/npsa_automat.pl?page=npsa_sopma.html). Besides, the chromosome location information of ZmNACs was downloaded from the Phytozome database. MapInspect software (http://www.plantbreeding.wur.nl/uk/software_mapinsp ect.html) was performed to generate the chromosomal distribution image of these ZmNAC genes [43].

Phylogenetic analysis

Firstly, the ClustalX version 2.1 was performed to align the ZmNAC members. Then the Jones, Taylor, and Thorton (JTT) model was selected as the best-fitting amino acid substitution model through the ProtTest version 2.4 [44]. Furthermore, the maximum-likelihood analysis was run with the PhyML version 3.1 based on JTT model [45]. Besides, MrBayes version 3.1.2 was used to conduct the Bayesian analysis: two independent computations kept running until the standard deviation of split frequencies was less than 0.01 [46]. All trees were visualized through the Figtree version 1.4.0.

After phylogenetic analysis, all of the ZmNAC proteins were subjected to online MEME program to investigate conserved motifs [47] (http://meme.nbcr.net/meme/). Parameters were set as follows: optimum motif width was set to ≥6 and ≤200; Maximum number of motifs was set to 20 [43]. Sequence logos of the conservative NAC domain and TAR region were generated through WebLogo [48] (http://weblogo.berkeley.edu/).

Gene duplication and OG identification

Gene duplication was defined according to the previous report [43]. Furthermore, tandem and segmental duplication were recognized by the chromosomal locations. Besides, the Ka/Ks ratio was used to show the selection pressure for the duplicate genes [49]. According to the synonymous substitutions per year (λ) of 6.5×10^{-9} for *Z. mays*, the approximate time of the duplication events about the duplicated pair was estimated ($T = Ks/2\lambda\times10^{-6}$ Mya) by substituting the calculated Ks values [50].

Additionally, the orthologous genes of the duplicate ZmNACs were searched in *S. italica*, *O. sativa* and *B. distachyon*. Firstly, the NAC proteins in *S. italica*, *O. sativa* and *B. distachyon* were obtained according to similar approach of ZmNAC identification. Then, the whole set of ZmNAC, SiNAC, OsNAC and BdNAC proteins was clustered using OrthoMCL program to identify the orthologous groups (OG) [51]. Moreover, the orthologous genes were also defined: (1) the length of aligned sequence cover >50% of the longer gene; and (2) the identity of the aligned regions > 50% [52].

Public microarry-based data analysis

The expression patterns of the ZmNAC genes in various tissues and development stages were examined from Genevestigator [53] (https://www.genevestigator.com/gv/). The ZmNAC probes were adopted according to the previous standard [54]. The expression data were hierarchically clustered and gene-wise normalized through Euclidean Distance program [55]. Besides, the EST sequences of *Z. mays* were downloaded from the GenBank EST database (http://www.ncbi.nlm.nih.gov/). The EST sequences, which were less than 0.01 in E-value and more than 80% in the identity to the ZmNAC proteins, were collected for further analysis.

Plant materials, RNA extraction and quantitative real-time PCR

The *Z. mays* cv B73 were used to construct the expression profiles of ZmNAC family. Roots, stems and leaves were collected from three-week-old seedlings, and the fresh flowers were harvested from maize plants. Then, total RNAs were isolated from the collected samples through RNAiso Plus (TakaRa), and the first-strand cDNA was synthesized from DNase-treated RNA using PrimerScript 1st Strand cDNA synthesis kit (TaKaRa). Gene-specific qRT-PCR primers were designed according to their CDSs (Table S8) and then synthesized commercially (Generay). The quantitative real-time PCR was performed in 96-well blocks with CFX96 Realtime System (BioRad) using SYBR premix Ex Taq (TakaRa). The qRT-PCR machine was set with 40 cycles and an annealing temperature of 60°C. The *ZmHMG* gene was used as an endogenous control for all the qRT-PCR analyses. The $2^{-\Delta\Delta Ct}$ method was used to determine relative transcription levels and the relative expression level in the root was normalized to 1. Three biological replications were performed in all reactions. The expression patterns of ZmNACs were clustered using the Cluster 3.0 software.

Results

Identification of the NAC family in *Z. mays*

The NAC transcription factor is one of the largest families in plant kingdom. As a model plant, the ANAC proteins from Arabidopsis have been comprehensively investigated [8]. Thus, the ANAC sequences were regarded as a query to search against the *Z. mays* genome database with the Blastp program. After Blastp search, 211 NAC-containing sequences were identified in maize. Then the putative NAC domain in their N-terminal region was confirmed by Pfam and SMART tool. Moreover, the amino acid sequences with at least four conserved NAC subdomains (from A to E) were identified for the further analysis. After this removal, 124 sequences in maize were isolated and sequentially named as *Z. mays* NAC (ZmNAC) (Table S1). The total number of NAC genes was a little greater in *Z. mays* than in *O. sativa*, *B. distachyon* or *S. italica* (Figure S1).

Phylogenetic investigation of ZmNAC family

The phylogenetic analysis among the identified ZmNAC and ANAC proteins was processed by the PhyML and MrBayes tool. There were similar results with high support values from each method (Figure 1A, Figure S2). According to previous studies [8,42], the ZmNAC family was divided into 13 subfamilies through the phylogenetic analysis. The member proportion was different in each ZmNAC subfamily (Figure 2A). The ONAC022 subfamily (17%) occupied the most members, followed by OsNAC7 subfamily (14%), ONAC003 subfamily (13%) and NAM subfamily (11%). The least was OsNAC8 and TIP subfamily (2%). Besides, ZmNAC sequences were submitted to OrthoMCL clustering. With the default stringency, 37 orthologous groups (OGs) were shown and they covered all the ZmNAC members in maize (Table S2). Each subfamily contained one or more OGs, and almost the different subfamilies had distinct OGs. The OG distributions were similar to the phylogenetic classifications of the NAC family in maize.

Although all the ZmNAC proteins contain the NAC domain, their protein structures are highly diverse (Table S3). The amino acid length in the TIP subfamily, NAC2 subfamily and OsNAC8 subfamily were relatively longer, while the NAC1 and SENU5 subfamily had relatively shorter amino acid length. The similar distribution of ZmNAC family also existed in the molecular weight and protein structure.

Conserved domain analysis in ZmNAC proteins

The MEME motif search tool was employed to identify 20 distinct conserved motifs in ZmNAC proteins (Figure 1B). According to the distribution of the 20 predicted motifs, 124 ZmNAC members can be classified into 13 distinct subfamilies, consistent with the categorization from the phylogenetic analysis (Figure 1A). The motif 2 was shared in all ZmNAC members and it corresponded to subdomain A in the NAC domain. The motif 4 or 6 is mapped to subdomain B in the ZmNAC proteins. All of the 124 ZmNACs had either motif 1, 7 or 14, which represented subdomain C. The motif 3, 7, 8 or 9 corresponding to subdomain D was also discovered in the ZmNAC family. Additionally, subdomain E matched with the motif 5 or 7 in the ZmNAC members. Besides, some subfamily-specific motifs were also found in some ZmNAC subfamilies. For example, the motif 13 was shared in TIP subfamily, and the NAC2 subfamily contained the motif 18.

All the conserved residues from subdomain A to subdomain E were shown in NAC domain through sequence alignment of NAC proteins in *Z. mays* (Figure S3). Although the TAR region was relatively divergent, some conserved subdomains existed in TAR region among some NAC subfamilies. Through WebLogo program, subdomain A, C and D showed a high conservation among the ZmNAC members, whereas subdomain B and E were relatively divergent. In addition, some conserved subdomains were also discovered in TAR region of some ZmNAC subfamilies (ATAF, NAC2, NAM, ONAC003, ONAC022 and OsNAC7 subfamily). But none of the 6 conserved subdomains corresponded to any known domain. The DNA binding domain (DBD) was highly conserved in the subdomain C, and a degenerate bipartite nuclear localization signal (NLS) was also detected in the subdomain D.

Intron dynamic detection of ZmNAC family

Gene structure and intron phase were investigated among the ZmNAC family. According to the number of exons/introns, the ZmNAC family could be divided into five types (Table S3). The first type contained 67 ZmNACs and had three exons. The second type had no intron as 6 ZmNAC belonged to this type. 19 genes formed the third type with only one intron. The fourth type had 24 genes which contained three, four or five introns. The fifth type mainly covered genes with the most exons (the number is 7, 8, 9 or 14). The largest number of the exons was found in ZmNAC39 with 14 exons. The first type of gene structure included most of ZmNAC subfamily, while the ANAC063 subfamily covered the second type. The third type appeared in the 7 subfamilies, and 8 subfamilies had the fourth type of gene structure.

Intron position and phase were examined to unravel evolutionary process of the ZmNAC gene structure (Figure 3). The primary gene structure contained three exons and two introns in ZmNAC family. The first (140 bp–230 bp) and second exon (230 bp–340 bp) showed the relatively conserved exon length. However, the divergent length and number mainly existed after the second exon, especially for ONAC003, TIP, ANAC011 and NAC2 subfamily. Besides, the gain and loss of introns resulted in the different gene structures. Some SENU5, NAC1 and OsNAC7 members lost an intron in the second and third exon, while the similar loss in the first and second exon mainly existed in the ATAF, TIP and NAC2 subfamily. Moreover, the ONAC022 and NAM subfamily contained the above-mentioned two intron loss. In addition, the intron gain usually existed in the third exon, especially for the ONAC003, TIP, ANAC011 and NAC2 subfamily. In contrast, the NAP subfamily did not undergo any changes in the gene structure, and still had three exons. Furthermore, there were not any introns in the ANAC063 members. Meanwhile, the first and second intron phase is 1 and 0 in most ZmNAC subfamilies. However, the ONAC003 subfamily had totally different intron phase, that is 0 and 1 in the first and second intron phase; and no intron phase were found in the ANAC063 subfamily.

Genomic locations and duplication of ZmNAC genes

A total of 124 ZmNAC members were mapped into 10 chromosomes in maize (Figure 4; Table S4). Every one of the 10 chromosomes contained the ZmNAC genes. The 124 ZmNAC genes distributed unevenly across the *Z. mays* chromosomes (Figure 1B). Chromosome 6 had the largest number of ZmNAC genes with 16 members, followed by chromosome 1 and chromosome 2 with 15 genes each. In contrast, only 7 members were located on chromosome 10. Besides, several ZmNAC genes clustered within a short distance such as the top of chromosome 2 and the bottom of chromosome 6.

15 gene duplication events were identified in maize (Figure 4, Table 1). Among them, 4 duplication events occurred in the ONAC022 subfamily, and OsNAC7 and ONAC003 subfamily had 3 duplication ones. Based on the sequence analysis and the chromosomal distribution, 13 gene pairs were identified to be involved in the segmental duplication events, while other 2 pairs were related to the tandem duplication events. Meanwhile, the expression patterns of some duplicated genes were also analyzed in the different tissues. Out of the 15 pairs of duplicated genes, 9 pairs were selected to reveal the expression relationship. The average signal values for the 9 pairs of duplicated genes were shown as an area-diagram. In the tandem duplication events, one pair of genes (ZmNAC22/113) had highly similar expression level, whereas another one pair (ZmNAC45/104) showed the divergent expression pattern (Figure 5C, Figure 5D). In the segmental duplication events, two pairs of genes (ZmNAC16/63 and ZmNAC65/100) had different expression profiles in the tissue tested (Figure 5E, Figure 5F). In each of other 8 pairs, their member share very similar expression model in different tissues, although the amplitude of expression is a little different in paired partners

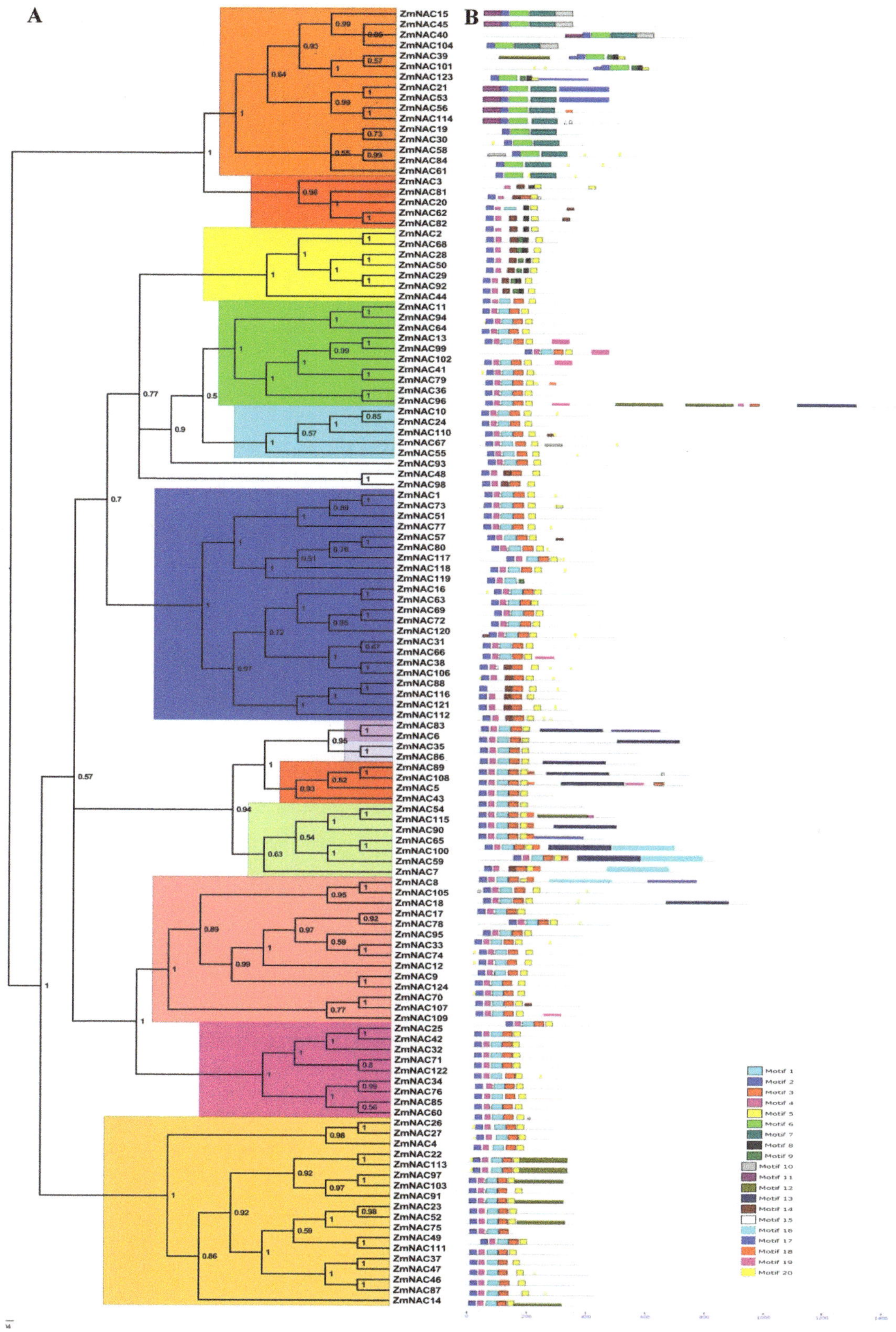

Figure 1. Phylogenetic relationships and putative conserved domain distributions of ZmNAC family. (A) The phylogenetic tree was generated using the Bayesian method based on the multiple alignments of ZmNAC protein sequences. The numbers in the clades are posterior probability values. The NAC subfamilies were indicated by different colors. (B) The conserved motifs were identified through MEME web server. Different motifs were represented by various colored boxes. The location of each motif can be estimated using the scale at the bottom.

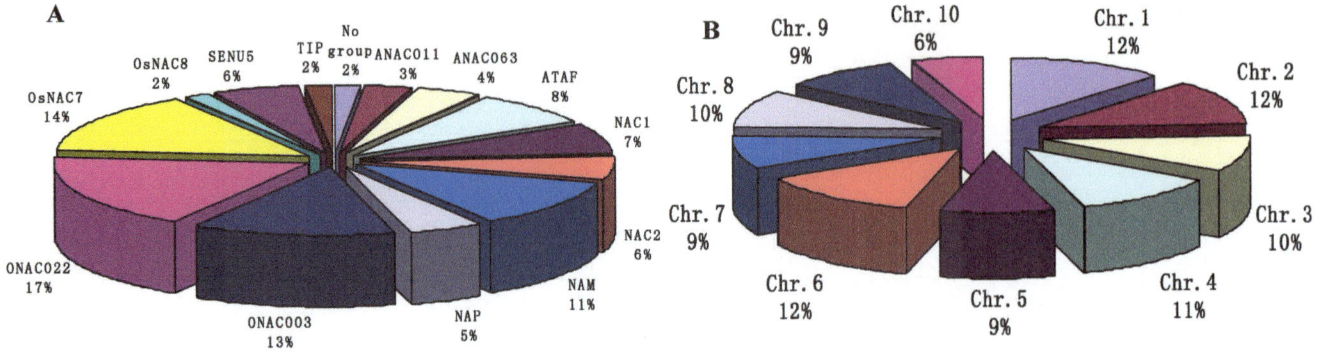

Figure 2. The percentage of ZmNAC proteins' subfamily (A) and ZmNAC chromosomes localizations (B) in *Zea mays*.

(Figure 5G–Figure 5K). Besides, the Ka/Ks ratio of each duplicated gene pairs was calculated to estimate the molecular evolutionary rates (Table 1). The Ka/Ks ratios from 12 duplicated gene pairs were less than 1, while in other 3 duplicated gene pairs are more than 1. Moreover, Ks age distribution of ZmNAC paralogs was constructed to predict the burst of duplication (Figure 5A). Most of Ks value were less than 0.4, and their corresponding duplication age might have been less than 30 million years ago (Mya). Furthermore, there are 4 duplicated pairs of which the Ks values are between 0.9 and 1.1, and their duplications might occur in 70–87 Mya. Additionally, the Ks values were less than 0.06 in the tandem duplicated pairs, and its corresponding events might occur in less than 0.5 Mya.

Furthermore, orthologous genes of some 15 duplicated gene pairs were detected in *S. italica*, *O. sativa* and *B. distachyon* (Table S5). For most of these duplicated genes, their corresponding orthologous genes can be discovered in three monocots. Then phylogenetic tree including these paralogous and orthologous genes was employed to estimate the relative time of the duplication events (Figure S4). Phylogenetic analysis indicated that the dominant topology between two *Z. mays* paralogs and the *S. italica* ortholog is ((*Z. mays*, *Z. mays*) *S. italica*) (Figure S4A–N, Figure 5B). But a different topology existed in one duplication event (ZmNAC 49/111), which might occur before the divergence between the *Z. mays* and *S. italica* (Figure S4O).

Expression profiles of ZmNAC members

The *in silico* frequencies of ZmNAC ESTs provide information to the basic statistical analysis of gene expression profiles in different tissues (Figure S5, Table S6). The *Z. mays* ESTs sequenced at Stanford Universtiy mainly contained three large cDNA libraries, which covered root (10611 ESTs), immature ear (9277 ESTs) and leaf (5871 ESTs). Through EST screening, 15 ZmNACs were found to express in the three libraries. 7 ZmNAC genes were expressed in immature ear and 4 members of these genes belonged to NAM subfamily. 3 NAC2 members were preferentially expressed in the root, and shoot EST library contained 2 ZmNAC genes from the ONAC022 subfamily.

Many microarray chips of maize were available on the Genevestigator database. 29 ZmNAC genes have their corresponding probe sets in the ZM-15K dataset (Table S7). First, the tissue-specific expression pattern was performed to show spatial feature of these ZmNACs (Figure 6A). The NAC2 subfamily (ZmNAC 115, ZmNAC 54, ZmNAC 59 and ZmNAC 90) and SENU5 subfamily (ZmNAC 28 and ZmNAC 68) were extensively expressed in almost of all the tested tissues. The expression level of the NAC2 subfamily was very high in the root cell, lateral root and root tip, while the SENU5 subfamily expressed abundantly in the leaf and shoot. Moreover, most of the ATAF subfamily (ZmNAC 13, ZmNAC 79, ZmNAC 41, ZmNAC 36, ZmNAC 102, ZmNAC 11 and ZmNAC 64) had relatively high transcription levels in root cell, glant cell, glume and foliar leaf. Furthermore, the expression of NAM subfamily (ZmNAC74, ZmNAC107, ZmNAC109, ZmNAC8, ZmNAC95 and ZmNAC12) was only restricted to very few tissues that were mainly involved in the flower development. Meanwhile, the ZmNAC expression was also investigated under 7 developmental stages in maize (Figure 6B). Most of the NAM members were found in low transcription level over all stages, while relatively high expression level was detected in the NAC2 and SENU5 subfamily. The NAC1 subfamily and the ATAF subfamily had relatively high expression level in the inflorescence formation and anthesis stage. The expression level of some ATAF members was relatively high at the germination stage.

qRT-PCR analysis was performed to examine the expression pattern of 27 selected ZmNAC in root, stem, leaf and flower. As it is shown, two NAC2 members (ZmNAC65 and ZmNAC100), two OsNAC8 members (ZmNAC35 and ZmNAC86), two TIP members (ZmNAC6 and ZmNAC83) and two SENU5 members (ZmNAC29 and ZmNAC50) exhibited high expression level in the leaf and stem. On the contrary, the OsNAC7 subfamily (ZmNAC22 and ZmNAC113) had the opposite expression pattern. Furthermore, the NAM subfamily (ZmNAC17 and ZmNAC78) and one NAC1 member (ZmNAC34) showed relatively lower transcription level in root than in other three tissues. The expression profile of some ATAF members (ZmNAC13, ZmNAC41, ZmNAC79 and ZmNAC102) and the ONAC003 subfamily (ZmNAC45, ZmNAC58 and ZmNAC104) in the flower is the lowest in all the tissues tested. However, the leaf showed the relatively higher transcription level of the ONAC022 subfamily (ZmNAC16, ZmNAC31, ZmNAC63, ZmNAC66 and ZmNAC69). In addition, the ZmNAC81 (ANAC063 subfamly) and ZmNAC 67 (NAP subfamily) had the high transcription level in root, and ZmNAC 67 also showed high expression level in flower.

Discussion

As an important monocot, the announcement of maize genome sequencing offers a good opportunity to further investigate the monocot and plant evolution in general. The present research mainly analyzed the ZmNAC molecular evolution, and its corresponding expansion patterns in monocot.

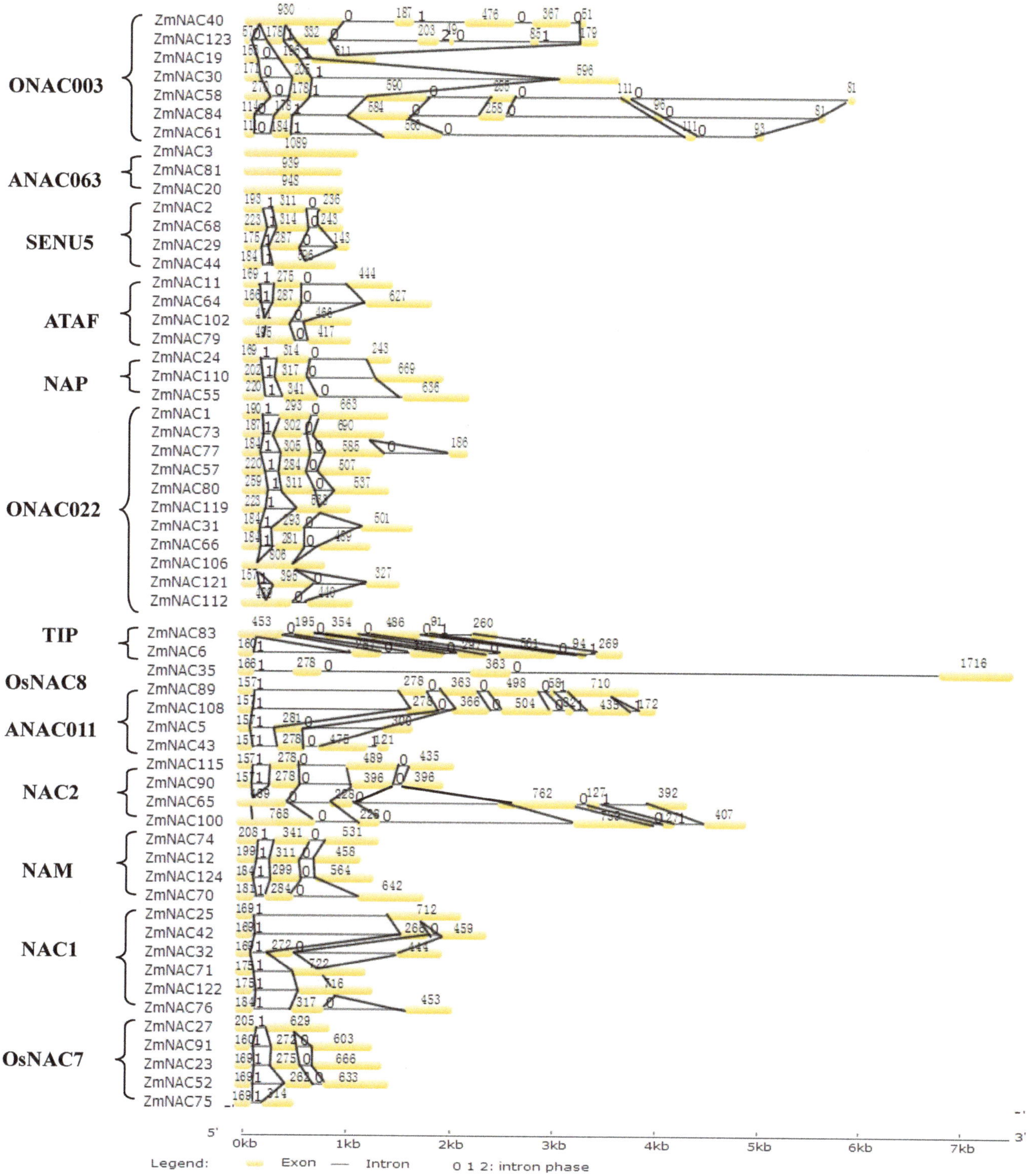

Figure 3. Gene structure dynamics of some ZmNAC members. Gene structures of ZmNAC were performed by the GSDS software. Subgroup designations are indicated by brackets.

The identification and function of ZmNAC family

In the present study, 124 NAC members were isolated in *Z. mays* using Arabidopsis NAC proteins as query (Table S1). This number, however, can be a conservative estimate for ZmNACs in maize, because the selected ZmNACs contained at least four conserved NAC subdomains (A to E) [15]. According to previous studies [8], 13 NAC subfamilies were phylogenetically clustered in the ZmNAC family (Figure 1A, Figure S2). The ONAC022, OsNAC7, ONAC003 and NAM subfamily had the most members, while other subfamilies contained relatively few

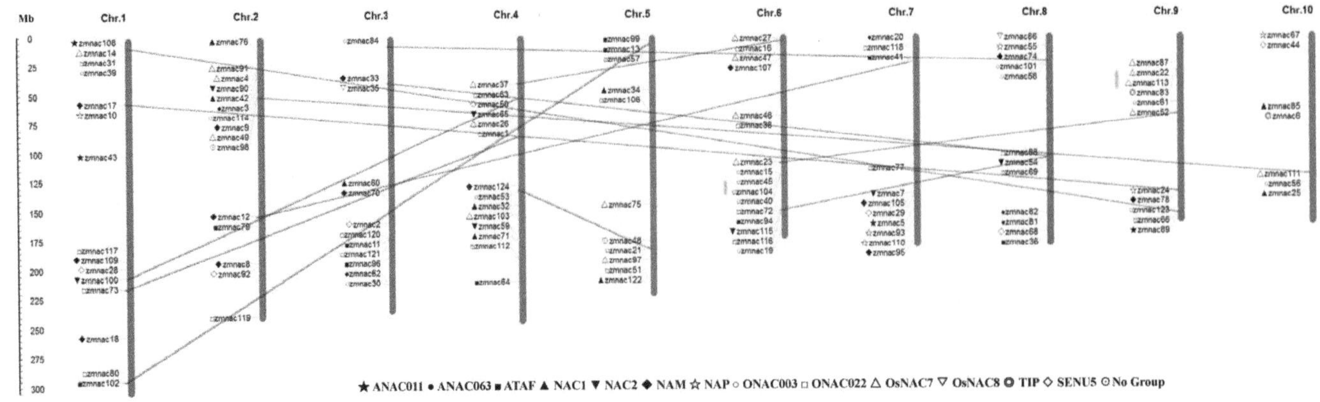

Figure 4. Chromosomal locations of ZmNAC genes on all 10 chromosomes. The scale is megabases (Mb). Markers before the gene names indicate the NAC subfamily. The red lines mark the tandem duplication of the ZmNAC genes, and genes related to segmental duplication are joined by gray lines.

ZmNACs (Figure 2A). The similar distribution of NAC family also can be found in banana [14], chinese cabbage [57] and so on. Then the protein structure, orthologous group, gene structure and conserved motifs confirmed the similar classification of NAC family in maize (Figure 1B, Figure 3, Table S2 and Table S3). In addition, NAC family plays significant roles in plant-specific processes [7,19,21,58]. The microarray chips were used to investigate ZmNAC's functions in the developmental stages and tissues (Figure 6). However, only 29 ZmNACs have their corresponding probes in the Genevestigator database (Table S7). These arrays from ESTs, not whole genome coverage, maybe partially lead to the relatively low coverage of ZmNAC members. Similar phenomenon also existed in the soybean [54]. Furthermore, combing with the EST frequency, NAC2 subfamily was discovered to express highly in the root, which indicated that NAC2 subfamily in maize may be related to root development. But two NAC2 members (ZmNAC65 and ZmNAC100) showed high expression level in the leaf and stem by qRT-PCR analysis. The different result maybe partially due to the extensively expression of NAC2 subfamily in almost of all the tissues and all the developmental stages. Meanwhile, the NAM subfamily in maize may be involved in the flower development, which has been confirmed in the NAM subfamily of Arabidopsis and petunia [9,59]. Moreover, ZmNAC102, one of ATAF members, has been previously found to control the lateral root development through miRNA164-directed cleavage [56], and due to relatively high expression in root, other ATAF members might have similar function in maize like ZmNAC102. Thus, the functions within the NAC subfamily were relatively conserved throughout maize and plant kingdom; and different subfamilies might have different biological functions, mainly because of distinct structure, especially in the TAR region. Additionally, it is suggested that the NAC family is a perfect candidate to regulate the plant growth and development.

Expansion of the NAC family in *Z. mays*

Z. mays is an important model plant for fundamental research on evolutionary history. In the current research, we isolated 124 NAC transcription factors in maize (Table S1). Then the NAC members were also identified in *O. sativa*, *B. distachyon* and *S. italica* through the same method (Figure S1). The number of NAC members revealed that the NAC family was a little larger in maize than in other three monocots, and it may be due to the larger genome. Meanwhile, the expansion of NAC members in maize confirmed some previous observations which the same subfamily from different plants shared the similar functions (Figure 6, Figure 7). Through sequence and OG analysis, the duplicated genes of NAC family were discovered in maize (Table 1, Table S2). According to the chromosomal distributions, most of the duplicated genes located in different chromosomes (Figure 4, Table S4). Meanwhile, the duplication events mainly concentrated on some specific subfamilies including the ONAC022, NAC2, ATAF, NAM, ONAC003, OsNAC8 and OsNAC7 subfamily, whereas no duplicated ones were detected in other subfamilies. Thus, it revealed the strong expansion preference for some NAC subfamilies. As an important paleopolyploid plant, maize has experienced a meaningful process from an ancient allotetraploid to a genetically diploid state [38]. During this process, numerous chromosomal breakages and fusions resulted in gene losses and gene retentions as duplicate orthologs. Based on the analysis of NAC family in maize, retention of ZmNAC members as duplicates is not random, e.g. the expansion preference of some ZmNAC subfamilies. Thus, it is indicated that the ZmNAC subfamily might have various retention rate during the plant evolution and the ZmNAC reference contributes to basic researches about plant evolution, especially polyploid plants e.g. *Gossypium hissutum*, *Brassica campestris* and *Triticum aestivum*. In addition, the gene losses and retentions might be also associated to the related functions during the plant evolution. In addition, NAC family can control a variety of plant growth and development processes [7,8]. Thus, according to the various retention rate of the ZmNAC subfamily, some putative key NAC members can be selected to regulate some phenotypes in the polyploid plants.

Previous studies revealed that the maize genome has undergone several rounds of genome duplication, including a paleopolyploid duplication event (about 70 Mya) [60] and an additional whole-genome duplication (WGD) event (about 12 Mya) [61]. In the ZmNAC family, most of duplication events (53%) might occur about less than 30 Mya, while 27% duplicated ones between 70 Mya and 87 Mya. Moreover, the othologous genes of these duplicated genes were isolated in *O. sativa*, *B. distachyon* and *S. italica* (Table S5). Through phylogenetic analysis, the dominant topology in four monocots is ((*O. sativa*, *B. distachyon*), ((*Z. mays*, *Z. mays*) *S. italica*)) (Figure 5B, Figure S4). Only one pair of duplicated genes (ZmNAC 49/111) showed the different topology (Figure S4O). The divergence may be attributed to the extensive rate variation among species [62]. Thus, it was concluded that the ZmNAC duplication events might mainly occur after the

Table 1. Ka/Ks analysis and estimated divergence time for the ZmNAP duplicated genes.

Duplicated gene 1	Duplicated gene 2	Subfamily	Ka	Ks	Ka/Ks	Purifying selection	Age(MYA)
ZmNAC1	ZmNAC73	ONAC022	0.075446	0.383504	0.196728	yes	29.500308
ZmNAC13	ZmNAC102	ATAF	0.927048	1.57896	0.587125	yes	121.45846
ZmNAC16	ZmNAC63	ONAC022	1.00354	0.987456	1.01629	no	75.958154
ZmNAC17	ZmNAC78	NAM	0.043229	0.292898	0.14759	yes	22.530615
ZmNAC21	ZmNAC53	ONAC003	0.054503	0.195477	0.278822	yes	15.036692
ZmNAC22	ZmNAC113	OsNAC7	1.73E-07	1.10E-06	0.157044	yes	8.48E-05
ZmNAC23	ZmNAC52	OsNAC7	0.97065	1.12054	0.866235	yes	86.195385
ZmNAC31	ZmNAC66	ONAC022	0.956598	1.24751	0.766806	yes	95.962308
ZmNAC35	ZmNAC86	OsNAC8	0.087589	0.217746	0.402251	yes	16.749692
ZmNAC41	ZmNAC79	ATAF	0.037907	0.603107	0.062852	yes	46.392846
ZmNAC45	ZmNAC104	ONAC003	0.013719	0.005979	2.29461	no	0.4599123
ZmNAC49	ZmNAC111	OsNAC7	0.019899	0.33814	0.058847	yes	26.010769
ZmNAC58	ZmNAC84	ONAC003	0.993472	1.02646	0.967863	yes	78.958462
ZmNAC65	ZmNAC100	NAC2	1.01946	0.937213	1.08776	no	72.093308
ZmNAC69	ZmNAC72	ONAC022	0.085138	0.221639	0.384129	yes	17.049154

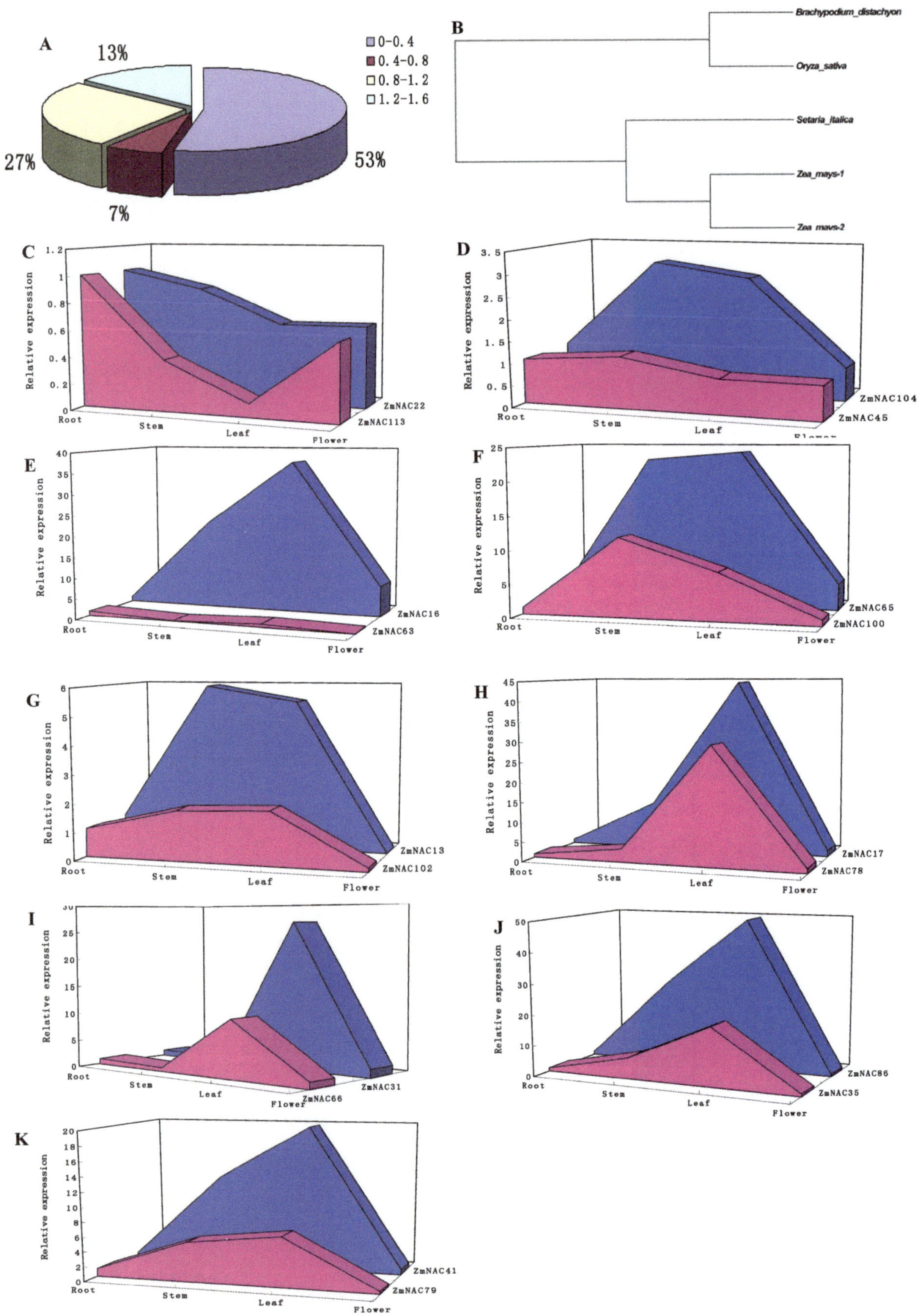
A
13%
27%
7%
53%
0-0.4
0.4-0.8
0.8-1.2
1.2-1.6
B
Brachypodium_distachyon
Oryza_sativa
Setaria_italica
Zea_mays-1
Zea_mays-2
C
D
E
F
G
H
I
J
K
Relative expression
Root
Stem
Leaf
Flower
ZmNAC22
ZmNAC113
ZmNAC104
ZmNAC45
ZmNAC16
ZmNAC63
ZmNAC65
ZmNAC100
ZmNAC13
ZmNAC102
ZmNAC17
ZmNAC78
ZmNAC66
ZmNAC31
ZmNAC86
ZmNAC35
ZmNAC41
ZmNAC79

Figure 5. Age distribution, evolutionary history and expression analysis of the duplicated ZmNAC genes. (A) Age distribution of the duplicated ZmNAC genes based on Ks values. (B) Phylogenetic relationships among the duplicated ZmNACs and their orthologous genes in other three monocots. (C–K) Expression pattern of some duplicated ZmNAC genes. The expression values of duplicated genes obtained from quantitative real-time PCR were compared in different tissues.

divergence of the lineages of *Z. mays* and *S. italica*. Above all, the expansion of NAC family in *Z. mays* arose from the whole genome duplication events, might mainly occur after the divergence of the lineages of *Z. mays* and *S. italica*, and the additional WGD event might have more influence on the expansion of ZmNAC members. The expansion of NAC family in other plants (e.g. *G. raimondii* and *P. trichocarpa*) has been similarly related to the whole genome duplication events [63,64].

Compared with other monocots, maize has undergone relatively more chromosomal breakages and fusions [38]. Meanwhile, the tandem duplication events in ZmNAC family might occur in less than 0.5 Mya (Table 1). Therefore, after an additional WGD event, interchromosomal rearrangements also have a certain effect on the expansion of ZmNAC members during the return to a genetically diploid state (Figure 4). Coincidently, this finding is consistent with the expansion of other gene families including ERF family and F-box family [65,66].

The conserved motifs' role during ZmNAC expansion

The NAC family has a conserved structure including a highly conserved NAC domain in the N-terminal domain and a relatively divergent C-terminal domain [7,8]. Without exception, the ZmNAC family also contained the NAC domain and TAR region (Figure 1B, Figure S3). Then through sequence alignment, we found, subdomain A, C and D were tightly conserved, while subdomain B and E were relatively divergent. The similar phenomena also exist in NAC family of other plants [8]. In addition, subdomain C was found to have a DNA binding domain (DBD), which indicated that subdomain C may be involved in the DNA binding [20]. Moreover, the DBD sequence (WKATGXD [K/R]) resembled the DBD sequence (WRKYGQK) in the WRKY family [67]. This sequence similarity revealed that the NAC family and WRKY family might share the similar evolutionary history in maize, and the ZmNAC family may originate from the protist WRKY family in WRKY-GCM1 superfamily [67]. Furthermore, the putative nuclear location signal (NLS) was found in the subdomain D, which indicated their nuclear localization. The transient expression of some NAC members from other plants also suggested that most of the NACs are a nuclear protein [25,42]. Thus, the NAC domain played an important role in some conserved functions such as stress responses.

Although the TAR region was divergent in the ZmNAC family, the TAR region within the same subfamily was relatively conserved, especially for the ATAF, NAC2, NAM, ONAC003, ONAC022 and OsNAC7 subfamily (Figure 1, Figure S3). This also may be the reason why the same NAC subfamily might regulate similar processes in different plants. Other plant-specific transcription factors have similar conditions, such as WRKY and MADS family [68,69].

Gene structure dynamic during ZmNAC expansion

Intron is a characteristic feature of the eukaryotic genes, and the relatively little selective pressure acts on intron [70]. Most of ZmNAC had three exons and two introns (Figure 3, Table S3). The first and second exon was relatively conserved in length, partially due to encoding the conserved NAC domain. However, there were different length and intron insertions in the third exon, especially for ONAC003, TIP, ANAC011 and NAC2 subfamily. The divergence of gene structure might mainly come from the exonization of intronic sequences or pseudoexonization of exonic sequences. Furthermore, compared with NAC family in other plants, the ZmNAC genes contained more large introns, partially because of insertion of repetitive elements [38]. Meanwhile, gene structure also confirmed the previous classification of ZmNAC subfamily (Figure 1A). Besides, the first and second intron phase were similar in ZmNAC subfamilies except the ONAC003 and ANAC063 subfamily. The ANAC063 subfamily did not have any intron phases, while first and second intron phase in the ONAC003 subfamily was entirely different from other subfamilies. This indicated that the expansion of ANAC063 and ONAC003 subfamily might be relatively independent from other subfamilies in maize. Above all, gain and loss of introns might take important parts in the expansion of ZmNAC members, and ANAC063 and ONAC003 subfamily perhaps have experienced the different evolutionary history from other ZmNAC subfamilies.

Functionalization of the ZmNAC duplicated genes during the maize evolution

During the evolutionary process, duplication genes might have experienced functionalization at the level of gene expression to retain in the genome [71]. In this study, 15 pairs of duplicated genes were identified in the ZmNAC family (Table 1). Most of Ka/Ks ratio was less than 1 in these duplicated gene pairs, indicating that the ZmNAC genes have mainly undergone purifying selection pressure with limited functional divergence after the duplications. Then, the expression patterns of six pairs of duplicated genes (Ka/Ks <1) were very similar in the different tissues through qRT-PCR analysis (Figure 5C, Figure 5G–K). These results indicated that most of the duplicated genes of ZmNAC family might retain some essential functions during sequent evolution. The retaining similar expression profiles may be related to highly similar amino acid sequence of these duplicated genes. Consistent with this observation, two duplicated ZmNAC paralogs (ZmNAC 36/96) were induced in response to *C. graminicola* infection [39]. However, this reported paralogs was not regarded as duplicated genes in our study, mainly because of the relatively low sequence similarity. Besides, there was three pairs of duplicated gene with a Ka/Ks ratio more than 1, which suggested strongly accelerated evolution with positive selection. Then three pairs of the duplicated gene showed the divergent expression pattern (Figure 5D, Figure 5E, Figure 5F). This result indicated that this pair of duplicated genes might have undergone significant diversification after duplication. Above all, functional divergence was limited in ZmNAC family after most of the gene duplication events.

In summary, 124 NAC transcription factors were identified in *Z. mays* in this study. Moreover, maize contained much more NAC members than other three monocots, and the duplication events (mainly segmental duplication event) might occur after the divergence of the lineages of *Z. mays* and *S. italica*. Meanwhile, the whole genome duplication, especially an additional WGD event, might take important roles during the ZmNAC expansion and this expansion has a certain NAC subfamily preference in maize. Furthermore, the expansion of ZmNAC members may be related to gain and loss of introns. Moreover, the restriction of

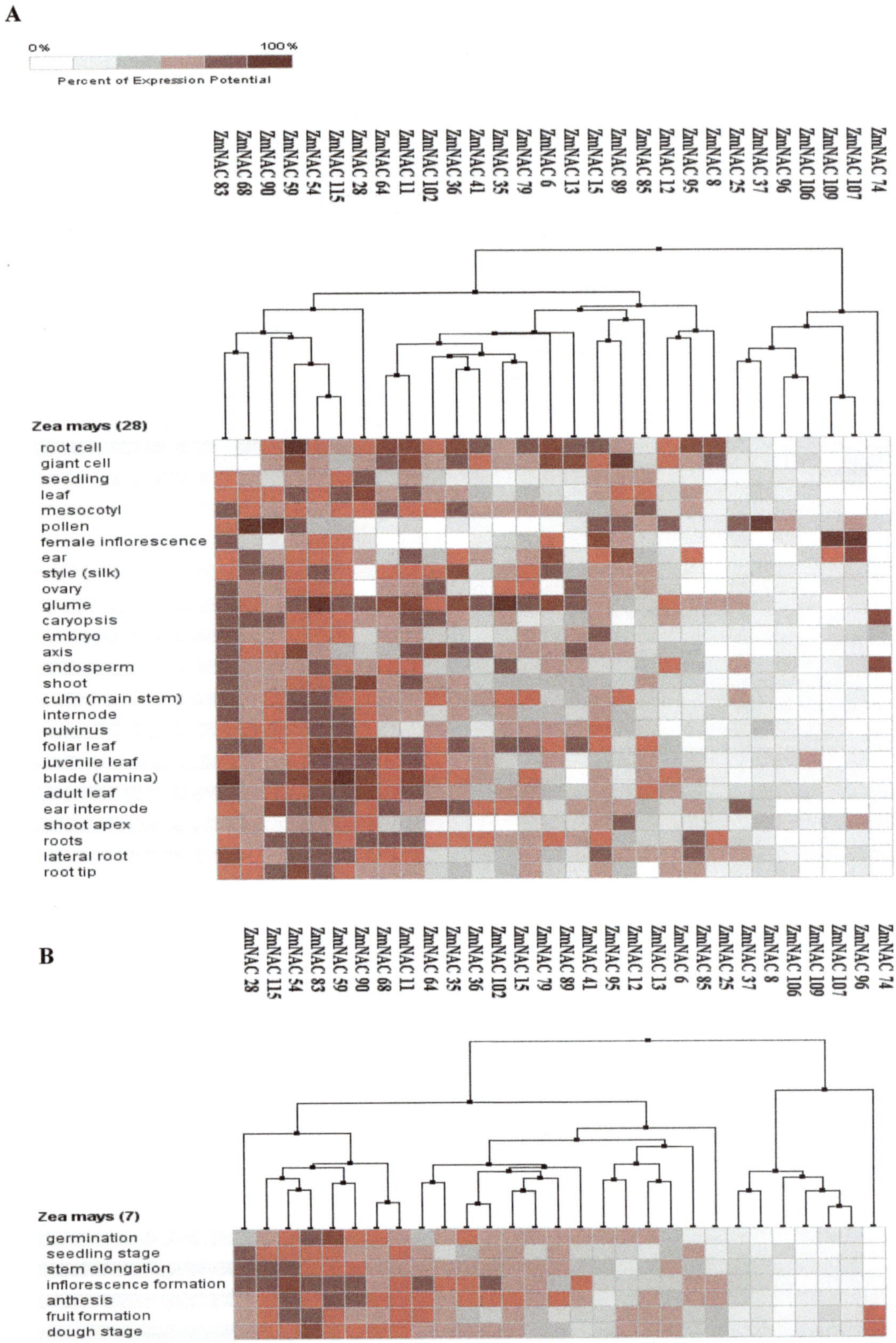

Figure 6. Hierarchical clustering of tissue-specific (A) and developmental (B) expression patterns of some ZmNAC genes by Genevestigator database. Color bar at top indicates percent of expression potential. The hierarchical clustering was calculated by Euclidean distance.

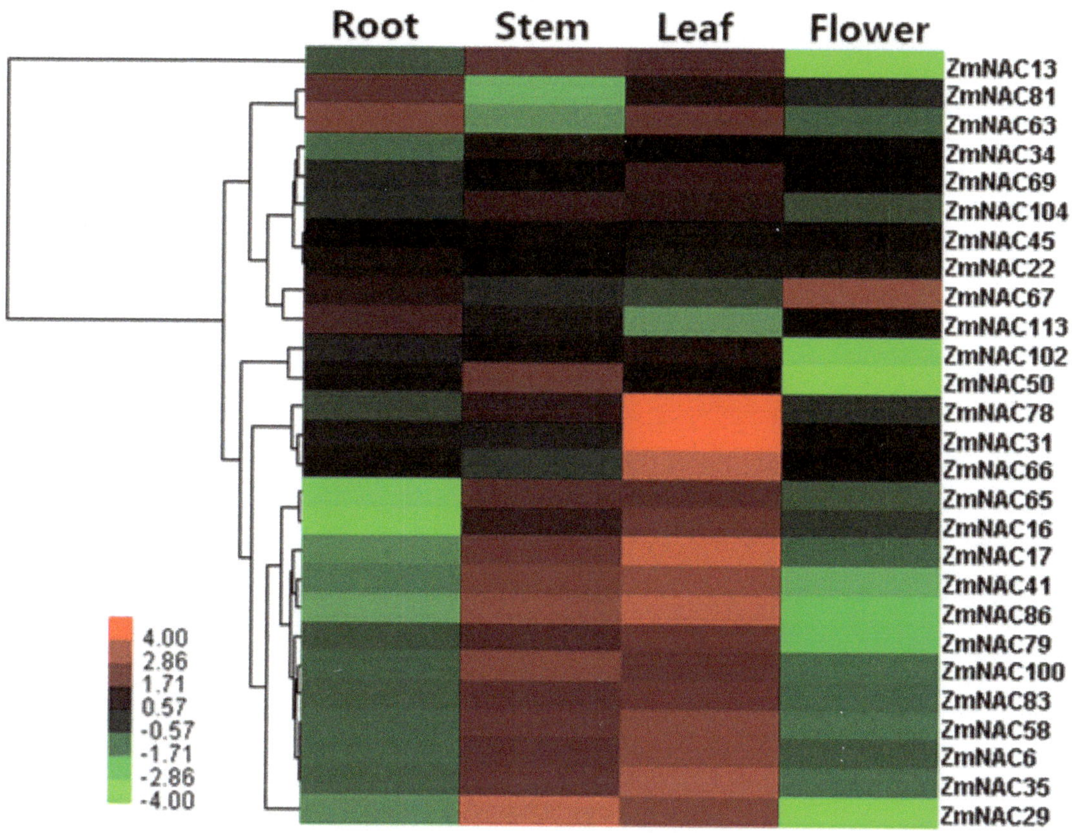

Figure 7. Heat map representation and hierarchical clustering of ZmNAC genes in *Zea mays* across different tissues. The color bar represents the relative signal intensity value.

functional divergence was concluded after most of the gene duplication events. Besides, the specific motifs and functions were highly conserved within the same subfamilies. The findings here provide the researchers a novel draft about molecular evolution and expansion history of NAC family in *Z. mays* and offer a good opportunity to further investigate NAC family in plants.

Supporting Information

Figure S1 A simplified phylogeny of three monocots. The total number of NAC family is showed in each species.

Figure S2 Phylogenetic tree of NAC proteins from *Z. mays* and Arabidopsis. Amino acid sequences were aligned using ClustalW and the maximum likelihood was generated through PhyML software. Names beginning with "ZmNAC" are NAC domains in *Z. mays*. All of ANACs in *A. thaliana* were achieved from TAIR. The NAC proteins of *Z. mays* are isolated as listed in Table S1. The subfamilies within the NAC family, as designated by Ooka et al. (2003), are grouped by colors.

Figure S3 The conserved domain analysis in ZmNAC protein using WebLogo program. Sequence logos of NAC domain (A) and TAR region (B) among ZmNAC family. The height of letter designating the amino acid residue at each position represents the degree of conservation. The numbers on the x-axis represent the sequence positions in its corresponding conservative domains. The y-axis represents the information content measured in bits.

Figure S4 Phylogenetic relationships among the 15 pairs of duplicated ZmNAC genes and its corresponding orthologous genes in other three monocots. The Bayesian method was used to construct the phylogenetic tree. The numbers in the clades are posterior probability values.

Figure S5 *In silico* frequency of *Z. mays* NAC genes ESTs. The EST frequencies of 14 ZmNAC genes were calculated through screening three EST libraries representing three different tissues.

Table S1 ZmNAC proteins identified in *Z. mays*.

Table S2 List of the orthologous groups of ZmNAC proteins through OtrhoMCL clustering.

Table S3 The structural analysis of ZmNAC identified in this study.

Table S4 Genomic locations of NAC genes in *Z. mays*.

Table S5 The duplicated ZmNAC genes and its corresponding orthologous genes in other three monocots.

Table S6 EST frequencies of ZmNAC genes.

Table S7 Selected some ZmNAC genes used for coexpression analysis in Genevestigator database.

Table S8 PCR primers used in this study.

Acknowledgments

We are greatly grateful to Gaixia Zhang, Hao Shen, Yijing Zhao (Zhejiang University, China) and Waqas Malik (Bahauddin Zakariya University) for their support in this study.

Author Contributions

Conceived and designed the experiments: XW KF MW YM MN. Performed the experiments: KF MW NB SY. Analyzed the data: KF MW YM FL. Contributed reagents/materials/analysis tools: XW KF. Contributed to the writing of the manuscript: KF XW YM MN NB.

References

1. Badis G, Berger MF, Philippakis AA, Talukder S, Gehrke AR, et al. (2009) Diversity and complexity in DNA recognition by transcription factors. Science 324: 1720–1723.
2. Luscombe NM, Thornton JM (2002) Protein-DNA interactions: amino acid conservation and the effects of mutations on binding specificity. J Mol Bio 320: 991–1009.
3. Hobert O (2008) Gene regulation by transcription factors and microRNAs. Science 319: 1785–1786.
4. Walford SA, Wu Y, Llewellyn DJ, Dennis ES (2011) GhMYB25-like: a key factor in early cotton fibre development. Plant J 65: 785–797.
5. Ricachenevsky FK, Sperotto RA, Menguer PK, Fett JP (2010) Identification of Fe-excess-induced genes in rice shoots reveals a WRKY transcription factor responsive to Fe, drought and senescence. Mol Biol Rep 37: 3735–3745.
6. Van Camp W (2005) Yield enhancement genes: seeds for growth. Curr Opin Biotech 16: 147–153.
7. Olsen AN, Ernst HA, Leggio LL, Skriver K (2005) NAC transcription factors: structurally distinct, functionally diverse. Trends Plant Sci 10: 79–87.
8. Ooka H, Satoh K, Doi K, Nagata T, Otomo Y, et al. (2003) Comprehensive analysis of NAC family genes in *Oryza sativa* and *Arabidopsis thaliana*. DNA Res 10: 239–247.
9. Souer E, van Houwelingen A, Kloos D, Mol J, Koes R (1996) The *No Apical Meristem* Gene of Petunia Is Required for Pattern Formation in Embryos and Flowers and Is Expressed at Meristem and Primordia Boundaries. Cell 85: 159–170.
10. Jensen M, Kjaersgaard T, Nielsen M, Galberg P, Petersen K, et al. (2010) The *Arabidopsis thaliana* NAC transcription factor family: structure-function relationships and determinants of ANAC019 stress signalling. Biochem J 426: 183–196.
11. Nuruzzaman M, Manimekalai R, Sharoni AM, Satoh K, Kondoh H, et al. (2010) Genome-wide analysis of NAC transcription factor family in rice. Gene 465: 30–44.
12. Le DT, Nishiyama R, Watanabe Y, Mochida K, Yamaguchi-Shinozaki K, et al. (2011) Genome-wide survey and expression analysis of the plant-specific NAC transcription factor family in soybean during development and dehydration stress. DNA Res 18: 263–276.
13. Singh AK, Sharma V, Pal AK, Acharya V, Ahuja PS (2013) Genome-Wide Organization and Expression Profiling of the NAC Transcription Factor Family in Potato (*Solanum tuberosum* L.). DNA Res 20: 403–423.
14. Shan W, Kuang J, Chen L, Xie H, Peng H, et al. (2012) Molecular characterization of banana NAC transcription factors and their interactions with ethylene signalling component EIL during fruit ripening. J Exp Bot 63: 5171–5187.
15. de Oliveira TM, Cidade LC, Gesteira AS, Coelho Filho MA, Soares Filho WS, et al. (2011) Analysis of the NAC transcription factor gene family in citrus reveals a novel member involved in multiple abiotic stress responses. Tree Genet Genomes 7: 1123–1134.
16. Puranik S, Sahu PP, Mandal SN, Parida SK, Prasad M (2013) Comprehensive genome-wide survey, genomic constitution and expression profiling of the NAC transcription factor family in Foxtail Millet (*Setaria italica* L.). Plos one 8: e64594.
17. Su H, Zhang S, Yuan X, Chen C, Wang X, et al. (2013) Genome-wide analysis and identification of stress-responsive genes of the NAM-ATAF1, 2-CUC2 transcription factor family in apple. Plant Physiol Bioch 71: 11–21.
18. Wang N, Zheng Y, Xin H, Fang L, Li S (2013) Comprehensive analysis of NAC domain transcription factor gene family in *Vitis vinifera*. Plant Cell Rep 32: 61–75.
19. Puranik S, Sahu PP, Srivastava PS, Prasad M (2012) NAC proteins: regulation and role in stress tolerance. Trends Plant Sci 17: 369–381.
20. Ernst HA, Olsen AN, Skriver K, Larsen S, Leggio LL (2004) Structure of the conserved domain of ANAC, a member of the NAC family of transcription factors. EMBO Rep 5: 297–303.
21. Wang Z, Dane F (2013) NAC (NAM/ATAF/CUC) transcription factors in different stresses and their signaling pathway. Acta Physiol Plant 35: 1397–1408.
22. Sablowski RW, Meyerowitz EM (1998) A Homolog of *NO APICAL MERISTEM* Is an Immediate Target of the Floral Homeotic Genes *PETALA3/PISTILLATA*. Cell 92: 93–103.
23. Guo H, Xie Q, Fei J, Chua N (2005) MicroRNA directs mRNA cleavage of the transcription factor NAC1 to downregulate auxin signals for Arabidopsis lateral root development. Plant Cell 17: 1376–1386.
24. Duval M, Hsieh T, Kim SY, Thomas TL (2002) Molecular characterization of AtNAM: A member of the Arabidopsis NAC domain superfamily. Plant Mol Biol 50: 237–248.
25. Guo Y, Gan S (2006) AtNAP, a NAC family transcription factor, has an important role in leaf senescence. Plant J 46: 601–612.
26. Mitsuda N, Seki M, Shinozaki K, Ohme-Takagi M (2005) The NAC transcription factors NST1 and NST2 of Arabidopsis regulate secondary wall thickenings and are required for anther dehiscence. Plant Cell 17: 2993–3006.
27. Kim JH, Woo HR, Kim J, Lim PO, Lee IC, et al. (2009) Trifurcate feed-forward regulation of age-dependent cell death involving miR164 in Arabidopsis. Science 323: 1053–1057.
28. Meng Q, Zhang C, Gai J, Yu D (2007) Molecular cloning, sequence characterization and tissue-specific expression of six NAC-like genes in soybean (*Glycine max*(L.) Merr.). J Plant Physiol 164: 1002–1012.
29. Fujita M, Fujita Y, Maruyama K, Seki M, Hiratsu K, et al. (2004) A dehydration-induced NAC protein, RD26, is involved in a novel ABA-dependent stress-signaling pathway. Plant J 39: 863–876.
30. Wang X, Basnayake BVS, Zhang H, Li G, Li W, et al. (2009) The Arabidopsis ATAF1, a NAC transcription factor, is a negative regulator of defense responses against necrotrophic fungal and bacterial pathogens. Mol Plant Microbe In 22: 1227–1238.
31. Faria J, Reis P, Reis M, Rosado G, Pinheiro G, et al. (2011) The NAC domain-containing protein, GmNAC6, is a downstream component of the ER stress-and osmotic stress-induced NRP-mediated cell-death signaling pathway. BMC Plant Biol 11: 129.
32. Taji T, Seki M, Satou M, Sakurai T, Kobayashi M, et al. (2004) Comparative genomics in salt tolerance between Arabidopsis and Arabidopsis-related halophyte salt cress using Arabidopsis microarray. Plant Physiol 135: 1697–1709.
33. Kim Y, Kim S, Park J, Park H, Lim M, et al. (2006) A membrane-bound NAC transcription factor regulates cell division in Arabidopsis. Plant Cell 18: 3132–3144.
34. Lu P, Chen N, An R, Su Z, Qi B, et al. (2007) A novel drought-inducible gene, *ATAF1*, encodes a NAC family protein that negatively regulates the expression of stress-responsive genes in Arabidopsis. Plant Mol Biol 63: 289–305.
35. Yoon H, Kim S, Kim S, Park C (2008) Regulation of leaf senescence by NTL9-mediated osmotic stress signaling in Arabidopsis. Mol Cell 25: 438.
36. Jeong JS, Kim YS, Redillas MC, Jang G, Jung H, et al. (2013) *OsNAC5* overexpression enlarges root diameter in rice plants leading to enhanced drought tolerance and increased grain yield in the field. Plant Biotechnol J 11: 101–114.
37. Uauy C, Distelfeld A, Fahima T, Blechl A, Dubcovsky J (2006) A NAC gene regulating senescence improves grain protein, zinc, and iron content in wheat. Science 314: 1298–1301.
38. Schnable PS, Ware D, Fulton RS, Stein JC, Wei F, et al. (2009) The B73 maize genome: complexity, diversity, and dynamics. Science 326: 1112–1115.
39. Voitsik A, Muench S, Deising HB, Voll LM (2013) Two recently duplicated maize NAC transcription factor paralogs are induced in response to *Colletotrichum graminicola* infection. BMC Plant Biol 13: 85.
40. Lu M, Ying S, Zhang D, Shi Y, Song Y, et al. (2012) A maize stress-responsive NAC transcription factor, ZmSNAC1, confers enhanced tolerance to dehydration in transgenic Arabidopsis. Plant Cell Rep 31: 1701–1711.
41. Zhong R, Lee C, McCarthy RL, Reeves CK, Jones EG, et al. (2011) Transcriptional activation of secondary wall biosynthesis by rice and maize NAC and MYB transcription factors. Plant Cell Physiol 52: 1856–1871.
42. Pinheiro GL, Marques CS, Costa MD, Reis PA, Alves MS, et al. (2009) Complete inventory of soybean NAC transcription factors: sequence conservation and expression analysis uncover their distinct roles in stress response. Gene 444: 10–23.
43. Zhao Y, Zhou Y, Jiang H, Li X, Gan D, et al. (2011) Systematic analysis of sequences and expression patterns of drought-responsive members of the *HD-Zip* gene family in maize. Plos one 6: e28488.
44. Abascal F, Zardoya R, Posada D (2005) ProtTest: selection of best-fit models of protein evolution. Bioinformatics 21: 2104–2105.

45. Guindon S, Dufayard J, Lefort V, Anisimova M, Hordijk W, et al. (2010) New algorithms and methods to estimate maximum-likelihood phylogenies: assessing the performance of PhyML 3.0. Systematic Biol 59: 307–321.
46. Huelsenbeck JP, Ronquist F (2001) MRBAYES: Bayesian inference of phylogenetic trees. Bioinformatics 17: 754–755.
47. Bailey TL, Williams N, Misleh C, Li WW (2006) MEME: discovering and analyzing DNA and protein sequence motifs. Nucleic Acids Res 34: W369–W373.
48. Crooks GE, Hon G, Chandonia J, Brenner SE (2004) WebLogo: a sequence logo generator. Genome Res 14: 1188–1190.
49. Zhang Z, Li J, Zhao X, Wang J, Wong GK, et al. (2006) KaKs_Calculator: calculating Ka and Ks through model selection and model averaging. Genomics, Proteomics & Bioinformatics 4: 259–263.
50. Gaut BS, Morton BR, McCaig BC, Clegg MT (1996) Substitution rate comparisons between grasses and palms: synonymous rate differences at the nuclear gene *Adh* parallel rate differences at the plastid gene *rbcL*. Proc Natl Acad Sci USA 93: 10274–10279.
51. Li L, Stoeckert CJ, Roos DS (2003) OrthoMCL: identification of ortholog groups for eukaryotic genomes. Genome Res 13: 2178–2189.
52. Cenci A, Guignon V, Roux N, Rouard M (2014) Genomic analysis of NAC transcription factors in banana (*Musa acuminata*) and definition of NAC orthologous groups for monocots and dicots. Plant Mol Biol: 1–18.
53. Zimmermann P, Hennig L, Gruissem W (2005) Gene-expression analysis and network discovery using Genevestigator. Trends Plant Sci 10: 407–409.
54. Chi X, Yang Q, Lu Y, Wang J, Zhang Q, et al. (2011) Genome-wide analysis of fatty acid desaturases in soybean (*Glycine max*). Plant Mol Biol Rep 29: 769–783.
55. Danielsson P (1980) Euclidean distance mapping. Computer Graphics and Image Processing 14: 227–248.
56. Li J, Guo G, Guo W, Guo G, Tong D, et al. (2012) miRNA164-directed cleavage of ZmNAC1 confers lateral root development in maize (*Zea mays* L.). BMC Plant Biol 12: 220.
57. Ma J, Wang F, Li M, Jiang Q, Tan G, et al. (2014) Genome wide analysis of the NAC transcription factor family in Chinese cabbage to elucidate responses to temperature stress. Sci Hortic165: 82–90.
58. Huang G, Li W, Zhou W, Zhang J, Li D, et al. (2013) Seven cotton genes encoding putative NAC domain proteins are preferentially expressed in roots and in responses to abiotic stress during root development. Plant Growth Regul 71: 101–112.
59. Aida M, Ishida T, Fukaki H, Fujisawa H, Tasaka M (1997) Genes involved in organ separation in Arabidopsis: an analysis of the cup-shaped cotyledon mutant. Plant Cell 9: 841–857.
60. Paterson AH, Bowers JE, Chapman BA (2004) Ancient polyploidization predating divergence of the cereals, and its consequences for comparative genomics. Proc Natl Acad Sci USA 101: 9903–9908.
61. Blanc G, Wolfe KH (2004) Widespread paleopolyploidy in model plant species inferred from age distributions of duplicate genes. Plant Cell 16: 1667–1678.
62. Tang H, Wang X, Bowers JE, Ming R, Alam M, et al. (2008) Unraveling ancient hexaploidy through multiply-aligned angiosperm gene maps. Genome Res 18: 1944–1954.
63. Shang H, Li W, Zou C, Yuan Y (2013) Analyses of the NAC transcription factor gene family in *Gossypium raimondii* Ulbr.: chromosomal location, structure, phylogeny, and expression patterns. J Integ Plant Biol 55: 663–676.
64. Hu R, Qi G, Kong Y, Kong D, Gao Q, et al. (2010) Comprehensive analysis of NAC domain transcription factor gene family in *Populus trichocarpa*. BMC Plant Biol 10: 145.
65. Nakano T, Suzuki K, Fujimura T, Shinshi H (2006) Genome-wide analysis of the ERF gene family in Arabidopsis and rice. Plant Physiol 140: 411–432.
66. Gagne JM, Downes BP, Shiu S, Durski AM, Vierstra RD (2002) The F-box subunit of the SCF E3 complex is encoded by a diverse superfamily of genes in Arabidopsis. Proc Natl Acad Sci USA 99: 11519–11524.
67. Ditte HW, Soren L, Niels EM, Addie NO, Charlotte H, et al. (2012) DNA binding by the plant-specific NAC transcription factors in crystal and solution: a firm link to WRKY and GCM transcription factors. Biochem J 444: 395–404.
68. Zahn LM, Kong H, Leebens-Mack JH, Kim S, Soltis PS, et al. (2005) The Evolution of the *SEPALLATA* Subfamily of MADS-Box Genes. Genetics 169: 2209–2223.
69. Eulgem T, Rushton PJ, Robatzek S, Somssich IE (2000) The WRKY superfamily of plant transcription factors. Trends Plant Sci 5: 199–206.
70. Park K, Kwon S, Kim N (2010) Intron loss mediated structural dynamics and functional differentiation of the polygalacturonase gene family in land plants. Genes Genom 32: 570–577.
71. Duarte JM, Cui L, Wall PK, Zhang Q, Zhang X, et al. (2006) Expression pattern shifts following duplication indicative of subfunctionalization and neofunctionalization in regulatory genes of Arabidopsis. Mol Biol Evol 23: 469–478.

Diverse and Widespread Contamination Evident in the Unmapped Depths of High Throughput Sequencing Data

Richard W. Lusk*

Department of Ecology and Evolutionary Biology, University of Michigan, Ann Arbor, Michigan, United States of America

Abstract

Trace quantities of contaminating DNA are widespread in the laboratory environment, but their presence has received little attention in the context of high throughput sequencing. This issue is highlighted by recent works that have rested controversial claims upon sequencing data that appear to support the presence of unexpected exogenous species. I used reads that preferentially aligned to alternate genomes to infer the distribution of potential contaminant species in a set of independent sequencing experiments. I confirmed that dilute samples are more exposed to contaminating DNA, and, focusing on four single-cell sequencing experiments, found that these contaminants appear to originate from a wide diversity of clades. Although negative control libraries prepared from 'blank' samples recovered the highest-frequency contaminants, low-frequency contaminants, which appeared to make heterogeneous contributions to samples prepared in parallel within a single experiment, were not well controlled for. I used these results to show that, despite heavy replication and plausible controls, contamination can explain all of the observations used to support a recent claim that complete genes pass from food to human blood. Contamination must be considered a potential source of signals of exogenous species in sequencing data, even if these signals are replicated in independent experiments, vary across conditions, or indicate a species which seems *a priori* unlikely to contaminate. Negative control libraries processed in parallel are essential to control for contaminant DNAs, but their limited ability to recover low-frequency contaminants must be recognized.

Editor: Tom Gilbert, Natural History Museum of Denmark, Denmark

Funding: RWL was supported by NIH postdoctoral fellowship F32-GM100685 (http://www.nih.gov). The funders had no role in study design, data collection and analysis, decision to publish, or preparation of the manuscript.

Competing Interests: The author has declared that no competing interests exist.

* Email: rlusk@umich.edu

Introduction

While contamination by foreign DNA is a concern for many experiments, it requires particular attention for those that rely on sensitive methods to describe samples that are themselves dilute or degraded. Low quality samples poorly compete with contaminant DNAs over the course of an experiment, and methods to characterize and eliminate contaminants have been rigorously evaluated in fields where such samples are common. These studies have found that free DNA at trace but detectable levels is widespread, often being found in "clean" new PCR tubes [1], dNTPs [2], and a variety of other sources [3-8]. Even extreme precautions, such as UV-treatment of reagents, the use of positive pressure laboratory ventilation systems, etc., appear sufficient only to reduce, rather than eliminate, the abundance of contaminant DNAs [9-11].

These results were established using PCR, but they have received surprisingly little attention in the context of high throughput sequencing, which shares with PCR the ability to potentially detect single DNA molecules. Moreover, sequencing samples are processed through a longer and more complex experimental pipeline, which typically includes a PCR amplification step, increasing their exposure to contaminants. The power of sequencing to recover, but not necessarily to identify, contaminant DNAs should grow as read depths increase and as library preparation methods require progressively smaller amounts of input material. Nevertheless, the presentation of contaminants in sequencing data is poorly understood. Although sequences from humans and a handful of other species have been found to contaminate sequencing samples and databases [12-17], the roster of potential contaminant species is unknown, as is their distribution within and between experiments, making it difficult to infer whether or not any given read originated from the intended sample.

To help create a framework for this inference, I sought to describe the diversity of contaminant reads in four independent sequencing experiments and how these reads are replicated across samples and negative controls. I used this information to evaluate whether contaminants may have evaded the heavy replication and plausible controls described in a recent paper to provide an alternative explanation for its claim that complete genes pass from food to human blood.

Materials and Methods

Inference of human contaminants in E. coli sequencing data

I downloaded data from Parkinson et al [18] from the European Nucleotide Archive. Run accession numbers are listed in

Additional File 1, Table S1. Splitting apart the paired-end reads, I used bowtie version 1.0.0 [19] to attempt to map each end to the *E. coli* K12 reference genome, allowing up to three mismatches in the seed region ('-n 3') but otherwise using default parameters. Unmapped reads were then aligned to the human hg18 reference genome [20] with no mismatches permitted in the seed ('-n 0'). Reads were considered a 'hit' if both ends from each pair mapped to within one kilobase of each other. The fraction of human reads was considered the total number of paired end reads meeting these criteria over the total number of pair-end reads. A generalized linear model was used to describe the relationship between the log of the sample DNA concentration and the frequency, modeled with the binomial distribution, of human-matching reads within the total library. This analysis was performed using the GLM2 package in R [21,22].

Inference of contaminant reads and species composition in four sequencing experiments

I downloaded data from the short read archive corresponding to study accession numbers SRP002535 (referred to in the main text as the "Tumor" dataset [23]), SRP014866 ("Strandseq" [24]), SRP006834 ("RNAseq" [25]), and SRP017186 ("Sperm" [26]). The specific SRA accessions for each dataset are listed in Additional File 2, Table S2. Due to differences in scope and sequencing technology, the numbers of reads sequenced by these studies differed over orders of magnitude, and so samples from the RNAseq and Strandseq datasets were pooled to provide a more comparable number of total reads, whereas samples having close to the median number of reads in the Sperm and Tumor datasets were analyzed individually. Given the unusual library preparation of the RNAseq dataset [25], in which each read contains a barcode and an uncertain number of guanines, I determined the information content of each position in the library (Additional File 3, Figure S1) and, based upon these data, removed the first fifteen base pairs of each read. The Strandseq and Sperm studies used paired-end libraries; I split paired ends and processed each end independently.

For each dataset, I used bowtie as described above to screen out reads that aligned to the appropriate mouse or human reference genome (hg18 [20] for Sperm and Tumor; mm9 [27] for RNAseq and Strandseq), allowing three mismatches in the seed region. Using the Amazon EC2 image ami-ef42d586, which contains snapshots of the NCBI nucleotide databases from 02/04/2013, I used BLAST to search for hits to each unaligned read to the NR database, using flags –task megablast, -outfmt 6, and –num_alignments 5 (500 for the comparison of positive and blank samples from the Strandseq and RNAseq experiments).

Each experiment having different read lengths and qualities, I considered hits of 100% identity over 48, 75, 40, and 101 base pairs for the Tumor, Strandseq, RNAseq, and Sperm datasets, respectively, to be a perfect match. Due to its large size, the number of reads used for the Strandseq dataset was capped at 7.5 million.

For each BLAST match, I used the GI sequence identification number to locate the corresponding taxID from the NCBI taxonomy database, discarding reads that did not cross-reference. For each taxID, I ascended the tree to the fourth-lowest node and assigned the BLAST hit to that taxonomic category. Reads were discarded if they had BLAST hits to more than one of these categories, if they matched sequences from the species of the reference genome, or if they only matched sequences in the database that were taxonomically unclassifiable (e.g. taxID 155900, "uncultured organism"). Reads matching these criteria were also used for the comparison of positive and blank samples from the Strandseq and RNAseq experiments, where the taxonomy tree was ascended to the "genus" level and reads that matched to more than one genus were discarded.

Diversity of chloroplast DNAs in individual libraries

Using the sequencing data described in Additional File 2, Table S2 and the methods described above, I screened out reads in each of these 100 samples that matched the human hg18 genome. I then used bowtie to attempt to match each unmapped read to a database of chloroplast genomes (Additional Table S4, Table S3), allowing no mismatches in the seed region ('-n 0'). These settings were specified to closely follow those used by Spisak et al [28]. As a further precaution against incorrect assignment, reads that matched chloroplasts using the above criteria were screened using BLAST against the NR database, removing those that had higher similarity to non-chloroplast sequences.

Other contaminant species detected in cell-free DNA sequencing samples

Downloading the data generated by Spisak et al [28] (European Nucleotide Archive accession number ERP002472), I created a representative subsample by combining one million reads from each sequencing sample. I screened these data against the *Escherichia coli, Malassezia globusa* and *Propionibacterium acnes* genomes and the *Solanum lycopersicum* chloroplast genome using bowtie as described above and in Spisak et al. To more closely follow the methods used by Spisak et al, matching reads were not subsequently screened against the NR database.

Results and Discussion

Contaminant read count is inversely related to sample concentration

High throughput sequencing quantifies the relative, not absolute, quantities of different DNAs in a library. If we assume that the amount of contaminant DNA contributed to each of a set of identically-prepared libraries is constant, we should expect that the number of reads matching contaminant genomes will increase as the concentration of the sample decreases. To illustrate this relationship, I reanalyzed data from a study that sequenced different dilutions of a single sample of *E. coli* DNA [18], examining how changes in the concentration of *E. coli* DNA affected the number of reads matching sequences in the human genome, a common laboratory contaminant [12,14].

The authors of this study sought to demonstrate the efficacy of a novel low-concentration library preparation protocol. To this end, they used it to create libraries from 1ng, 100pg, and 10pg of a shared *E. coli* DNA sample; they also used 1 μg of this DNA to prepare a sequencing library using the standard Illumina protocol. As expected, there is an inverse relationship between the concentration of the sample and the frequency of contaminant reads (Figure 1A). For the four libraries prepared from the highest concentration of DNA (1ng), I found only three reads in total that mapped to the human but not to the *E. coli* genome. However, the frequency of these reads increased with decreasing sample concentration, to approximately 175 and 2,500 reads per million in the 100 and 10pg samples, respectively ($z = -108.4$, $p < 2e-16$, see Methods). The protocol used to prepare the libraries can influence the amount of contamination, however: the protocol described above used a single tube to fragment input DNA, ligate adapters, and amplify fragments, potentially reducing the sample's exposure to contaminants. I found an intermediate frequency of contaminating reads in the library prepared from the standard

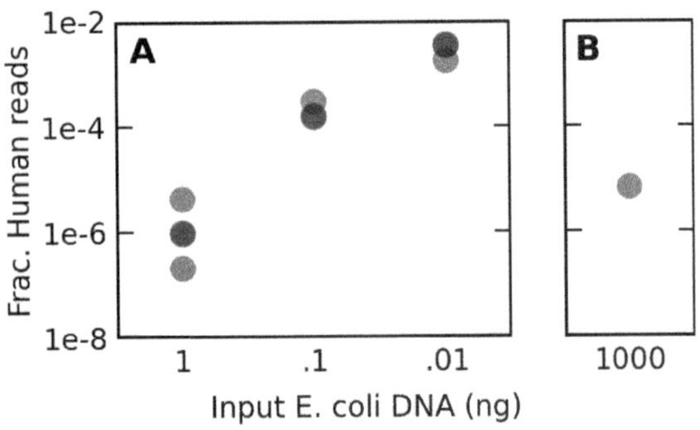

Figure 1. Reads matching the human genome are more prevalent in libraries prepared from dilute samples. (a) The fraction of paired-end reads which preferentially map to the contaminant human genome instead of the *E. coli* K-12 genome, measured against the total number of reads in the library, is plotted against the amount of *E. coli* K-12 DNA used per tagmentation procedure as described by Parkinson et al [18]. Shading is used to highlight closely overlapping points (n = 4, 3, and 3 for the 1ng, 100pg, and 10pg libraries, respectively). Libraries listed at each concentration were not identically prepared, each using a different restriction enzyme or set of restriction enzymes at an intermediate step in the protocol (Additional File 1, Table S1), but the number and composition of enzymes used did not appreciably change the number of contaminant reads recovered. (b) The same fraction is plotted for a library prepared in the same experiment using a standard Illumina library preparation protocol. Despite a higher concentration of input DNA, an intermediate number of contaminant reads was detected.

Illumina protocol, despite its higher concentration of sample DNA (Figure 1B).

This relationship between the concentration of the sample and the frequency of contaminant reads may differ in magnitude within and between experiments due to, for instance, differing preparation protocols, laboratory environments, and human error. Nevertheless, we can expect in general that libraries prepared from dilute samples will be more vulnerable to contaminating DNAs than will libraries prepared from more concentrated samples.

Unmapped reads from independent experiments match a wide range of species

Although contamination by DNA from humans [12,14], plants [17], and a handful of bacterial species [13,15,16] has been described in sequencing data, there has been to my knowledge no systematic effort to describe the diversity of contaminants evident in sequencing data, making it difficult to gauge the likelihood that any given species inferred to be present from sequencing data originated from contamination. To address this issue, I performed a metagenomic analysis of potential contaminants in a set of four independent sequencing experiments [23-26].

In order to recover the greatest possible yield of contaminant DNAs, I chose experiments that worked with low quantities of input material. All of these experiments prepared libraries from individual cells, but they otherwise had different goals and used different species and experimental techniques (Table 1). I took several precautions to limit the impact of sequencing errors and to ensure that inferences of contaminant species were as conservative as possible. For each of these experiments I used permissive settings to screen out reads that could potentially map to the appropriate reference genome, and I then used BLAST to search for perfect matches to the remaining reads in the NR database across the entire or, depending on the dataset, close to the entire length of each read (see Methods). The use of BLAST for phylogenetic assignment, although once standard [29], has important shortcomings, including loss of information from non-global alignment, ignorance of population genetic and phylogenetic issues, and the use of sequence-level rather than clade-level confidence metrics [30]. However, alternative methods that take these issues into account, e.g. SAP [30] and EcoTag [31], are not computationally feasible at this scale. As highly conserved regions are less informative for taxonomic classification, I further screened out reads with reported BLAST hits to more than one broad taxonomic category.

In all of these experiments I found reads matching sequences belonging to a wide diversity of groups (Figure 2), and the observed abundance of each group was broadly correlated across experiments. These groups include some which might seem unlikely contributors to contamination, such as Streptophyta, a phylum encompassing the land plants and some green algaes.

There are several potential sources for reads such as these that preferentially map to alternate genomes in sequencing experiments. In some cases, they may truly be evidence for an alternate organism in the sample, e.g. the serendipitous sequencing of new genomes of the bacterial endosymbiont *Wolbachia* in several *Drosophila* genome sequencing projects [32]. This seems unlikely for the reads described in Figure 2, because all of these experiments sequenced individual cells. Many of these reads map to organisms that seem likely to contaminate samples, such as *Propionibacterium acnes*, a pervasive skin bacterium, and the simplest explanation for the diversity detected may be that traces of many species are widespread in the laboratory environment. It is also possible for multiplexed libraries to cross-contaminate each other during sequencing [33], although this would not explain the diversity of species observed here.

Many reads observed here appear to map to alternate genomes due to sequencing errors. While most reads containing errors will presumably not perfectly match any species' genome, the large number of reads analyzed here ensures that many will coincidentally match genomes of species only related to those actually present in the sample. For instance, many reads in each experiment matched the chimpanzee genome, even though the sample organism was either mouse or human. Whether a given match to an exogenous species originated from a true contaminant or from a sequencing error can be difficult to distinguish. Indeed, some of the matches to the chimpanzee genome, particularly those from the mouse experiments, may be due to errors made during sequencing of contaminant human DNA. The choice of aligner may also play a role in species assignment. Here I have used an ungapped aligner, bowtie v.1.0, and so a missed base in the reference sequence would prevent mapping to the correct sequence and would preferentially match homologous sequences from related species. Similarly, incomplete reference genomes may lead to false inference of contamination, and differences in completeness of alternate genomes may bias the inference of the origins of contamination. However, these issues, driven by close homology, would be unlikely to shift the distribution of broad taxonomic categories shown in Figure 3.

It is possible that examination of error patterns and rates with different sequencing technologies would allow probabilistic attribution of each potential contaminant read to either experimental or sequencing errors, but for the purposes of this study the precise origins of each read will remain ambiguous. To reflect this ambiguity, I will refer to reads that match exogenous genomes as "contaminant matching reads."

Table 1. Description of the sequencing experiments used in this study.

Experiment	Reference	Organism	RNA/DNA	Platform	Cell isolation	Pooled	Read length	Number of reads
RNAseq	[25]	Mouse	RNA	GA IIx	Cell picker	Yes	56	93,471,748
Tumor	[23]	Human	DNA	GA II	FACS	No	48	8,568,573
Strandseq	[24]	Mouse	DNA	GAIIx/Hiseq2000	FACS	Yes	75	100,026,526
Sperm	[26]	Human	DNA	HiSeq2000	Pipetting	No	101	17,066,385

'Blank' negative controls effectively describe the distribution of contaminants per sample

The "RNAseq" and "Strandseq" (Table 1) experiments prepared negative control libraries from "blank" buffer samples into which no sample cell had been deposited. However, only a small number of such libraries were prepared, and each produced a relatively small number of reads, having in total 0.56% and 0.21% of the corresponding total number of reads in the positive samples from the RNAseq and Strandseq studies, respectively. It is possible that this low yield compromised their ability to represent the spectrum of contaminants found in the positive samples.

To address this possibility, I compared the frequency distribution of genera represented by contaminant matching reads in the positive and blank samples, discarding reads that promiscuously matched more than one genus. I found that the frequency distribution of recovered genera correlated well between positive and blank samples from the same experiment, suggesting that, despite their low yield, blank samples can serve as effective negative controls (Figure 3). Furthermore, the correlation is poorer between mismatched positive and blank samples taken from different experiments, consistent with the spectrum of contaminants being experiment-specific. For instance, reads matching the skin bacterium *Propionibacterium acnes* are more than 100-fold more frequent in the genomic DNA sample than they are in the RNA-seq dataset, perhaps reflecting the stricter guidelines for handling RNA.

Despite the strong correlation between the positive and blank libraries, the low yield of the blank libraries limited their ability to control for low frequency genera in the positive libraries, many of which were not represented in the blanks. As the lack of input DNA pushes the boundaries of preparation protocols, this low yield may be inevitable in true blanks. To increase overall yield and thereby lower the detection threshold for potential contaminants, it may be sufficient to increase the number of blank samples or to use a distinguishable carrier DNA (e.g. [34]). Sequencing data generated from libraries prepared in the same laboratory and with the same equipment may also be informative [17]. In cases where exclusion of contaminants is paramount, the use of techniques from ancient DNA sequencing may be considered (e.g. [9]).

Libraries prepared from a single tissue appear contaminated by different species

If a reagent or piece of equipment is heavily contaminated by a given species, then all exposed samples should share evidence of being contaminated by that species. In this scenario, one way to control for contamination would be to examine how exogenous species are distributed, considering as contaminants species that are recovered from most or all samples. However, while this assumption is reasonable if the contaminant concentration is high, we should expect a weaker correlation between exogenous species if the contaminant concentration is low. Indeed, at the lower limit, where each contaminant species contributes one DNA molecule, samples will necessarily be contaminated by different species. To examine whether these low-frequency contaminants are detectable and to what extent they differ between samples, I reanalyzed data from the "Tumor" experiment in which independent sequencing samples were prepared from the same tissue rather than from independent individuals. In this case, all differences between samples outside changes to the target genome can be attributed to contamination.

In this experiment, Navin et al [23] used FACS to isolate 100 individual nuclei from a section of a single tumor, and, processing

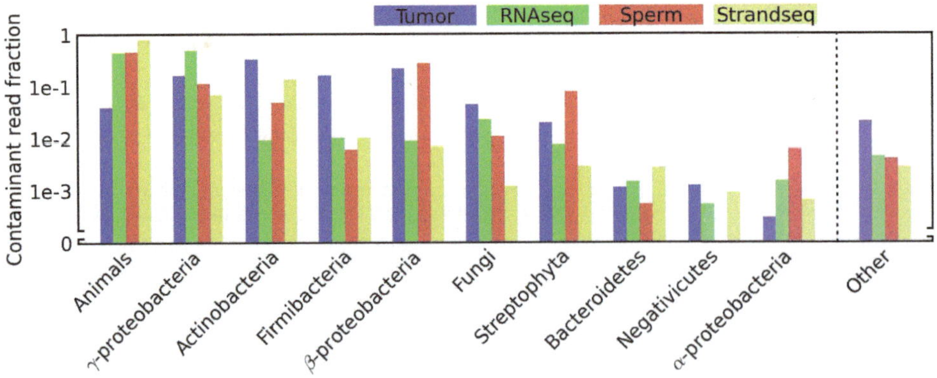

Figure 2. Reads that do not map to the reference genome match a diverse array of clades. For each experiment ("Tumor" [23], "RNAseq" [25], "Sperm" [26], and "Strandseq" [24]), all reads were individually mapped to the appropriate reference genome using permissive parameters before being used to query the NR database using BLAST. BLAST hits were considered "perfect" if they matched with 100% identity over a dataset-specific length threshold (see Methods). A read was assigned to one of the depicted phylogenetic categories if it did not map or have a perfect BLAST hit to the reference genome, had a perfect BLAST hit to a species in that category, and had no BLAST hits to species outside that category. For each category and experiment, the fraction of reads meeting this criteria against the total number of reads in the experiment is depicted.

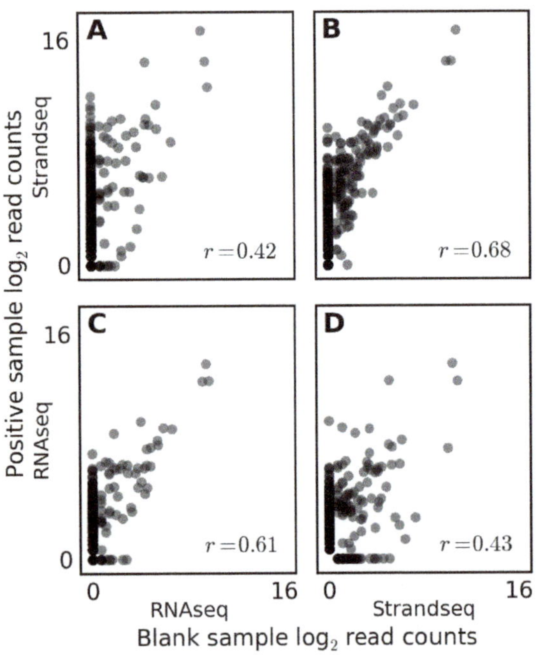

Figure 3. Experiment-specific correlation between distributions of genera recovered from 'blank' and positive samples. The "RNAseq" [25] and "Strandseq" [24] experiments sequenced libraries prepared from blank samples into which no cells had been introduced. Reads from these blank samples and from all other samples were separately pooled, screened against the mouse reference genome, and queried against the BLAST NR database. Reads were screened using the same criteria as described for Figure 2, but adjusted to the genus taxonomic level. The number of reads matching each genus in each dataset was counted, incremented by one, and log transformed. Values for the pooled positive samples (StrandSeq and RNAseq in rows A-B and C-D, respectively) are plotted with their Pearson correlation against values for the pooled negative samples (Strandseq and RNAseq in columns A-C and B-D, respectively). Matched positive and negative samples in (B) and (C) exhibit more correlated read counts than do mismatched positive and negative samples in (A) and (D).

them similarly through 96-well plates, used whole genome amplification to accumulate sufficient DNA for library preparation and sequencing. For each of these libraries, I first screened out loose matches to the human genome before searching for strong matches to a database of chloroplast genomes. Reads that matched chloroplast genomes were further screened against the NR database using BLAST, and chloroplast genomes with uneven read coverage, presumably caused by matches to low-complexity sequences, were screened out as well (Additional File 5, Figure S2). I found reads matching chloroplast DNAs in all samples, and different samples had different rosters of inferred contaminants (Figure 4; Additional File 6, Table S4), suggesting that contaminants can contribute unevenly to samples within a single experiment. Many of the species recovered are edible and commonly eaten, such as tomato and lettuce, although a slight majority (53%) of the reads from Figure 4 matched to oak and chestnut. Many reads also matched to *Heliconia,* but these could be due to *Heliconia*'s close relation to the banana, which was not represented in the chloroplast database.

Contaminants provide an alternative explanation for observations of plant DNAs in human blood

Highlighting the potential impact of contaminants in high throughput sequencing data is a controversial paper that claimed that complete genes pass from food to human blood [28]. The authors of this study based their claim on their observations of chloroplast sequences from edible plants in sequencing data from samples of blood plasma. They were able to replicate this observation in independent datasets, and, contrary to what might be expected from a single contamination event, found that the plant species recovered differed from sample to sample within a single experiment. Furthermore, while these reads could not be detected in a negative control sample of fetal blood, which circulates independently, they found many in the corresponding maternal plasma sample, which is presumably more exposed to the digestive system.

On the basis of their observation's heavy replication, heterogeneity among samples, absence from a plausible negative control, and, finally, their assumption that plant DNA would be unlikely to infiltrate stringent laboratory practices, the authors dismissed the possibility of contamination. However, as I have shown above, all of these assumptions about the nature of contamination are

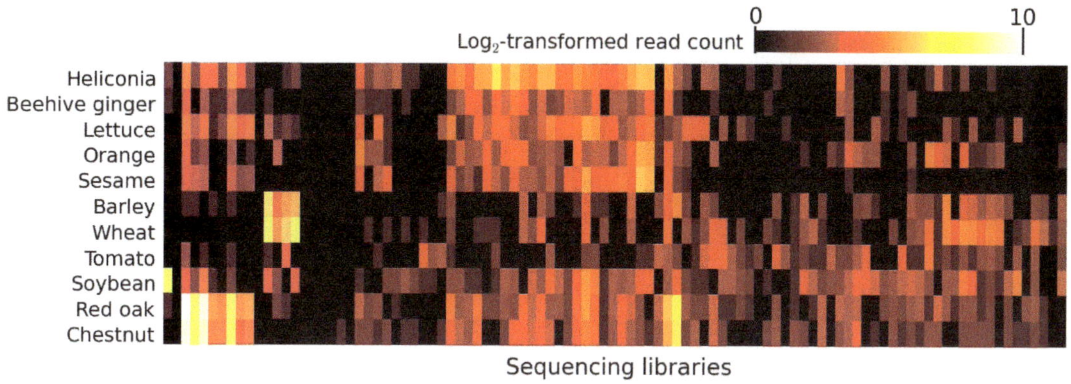

Figure 4. Heterogeneous species appear to contaminant samples from the same tissue and experiment. The "Tumor" [23] experiment dissociated 100 individual cells from a sample of a single tumor and sequenced libraries from each. Following the analysis pipeline of a study that claimed to find different plant species in different blood plasma samples from a single experiment, I used bowtie to screen each read in each library against the human reference genome before using it to query a database of chloroplast genomes. The number of such hits to each genome is depicted here, each count incremented by one and log-transformed. Only chloroplast genomes with at least 200 hits are shown. Rows and columns were clustered using a neighbor-joining algorithm.

incorrect. DNA from plants is a common contaminant, and it should be expected to appear to replicate across independent experiments (Figure 2). Furthermore, the particular species inferred can vary considerably from sample to sample within an experiment (Figure 4). Finally, although the authors did find hundreds of reads matching plants in a maternal plasma sample, and none in a matched sample of whole fetal blood, the concentration of DNA in whole blood is tens-of-thousands-fold higher than that in blood plasma (approximately 60 µg/ml vs. 21ng/ml [35,36]). Given the inverse relationship between sample DNA concentration and contaminant frequency (Figure 1), we should expect plasma samples to exhibit more evidence of contamination than full blood samples. Indeed, the corresponding maternal whole blood sample contained only one plant-matching read, a number statistically indistinguishable from zero, suggesting that using fetal whole blood as a negative control in this case was inappropriate. Contamination thus provides an alternative explanation for all of the observations originally used to support the authors' claim.

To further investigate this possibility, I used the same criteria used by Spisak et al [28] to assign reads to the tomato genome, their most frequent inferred contaminant, to search for reads matching three other potential contaminants in a representative sample of the same datasets (see Methods): *E. coli*, *P. acnes* and *M. globosa*. *P. acnes* and *M. globosa* are widely associated with the human skin flora [37,38], although both have been detected in stool samples as well [39,40]. Reads matching the genomes of each of these species were more frequent than reads matching the tomato chloroplast genome (Figure 5). While it cannot be proven that these reads did not originate from genes passing through the epithelium of the gastrointestinal tract, contamination appears to be their more likely source.

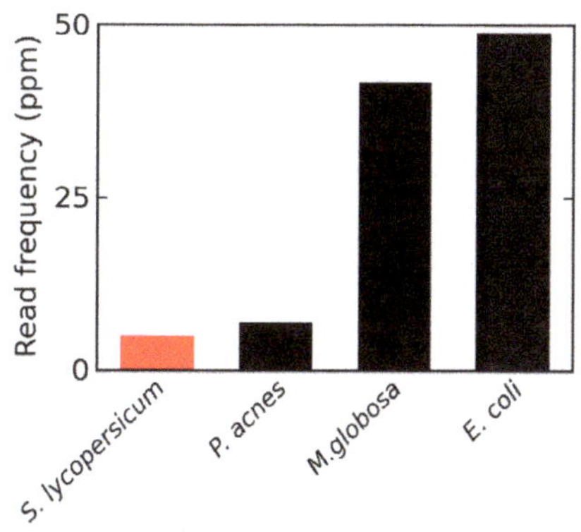

Figure 5. Reads matching the tomato chloroplast genome are less frequent than other contaminant matching reads in samples of cell-free DNA. Spisak et al [28] used the frequency of reads matching chloroplast genomes as evidence that genes pass intact from food to the bloodstream, and found *S. lycopersicum* (tomato) to be the most common contaminant. Evenly sampling from all of the sequencing samples generated by Spisak et al, I used identical criteria to investigate potential matches to three other contaminant species, *E. coli*, *P. acnes* and *M. globosa*. *P. acnes* and *M. globosa* are associated with the human skin flora. The frequency of reads matching each of these contaminant reads, per million reads in the pooled samples, is depicted.

Conclusions

Contaminants can in some cases effectively mimic behavior intuitively expected from true signals, including replication across independent experiments and variation between samples, and can include species that are not typically considered potential contaminants. This calls into question several controversial works that have rested their claims on observations of rare matches to exogenous species in sequencing data [28,41-43], and suggests that blank negative controls, prepared in parallel with experimental samples, are essential. Nevertheless, the rarest contaminants, being difficult to recover in these controls, may be intrinsically difficult to control for.

Supporting Information

Figure S1 Information per position in RNAseq dataset. The entropy of each position in each of 460 libraries was calculated, and the distribution of these among datasets is depicted for each position. Due to the protocol used to generate these data, each read began with a six base pair barcode, followed by at least three guanines. On the basis of this analysis, the first fifteen bases, highlighted in gray, were trimmed from each read.

Figure S2 Coverage of chloroplast genomes. Reads from the "Tumor" experiment were mapped to a database of chloroplast genomes, and the coverage of genomes with more than 200 matches is depicted here. The *B. hypnoides* genome was removed from further analysis due to the observed uneven coverage.

Table S1 The European Nucleotide Archive accession numbers associated with each sample analyzed in Figure 1.

Table S2 The SRA accession numbers associated with the samples analyzed in Figures 3 and 4.

Table S3 The chloroplast genomes which made up the database used in Figure 4.

Table S4 The total number of hits to each chloroplast genome in each library from the Tumor dataset.

Acknowledgments

I would like to thank Patricia Wittkopp, Brian Metzger, Kraig Stevenson, Joseph Coolon, and Fabien Duveau for critical comments on the manuscript. I would also like to thank those listed above and other members of the Wittkopp laboratory for valuable discussions.

Author Contributions

Conceived and designed the experiments: RWL. Performed the experiments: RWL. Contributed reagents/materials/analysis tools: RWL. Wrote the paper: RWL.

References

1. Schmidt T, Hummel S, Herrmann B (1995) Evidence of contamination in PCR laboratory disposables. Naturwissenschaften 82(9): 423-31.
2. Léonard JA, Shanks O, Hofreiter M, Kreuz E, Hodges L, et al. (2007) Animal DNA in PCR reagents plagues ancient DNA research. Journal of Archaeological Science 34(9): 1361-6.
3. Peters RP, Mohammadi T, Vandenbroucke Grauls CM, Danner SA, van Agtmael MA, et al. (2004) Detection of bacterial DNA in blood samples from febrile patients: underestimated infection or emerging contamination? FEMS Immunol Med Microbiol 42(2): 249-53.
4. Ehricht R, Hotzel H, Sachse K, Slickers P (2007) Residual DNA in thermostable DNA polymerases - a cause of irritation in diagnostic PCR and microarray assays. Biologicals 35(2): 145-7.
5. Evans GE, Murdoch DR, Anderson TP, Potter HC, George PM, et al. (2003) Contamination of Qiagen DNA extraction kits with *Legionella* DNA. J Clin Microbiol 41(7): 3452-3.
6. Hughes MS, Beck LA, Skuce RA (1994) Identification and elimination of DNA sequences in Taq DNA polymerase. J Clin Microbiol 32(8): 2007-8.
7. Erlwein O, Robinson MJ, Dustan S, Weber J, Kaye S, et al. (2011) DNA extraction columns contaminated with murine sequences. PLoS One 6(8): e23484.
8. Persing DH (1991) Polymerase chain reaction: trenches to benches. J Clin Microbiol 29(7): 1281-5.
9. Champlot S, Berthelot C, Pruvost M, Bennett EA, Grange T, et al. (2010) An efficient multistrategy DNA decontamination procedure of PCR reagents for hypersensitive PCR applications. PLoS One 5(9).
10. Gill P, Whitaker J, Flaxman C, Brown N, Buckleton J (2000) An investigation of the rigor of interpretation rules for STRs derived from less than 100 pg of DNA. Forensic Sci Int 112(1): 17-40.
11. Corless CE, Guiver M, Borrow R, Edwards-Jones V, Kaczmarski EB, et al. (2000) Contamination and sensitivity issues with a real-time universal 16S rRNA PCR. J Clin Microbiol 38(5): 1747-52.
12. Longo MS, O'Neill MJ, O'Neill RJ (2011) Abundant human DNA contamination identified in non-primate genome databases. PLoS One 6(2): e16410.
13. Binns M (1993) Contamination of DNA database sequence entries with *Escherichia coli* insertion sequences. Nucleic Acids Res 21(3): 779.
14. Schmieder R, Edwards R (2011) Fast identification and removal of sequence contamination from genomic and metagenomic datasets. PLoS One 6(3): e17288.
15. Astua-Monge G, Lyznik A, Jones V, Mackenzie SA, Vallejos CE (2002) Evidence for a prokaryotic insertion-sequence contamination in eukaryotic sequences registered in different databases. Theor Appl Genet 104(1): 48-53.
16. Percudani R (2013) A Microbial Metagenome (*Leucobacter sp.*) in *Caenorhabditis* Whole Genome Sequences. Bioinform Biol Insights 7: 55-72.
17. Tosar JP, Rovira C, Naya H, Cayota A (2014) Mining of public sequencing databases supports a non-dietary origin for putative foreign miRNAs: underestimated effects of contamination in NGS. RNA 20: 754-7.
18. Parkinson NJ, Maslau S, Ferneyhough B, Zhang G, Gregory L, et al. (2012) Preparation of high-quality next-generation sequencing libraries from picogram quantities of target DNA. Genome Res 20: 754-7.
19. Langmead B, Trapnell C, Pop M, Salzberg SL (2009) Ultrafast and memory-efficient alignment of short DNA sequences to the human genome. Genome Biol 10(3): R25.
20. International Human Genome Sequencing Consortium (2001) Initial sequencing and analysis of the human genome. Nature 409(6822): 860-921.
21. Marschner IC (2011) glm2: fitting generalized linear models with convergence problems. The R journal 3(2): 12-5.
22. R Core Team (2014) R: A language and environment for statistical computing. Vienna: the R Foundation for Statistical Computing. 3397 p.
23. Navin N, Kendall J, Troge J, Andrews P, Rodgers L, et al. (2011) Tumour evolution inferred by single-cell sequencing. Nature 472(7341): 90-4.
24. Falconer E, Hills M, Naumann U, Poon SS, Chavez EA, et al. (2012) DNA template strand sequencing of single-cells maps genomic rearrangements at high resolution. Nat Methods 9(11): 1107-12.
25. Islam S, Kjällquist U, Moliner A, Zajac P, Fan JB, et al. (2011) Characterization of the single-cell transcriptional landscape by highly multiplex RNA-seq. Genome Res 21(7): 1160-7.
26. Lu S, Zong C, Fan W, Yang M, Li J, et al. (2012) Probing meiotic recombination and aneuploidy of single sperm cells by whole-genome sequencing. Science 338(6114): 1627-30.
27. Mouse genome sequencing consortium (2002) Initial sequencing and analysis of the mouse genome. Nature 420(6915): 520-62.
28. Spisák S, Solymosi N, Ittzés P, Bodor A, Kondor D, et al. (2013) Complete genes may pass from food to human blood. PLoS One 8(7): e69805.
29. Altschul SF, Madden TL, Schaffer AA, Zhang J, Zhang Z, et al. (1997) Gapped Blast and psi-Blast: A new generation of protein database search programs. Nucleic Acids Res 25: 3389-402.
30. Munch K, Boomsma W, Huelsenbeck JP, Willerslev E, Nielsen R (2008) Statistical assignment of DNA sequences using Bayesian phylogenetics. Syst Biol 57(5): 750-7.
31. OBITool Development Team. Available: http://www.grenoble.prabi.fr/trac/OBITools. Accessed 2014 August 17.
32. Salzberg SL, Dunning Hotopp JC, Delcher AL, Pop M, Smith DR, et al. (2005) Serendipitous discovery of *Wolbachia* genomes in multiple *Drosophila* species. Genome Biol 6(3): R 23.
33. Kircher M, Sawyer S, Meyer M (2012) Double indexing overcomes inaccuracies in multiplex sequencing on the Illumina platform. Nucleic Acids Rsch 40(1): e3.
34. Combs PA, Eisen MB (2013) Sequencing mRNA from cryo-sliced Drosophila embryos to determine genome-wide spatial patterns of gene expression. PLoS One 8(8): e71820.
35. Salazar LA, Hirata MH, Cavalli SA, Machado MO, Hirata RD (1998) Optimized procedure for DNA isolation from fresh and cryopreserved clotted human blood useful in clinical molecular testing. Clin Chem 44(8): 1748-50.
36. Cherepanova AV, Tamkovich SN, Bryzgunova OE, Vlassov VV, Laktionov PP (2008) Deoxyribonuclease activity and circulating DNA concentration in blood plasma of patients with prostate tumors. Ann N Y Acad Sci 1137: 218-21.
37. Cogen AL, Nizet V, Gallo RL (2009) Skin microbiota: a source of disease or defence? Br J Dermatology 158(3): 442-55.
38. Yim SM, Kim JY, Ko JH, Lee YW, Choe YB, et al. (2010) Molecular analysis of Malassezia microflora on the skin of the patients with atopic dermatitis. Ann Dermatol 22(1): 41-7.
39. Gossling J, Slack JM (1974) Predominant Gram-Positive Bacteria in Human Feces: Numbers, Variety, and Persistence. Infection and Immunity 9(4): 719-29.
40. Gouba N, Raoult D, Drancourt M (2013) Plant and fungal diversity in gut microbiota as revealed by molecular and culture investigations. PLOS ONE 8(3): e59474.

41. Zhang L, Hou D, Chen X, Li D, Zhu L, et al. (2012) Exogenous plant MIR168a specifically targets mammalian LDLRAP1: evidence of cross-kingdom regulation by microRNA. Cell Res 22(1): 107-26.
42. Shtarkman YM, Koçer ZA, Edgar R, Veerapaneni RS, D'Elia T, et al. (2013) Subglacial Lake Vostok (Antarctica) accretion ice contains a diverse set of sequences from aquatic, marine and sediment-inhabiting bacteria and eukarya. PLoS One 8(7): e67221.
43. Wang K, Li H, Yuan Y, Etheridge A, Zhou Y, et al. (2012) The complex exogenous RNA spectra in human plasma: an interface with human gut biota? 7(12): e51009.

PERMISSIONS

All chapters in this book were first published in PLOS ONE, by The Public Library of Science; hereby published with permission under the Creative Commons Attribution License or equivalent. Every chapter published in this book has been scrutinized by our experts. Their significance has been extensively debated. The topics covered herein carry significant findings which will fuel the growth of the discipline. They may even be implemented as practical applications or may be referred to as a beginning point for another development.

The contributors of this book come from diverse backgrounds, making this book a truly international effort. This book will bring forth new frontiers with its revolutionizing research information and detailed analysis of the nascent developments around the world.

We would like to thank all the contributing authors for lending their expertise to make the book truly unique. They have played a crucial role in the development of this book. Without their invaluable contributions this book wouldn't have been possible. They have made vital efforts to compile up to date information on the varied aspects of this subject to make this book a valuable addition to the collection of many professionals and students.

This book was conceptualized with the vision of imparting up-to-date information and advanced data in this field. To ensure the same, a matchless editorial board was set up. Every individual on the board went through rigorous rounds of assessment to prove their worth. After which they invested a large part of their time researching and compiling the most relevant data for our readers.

The editorial board has been involved in producing this book since its inception. They have spent rigorous hours researching and exploring the diverse topics which have resulted in the successful publishing of this book. They have passed on their knowledge of decades through this book. To expedite this challenging task, the publisher supported the team at every step. A small team of assistant editors was also appointed to further simplify the editing procedure and attain best results for the readers.

Apart from the editorial board, the designing team has also invested a significant amount of their time in understanding the subject and creating the most relevant covers. They scrutinized every image to scout for the most suitable representation of the subject and create an appropriate cover for the book.

The publishing team has been an ardent support to the editorial, designing and production team. Their endless efforts to recruit the best for this project, has resulted in the accomplishment of this book. They are a veteran in the field of academics and their pool of knowledge is as vast as their experience in printing. Their expertise and guidance has proved useful at every step. Their uncompromising quality standards have made this book an exceptional effort. Their encouragement from time to time has been an inspiration for everyone.

The publisher and the editorial board hope that this book will prove to be a valuable piece of knowledge for researchers, students, practitioners and scholars across the globe.

LIST OF CONTRIBUTORS

Marc Stegger and Paal S. Andersen
Microbiology and Infection Control, Statens Serum Institut, Copenhagen, Denmark
Translational Genomics Research Institute, Pathogen Genomics Division, Flagstaff, Arizona, United States of America

Maliha Aziz
Translational Genomics Research Institute, Pathogen Genomics Division, Flagstaff, Arizona, United States of America

Tomasz Chroboczek and Frederic Laurent
National Reference Center for Staphylococci, Laboratory of Bacteriology, Ho^ pital de la Croix Rousse, Inserm U851, Lyon, France

Lance B. Price
Translational Genomics Research Institute, Pathogen Genomics Division, Flagstaff, Arizona, United States of America
Department of Environmental and Occupational Health, George Washington University, Washington, D. C., United States of America

Troels Ronco, Kristoffer Kiil and Robert L. Skov
Microbiology and Infection Control, Statens Serum Institut, Copenhagen, Denmark

Mehanathan Muthamilarasan, Rohit Khandelwal, Chandra Bhan Yadav, Venkata Suresh Bonthala, Yusuf Khan and Manoj Prasad
National Institute of Plant Genome Research, New Delhi, India

Alexander E. Lipka and Fei Lu
Institute for Genomic Diversity, Cornell University, Ithaca, New York, United States of America

Jerome H. Cherney
Department of Crop and Soil Sciences, Cornell University, Ithaca, New York, United States of America

Edward S. Buckler
Institute for Genomic Diversity, Cornell University, Ithaca, New York, United States of America
Agricultural Research Service, United States Department of Agriculture, Ithaca, New York, United States of America
Department of Plant Breeding and Genetics, Cornell University, Ithaca, New York, United States of America

Michael D. Casler
Agricultural Research Service, United States Department of Agriculture, Madison, Wisconsin, United States of America
Department of Agronomy, University of Wisconsin–Madison, Madison, Wisconsin, United States of America

Denise E. Costich
Institute for Genomic Diversity, Cornell University, Ithaca, New York, United States of America
Agricultural Research Service, United States Department of Agriculture, Ithaca, New York, United States of America

Huei-Jiun Su
Institute of Ecology and Evolutionary Biology, National Taiwan University, Taipei, Taiwan

Saskia A. Hogenhout
Department of Cell and Developmental Biology, John Innes Centre, Norwich, United Kingdom

Abdullah M. Al-Sadi
Department of Crop Sciences, Sultan Qaboos University, Al Khoud, Oman

Chih-Horng Kuo
Institute of Plant and Microbial Biology, Academia Sinica, Taipei, Taiwan
Molecular and Biological Agricultural Sciences Program, Taiwan International Graduate Program, National Chung Hsing University and Academia Sinica, Taipei, Taiwan
Biotechnology Center, National Chung Hsing University, Taichung, Taiwan

Marta Brozynska, Agnelo Furtado and Robert James Henry
Queensland Alliance for Agriculture and Food Innovation, The University of Queensland, Brisbane, Queensland, Australia

Christine D. Zanke, Jie Ling, Kerstin Neumann and Marion S. Röder
Leibniz Institute of Plant Genetics and Crop Plant Research (IPK), Gatersleben, Germany

Jörg Plieske and Martin W. Ganal
TraitGenetics GmbH, Gatersleben, Germany
Sonja Kollers, Erhard Ebmeyer and Viktor Korzun
KWS LOCHOW GMBH, Bergen, Germany

Odile Argillier
Syngenta France S.A.S., Orgerus, France

Gunther Stiewe and Maike Hinze
Syngenta Seeds GmbH, Bad Salzuflen, Germany

Luca Bianco, Alessandro Cestaro, Daniel James Sargent, Elisa Banchi, Roberto Viola, Riccardo Velasco and Michela Troggio
Research and Innovation Centre, Fondazione Edmund Mach, San Michele all'Adige, Trento, Italy

Sophia Derdak and Ivo Gut
CNAG - Centro Nacional de Análisis Genómico, Parc Científic de Barcelona, Barcelona, Spain

Mario Di Guardo
Research and Innovation Centre, Fondazione Edmund Mach, San Michele all'Adige, Trento, Italy
Wageningen UR Plant Breeding, Wageningen University and Research Centre, Wageningen, The Netherlands

Eric van de Weg
Wageningen UR Plant Breeding, Wageningen University and Research Centre, Wageningen, The Netherlands

Silvio Salvi
DipSA - University of Bologna, Bologna, Italy

Johannes Jansen
Biometris, Wageningen University and Research Centre, Wageningen, The Netherlands

Francois Laurens
INRA, UMR1345 Institut de Recherche en Horticulture and Semences, Beaucouzé, France

David Chagné
Plant & Food Research, Palmerston North Research Centre, Palmerston North, New Zealand

Ing-Nang Wang
Institute of Plant and Microbial Biology, Academia Sinica, Taipei, Taiwan, Republic of China
Department of Biological Sciences, University at Albany, Albany, New York, United States of America

Chung-Chi Hu, Ching-Wei Lee, Sih-Min Yen and Yau-Heiu Hsu
Graduate Institute of Biotechnology, National Chung Hsing University, Taichung, Taiwan, Republic of China

Wen-Bing Yeh
Department of Entomology, National Chung Hsin University, Taichung, Taiwan, Republic of China

Na-Sheng Lin
Institute of Plant and Microbial Biology, Academia Sinica, Taipei, Taiwan, Republic of China

Jordon Pace and Thomas Lübberstedt
Department of Agronomy, Iowa State University, Ames, Iowa, United States of America

Nigel Lee, Hsiang Sing Naik and Baskar Ganapathysubramanian
Department of Mechanical Engineering, Iowa State University, Ames, Iowa, United States of America

Martha E. Trujillo, Rodrigo Bacigalupe, Patricia Benito and Raúl Riesco
Departamento de Microbiología y Genética, Edificio Departamental, Campus Miguel de Unamuno, Universidad de Salamanca, Salamanca, Spain
Petar Pujic and Philippe Normand
Université Lyon 1, Université de Lyon, CNRS-UMR5557 Ecologie Microbienne, Villeurbanne, France

Yasuhiro Igarashi
Biotechnology Research Center, Toyama Prefectural University, Kurokawa, Imizu, Toyama, Japan

Claudine Médigue
Genoscope, CNRS-UMR 8030, Atelier de Ge´nomique Comparative, Evry, France

Da-Song Chen, Peng-Yu Jin, Xiu-Lei Ding, Si-Xia Yang, Jia-Fei Ju, Jing-Yu Zhao and Xiao-Yue Hong
Department of Entomology, College of Plant Protection, Nanjing Agricultural University, Nanjing, Jiangsu, China

Kai-Jun Zhang
Department of Entomology, College of Plant Protection, Southwest University, Chongqing, China

Justin N. Vaughn, Srinivasa R. Chaluvadi and Jeffrey L. Bennetzen
Department of Genetics, University of Georgia, Athens, Georgia, United States of America

Tushar and Latha Rangan
Department of Biotechnology, Indian Institute of Technology Guwahati, Assam, India

Adam D. Idoine, Alix Boulouis, Jens Rupprecht and Ralph Bock
Max-Planck-Institut für Molekulare Pflanzenphysiologie, Potsdam, Golm, Germany

Zhang Wang and Martin Wu
Department of Biology, University of Virginia, Charlottesville, Virginia, United States of America

Tiange Lang and Jinyu Liu
Key Laboratory of Tropical Forest Ecology, Xishuangbanna Tropical Botanical Garden, Chinese Academy of Sciences, Menglun, Mengla, Yunnan Province, China

Kangquan Yin
College of Forestry, Beijing Forestry University, Beijing, China
School of Life Science, Tsinghua University, Beijing, China

Kunfang Cao
Key Laboratory of Tropical Forest Ecology, Xishuangbanna Tropical Botanical Garden, Chinese Academy of Sciences, Menglun, Mengla, Yunnan Province, China
State Key Laboratory for Conservation and Utilization of Subtropical Agro-Bioresources, and College of Forestry, Guangxi University, Nanning, Guangxi, China

Charles H. Cannon
Key Laboratory of Tropical Forest Ecology, Xishuangbanna Tropical Botanical Garden, Chinese Academy of Sciences, Menglun, Mengla, Yunnan Province, China
Department of Biological Sciences, Texas Tech University, Lubbock, Texas, United States of America

Fang K. Du
College of Forestry, Beijing Forestry University, Beijing, China

Yang Yang, Dang Yuanye, Lu Jinjian and Wang Yitao
State Key Laboratory of Quality Research in Chinese Medicine, Institute of Chinese Medical Sciences, University of Macau, Macau, China

Li Qing
Department of Pharmacy, Shanghai Changzheng Hospital, Second Military Medical University, Shanghai, China

Li Xiwen
State Key Laboratory of Quality Research in Chinese Medicine, Institute of Chinese Medical Sciences, University of Macau, Macau, China
Institute of Chinese Materia Medica, China Academy of Chinese Medical Sciences, Beijing, China

Elena Vinay-Lara and James L. Steele
Department of Food Science, University of Wisconsin-Madison, Madison, Wisconsin, United States of America

Joshua J. Hamilton and Jennifer L. Reed
Department of Chemical and Biological Engineering, University of Wisconsin-Madison, Madison, Wisconsin, United States of America

Buffy Stahl
DuPont Nutrition and Health, Madison, Wisconsin, United States of America

Jeff R. Broadbent
Utah State University Department of Nutrition, Dietetics, and Food Sciences, Logan, Utah, United States of America

Ning Yang
School of Automation, Northwestern Polytechnical University, Xi'an, shaanxi, China

Fei Hu, Lingxi Zhou and Jijun Tang
Tianjin Key Laboratory of Cognitive Computing and Application, Tianjin University, Tianjin, China
Department of Computer Science and Engineering, University of South Carolina, Columbia, South Carolina, United States of America

Ian P. Adams, Anna Skelton, Roy Macarthur, Laura Flint, Neil Boonham and Adrian Fox
Centre for Crop Protection, Food and Environment Research Agency, Sand Hutton, York, United Kingdom

Tobias Hodges
Centre for Crop Protection, Food and Environment Research Agency, Sand Hutton, York, United Kingdom
University of York, York, United Kingdom

Howard Hinds
RootCrop Ltd., Hoveringham, Nottinghamshire, United Kingdom

Palash Deb Nath
Department of Plant Pathology, Assam Agricultural University, Jorhat, India

Kai Fan, Ming Wang, Mi Ni, Noreen Bibi, Shuna Yuan, Feng Li and Xuede Wang
Institute of Crop Science, College of Agriculture and Biotechnology, Zhejiang University, Hangzhou, P.R. China

Ying Miao
Center for Molecular Cell and Systems Biology, College of Life Science, Fujian Agriculture and Forestry University, Fuzhou, P.R. China

Richard W. Lusk
Department of Ecology and Evolutionary Biology, University of Michigan, Ann Arbor, Michigan, United States of America

Index